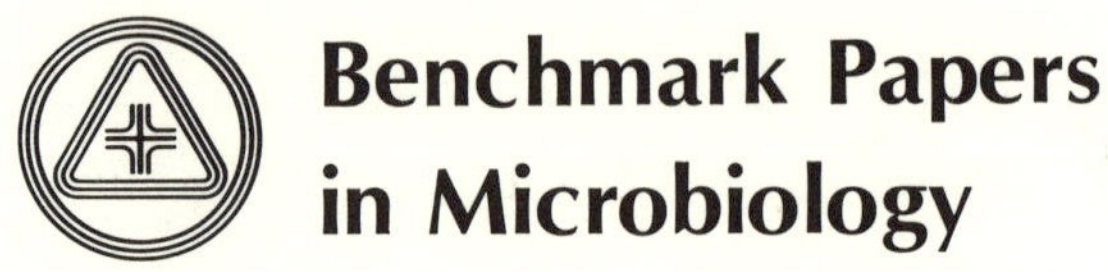

Benchmark Papers in Microbiology

Series Editor: Wayne W. Umbreit
Rutgers—The State University

PUBLISHED VOLUMES

MICROBIAL PERMEABILITY / *John P. Reeves*
CHEMICAL STERILIZATION / *Paul M. Borick*
MICROBIAL GENETICS / *Morad Abou-Sabé*
MICROBIAL PHOTOSYNTHESIS / *June Lascelles*
MICROBIAL METABOLISM / *H. W. Doelle*
ANIMAL CELL CULTURE AND VIROLOGY / *Robert J. Kuchler*
PHAGE / *Sewell P. Champe*
MICROBIAL GROWTH / *P. S. S. Dawson*
MICROBIAL INTERACTION WITH THE PHYSICAL ENVIRONMENT / *D. W. Thayer*

Additional Volumes in Preparation

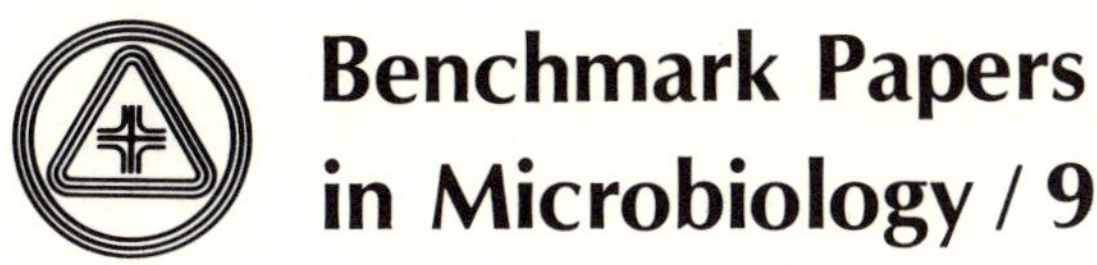

A BENCHMARK® Books Series

MICROBIAL INTERACTION WITH THE PHYSICAL ENVIRONMENT

Edited by
D. W. THAYER
Texas Tech University

Distributed by
HALSTED PRESS *A Division of John Wiley & Sons, Inc.*

Benchmark Papers in Microbiology, Volume 9
Library of Congress Catalog Card Number: 75–20439
ISBN: 0–470–85842–7

77 76 75 1 2 3 4 5
Manufactured in the United States of America.

LIBRARY OF CONGRESS CATALOGING IN PUBLICATION DATA
Main entry under title:

Microbial interaction with the physical environment.

(Benchmark papers in microbiology ; 9)
Includes indexes.
1. Microbial ecology. 2. Micro-organisms--Physiology. 3. Microbiological chemistry. I. Thayer, Donald Wayne, 1937- [DNLM: 1. Environment--Collected works. 2. Microbiology--Collected works. QW52 M6265]
QR100.M52 576'.11 75-20439
ISBN 0-470-85842-7 (Halsted)

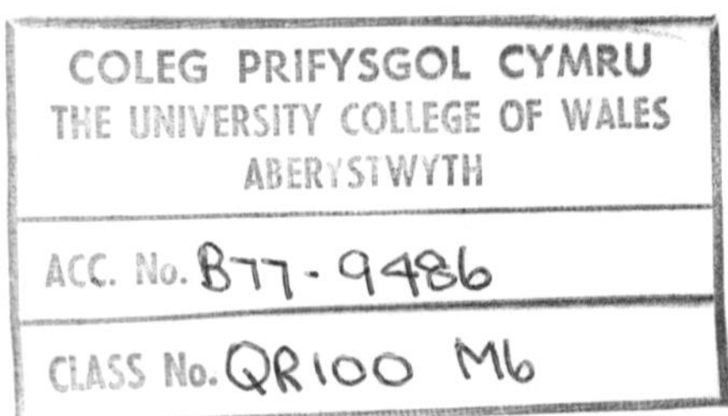

Exclusive Distributor: **Halsted Press**
A Division of John Wiley & Sons, Inc.

ACKNOWLEDGMENTS AND PERMISSIONS

ACKNOWLEDGMENTS

DANISH SCIENCE PRESS LTD.—*Galathea Report*
Deep-Sea Bacteria

PERMISSIONS

The following papers have been reprinted or translated with the permission of the authors and copyright holders.

AMERICAN SOCIETY FOR MICROBIOLOGY
Bacteriological Reviews
Current Status of Knowledge of Halophilic Bacteria
Effects of Hydrostatic Pressure on Microbial Systems
Journal of Bacteriology
Effects of Hydrostatic Pressure and Temperature on the Uptake and Respiration of Amino Acids by a Facultatively Psychrophilic Marine Bacterium
Electromechanical Interactions in Cell Walls of Gram-Positive Cocci
Fermentation of Glucose by Suspensions of *Escherichia coli*
Influence of pH on the Dissimilation of Glucose by *Aerobacter indologenes*
The Isoelectric Point of Bacterial Cells
Membrane Permeability and the Loss of Germination Factor from *Neurospora crassa* at Low Water Activities
Osmotic Pressure in *Escherichia coli* as Rendered Detectable by Lysozyme Attack
Osmotic Regulation of Invertase Formation and Secretion by Protoplasts of *Saccharomyces*
Purification and Reversible Inactivation of the Isocitrate Dehydrogenase from an Obligate Halophile
Ultrasonic Disintegration as a Method of Extracting Bacterial Enzymes

ASP BIOLOGICAL AND MEDICAL PRESS—*Biochimica et Biophysica Acta*
Hydrostatic Pressure Effects on the Translation Stages of Protein Synthesis in a Cell-Free System from *Escherichia coli*
Studies of the Electron Transport Chain of Extremely Halophilic Bacteria: V. Mode of Action of Salts on Cytochrome Oxidase

THE BIOLOGICAL LABORATORY, LONG ISLAND BIOLOGICAL ASSOCIATION, INC.—*Cold Spring Harbor Symposia on Quantitative Biology*
Electrophoresis and the Chemistry of Cell Surfaces

CAMBRIDGE UNIVERSITY PRESS FOR THE SOCIETY OF GENERAL MICROBIOLOGY
The Journal of General Microbiology
The Freezing Points of Bacterial Cells in Relation to Halophilism

Symposia of the Society for General Microbiology
The Action of Pressure and Temperature

THE CHEMICAL SOCIETY, LONDON—*Journal of the Chemical Society, London*
The Chemical Action of *Bacillus coli communis* and Similar Organisms on Carbohydrates and Allied Compounds

JOURNAL OF MARINE RESEARCH (SEARS FOUNDATION FOR MARINE RESEARCH)—*Journal of Marine Research*
The Growth and Viability of Sixty-Three Species of Marine Bacteria as Influenced by Hydrostatic Pressure

NATIONAL RESEARCH COUNCIL OF CANADA
Canadian Journal of Microbiology
Effects of Sodium and Potassium Chloride on Certain Enzymes of *Micrococcus halodenitrificans and Pseudomonas salinaria*
Canadian Journal of Technology
Dissimilation of Glucose by Yeast at Poised Hydrogen Ion Concentrations

THE ROYAL SOCIETY, LONDON—*Proceedings of the Royal Society*
The Limits of Microbial Existence

THE SOCIETY FOR APPLIED BACTERIOLOGY—*The Journal of Applied Bacteriology*
Combined Effect of Water Activity, pH and Temperature on the Growth of *Clostridium botulinum* from Spore and Vegetative Cell Inocula
The Water Relations of Staphylococci and Micrococci

SOCIETY FOR EXPERIMENTAL BIOLOGY AND MEDICINE—*Proceedings of the Society for Experimental Biology of Medicine*
The Mechanism of the Lethal Effects of Ultrasonic Radiation

SPRINGER-VERLAG—*Biochemische Zeitschrift*
The Third Form of Sugar Fermentation

TAYLOR & FRANCIS LTD.—*The Philosophical Magazine*
The Physical and Biological Effects of High-Frequency Sound-Waves of Great Intensity

JOHN WILEY & SONS, INC.—*Biotechnology and Bioengineering*
The Influence of Environmental Conditions on the Macromolecular Composition of *Candida utilis*

SERIES EDITOR'S PREFACE

While it is probably an outworn cliche to say that microorganisms are creatures of their environment, it still serves to emphasize that in microorganisms, in contrast to more complex organisms, the mechanism of homeostasis are rudimentary at best. In the case of temperature, not considered in this volume, homeostatic mechanisms are essentially non-existent and the microbial cell has no mechanism for cooling itself in a warm environment or of warming itself in a cold one, that is, it rapidly loses what heat it may generate. The other factors always in the environment, in contrast to those which might be introduced, do have some measure of homeostasis. The internal pH is not necessarily that of the external environment. Indeed, even in common laboratory organisms such as *Escherichia coli,* the external pH may be 4, but the internal pH is still at 6, even though the external pH is such that no further growth occurs. Somewhat the same relations exist between internal and external osmotic pressure, and there are organisms (halophiles, osmophiles) which have adapted to conditions which limit the growth of the more normal forms. It is of interest to determine how they have managed such adaptation. But, perhaps not surprising, there are few homeostatic mechanisms for factors introduced into the environment, such as radiation, or ultrasonic sound.

Yet, as Dr. Thayer points out, none of these environmental factors can be really understood without some reference to the curious, unique and remarkable properties of water, a substance which, if it did not exist, would have to be invented before life, as we know it, would be possible. Indeed, it is almost impossible to think of life without water. One could perhaps think of a silica rather than a carbon based structural and metabolic living system, but there is no substance that could substitute for water. We think, of course, that life started in water, that water is abundantly available, and that rarely does the lack of water limit life, except in very dry environments. However, even in environments in which fluid water exists, not all of it is biologically available and it is the "water activity" (surely a poor choice of words) which is important.

Assuming that water is available, the microorganism is still at the mercy of its environment, and we do, indeed, manipulate the environment both to stimulate and to inhibit growth of microorganisms. What happens and why is the subject of this collection of classical and pertinent papers on the relation between the environment and microbial life.

WAYNE W. UMBREIT

PREFACE

The editor takes full responsibility for the selection of articles in this manuscript. The diversity of topics covered by this manuscript forced the exclusion of many fine articles which otherwise would have been included. The editor also takes full credit or blame for the two translations. They are attempts to produce readable and scientifically accurate rather than strictly literal translations.

The editor would like to express his thanks to the many authors and publishers who have granted permission for the use of their articles. He is especially grateful to his wife, Suzanne, who typed and helped edit the manuscript, to W. W. Umbreit for his review of the manuscript and to R. W. Strandtmann for his review of the German translation.

D. W. THAYER

CONTENTS

PART I: HYDROGEN-ION CONCENTRATION

PART II: WATER ACTIVITY AND OSMOTIC PRESSURE

PART III: HALOPHILIC INTERACTIONS

PART IV: HYDROSTATIC PRESSURE

PART V: ELECTROKINETIC PROPERTIES

PART VI: SOUND

CONTENTS BY AUTHOR

INTRODUCTION: THE LIMITS OF MICROBIAL EXISTENCE

This book presents selected papers on the interaction of the microbe with its physical environment. For our purposes, the physical environment is defined as temperature, pH, osmotic pressure and/or water activity, ionic environment, baric and hydrostatic pressure, electrokinetic effects, surface tension, and radiant energy. The subjects of temperature and radiation are covered in other volumes in this series. The book is divided into six parts, each of which emphasizes one environmental parameter. Paper 1, by F. A. Skinner, discusses the limits of microbial existence and thus serves as an introduction to the literature and to the papers that follow. Skinner provides an extensive bibliography on the subject. Few papers defining the limits of microbial existence will be included in this collection; instead, papers dealing with mechanisms of microbial interaction with the physical environment will be emphasized.

There is really no way to understand the microbe's environment without understanding the physical and chemical properties of water. This volume cannot deal with these in detail, but it can serve as a guide to the literature, especially for the graduate student. A bibliography is presented on each subject as a guide to the literature. The article by Buswell and Rodebush is especially recommended as a general review of the properties of water. The many books, monographs, and articles by P. W. Bridgman are the most important of the older articles.

The limits of microbial existence are defined by the properties of water, because the microbe's environment is aqueous. The freezing and boiling points of pure water at sea level are 0°C and 100°C, respectively. Water is unique in these properties, behaving in a manner that differs from similar organic or inorganic compounds. As examples, hydrogen sulfide boils at −59.6°C and methanol boils at 64.7°C. These properties are due, in part, to the formation of hydrogen bonds between individual water molecules, forming, in effect, one giant molecule. The high surface tension of water also results from the hydrogen bonding. In general, few metabolically active microorganisms are found near the boiling or freezing points of water. The freezing point can be lowered by the addition of solutes, and in strongly saline pools in Antarctica, for example, active microorganisms have been found at −23°C (Paper 1). Skinner discusses the limits of microbial existence.

The hydrogen-ion concentration of the medium is a direct result of the dielectric constant of water, which is the highest of any known liquid. Even the term *pH* is defined according to the ionization constant of water, pH being the negative logarithm of the hydrogen-ion concentration. Since the ionization constant of water is 1×10^{-14}, the number of hydrogen ions in pure water is 1×10^{-7} Almost all organisms live in a very narrow pH range, around 7.0. Even slight alterations of the hydrogen-ion concentration often result in drastic changes in the metabolic products and in the macromolecular composition of the cell. In 1901 Harden developed a very complete and accurate fermentation balance for *Escherichia coli*. His work laid a solid foundation upon which others were to build. Neuberg and his coworkers analyzed the effects of pH on the fermentation of sugars by yeast, and some of these studies led to new industrial processes. Mickelson, Neish, and Blackwood were among the first to control pH during fermentation studies with growing cells. Stokes separated the catabolic from the anabolic functions by working with resting cell suspensions. Alroy and Tannenbaum utilized the chemostat to determine the effects of pH on the macromolecular composition of yeast.

Water is the best-known solvent. It will ionize electrolytes and form hydrates of many nonelectrolytes. Its dielectric constant makes it the best solvent for electrolytes. Membranes that permit the passage of solvent molecules but not those of the solute are known as *semipermeable*. When a solution containing a solute such as sucrose is placed within a membrane tubing surrounded by distilled water, the pressure that must be exerted to prevent an increase in volume is called the *osmotic pressure*.

In dilute solutions the ideal nonelectrolyte behaves in a manner analogous to an ideal gas. Thus 1 gram molecular weight of an ideal gas at 0°C occupies 22.41 liters at 1 atmosphere pressure, or 22.41 atmospheres pressure in a volume of 1 liter. A solution of 1 gram molecular weight of an ideal nonelectrolyte in water will exert an osmotic pressure of 22.41 atmospheres. Osmotic pressure is a colligative property; therefore, an electrolyte will exert a greater osmotic pressure than will a nonelectrolyte.

The importance of water to the living cell is not disputed and cells grow only above certain limits of available water. Even deuterium oxide cannot substitute for water physiologically. Skinner mentions osmophilic yeasts which are active in solutions containing 60 percent sucrose. This corresponds to an osmotic pressure in excess of 50 atmospheres. The influence of water activity on cell composition was reviewed by Christian and Waltho (Paper 8). Water activity, not solute concentration, was the controlling factor in the growth of staphylococci. The combined effects of pH, temperature, and water activity were examined by Baird-Parker and Freame (Paper 9). Kuo and Lampen (Paper 10) linked enzyme synthesis, secretion, and osmotic pressure. Scheie measured cellular osmotic pressures. Other studies (Paper 12) have demonstrated the release of components essential to germination from *Neurospora* exposed to media having low water activity.

Halophiles are uniquely adapted to exist in water containing high concentrations of salt. They have developed unusual enzymes and cell-wall components which are dependent on high salt concentrations. The ability of various salts to substitute for NaCl diminishes with the Hofmeister series, Na > K > Li > Mg > Ca. In aqueous solution the effective radii of the hydrated ions are 10.8Å, 5.6Å, and 3.8Å for Mg^{2+}, Na^{+}, and K^{+}, respectively (Florey, 1966). The hydration number for Li^{+} is 4. Again the physiological properties are closely correlated with the hydration of ions and consequent effects on ionic mobilities. Flannery reviewed the knowledge of halophiles prior to 1956. Baxter and Gibbons (Paper 14) compared the effects of potassium and sodium chloride on several enzymes from two halophiles. The freezing points of the halophiles indicate that they do not maintain greater intracellular water activities than that of the media. Hubbard and Miller (Paper 16) purified isocitrate dehydrogenase from an obligate halophile and studied the salt-dependent properties of this enzyme. Even the electron transport system of halophiles is salt-dependent (Paper 17).

Bacteria are remarkable for their diversity, including their reaction to various hydrostatic pressures. Certes recovered bacteria at ocean depths of 5100 m in 1883 and succeeded in cultivating these organisms while at sea. The known limits were extended to 1000 atmospheres when active bacteria were recovered from the Philippine Trench at depths of 10,000 m. Organisms soon were found which required high hydrostatic pressures. Johnson (Paper 20) discusses the effects of pressure and temperature on biological processes. Heden investigated the effects of hydrostatic pressure and temperature on transforming DNA. Amino acid uptake and oxidation by marine bacteria were altered by hydrostatic pressure. Arnold and Albright traced the effect of pressure to the translation stages of protein synthesis.

During the period 1925–1936, practical methods were developed for the study of the surface charge of cells. Abramson was one of the pioneers in these studies and reviewed the early work in a paper published in 1931. In Paper 26, Abramson et al. discussed the effects of the surface geometry of particles on their electrophoretic mobility. (Throughout the literature one constantly finds Abramson's work cited. The *Journal of Asthma Research* reprinted several of his articles in 1972.) One of several comparisons of electrophoretic mobility of bacteria to the staining reaction is given in Paper 27. Recent studies have concentrated on properties of the cell-wall polymers (Paper 28).

The use of ultrasonic sound for disruption of bacteria and other cells developed very rapidly after Wood and Loomis described some of the biological effects in 1927. From that period both the physical and biological effects were studied extensively. Schmitt and his coworkers described the coagulation of protoplasm and verified the concept of cavitation as one of the prime causes of cell rupture during sonic irradiation. The first systematic study of ultrasonic sound for the extraction of bacterial enzymes was published in 1946.

SELECTED READINGS

The Limits of Microbial Existence

Brock, T. D. 1969. Microbial growth under extreme conditions. *Symp. Soc. Gen. Microbiol.* *19*: 15–41.

———, and G. K. Darland. 1970. Limits of microbial existence: temperature and pH. *Science* *169*(3952): 1316–1318.

Sherman, J. M., and G. M. Cameron. 1934. Lethal environmental factors within the natural range of growth. *J. Bacteriol.* *27*(4): 341–348.

Tansey, M. R., and T. D. Brock. 1972. The upper temperature limit for eukaryotic organisms. *Proc. Nat. Acad. Sci. U.S.A. 69*(9): 2426–2428.

Water

Arrhenius, S. 1887. Über die Dissociation der in Wasser geloesten Stoffe. *Z. Physik. Chem. 1:* 631–648.

Bell, R. P. 1958. The hydration of ions in solution. *Endeavour 17*(65): 31–35.

Benson, S. W., and J. A. Berson. 1962. The effect of pressure on the rate and equilibrium constants of chemical reactions. The calculation of activation volumes by application of the tait equation. *J. Amer. Chem. Soc. 84*(2): 152–158.

Bockris, J. O'M. 1949. Ionic solvation. *Quart. Rev. (London) 3*: 173–180.

Bridgman, P. W. 1912. Water, in the liquid and five solid forms, under pressure. *Proc. Amer. Acad. Arts Sci. 47*(13): 439–559.

———. 1930. The volume of eighteen liquids as a function of pressure and temperature. *Proc. Amer. Acad. Arts Sci. 66*: 184–233.

———. 1931. *The Physics of High Pressure.* Macmillan, New York.

Buswell, A. M., and W. H. Rodebush. 1956. Water. *Sci. Amer. 194*(4): 76–89.

Enns, T., P. F. Scholander, and E. D. Bradstreet. 1965. Effect of hydrostatic pressure on gases dissolved in water. *J. Phys. Chem. 69*(2): 389–391.

Eyring, H. 1935. The activated complex in chemical reactions. *J. Chem. Phys. 3*(2): 107–115.

Florey, E. D. 1966. *An Introduction to General and Comparative Animal Physiology.* W. B. Saunders, Philadelphia.

Frank, H. S., and M. W. Evans. 1945. Free volume and entrophy in condensed systems. III. Entropy in binary liquid mixtures; partial molar entropy in dilute solutions; structure and thermodynamics in aqueous electrolytes. *J. Chem. Phys. 13*(11): 507–532.

———, and W.-Y. Wen. 1957. Structural aspects of ion-solvent interaction in aqueous solutions: A suggested picture of water structure. *Disc. Faraday Soc. 24*: 133–140.

Hamann, S. D. 1963. The ionization of water at high pressures. *J. Phys. Chem. 67*(10): 2233–2235.

Horne, R. A., and D. S. Johnson. 1966. The viscosity of water under pressure. *J. Phys. Chem. 70*(7): 2182–2190.

Luzzati, V., and F. Husson. 1962. The structure of the liquid crystalline phases of lipid-water systems. *J. Cell. Biol. 12*(1): 207–219.

Oppenheimer, C. H., and W. Drost-Hansen. 1960. A relationship between multiple temperature optima for biological systems and the properties of water. *J. Bacteriol. 80*(1): 21–24.

Szent-Gyorgyi, A. 1956. Bioenergetics. *Science 124*(3227): 873–875.

Webb, S. J. 1965. *Bound-Water in Biological Integrity.* Charles C Thomas, Springfield, Ill.

ZoBell, C. E. 1959. Thermal changes accompanying the compression of aqueous solutions to deep-sea conditions. *Limnol. Oceanogr. 4*(4): 463–471.

1

Reprinted from *Proc. Roy. Soc. (London),* **B171**(1022), 77–89 (1968)

The limits of microbial existence

By F. A. Skinner

Soil Microbiology Department, Rothamsted Experimental Station, Harpenden, Herts.

Introduction

Many parts of the world, such as the ocean depths, polar and geothermal regions, where the environment severely restricts most forms of life, are nevertheless colonized to some extent by specially adapted micro-organisms. There are also man-made environments, some especially designed to inhibit microbial growth, such as the strong brines used to pickle fish and meat, and cold stores, in which some micro-organisms can survive and grow.

Although such organisms have been known since the early days of microbiology only recently have they been systematically studied. Modern interest in the ability of micro-organisms to tolerate such unfavourable conditions stems largely from the need of the food industries to prevent microbial growth in their products. A further stimulus to understand the limiting conditions for microbial life, and to collect and systematize the available information has been provided by the requirements of space research (see Vallentyne 1963, 1965).

Growth at extremes of temperature

Most species of micro-organisms grow between 20 and 45 °C and are said to be mesophilic. The thermophiles, comprising fewer species, grow from 45 to *ca.* 75 °C with optima between 50 and 55° C: some of these are obligate, requiring high temperature for growth; others only tolerate it and have comparatively low-temperature optima. The psychrophiles grow well at low temperatures, sometimes even below 0 °C, and are inhibited above *ca.* 30 °C.

Thermophiles

The first known thermophile, a bacterium that could grow at 73 °C but not at low temperatures, was isolated by Miquel (1888) from the River Seine. Since that time thermophiles, mostly bacteria, have been found in most samples of soil, mud or water from regions of widely differing climate. They also occur in composts, sewage, in deep ocean floor deposits (Bartholomew & Paik 1966) and in hydrocarbon-containing rocks (Bairiev & Mamedov 1963). Thus, though thermophiles are ubiquitous they mostly occur in habitats that offer little or no scope for growth at high temperature. Indeed, it is not uncommon for micro-organisms to be adapted to conditions they do not normally encounter.

Microbes are at the temperature of their environment (see Kempner 1963), but when growing they produce heat and so can raise the temperature of that environment if its thermal properties impede heat loss. Thus, rapid mesophilic growth can provide conditions in which thermophiles can flourish, and by so doing raise the temperature even further. An example of this is provided by the heating of hay stacks or bales when prepared from hay too wet for proper storage. Rapid growth of an initially very mixed microflora of fungi, actinomycetes and bacteria quickly raises the temperature to a level at which thermophilic representatives of these groups proliferate. A few species of thermophilic fungi such as *Mucor pusillus*, *Absidia ramosa*, *Aspergillus fumigatus* and *Humicola lanuginosa* develop vigorously in heating hay as also do large numbers of thermophilic actinomycetes (Gregory, Lacey, Festenstein & Skinner 1963; Festenstein *et al.* 1965).

Upper layers of the soil are usually too cool to permit extensive development of thermophiles. However, thermophilic fungi have been isolated from the zone of vegetation litter just above soil level where they seem to be well established. This zone, which is often rather warmer than the surrounding air, may be a natural habitat in which thermophiles can grow, though probably below their temperature optima (Apinis 1963*a*, *b*).

Few bacteria are really obligate thermophiles; most are facultative such as the aerobic sporeformers, *Bacillus coagulans* and *B. circulans*. Economically, the most important thermophile is *B. stearothermophilus*; this is an obligate thermophile though some merely thermotolerant strains exist. It is particularly troublesome in the food industries because its spores are so resistant to heat; some strains will grow at 80 to 85 °C. There are some anaerobic thermophilic bacteria such as the cellulose decomposers (McBee 1950) and sulphate reducers (Postgate & Campbell 1966).

The transient conditions that favour thermophiles in heating hay are in striking contrast to the hot springs that provide unusual but constant habitats for a few specialized thermophiles, especially algae and bacteria (Kempner 1963; Brock 1967).

The main problems concerning life at high temperatures are to explain how these organisms can grow at temperatures at which most proteins are denatured. It is possible that the proteins and other important macromolecules of thermophiles are intrinsically more stable at high temperature than the corresponding compounds or structures of mesophiles, or that they are protected against denaturation in the cell. Alternatively, proteins may be denatured just as easily in the thermophile as in the mesophile, but that the former can repair the damage more readily (Allen 1950, 1953; Recknitz & Janota-Bassalik 1967).

There is some evidence for the first hypothesis. Oprescu (1898) demonstrated the existence of thermostable hydrolytic enzymes that could be isolated from culture filtrates of thermophiles, and since then, thermostability of enzymes has been frequently reported (e.g. Militzer, Sonderegger, Tuttle & Gorgi 1949; Marsh & Militzer 1956*a*, *b*; O'Brien & Campbell 1957; Amelunxen 1967; Daron 1967). Heat stability of enzymes is often related to the growth temperature of the organism; thus, Campbell (1955) found that α-amylases produced by a strain of facultatively thermophilic bacteria grown at 35 and 55 °C respectively differed in that the latter was more stable at 90 °C (see also Militzer *et al.* 1949; Marsh & Militzer 1956*b*; Brown, Militzer & Georgi 1957). Differences in the stability of bacterial flagellar proteins to heat have also been demonstrated. Whereas the flagella of mesophiles usually disintegrated easily above 50 °C, those of thermophiles remained intact even at 70 °C (Kofler 1957; Kofler, Mallet & Adye 1957).

Ribosomes from *Bacillus stearothermophilus* (the most frequently used organism in thermophilic studies) are much more resistant to high temperatures than those from the mesophile *Escherichia coli*, as judged by experiments on the incorporation of amino acids (Algranati & Lengyel 1966; Friedman & Weinstein 1966; Friedman, Axel & Weinstein 1967). The thermal denaturation point is also higher for ribosomes from thermophilic than for those from mesophilic bacteria, but this greater resistance to heat seems not to arise from intrinsic differences in the *RNA* (Mangiantini, Tecce, Toschi & Trentalance 1965; Saunders & Campbell 1966*a*; Friedman *et al.* 1967; Stenesh & Yang 1967) or the ribosomal proteins (Saunders & Campbell 1966*a*). The greater resistance of ribosomes to heat is correlated positively with maximal growth temperatures (Pace & Campbell 1967), but the reason for this stability is not known.

Magnesium ions may play a part in enzyme stability, for increasing the concentration of magnesium increases the resistance of some enzyme preparations to inactivation by heat (Brown, Militzer & Georgi 1957). Friedman & Weinstein (1966), working with enzyme systems from *Bacillus stearothermophilus* and the mesophile *Escherichia coli*, also emphasized the importance of magnesium ions in stabilizing the reaction between ribosomes, *s-RNA* and *m-RNA*, and suggested that thermophiles might make their protein synthetic apparatus stable to heat by maintaining a high concentration of magnesium or other polyvalent cations.

Intrinsic resistance of macromolecules to heat is also demonstrated by the existence of phages not easily inactivated at high temperature (Lark & Adams 1953). Saunders & Campbell (1966*b*) described a bacteriophage able to lyse strains of *Bacillus stearothermophilus* over the range of 43 to 76 °C; it was inactivated rapidly at 65 °C in citrate buffer, but not in broth during a 12 h period. White, Georgi & Militzer (1954) also described a bacteriophage for thermophilic bacteria; this strain remained infective after 2 h at 95 °C in broth.

The thermophilic actinomycete *Thermoactinomyces* (*Micromonospora*) *vulgaris* also has its phages (Agre 1961), and current work with these at Rothamsted shows that they are fully active after 30 min at 70 °C but many particles are not infective at 75 °C and inactivation is almost complete at 80 °C (J. J. Patel, personal communication).

Brock (1967) measured the photosynthetic efficiency at different temperatures of algae taken from growths in hot springs at several temperatures in the Yellowstone National Park. Photosynthetic efficiency of each sample was greatest at the temperature of the water from which that particular sample of algae was taken, even though the biomass at the limiting temperature of 73 to 75 °C was smaller than at the optimal range of 50 to 55 °C. Possibly, some part of the photo-synthetic mechanism does not operate above *ca.* 75° C, because non-photosynthetic bacteria grow well in these springs at 85 to 88 °C and make slight growth at 91 °C, which is only two degrees below the boiling point of water at that altitude. Growth has not been reported at a higher temperature than 91 °C at atmospheric pressure. Hot spring bacteria also have different temperature optima according to the position along the stream (Brock & Brock 1966).

Psychrophiles

Bacteria that can grow well at or near 0 °C are widely distributed and can be readily isolated from many sources, especially the sea, most of which is always colder than 5 °C. Mostly they are Gram-negative non-sporeforming rods (e.g. *Pseudomonas* spp.). Although these bacteria tolerate cold, most of them have optima between 20 and 40 °C so it has been suggested that they should be called 'psychrotrophic' and the term 'psychrophilic' reserved for those organisms that grow optimally below *ca.* 35 °C (Eddy 1960). Few species have optima below 20 °C. There has been much disagreement among microbiologists about this terminology (see Ingraham & Stokes 1959).

Psychrophilic, facultatively anaerobic, bacteria can also be easily isolated from many sources; their fermentation activities are not necessarily the same at temperatures near 0 °C as they are at *ca.* 20 °C, presumably because enzymes are either not formed or are inhibited at low temperatures (Upadhyay & Stokes 1962). Strictly anaerobic psychrophiles seem to be rare in most habitats, though sulphate reducers occur in ocean deposits (ZoBell & Morita 1957) and some sporeforming anaerobes, such as *Clostridium carnofoetidum* var. *amylolyticum* and *Cl. perfringens* will grow well at 6 °C though their growth optima are at *ca.* 30 °C (Beerens, Sugama & Tahon-Castell 1965). Bacteriophage with a growth temperature maximum only slightly above that of the psychrophilic host has also been reported (Olsen 1967).

Some bacteria, including *Bacillus megaterium*, micrococci and coryneform bacteria, and the fungus *Sporobolomyces* have been found growing in strongly saline pools in Antarctica at *ca.* −23 °C, the lowest temperature at which active life has been observed: these organisms grew in laboratory media at room temperature, thereby displaying remarkable adaptability to changing conditions (Meyer *et al.* 1962). The flagellate alga, *Dunaliella salina*, has also been seen to swim in brine pools at −15 °C (Zernow 1944).

Some small algae, such as *Chlamydomonas nivalis*, live on snow surfaces (Lewin 1962). The resistance to freezing of these algae is probably partly explained by the high osmotic pressure of the cell contents and perhaps by the ability of the cytoplasm to survive damage caused by freezing and thawing (Levitt 1956; Kanwisher 1957). Typically, these algae are red with carotenoid pigments.

Other examples of microbial growth at very low temperatures have been given in the review by Geiger, Jaffe & Mamikunian (1965).

Barophiles

Many bacteria occur in the sediments of the ocean floor, even at the greatest known depths of water. A characteristic feature of this environment is the enormous hydrostatic pressure, which increases at the rate of about one atmosphere for each 10 m depth. Most bacteria can tolerate considerable pressure, but some grow only at very high pressures and not at atmospheric pressure. ZoBell & Johnson (1949) found that whereas terrestrial mesophilic bacteria did not grow at 30 °C under 600 atm pressure, and that some strains were killed, marine bacteria isolated from depths where this order of pressure obtains grew readily; they introduced the term 'barophilic' for such bacteria whose growth or metabolism was favoured by high pressure.

Many barophiles were recovered from samples of the uppermost layers of sediments taken from the Philippine Trench at depths greater than 10000 m. In culture at 2·5 °C (ocean floor temperature) 10 to 1000 times more bacteria developed at 1000 atm pressure than at atmospheric pressure. Mostly, the barophiles are rod-shaped cells 2 to 4 μm long, many encapsulated and many forming endospores. The genera *Pseudomonas*, *Vibrio*, *Spirillum*, *Bacillus* and *Clostridium* were all represented. Bacteria also occurred in the deeper sediments, down to about 1 m below the ocean floor, an environment likely to favour longevity of cells and activity of anaerobes (ZoBell 1952).

Attempts to grow these deep sea bacteria at various pressures and temperatures revealed that the two factors were not acting independently. For example, some cultures grew under 400 to 600 atm at 40 °C, a temperature too high for their growth at atmospheric pressure. The same high pressures retarded growth at lower temperature. Cold greatly accentuated the growth-inhibiting and killing effects of pressure. Conversely, the adverse effects of pressure were alleviated by heat.

Pressure does not cause damage by crushing the cells, but is though to act by changing molecular volumes. It so happens that high pressure acts on bacterial growth in the same direction as cooling. These temperature–pressure relationships are affected by the specific temperature characteristics of each organism. For example, high pressure might prevent a thermophile growing at its usually high optimal temperature but permit growth at temperatures higher than this. ZoBell (1958) found that one strain of thermophilic sulphate-reducing bacteria that did not grow above 85 °C at atmospheric pressure, grew at 104 °C under a pressure of 1000 atm, the highest growth temperature recorded.

It is considered that barophiles possibly adapt to high pressure by developing metabolic mechanisms that operate with little or no changes in molecular volumes: thus, pressure would have little effect. This could explain why some barophiles can be cultivated at atmospheric pressure, even for several years, and still retain their barophilic properties (ZoBell & Johnson 1949).

Pressure changes can have profound effects on the morphology of marine bacteria. For example, in sea-water broth at ordinary pressures, *Serratia marinorubra* grew as a small rod, 0·3 to 0·5 μm × 0·6 to 1·5 μm, but at 600 atm the cells grew to form filaments up to 200 μm long with no sign of division. When the long forms were transferred to atmospheric pressure they began to fragment after a few minutes (ZoBell & Oppenheimer 1950).

Halophiles

Micro-organisms tolerate greater or lesser concentrations of sodium chloride in their environment; some, the halophiles, require this salt. Flannery (1956) defines facultative halophiles as those that will grow with less than 2 % of salt but grow better with more, and obligate halophiles as those that grow only with more than 2 % of salt, 15 to 25 % being usually required for satisfactory growth.

Halophiles, especially bacteria, occur in natural briny waters such as the Dead Sea (Wilkansky 1936) and in man-made habitats wherever high concentrations of salt are employed. Typically, the extreme halophilic species are red-pigmented (carotenoids) and their presence is often indicated by the reddening of suitable substrates such as crude salt, salted fish and salted hides. The brine flagellate, *Dunaliella salina*, is also coloured bright red by carotenoids.

No Gram-positive extreme halophiles are known (Brown 1964) though some strains of *Staphylococcus* and *Micrococcus* tolerate very high salt concentrations (Christian & Waltho 1962): such tolerance is a reflexion, at least for some species, of a very small requirement for water (Scott 1953). Most extreme halophiles are Gram-negative rods, and some have growth temperature optima of 40 to 50 °C which is high for Gram-negative bacteria (Gibbons & Payne 1961). Some bacteria can adjust to higher salt concentrations; this adjustment can be reversible (Kluyver & Baars 1932) or irreversible (Hof 1935).

The fact that the freezing-points of halophiles are often lower than those of non-halophiles (Brown 1964) suggests that their cells contain more concentrated solutes. Certainly, the intracellular enzyme systems of halophilic bacteria are active in high concentrations of salts, although the concentration favouring optimal activity of cell-free preparations is often lower than that necessary for growth of the organism. For example, the nitritase and lactic dehydrogenase of *Micrococcus halodenitrificans* were not resistant to the sodium chloride content optimal for growth. Thus, the salt concentration within a halophilic cell, though possibly large compared with that of a non-halophile, may be less than in the surrounding medium (see Robinson 1952; Robinson & Gibbons 1952; Robinson, Gibbons & Thatcher 1952; Baxter & Gibbons 1954).

Though the cell envelopes of halophiles and non-halophiles have properties in common, for instance, they all permit cells to accumulate K^+ ions and exclude Na^+ ions against concentration gradients, there are chemical and structural differences between them. The cell membranes of extreme halophiles (e.g. *Halobacterium* spp.) contain little or no mucopeptide, but proportionately more protein and less carbohydrate than non-halophilic Gram-negative bacteria. Moreover, such cell membranes (e.g. of *H. halobium*, one of the few species examined) break up in solutions of low ionic strength; the membranes require a high concentration of cations for mechanical stability. This cation-dependent stability of the membranes is correlated with a larger proportion of aspartic and glutamic acid in the protein than of lysine and arginine. Isolated membranes of a non-halophilic pseudomonad acquired halophilic properties when acidic groups (—COO^-) were substituted for lysine-ϵ-NH_2^+ groups. It is suggested that all bacteria needing much salt have membranes of this type, which are stabilized by a high concentration of cations (Brown 1964).

Extreme halophiles are strict aerobes; perhaps very saline environments with restricted aeration and adequate supplies of organic nutrients are rare so that there is little evolutionary opportunity for anaerobes to develop. However, some sulphate-reducers tolerate concentrated salt solutions (ZoBell 1958), as do some facultative anaerobes such as staphylococci and a *Bacillus* sp. (Eddy & Ingram 1956). One species of obligate anaerobe, the sulphate-reducer *Desulfovibrio salixigens*, requires 2·5 to 5 % of salt (Postgate & Campbell 1966). However, here the requirement is for chloride ions and not, as is more usual, for cations.

Osmophiles

Only a very restricted microflora can develop in media of very high osmotic pressure such as that provided by 20 to 50 % sugar solutions. Some microbes have adapted to this environment, the outstanding examples being the osmophilic yeasts able to grow in honey and molasses. These yeasts have been referred to the genus *Saccharomyces*, mostly to *S. rouxii*. Some of these yeasts are obligate osmophiles; others can adapt themselves to grow in more dilute sugar solutions. Osmophilic yeasts form only a small proportion of the naturally occurring microflora on fruits but they predominate on the surfaces of dried grapes and dates (Scarr 1953).

Some other fungi have osmophilic properties: examples are *Xeromyces bisporus* (Bunker 1967) and *Aspergillus glaucus*, both of which will grow in 60 to 65 % sugar solutions (see Scarr 1953; Borgstrom 1961).

Acidity and alkalinity

Most micro-organisms grow in the pH range of 3·5 to 9·0, but a few tolerate extreme acidity: the thiobacilli being notable examples. *Thiobacillus thiooxidans* grows at pH values down to 0·5, although there is evidence that the internal pH of the cell is about 6 to 7. Certainly, enzymes isolated from such cells operate at neutrality rather than in acid solution. *T. thiooxidans* brings about its own extremely acid conditions by forming sulphuric acid. It can survive in 7 % sulphuric acid and will stand subculture directly into a medium of not less than pH 4 (H. J. Bunker, personal communication). The mould *Acontium velatum* also grows in media containing up to 2·5 N sulphuric acid, and in media saturated with copper sulphate (Starkey & Waksman 1943).

The remarkable alga *Cyanidium caldarium* lives in hot acid springs and has excited as much interest by virtue of its thermophilic as by its acidophilic properties. It has an optimal growth temperature range of 45 to 50 °C but will grow at up to 75 °C at a pH of *ca.* 1·2 (corresponding to about 0·1 N sulphuric acid) (see Allen 1959).

There are also examples of adaptation to high alkalinity. Bacteria such as *Streptococcus faecalis* and *Bacillus circulans* can grow at pH 10 to 11 and there are reports of algae occurring in alkaline lakes at pH 9 to 11. The blue-green alga *Plectonema nostocorum* grows at pH 13 which seems to be the highest at which active life has been recorded (Geiger *et al.* 1965).

This is a convenient point to refer to work with atmospheres enriched with ammonia gas. Some higher plants and animals tolerate gaseous ammonia and it has also been found that many micro-organisms grow from soil samples incubated in ammonia-rich atmospheres (Siegel & Giumarro 1966). When soil samples from Harlech, Wales, were inoculated on to a meat extract peptone agar, and the cultures incubated in an 'ammonia-air' mixture (containing 25% ammonia and 25% methane) an unusual micro-organism appeared (Siegel & Giumarro 1966). This was an umbrella-like structure of *ca.* 5 μm diam. with a stalk *ca.* 10 μm long, and although it had no obvious affinity with any known species of soil micro-organism, it was remarkably similar to the Pre-Cambrian fossil organism *Kakabekia umbellata* occurring in the Gunflint chert formation of Southern Ontario (Barghoorn & Tyler 1965*a*, *b*). It may be that the modern form flourishes only in atmospheres similar to those in which *Kakabekia* may have lived (Symposium 1965) and which are now to be found only in micro-environments such as soils where ammonia is produced and aeration is restricted. Other organisms of more familiar type also grew in these ammonia-rich mixtures; bacteria, fungi, actinomycetes, blue-green algae and protozoa.

Anaerobiosis

Although probably all multicellular and most unicellular organisms require oxygen, many species of bacteria are obligate anaerobes and cannot grow in its presence. Moreover, many other bacteria and yeasts are facultative anaerobes which usually grow better with oxygen but can grow without it provided suitable nutrients are available.

The obligate anaerobes are of great practical importance in clinical pathology and food preservation, and accounts of their activities form a substantial part of the literature of bacteriology. From less specialized viewpoints, these organisms pose some interesting and difficult problems, such as their significance in nitrogen fixation and the breakdown of organic matter in their natural environments, soils and muds.

The survival and growth of anaerobes is of particular interest to workers concerned with problems of contamination of extra-terrestrial environments, most of which are likely to be deficient in oxygen.

Growth and survival in simulated extra-terrestrial environments

Rather little attention has hitherto been given to the effect of two or more extreme environmental conditions acting together, though some examples of complex extreme habitats (very cold and saline; cold and high pressure; hot and acid) have been mentioned above. The study of such complex extreme sets of conditions is, however, an important part of space research programmes. The basis of simulation work is to study how terrestrial organisms react to environments likely to be found on other planets. Such experiments can indicate whether any particular environment, such as that of Mars, prohibits life as we know it, and they can also demonstrate whether terrestrial micro-organisms contaminating space vehicles are likely to survive or grow in their new environment. Emphasis is very much on micro-organisms in this work because they are the only living things likely to be carried in a space vehicle without being noticed, and they are also the most likely to find ecological niches suited to them. Many of these studies have also been aimed at establishing the effects of environmental factors acting in a cyclical manner. Such changing factors have been little investigated: the usual concern of the microbiologist is to keep experimental conditions as constant as possible (see Siegel *et al.* 1965).

Much recent information derives from numerous studies on the simulated 'Martian' environment. Experimental details vary considerably but the following conditions are typical: soil samples inoculated with the organisms under test are subjected to low-pressure atmospheres and a daily freeze–thaw cycle with about 4 h spent above 0 °C. Very small amounts of water are usually provided, and organic matter is sometimes included. Many such experiments have been recorded and reviewed (Jenkins 1966).

Bacteria survive these treatments extremely well, especially the sporeformers as would be expected. For example, in some experiments bacteria survived for at least 6 months in all soil samples tested in a low pressure, almost dry anoxic atmosphere and a 12 h freeze–thaw cycle ranging from −60 to +20 °C (Packer, Scher & Sagan 1963). Dryness and cold are not inconsistent with prolonged survival, and even very small and transient amounts of water permit some growth. Hawrylewicz *et al.* (1967) showed in environmental studies of this kind that bacterial species have different requirements for water, but with all the organisms studied, *Staphylococcus albus*, *S. aureus*, *Bacillus cereus*, *Pseudomonas aeruginosa*, *Lactobacillus plantarum* and PA 3679, the extreme diurnal temperature cycle aided conservation of water, and a limited amount of growth occurred.

It seems clear that terrestrial microbes might survive when introduced to the Martian environment, but would be unlikely to find favourable growth conditions. A large gap in our knowledge is the nature and amount of organic matter that might be available to terrestrial contaminants, and on which their growth must depend.

Conclusions

Micro-organisms grow in what seem to be remarkably forbidding environments, and to meet these conditions the cell becomes modified in many different ways. Cell enzymes and enzyme systems are formed with optimal activities at very different temperatures or at different ionic strengths, or are stabilized against heat denaturation by cations. Cell envelopes are modified to prevent their dissolution by extreme salinity, and there are cell repair mechanisms, possibly operating to counteract the damaging effect of heat, and certainly to offset radiation damage (Lett *et al.* 1967).

Considered above are some extreme physical and chemical factors characteristic of whole environments, and which manifest themselves by restricting the numbers of species that flourish. The more extreme the conditions, the fewer the species able to tolerate them, and the slower they grow. For instance, there are very few records of growth above 80 °C, and only one above 90 °C, yet in such situations as hot springs a suitably adapted organism may flourish untroubled by competitors.

It must be remembered, however, that these extreme factors are superimposed on others such as nutritional factors, that are limiting the growth of individual cells or species within that environment. Moreover, conditions for growth can be 'unfavourable' and thereby growth-limiting, without any one factor being obviously 'extreme'. The vigorous development of any one species of micro-organism is not so common in nature, except, paradoxically, in certain extreme conditions such as the hot springs mentioned above, and in some specialized saprophytic and parasitic associations. In soils, for example, any one species of micro-organism usually is not abundant even though the total numbers present are very large: many species exist together but growth of any one is limited.

No doubt, new man-made habitats, seemingly inimical to microbial life, will be developed and it will probably be found that some micro-organisms will be able to adapt to the new conditions. Moreover, not all natural environments have yet been exhaustively examined for the presence of unusual organisms: some have only recently been discovered. Thus, Miller (1964) and Charnock (1964) drew attention to the existence of pools of warm brine in depressions at the bottom of the Red Sea. The hottest of these pools, Atlantis II, is at 56 °C, is under 2000 m of normal sea water (hydrostatic pressure of *ca.* 200 atm), has 20 to 30 % total solids mostly sodium chloride, is rich in iron and manganese, and has no free oxygen (Munns, Stanley & Densmore 1967). This water is similar to that of oil field and deep well brines in which bacteria have been found (ZoBell 1958). So far, micro-organisms have not been reported from the Red Sea brines, though it is probably too soon for investigations to have been made, yet it would seem rash to regard these pools as sterile in view of what has been noted above about the ability of organisms to survive a hostile environment.

I am indebted to Mr H. J. Bunker for his helpful comments on unusual micro-organisms, and to Dr J. J. Patel for allowing me to refer to his unpublished work on thermophilic actinophage.

References (Skinner)

Agre, N. S. 1961 Phage of the thermophilic *Micromonospora vulgaris*. *Mikrobiologiya* **30**, 414–417.

Algranati, I. D. & Lengyel, P. 1966 Polynucleotide-dependent incorporation of amino acids in a cell-free system from thermophilic bacteria. *J. biol. Chem.* **241**, 1778–1783.

Allen, M. B. 1950 The dynamic nature of thermophily. *J. gen. Physiol.* **33**, 205–214.

Allen, M. B. 1953 The thermophilic aerobic sporeforming bacteria. *Bact. Rev.* **17**, 125–173.

Allen, M. B. 1959 Studies with *Cyanidium caldarium*, an anomalously pigmented chlorophyte. *Archiv. Microbiol.* **32**, 270–277.

Amelunxen, R. E. 1967 Some chemical and physical properties of thermostable glyceraldehyde-3-phosphate dehydrogenase from *Bacillus stearothermophilus*. *Biochem. biophys. Acta* **139**, 24.

Apinis, A. E. 1963*a* Thermophilous fungi of coastal grasslands. In *Soil organisms* (ed. J. Doeksen and J. van der Drift.) Amsterdam: North Holland Publishing Co.

Apinis, A. E. 1963*b* Occurrence of thermophilous microfungi in certain alluvial soils near Nottingham. *Nova Hedwigia* **5**, 57–78.

Bairiev, C. B. & Mamedov, S. M. 1963 A thermophilic bacterium isolated from ozokerite. *Fed. Proc.* (*Trans. Suppl.*) **22** (6II), 1224–1226.

Barghoorn, E. S. & Tyler, S. A. 1965*a* Microorganisms from the Gunflint Chert. *Science* **147**, 563–577.

Barghoorn, E. S. & Tyler, S. A. 1965*b* Microorganisms of middle Precambrian age from the Animikie series, Ontario, Canada. In *Current aspects of exobiology* (ed. G. Mamikunian and M. R. Briggs), pp. 93–118. New York: Pergamon Press.

Bartholomew, J. W. & Paik, G. 1966 Isolation and identification of obligate thermophilic sporeforming bacilli from ocean basin cores. *J. Bact.* **92**, 635–638.

Baxter, R. M. & Gibbons, N. E. 1954 The glycerol dehydrogenases of *Pseudomonas salinaria*, *Vibrio costicolus* and *Escherichia coli* in relation to bacterial halophilism. *Can. J. Biochem. Physiol.* **32**, 206–217.

Beerens, H., Sugama, S. & Tahon-Castel, M. 1965 Psychrotrophic clostridia. *J. appl. Bact.* **28**, 36–48.

Borgstrom, G. 1961 Unsolved problems in frozen food microbiology. *Proc. Low Temp. Microbiol. Symp.* pp. 197–250. Campbell Soup Co.

Brock, T. D. 1967 Microorganisms adapted to high temperatures. *Nature, Lond.* **214**, 882.

Brock, T. D. & Brock, M. L. 1966 Temperature optima for algal development in Yellowstone and Iceland hot springs. *Nature, Lond.* **209**, 733–734.

Brown, A. D. 1964 Aspects of bacterial response to the ionic environment. *Bact. Rev.* **28**, 296–329.

Brown, D. K., Militzer, W. & Georgi, C. E. 1957 The effect of growth temperature on the heat stability of a bacterial pyrophosphatase. *Arch. Biochem. Biophys.* **70**, 248–256.

Bunker, H. J. 1967 Dealing with microbiological trouble-makers. In *Biology and the manufacturing industries*, Symposium no. 16, Institute of Biology, 1966 (ed. M. Brook), pp. 63–71. London: Academic Press.

Campbell, L. L. 1955 Purification and properties of an α-amylase from facultative thermophilic bacteria. *Arch. Biochem. Biophys.* **54**, 154–161.

Charnock, H. 1964 Anomalous bottom water in the Red Sea. *Nature, Lond.* **203**, 591.

Christian, J. H. B. & Waltho, J. A. 1962 The water relations of staphylococci and micrococci. *J. appl. Bact.* **25**, 369–377.

Daron, H. H. 1967 Occurrence of isocitrate lyase in a thermophilic *Bacillus* species. *J. Bact.* **93**, 703–710.

Eddy, B. P. 1960 The use and meaning of the term 'psychrophilic'. *J. appl. Bact.* **23**, 189–190.

Eddy, B. P. & Ingram, M. 1956 A salt-tolerant denitrifying *Bacillus* strain which 'blows' canned bacon. *J. appl. Bact.* **19**, 62–70.

Festenstein, G. N., Lacey, J., Skinner, F. A., Jenkins, P. A. & Pepys, J. 1965 Self-heating of hay and grain in Dewar flasks and the development of farmer's lung antigens. *J. gen. Microbiol.* **41**, 389–407.

Flannery, W. L. 1956 Current status of knowledge of halophilic bacteria. *Bact. Rev.* **20**, 49–66.

Friedman, S. M., Axel, R. & Weinstein, I. B. 1967 Stability of ribosomes and ribosomal ribonucleic acid from *Bacillus steaorthermophilus*. *J. Bact.* **93**, 1521–1526.

Friedman, S. M. & Weinstein, I. B. 1966 Protein synthesis in a subcellular system from *Bacillus stearothermophilus*. *Biochim. Biophys. Acta* **114**, 593–605.

Geiger, P. J., Jaffe, L. D. & Mamikunian, G. 1965 Biological contamination of the planets. In *Current aspects of exobiology* (ed. G. Mamikunian and M. H. Briggs), pp. 283–322. New York: Pergamon Press.

Gibbons, N. E. & Payne, J. I. 1961 Relation of temperature and sodium shloride concentration to growth and morphology of some halophilic bacteria. *Can. J. Microbiol.* **7**, 483–489.

Gregory, P. H., Lacey, M. D., Festenstein, G. N. & Skinner, F. A. 1963 Microbial and biochemical changes during the moulding of hay. *J. gen. Microbiol.* **33**, 147–174.

Hawrylewicz, E. J., Hagen, C. A., Tolkacz, V., Anderson, B. T. & Ewing, M. 1967 Probability of growth (P_G) of viable microorganisms in Martian environments. Paper L. 63. Cospar. London, 1967.

Hof, T. 1935 Investigations concerning bacterial life in strong brines. *Recl Trav. bot. néerl.* **32**, 92–173.

Ingraham, J. L. & Stokes, J. L. 1959 Psychrophilic bacteria. *Bact. Rev.* **23**, 97–108.

Jenkins, D. W. 1966 *Significant achievements in space bioscience* 1958–1964. Publication NASA SP-92. Washington D.C.: National Aeronautics and Space Administration.

Kanwisher, J. 1957 Freezing and drying in intertidal algae. *Biol. Bull. mar. biol. Lab., Woods Hole* **113**, 275–285.

Kempner, E. S. 1963 Upper temperature limit of life. *Science* **142**, 1318.

Kluyver, A. J. & Baars, J. K. 1932 On some physiological artifacts. *Proc. K. ned. Akad. Wet. (Sect. Sci.)* **35**, 370–378.

Koffler, H. 1957 Protoplasmic differences between mesophiles and thermophiles. *Bact. Rev.* **21**, 227–240.

Koffler, H., Mallett, G. E. & Adye, J. 1957 Molecular basis of biological stability to high temperatures. *Proc. natn Acad. Sci. U.S.A.* **43**, 464–476.

Lark, K. G. & Adams, M. H. 1953 The stability of phages as a function of the ionic environment. *Cold Spring Harb. Symp. Quant. Biol.* **18**, 171–183.

Lett, J. T., Feldschreiber, P., Little, J. G., Steele, K. & Dean, C. J. 1967 The repair of X-ray damage to the deoxyribonucleic acid in *Micrococcus radiodurans*: a study of the excision process. *Proc. Roy. Soc.* B **167**, 184–201.

Levitt, J. 1956 *The hardiness of plants. Agronomy*, vol. VI. New York: Academic Press.

Lewin, R. A. 1962 *Physiology and biochemistry of algae.* New York: Academic Press.

Mangiantini, M. T., Tecce, G., Toschi, G. & Trentalance, A. 1965 A study of ribosomes and of ribonucleic acid from a thermophilic organism. *Biochim. Biophys. Acta* **103**, 252–274.

Marsh, C. & Militzer, W. 1956*a* Thermal enzymes VII. Further data on an adenosinetriphosphatase. *Arch. Biochem. Biophys.* **60**, 433–438.

Marsh, C. & Militzer, W. 1956*b* Thermal enzymes. VII. Properties of a heat-stable inorganic pyrophosphatase. *Arch. Biochem. Biophys.* **60**, 439–451.

McBee, R. H. 1950 Anaerobic thermophilic cellulolytic bacteria. *Bact. Rev.* **14**, 51–63.

Meyer, G. H., Morrow, M. B., Wyss, O., Berg, T. E. & Littlepage, J. L. 1962 Antarctica: The microbiology of an unfrozen saline pond. *Science* **138**, 1103–1104.

Miller, A. R. 1964 Highest salinity in the world ocean? *Nature, Lond.* **203**, 590–591.

Militzer, W., Sonderegger, T. B., Tuttle, L. C. & Georgi, C. E. 1949 Thermal enzymes. *Arch. Biochem.* **24**, 75–82.

Miquel, P. 1888 Monographie d'un bacille vivant au-delà de 70° centigrades. *Ann. Micrographie*, **1**, 3–10.

Munns, R. G., Stanley, R. J. & Densmore, C. D. 1967 Hydrographic observations of the Red Sea brines. *Nature, Lond.* **214**, 1215–1217.

O'Brien, R. T. & Campbell, L. L. 1957 Purification and properties of a proteolytic enzyme from *Bacillus stearothermophilus. Arch. Biochem. Biophys.* **70**, 432–441.

Olsen, R. H. 1967 Isolation and growth of psychophilic bacteriophage. *Appl. Microbiol.* **15**, 198.

Oprescu, V. 1898 Studien über thermophile Bacterien. *Arch. Hyg.* **33**, 164–186.

Pace, B. & Campbell, L. L. 1967 Correlation of maximal growth temperature and ribosome heat stability. *Proc. natn Acad. Sci. U.S.A.* **57**, 1110–1116.

Packer, E., Scher, S. & Sagan, C. 1963 Biological contamination of Mars. II. Cold and aridity as constraints on the survival of terrestrial microorganisms in simulated Martian environment. *Icarus* **2**, 293–316.

Postgate, J. R. & Campbell, L. L. 1966 Classification of *Desulfovibrio* species, the nonsporulating sulfate-reducing bacteria. *Bact. Rev.* **30**, 732–738.

Recknitz, B. & Janota-Bassalik, L. 1967 The effect of temperature on the glucose utilization by *Bacillus stearothermophilus. Acta microbiol. pol.* **16**, 13–18.

Robinson, J. 1952 The effects of salts on nitritase and lactic acid dehydrogenase activity of *Micrococcus halodenitrificans. Can. J. Bot.* **30**, 155–163.

Robinson, J. & Gibbons, N. E. 1952 The effect of salts on the growth of *Micrococcus halodenitrificans* (n.sp.). *Can. J. Bot.* **30**, 147–154.

Robinson, J., Gibbons, N. E. & Thatcher, F. S. 1952 A mechanism of halophilism in *Micrococcus halodenitrificans*. *J. Bact.* **64**, 69–77.

Saunders, G. F. & Campbell, L. L. 1966*a* Ribonucleic acid and ribosomes of *Bacillus stearothermophilus*. *J. Bact.* **91**, 332–339.

Saunders, G. F. & Campbell, L. L. 1966*b* Characterization of a thermophilic phage for *Bacillus stearothermophilus*. *J. Bact.* **91**, 340–348.

Scarr, M. P. 1953 Microbiology of sugar-preserved foods. *Proc. Soc. Appl. Bact.* **16**, 119–127

Scott, W. J. 1953 Water relations of *Staphylococcus aureus* at 30 °C. *Aust. J. biol. Sci.* **6**, 549–564.

Siegel, S. M. & Giumarro, C. 1966 On the culture of a micro-organism similar to the Precambrian microfossil *Kakabekia umbellata* Banghoorn in NH_3-rich atmospheres. *Proc. natn Acad. Sci. U.S.A.* **55**, 349–353.

Siegel, S. M., Renwick, G., Daly, O., Giumarro, C., Davis, G. & Halpern, L. 1965 The survival capabilities and the performance of earth organisms in simulated extra-terrestrial environments. In *Current aspects of exobiology* (ed. G. Mamikunian and M. A. Briggs), pp. 119–178. New York: Pergamon Press.

Stenesh, J. & Yang, C. 1967 Characterization and stability of ribosomes from mesophilic and thermophilic bacteria. *J. Bact.* **93**, 930–936.

Starkey, R. L. & Waksman, S. A. 1943 Fungi tolerant to extreme acidity and high concentrations of copper sulphate. *J. Bact.* **45**, 509–519.

Symposium on the evolution of the Earth's atmosphere 1965 *Proc. natn Acad. Sci. U.S.A.* **53**, 1169–1226.

Upadhyay, J. & Stokes, J. L. 1962 Anaerobic growth of psychophilic bacteria. *J. Bact.* **83**, 270–275.

Vallentyne, J. R. 1963 Environmental biophysics and microbial ubiquity. *Ann. N.Y. Acad. Sci.* **108**, 342–352.

Vallentyne, J. R. 1965 Why exobiology? In *Current aspects of exobiology* (ed. G. Mamikunian and M. A. Briggs.), pp. 1–12. New York: Pergamon Press.

White, R., Georgi, C. E. & Militzer, W. 1954 Heat studies on thermophilic bacteriophage. *Proc. Soc. Expt. Biol. Med.* **85**, 137–139.

Wilkansky, B. 1936 Life in the Dead Sea. *Nature, Lond.* **138**, 467.

Zernow, S. A. 1944 On limits of life at negative temperatures. *C. r. hebd. Séanc. Acad. Sci., Paris.* **44**, 76–77.

ZoBell, C. E. 1952 Bacterial life at the bottom of the Philippine Trench. *Science* **115**, 507–508.

ZoBell, C. E. 1958 Ecology of sulfate reducing bacteria. *Producers Mon.* **22**, 12–29.

ZoBell, C. E. & Johnson, F. H. 1949 The influence of hydrostatic pressure on the growth and viability of terrestrial and marine bacteria. *J. Bact.* **57**, 179–189.

ZoBell, C. E. & Morita, R. Y. 1957 Barophilic bacteria in some deep sea sediments. *J. Bact.* **73**, 563–568.

ZoBell, C. E. & Oppenheimer, C. H. 1950 Some effects of hydrostatic pressure on the multiplication and morphology of marine bacteria. *J. Bact.* **60**, 771–781.

Part I

HYDROGEN-ION CONCENTRATION

Editor's Comments on Papers 2 Through 7

2 **HARDEN**
The Chemical Action of Bacillus coli communis *and Similar Organisms on Carbohydrates and Allied Compounds*

3 **NEUBERG and HIRSCH**
The Third Form of Sugar Fermentation

4 **MICKELSON and WERKMAN**
Influence of pH on the Dissimilation of Glucose by Aerobacter indologenes

5 **STOKES**
Fermentation of Glucose by Suspensions of Escherichia coli

6 **NEISH and BLACKWOOD**
Dissimilation of Glucose by Yeast at Poised Hydrogen Ion Concentrations

7 **ALROY and TANNENBAUM**
The Influence of Environmental Conditions on the Macromolecular Composition of Candida utilis

Sir Arthur Harden continued the work of Pasteur and others on alcoholic fermentation and received the Nobel Prize in Chemistry in 1929. He was primarily interested in alcoholic fermentations and was a leading investigator. He published a primary review of the early concepts of fermentation in 1923. Much of his work is outside the scope of this volume, but in 1901 he obtained the first relatively complete fermentation balance for *Bacillus coli communis* (Paper 2). This study allowed a stoichiometric equation to be derived for the fermentation of glucose by several different strains of *Bacillus coli communis (Escherichia coli)* isolated from different sources.

$$2C_6H_{12}O_6 + H_2O = 2C_3H_6O_3 + C_2H_4O_2 + C_2H_6O + 2CO_2 + 2H_2$$

He compared the fermentation products obtained with *E. coli* with those produced by *Bacillus typhosus (Salmonella typhosa)*.

The results were similar except that formic acid was produced by *B. typhosus* instead of CO_2 and H_2. Fermentation of fructose by *E. coli* gave the same products as glucose and in the same proportion. His techniques and interpretations were not improved upon for many years. Studies of bacterial fermentation were extended by Harden and Walpole (1906), Harden and Norris (1912, 1913), Kay (1926), Tikka (1935), Mickelson and Werkman (Paper 4), Stokes (1942), Gale and Epps (1942), Gunsalus and Campbell (1944), and Stokes (Paper 5).

Carl Neuberg and his associates studied the effect of pH and bisulfite on ethanol fermentation by yeast. He reported the results of many of his studies in *Biochemische Zeitschrift* during 1917–1920. Several studies were completed prior to World War I, and his bisulfite fermentation was employed for the production of glycerol during the war. Carl Neuberg and associates described three basic types of fermentations by intact yeast cells (Neuberg, Hirsch, and Reinfurth, 1920) and a fourth type by dried cells (Neuberg and Kobel, 1930). The third form of fermentation is of particular interest; Neuberg and Hirsch (Paper 3) describe the effect of alkaline medium on yeast fermentation of sugar. In this form of fermentation glycerol is produced at the expense of ethanol in what he describes as a Cannizzaro reaction. The overall reaction was described as

$$2 \text{ glucose} = 2 \text{ glycerol} + \text{acetic acid} + \text{ethanol} + 2CO_2$$

Mickelson and Werkman (Paper 4) were among the first to carry out fermentations with constant pH control. They found that pH had a marked effect on the 2,3-butanediol fermentation by *Aerobacter aerogenes.* Above pH 6.3, fermentation resulted in the accumulation of acetic and formic acids. The production of hydrogen and carbon dioxide was suppressed. Below pH 6.3, acetic acid was converted into acetoin and 2,3-butylene glycol. Studies by Neish and Ledingham (1949) utilized automatic pH control. *Aerobacter* had a broad optimum pH range of 6.5 to 7.6

Stokes (Paper 5) studied the fermentation of glucose by resting cell suspensions of *Escherichia coli* at several pH values. Ethanol, acetic acid, lactic acid, and succinic acid were the major products. A pronounced effect of pH was found with phosphate but not with bicarbonate buffer. The amount of lactic acid produced at high pH values in phosphate buffers is greatly reduced and accompanied by increases in ethanol and formic acid.

Neish and Blackwood (Paper 6) utilized automatic pH control

with either NH_4OH or NaOH to study the catabolism of glucose by distiller's yeast. When NH_4OH was used to control pH, virtually all the glucose was catabolized in the pH range from 2.4 to 7.6 The yield of glycerol was increased approximately fivefold by raising the pH from 3 to 7.6. The concentration of glucose also affected the ratio of the end products. The authors present a complete analysis of the fermentation products. Although ethanol, glycerol, and acetic acid were the major products, as expected, traces of 2,3-butanediol, acetoin, butyric acid, formic acid, succinic acid, and lactic acid were found.

In Paper 7, Alroy and Tannenbaum consider the effect of pH on the macromolecular composition of *Candida utilis*. They utilized glucose-limited chemostat cultures cultivated at various pH levels, temperatures, and dilution rates with different nitrogen sources to study the influence of environmental conditions. Varying the growth pH from 3.05 to 7.5 caused no significant change in the level of nucleic acid or protein.

Gale and Epps (1942) determined the effect of the pH of the culture medium on the enzymatic activities of the organisms. The authors found that the investigated enzymes could be divided into two main groups according to their response to the pH of the growth medium. Formic dehydrogenase, catalase, urease, and fumarase fell into a group in which a compensatory formation of enzyme occurred as the growth pH deviated from the optimum for the enzyme, so that the effective activity remained approximately constant. A second group of enzymes were best formed when the growth pH approximated their optimum pH. The second group included succinic dehydrogenase, deaminases, decarboxylases, hydrogenase, and tryptophanase. The enzymes in the first group seemed to have a protective function and those in the second group an adaptive function. When grown under acidic conditions, the cells (*E. coli* and *Micrococcus lysodeikticus*) produced decarboxylases, which resulted in a decrease in the acidity of the medium. Under basic growth conditions the cells produced deaminases, which resulted in a decrease in pH. Gunsalus and Niven (1942) found that *Streptococcus liquefaciens* produced lactic acid almost exclusively below pH 6.5, but above pH 6.5 large quantities of formic and acetic acids and ethanol in the ratio of 2 : 1 were formed from glucose. These products accounted for 25 to 40 percent of the sugar fermented. Gunsalus and Campbell (1944) eliminated lactic acid as a fermentation product of *Streptococcus faecalis* by using an oxidized substrate and shifting the growth pH to an alkaline reaction. Thus lactic acid fermentation was not essential to the homofermentative streptococci.

REFERENCES

Gale, E. F., and H. M. R. Epps. 1942. The effect of pH of the medium during growth on the enzymic activities of bacteria (*Escherichia coli* and *Micrococcus lysodeikticus*) and the biological significance of the changes produced. *Biochem. J. 36*(7): 600–618.

Gunsalus, I. C., and J. J. R. Campbell. 1944. Diversion of the lactic acid fermentation with oxidized substrate. *J. Bacteriol. 48*(2): 455–561.

———, and C. F. Niven. 1942. Effect of pH on the lactic acid fermentation. *J. Biol. Chem. 145*(1): 131–136.

Harden, A. 1923. *Alcoholic Fermentation.* Longmans Green, London.

———, and D. Norris. 1912. The bacterial production of acetylmethylcarbinol and 2.3-butylene glycol from various substances. *Proc. Roy. Soc. (London) B84*(573): 492–499.

———, and D. Norris. 1913. The bacterial production of acetylmethylcarbinol and 2.3-butylene glycol from various substances II. *Proc. Roy. Soc. (London) B85*(576): 73–78.

———, and G. S. Walpole. 1906. Chemical action of *Bacillus lactis aerogenes* (Escherich) on glucose and mannitol: Production of 2:3-butylene glycol and acetylmethylcarbinol. *Proc. Roy. Soc. (London) B77*(519): 399–405.

Kay, H. D. 1926. XLII. Note on the variation in the end-products of bacterial fermentation resulting from increasing combined oxygen in the substrate. *Biochem. J. 20*(2): 321–329.

Neish, A. C., and G. A. Ledingham. 1949. Production and properties of 2.3-butanediol XXXII. 2.3-Butanediol fermentations at poised hydro gen ion concentrations. *Can. J. Res. B27*(1): 694–704.

Neuberg, C., and M. Kobel. 1930. Die Zerlegung von nicht phosphoryliertem Zucker durch Hefe unter Bildung von Glycerin und Brenztraubensäure. *Biochem. Z. 229*:446–454.

———, J. Hirsch, and E. Reinfurth. 1920. Die drei Vergärungsformen des Zuckers, ihre Zusammenhänge und Bilanz. *Biochem. Z. 105*: 307–338.

Stokes, J. L. 1942. Enzymatic aspects of gas formation by salmonella. *J. Bacteriol. 72*(2): 269–275.

Tikka, J. 1935. Über den Mechanismus der Glucoservergärung durch *B. coli. Biochem. Z. 279*: 264–288.

SELECTED READINGS

Brock, T. D. 1973. Lower pH limit for the existence of blue-green algae: Evolutionary and ecological implications. *Science 179*(4072): 480–483.

Caldwell, P. C. 1956. Intracellular pH. *Int. Rev. Cytol. 5*: 229–277.

Conway, E. J., and M. Downey. 1950. pH values of the yeast cell. *Biochem. J. 47*(3): 355–360.

Gale, E. F. 1940. The production of amines by bacteria. I. The decarboxylation of amino-acids by strains of *Bacterium coli. Biochem. J. 34*(3): 392–413.

———. 1943. Factors influencing enzymatic activities of bacteria. *Bacteriol. Rev. 7*(2): 139–173.

———, and W. E. Van Heyningen. 1942. The effect of the pH and the presence of glucose during growth on the production of $\alpha-$ and $\theta-$ toxins and hyaluronidase by *Clostridium welchii. Biochem. J. 36*(7): 624–630.
Gunsalus, I. C. 1947. Products of aerobic glycerol fermentation by *Streptococcus faecalis. J. Bacteriol. 54*(2): 239–244.
———, and M. Gibbs. 1952. The heterolactic fermentation II. Position of C^{14} in the products of glucose dissimilation by *Leuconostoc mesenteroides. J. Biol. Chem. 194*(2): 871–875.
———, and J. M. Sherman. 1943. Fermentation of glycerol by streptococci. *J. Bacteriol. 45*(2): 155–162.
———, and W. W. Umbreit. 1945. The oxidation of glycerol by *Streptococcus faecalis. J. Bacteriol. 49*(4): 347–357.
Harden, A., and W. J. Young. 1906. The alcoholic ferment of yeast juice. Part II. The coferment of yeast juice. *Proc. Roy. Soc. (London) B78*(526): 369–375.
———, and W. J. Young. 1908. The alcoholic ferment of yeast juice. Part III. The function of phosphates in the fermentation of glucose by yeast juice. *Proc. Roy. Soc. (London) B80*(540): 299–311.
———, and W. J. Young. 1909. The alcoholic ferment of yeast juice. Part IV. The fermentation of glucose, mannose and fructose by yeast juice. *Proc. Roy. Soc. (London) B81*(549): 336–347.
———, and W. J. Young. 1910. The alcoholic ferment of yeast juice. Part V. The function of phosphates in alcoholic fermentation. *Proc. Roy. Soc. (London) B82*(556): 321–331.
———, and W. J. Young. 1912. The mechanism of alcoholic fermentation. *Biochem. Z. 40*: 458–478.
Harvey, R. J. 1965. Damage to *Streptococcus lactis* resulting from growth at low pH. *Bacteriol. 90*(5): 1331–1336.
Hewitt, L. F. 1957. Influence of hydrogen-ion concentration and oxidation–reduction conditions on bacterial behavior. *Symp. Soc. Gen. Microbiol. 7*: 42–55.
Iandolo, J. J., Z. J. Ordal, and L. D. Witter. 1964. The effect of incubation temperature and controlled pH on the growth of *Staphylococcus aureus* MF at various concentrations of NaCl. *Can. J. Microbiol. 10*(5): 808–811.
Marquis, R. E., N. Porterfield, and P. Matsumura. 1973. Acid–base titration of streptococci and the physical states of intracellular ions *J. Bacteriol. 114*(2): 491–498.
Neuberg, C., and L. Czapski. 1914. Über Carboxylase im Saft aus obergäriger Hefe. *Biochem. Z. 67*: 9–11.
———, and E. Färber. 1917. Über den Verlauf der alkolischen Gärung bei alkalischer Reaktion I. Zellfreie Gärung in alkalischen Lösungen. *Biochem. Z. 78*: 238–263.
———, and J. Hirsch. 1919. Wirkungsweise der Abfangmethode bei der Acetaldehyd-Glycerin-Spaltung des Zuckers. *Biochem. Z. 98*: 141–158.
———, and E. Reinfurth. 1918. Die Festlegung der Aldehydstuf bei der alkoholischen Gärung. Ein experimenteller Beweis der Acetaldehyd-Brenztraubensäuretheorie. *Biochem. Z. 89*: 365–414.

———, and P. Rosenthal. 1913. Uber zuckerfreie Hefegarungen XI. Weiteres zur Kenntnis der Carboxylase. *Biochem. Z. 51*: 128–42.

———, and W. Ursum. 1920. Die dritte Vergärungsform des Zuckers als allgemeine Folge der Dismutationswirkung anorganischer und organisher Alkalisatoren. *Biochem. Z. 110*: 193–215.

Pardee, A. B. 1961. Response of enzyme synthesis and activity to environment. *Symp. Soc. Gen. Microbiol. 11*: 19–40.

Reed, G., and J. H. Orr. 1923. The influence of ion concentration upon structure. I. *H. influenzae. J. Bacteriol. 8*(2): 103–113.

Svanberg, O. 1919. Velocity of growth of lactic acid bacteria in different hydrogen-ion concentrations. *Z. Physiol. Chem. 108*: 120–146.

2

Reprinted from *J. Chem. Soc.*, **79**, 610–628 (1901)

The Chemical Action of Bacillus coli communis and Similar Organisms on Carbohydrates and Allied Compounds.

By ARTHUR HARDEN.

THE following experiments were undertaken with the object of examining quantitatively the mode of action of the organisms of the Colon group as fermentative agents, in the hope of finding some characteristics which might serve as a means of classification or discrimination. The organisms of this group can scarcely be distinguished microscopically, but are found to vary in their action on various sugars and proteid compounds. The *Bacillus coli communis*, at one end of the series, has the following characteristic properties:

(*a*) It decomposes glucose with production of (1) acid and (2) gas.

(*b*) It curdles milk.

(*c*) It produces indole from peptone.

The *Bacillus typhosus*, which may be taken as the other extreme of the series, does not possess properties *b* and *c*, and decomposes glucose with production of acid, but not of gas. Intermediate in the series are a number of organisms, which possess one or more, but not all, of the properties of *B. coli communis* or *B. typhosus*.

The action of *B. coli communis* and of *B. typhosus* on glucose has therefore been examined, as far as possible quantitatively, in order to obtain a standard with which that of the other organisms might be compared, and this comparison has been effected for three different organisms. The action of *B. coli communis* on *d*-fructose and mannitol has also been examined in detail.

The products of the fermentation of glucose by these organisms comprise lactic acid, succinic acid, acetic acid, ethyl alcohol, formic acid, carbon dioxide, and hydrogen, all of which have been quantitatively estimated.

The *Bacillus coli communis*, and similar organisms, grow well and bring about a vigorous fermentation of glucose in a large number of different nutrient media. Thus they are able to derive their nitrogen from inorganic salts of ammonium and of methylammonium, from simple amino-acids, such as asparaginic acid, from Witte's peptone, and from beef broth. Since the products of the fermentation are acid in character, it is necessary to carry out the process in presence of chalk in order to neutralise the acid as it is produced. When ammonium salts are employed, it is therefore necessary to sterilise the solution of the sugar containing the chalk separately from the solution of the ammonium salt, in order to avoid the loss of ammonia which would otherwise ensue. The solution of the ammonium salt must therefore be sterilised in a flask arranged like an ordinary washbottle, so that the liquid, before inoculation, can be driven over into the sugar solution by a current of sterile air. When a culture in such a medium is incubated at 37°, it is found that the fermentation of the sugar commences very slowly, a delay of as much as five days occurring in some cases, although after this initial period the action is as vigorous as in the presence of peptone alone. This delay is no doubt due to the alkalinity of the solution, caused by the liberation of ammonia by the action of the chalk on the ammonium salt, which renders the development of the organism extremely slow until sufficient acid has been produced to render the medium capable of promoting growth at the normal rate.

This is shown by the figures on p. 612, from which it is seen that although in the end nearly the whole of the sugar is decomposed, the evolution of gas commences much more slowly when ammonium or methylammonium salts are present than when peptone alone is employed.

As the action of the free ammonia might seriously affect the mode of growth and action of the organism, this form of nitrogenous nourishment is not well adapted for comparative experiments. Asparaginic acid is equally unsuitable, but from a different and very interesting cause. As will be shown, it undergoes reduction under these circumstances to ammonium succinate, a large proportion of the

Nitrogenous medium.	Gas evolved at the end of days						Total gas evolved c.c.	Total sugar decomposed. Grams.
	1.	2.	3.	4.	5.	6.		
Peptone	500	1560	2270	3400	3500	3600	3700	20
Ammonium sulphate...	—	—	—	100	690	2630	4980	20
Methylammonium sulphate...	—	60	960	2210	—	3430	3650	17·87

hydrogen which is evolved when other media are employed being thus absorbed. Moreover, the large amount of succinic acid produced renders the analytical separation of the products more difficult.

It is also difficult to obtain satisfactory analytical results with media containing beef broth or extract of meat, because this contains small quantities of sugar and *d*-lactic acid, and the products derived from these cannot easily be separated from those derived from the sugar undergoing examination. The sum of the products obtained in this medium is invariably greater than the weight of the sugar taken.

The most satisfactory medium, from this point of view, was found to be peptone water, prepared by boiling 10 grams of Witte's peptone with tap water, adding 20 grams of the sugar or other compound to be examined, together with 10 grams of pure calcium carbonate, and making up the whole to 1 litre. In some cases, 2 grams of calcium phosphate were added, but no beneficial effect could be observed. Common salt is purposely omitted from the solution, because in its presence hydrochloric acid invariably passes over both into the volatile acids and the ethereal extract of the non-volatile acids.

The use of Witte's peptone alone has the further advantage that organisms of the Colon group have been found by direct experiment to grow freely in its solutions, without producing volatile acids, acids extractable by ether, or gaseous products in quantities large enough to interfere with the estimation of those produced from the compound under examination.

The action of the organism on the peptone, however, is undoubtedly modified by the presence of sugar, so that complete certainty on this point cannot be readily attained. This is shown by two facts.

1. In the presence of glucose, indole is not produced from peptone, whereas in its absence it is freely formed.

2. In the absence of sugar, practically no gas is evolved from peptone water, whereas in the presence of glucose the gas evolved invariably contains nitrogen, which must have been derived from the peptone. Thus in two experiments in which 250 c.c. of peptone water

were inoculated with this organism, no gas was evolved in one case and only 0·1 c.c. in the other, during a whole week, whilst indole was produced in both cases. On the other hand, a similar solution containing 2 grams of glucose gave about 100 c.c. of gas containing 7·5 per cent. of nitrogen. This result, which is in agreement with that obtained by other observers, is of great importance, since it throws some light on the function of the glucose in the nutrition of the organism.

The fermentations were in every case carried out in an atmosphere of nitrogen; the exact disposition of the apparatus and the methods of collecting and examining the gaseous products have been previously described in detail (*Trans. Jenner Inst.*, 1899, **2**, 126).

The method of analysis of the remaining products, which follows the general plan proposed by Nencki (*Centr. Bakt.*, 1891, **9**, 305) has, however, undergone considerable modification, and is therefore given in full.

Analytical Methods.

(1) *Residual Carbohydrate.*—Glucose and fructose were estimated by taking 50 c.c. of the clear liquid, adding excess of ammonium oxalate to precipitate the calcium, diluting to 100 c.c., and estimating the sugar present by means of Pavy's copper solution. In some cases, especially when beef broth was employed, the end of the reaction was rendered indistinct by the production of a purple coloration, due to the unaltered peptone and albumoses in the liquid. In such cases, it was found advisable to add four volumes of alcohol to some of the solution, filter an aliquot portion, evaporate off the alcohol, make up to the original bulk and then titrate for the sugar.

Mannitol was determined by Müller's method of measuring the change of rotation produced by the addition of a known weight of borax to the solution. The other substances present were found not to interfere with this measurement.

(2) *Total Acids.*—Fifty c.c. of the liquid were boiled for a short time with pure calcium carbonate, to effectually neutralise any free acid which might be present, and at the same time to remove calcium carbonate held in solution by carbonic acid. The dissolved calcium was estimated in the filtrate by precipitating with ammonium oxalate, and weighing as lime. The peptone, which was usually employed as the nutrient medium, contained calcium and was found to yield 0·106 gram of lime per 10 grams of peptone. This amount was therefore subtracted from the quantity found, in order to obtain the lime equivalent to the acids produced by the fermentation.

(3) *Alcohol.*—About 800 c.c. of the liquid were taken, 10 grams of oxalic acid added, the solution measured, and about 700—800 c.c. filtered off and distilled. The first 250 c.c. of the distillate were

reserved for the estimation of alcohol, and for this purpose were neutralised with normal alkali in presence of phenolphthalein paper, and then redistilled through a fractionating column; 100 c.c. of the distillate were collected and the alcohol calculated from the specific gravity.

(4) *Volatile Acids.*—The distillation of the acidified liquid was continued until only about 100 c.c. remained in the flask, and this was then distilled in steam until 100 c.c. of the distillate only required about 0·2 c.c. of normal alkali for neutralisation. In the presence of lactic acid, it is impossible to attain perfect neutrality, as this acid is itself slightly volatile with steam, the error thus introduced amounting to about 1 c.c. per litre of distillate. The distillate was then neutralised with normal alkali in presence of phenolphthalein, the residue from the alcohol distillation added, and the whole evaporated to dryness on the water-bath, redissolved in water, and diluted to 100 c.c.

In order to ascertain the nature of the acids present, three fractions were obtained by the addition of successive quantities of sulphuric acid and subsequent distillation, and the barium salts prepared from these were analysed. As numerous experiments showed that no higher acid than acetic was produced, this extended examination was not in every case carried out, but the formic acid was directly estimated by oxidation with mercuric chloride, and the acetic acid found by difference. The small amount of lactic acid present was neglected, as it was found that the only satisfactory method of estimating lactic acid—oxidation with alkaline permanganate and estimation of the oxalic acid produced (Ulzer and Seidel)—is inapplicable in the presence of acetic acid.

(5) *Lactic and Succinic Acids.*—The residue from the steam distillation was diluted to 150 c.c., and 25 c.c. were extracted with ether, which removes the lactic and succinic acids, together with some oxalic acid.

Since lactic acid is difficult to extract from aqueous solution, and the nutrient solutions tend to form an emulsion with the ether when shaken up by hand, it is convenient to employ some form of continuous extraction apparatus, and a modification of that described by Foerster (*Chem. Zeit.*, 1898, **22**, 421) has been found very suitable.

The extraction was continued for 10 hours, at the close of which time experiment has shown that all the lactic acid is removed; the ether was then evaporated, and the liquid heated on the water-bath with excess of calcium carbonate for several hours. The lactic anhydride which is invariably formed when the ethereal extract is evaporated (Wislicenus) is thus completely converted into calcium lactate. The liquid was then filtered from the calcium oxalate and excess of carbonate and diluted to 400 c.c.; 150 c.c. of this solution were taken for the estimation of the total acid present by precipitation of the calcium as oxalate.

Separation of Calcium Lactate from Calcium Succinate.—It was found that the principle of the method proposed by Müller for the separation of lactic from succinic acid by means of the difference in the solubility of their barium salts in dilute alcohol could also be applied to the calcium salts. Traces only of calcium succinate dissolve in 90 per cent. alcohol, whereas calcium lactate dissolves to the extent of about 3 grams in 100 c.c. of the liquid. The two salts can therefore be completely separated either by digesting the mixture of the dry salts with 90 per cent. alcohol, or, more conveniently, by dissolving the mixture in a measured volume of water and adding 9 volumes of alcohol. The following experiment, made to test the method, shows that the separation is practically complete.

0·7056 gram of calcium lactate yielded 0·1292 gram of lime.

0·7976 gram of the same salt was mixed with 0·1646 gram of pure calcium succinate, the whole treated with 90 per cent. alcohol, and the calcium estimated in the filtrate. This was found to yield 0·1468 gram of lime, whilst the amount calculated from the foregoing analysis of the lactate is 0·1460 gram.

The following method was therefore adopted for the separation and estimation of the succinic acid. One hundred and fifty c.c. of the solution of the calcium salts were evaporated to dryness on the water-bath in a beaker, the residue was dissolved in 10 c.c. of boiling water, and 90 c.c. of absolute alcohol were added. The liquid was allowed to stand for some time and then filtered, the precipitate washed with 90 per cent. alcohol, and the calcium estimated in the filtrate after the removal of the alcohol by distillation. This gives the equivalent of the lactic acid present, whilst the difference between this and the total acid gives the succinic acid. The precipitated calcium succinate can also be dissolved in water and the calcium in it estimated directly.

(6) *Determination of the nature of the Lactic Acid produced.*—The method usually employed for this purpose is to convert the acid into the zinc salt and estimate the amount of water of crystallisation and the optical rotation of this, the zinc salts of the active acids crystallising with $2H_2O$ (12·9 per cent.) and that of the inactive acid with $3H_2O$ (18·1 per cent.).

The zinc salts of the inactive and active acids, however, differ in solubility, and it is therefore impossible to prepare the salt from the acids obtained by extracting with ether in a state of purity, without to some extent altering the proportions in which the active and inactive acids are present. The presence of succinic acid, moreover, renders it essential to remove this substance from the lactic acid before the conversion of the latter into the zinc salt.

Attempts were therefore made to take advantage of the fact that the calcium salts of the active lactic acids have an optical rotation which is the opposite to that of the acids from which they are derived, and to estimate the amount of the active acid present from the change of rotation which occurs when the salt is converted into the acid.

This method suffers from two disadvantages (1) the lactic acids and their calcium salts have a very low specific rotation ; (2) even in an aqueous solution of the acid there is a tendency towards the formation of an anhydride which has a much higher rotation than the acid, and in an opposite sense.

It was, however, found that by working in the manner described below, fairly consistent numbers could be obtained. The results given in the present paper are founded on the numbers obtained by the examination of a sample of calcium *d*-lactate derived from extract of meat, and the rotations were all measured on a Schmitt and Haensch saccharimeter which can be employed with lactic acid for relative determinations of this kind. The observations are being extended to specimens of pure *l*- and *d*-lactates resulting from the fractional crystallisation of zinc ammonium lactate, a supply of which I owe to the kindness of Prof. Purdie, and a full account of the absolute measurements and conditions and limits of accuracy will shortly be published. The numbers given in the present paper are probably accurate to about 5—10 per cent.

This method has the two advantages that a fair estimate of the nature of the whole of the lactic acid is obtained, and that the estimation can be carried out even in the presence of small amounts of optically active substances, such as proteid matter from the nutrient medium, or unfermented sugar, traces of which very frequently pass over into the ether extract in spite of the greatest care, owing to the character of the liquid to be extracted.

To carry out the estimation, the remainder of the 150 c.c. of the liquid left after the removal of the volatile acids by distillation is extracted with ether and the acid thus obtained converted into the calcium salt, which is then freed from calcium succinate in the manner already described, 10 c.c. of water and 90 c.c. of alcohol being used for each 1·5 grams of salt. The alcoholic filtrate is evaporated, decolorised if necessary with animal charcoal, and diluted so as to yield at least 30 c.c. of liquid. Five c.c. of the solution are taken for the estimation of the calcium, and from this the concentration of the solution is found. Two quantities of 12 c.c. are measured out and to one is added 1 c.c. of water, to the other 1 c.c. of aqueous hydrochloric acid containing sufficient acid to liberate the whole of the lactic acid. Two solutions are thus obtained of the same concentration, one containing the acid and the other the calcium salt. These are then examined in the polarimeter and the difference between their rotations observed. The amount of lime as active acid corresponding to this difference is then found from the table and the relation of this to the total amount of lime present gives the percentage of active acid.

Action of Normal B. coli communis *on* d-*Glucose.*

The following table contains the results of five experiments, each carried out with peptone water, as already described, in an atmosphere of nitrogen, 20 grams of pure glucose being used.

Experiment No. 3 was carried out with a solution of ammonium sulphate containing 6·6 grams of this salt per litre, instead of peptone.

The gaseous products are expressed in c.c. per gram of sugar, and the other products in percentages of the weight of sugar fermented.

The organism employed in experiments 1, 2, and 3 was isolated from the stool of an adult, that employed in No. 6 from that of an unweaned child; and those used for Nos. 4 and 5 were isolated from the fæces of a dog and a rabbit respectively. The culture was made in every case into Parietti broth; the resulting growth was plated out, and cultures taken from a single colony. All the organisms gave the usual reactions for the normal bacillus.

Source of organism.	Sugar fermented. Grams.	Lactic acid.	Percentage of excess of *l*-acid in the lactic acid.	Acetic acid.	Alcohol.	Succinic acid	Formic acid.	Carbon dioxide, per cent.	C.c. per 1 gram.	
									Carbon dioxide.	Hydrogen.
1. Adult......	12·00	40·1	70	10·05	16·0	5·5	trace	—	—	—
2. ,, ...	20	31·9	not determined	18·84	12·85	(5·2)	none	18·09	91·8	110
3. ,, ...	20	40·1	,,	15·63	9·84	(6·7)	none	16·25	83	86
4. Dog......	14·68	45·2	,,	14·14	8·9	(5·7)	none	12·55	65·2	79·9
5. Rabbit ...	11·63	44·4	75	19·69	10·2	4·6	1·4	12·1	61·6	80·7
6. Child	13·33	46·0	95	12·02	10·9	5·7	2·5	—	—	—

These results illustrate the degree of variation which is found in the action of these organisms. Even parallel experiments, carried out with the same culture under apparently identical conditions, have frequently been found to yield the various products in somewhat different proportions. The weight of lactic acid produced varies from 31·9 to 46 per cent., but never exceeds 50 per cent. of the sugar employed. In every case, this acid consists of a mixture of inactive acid with a preponderating amount of *l*-acid. In spite of this result however, the production of *l*-lactic acid cannot be taken as an essential function of this organism, since the production of the *d*-acid has been frequently observed by various authors, employing this bacillus derived from various sources. It has also been found by Péré (*Ann. Inst. Past.*, 1898, **12**, 63; 1893, **7**, 737) that some of these organisms may be made to yield either *d*- or *l*-lactic acid, according to the conditions of nourishment, whilst others yield the same acid under all conditions.

These numbers do not, in every case, account for the whole of the sugar, there being a deficit amounting to from 1/12th to 1/8th of the total carbon, but a careful search has not yet revealed the nature of the change which this remaining portion has undergone. It is possible that a portion of the sugar is reduced either by the hydrogen which is evolved, or by a reaction in which a molecule of water, or perhaps of formic acid, intervenes. Sorbitol, which might be thus produced, is not easily fermented by the organism, and might remain unaltered and escape notice. Further experiments are still in progress with the object of deciding this question.

A clearer view of the nature of the chemical change is afforded by expressing the results in terms of the carbon atoms of the original sugar. This is done in the following table:

Source of organism.	Carbon atoms as					Hydrogen atoms.
	Lactic acid.	Acetic acid.	Alcohol.	Succinic acid.	Carbon dioxide and formic acid.	
1. Adult	2·40	1·2	1·32	0·34	—	—
2. ,, ...	1·91	1·13	1·01	0·32	0·74	1·77
3. ,, ...	2·40	0·94	0·98	0·41	0·67	1·39
4. Dog	2·71	0·84	0·70	0·35	0·53	1·29
5. Rabbit	2·67	1·18	0·80	0·29	0·50	1·30
6. Child	2·76	0·81	0·85	0·35	—	—

This reveals the fact that whilst the lactic acid accounts for 2—3 of the carbon atoms of the original molecule, the alcohol and acetic acid are formed in nearly equivalent amounts, and each accounts for only one carbon atom of the sugar molecule. The amount of carbon dioxide obtained in different experiments represents less than 1 carbon atom, but never falls below 0·5. The ratio of hydrogen to carbon dioxide is not exactly 1 : 1 by volume, but does not depart widely from this ratio. The production of about half the weight of the sugar as lactic acid, and of acetic acid and alcohol in equivalent amounts, each corresponding to about 1 carbon atom of the sugar molecule, may be represented by a very simple equation, which, however, somewhat overestimates the amounts of carbon dioxide and hydrogen which are actually found:

$$2C_6H_{12}O_6 + H_2O = 2C_3H_6O_3 + C_2H_4O_2 + C_2H_6O + 2CO_2 + 2H_2.$$

It seems probable, however, that this represents the main reaction, but is subject to modification by secondary changes (see p. 626).

In solutions containing beef broth, as already mentioned, the products derived from the sugar cannot be distinguished accurately from those previously present in the broth and formed from it by the action of the organism. The lactic acid obtained in this medium is invariably much less active, and sometimes quite inactive, as has been pointed out in a previous paper (*Trans. Jenner Inst.*, 1899, **2**, 126). This is, at all events in part, to be ascribed to the presence of *d*-lactic acid in the broth, since this acid is not affected by the organism (p. 625). Precisely similar relations are found with lævulose and mannitol.

The Fermentation of Glucose by Bacillus typhosus.

The products of fermentation of glucose by *B. typhosus* are very similar in character to those produced by *B. coli communis*, the only essential difference being that formic acid is produced in large quantity, instead of only in traces. This formic acid doubtless represents the carbon dioxide and hydrogen which are produced by *B. coli communis*, the *B. typhosus* being unable to decompose formates, whilst *B. coli communis*, as will be shown later on, has this power (see also Pakes and Jollyman, this vol., p. 386). The fermentation is not a very vigorous one, and in all cases terminated when about half of the sugar had been decomposed.

The two following experiments were carried out with cultures of virulent typhoid in an atmosphere of nitrogen. The solutions were sterilised at 80° for an hour before being examined:

	I.		II.	
	Per cent.	Carbon atoms.	Per cent.	Carbon atoms.
Lactic acid	31·1	1·87	49·5	2·96
Acetic acid	16·1	0·97	12·7	0·76
Alcohol	10·6	0·83	9·1	0·70
Succinic acid	2·1	0·13	trace	trace
Formic acid	17·2	0·67	17·7	0·69

In both cases, a mixture of inactive with *l*-lactic acid was formed; that produced in experiment I was found to contain 47·5 per cent. of the active acid, this being a smaller proportion than is produced by the *B. coli communis.*

Action of Abnormal Forms of B. coli communis *on Glucose.*

Only three of these organisms have so far been employed: 1. Two organisms which differed from the normal only in not curdling milk until after the expiration of 7 or 8 days. One of these (A) was isolated from the fæces of a guinea-pig, and the second (B) from those of a pigeon.

2. An organism (C) which I owe to the kindness of Dr. Gordon, who had isolated it from sewage. This organism gave no gas with sugar solutions, but curdled milk and produced indole in the normal manner.

The results are stated both in percentages and in terms of the carbon atoms of the sugar molecule.

	A.		B.		C.	
	Per cent.	Carbon atoms.	Per cent.	Carbon atoms.	Per cent.	Carbon atoms.
Lactic acid	33·2	1·99	25·4	1·52	76·6	4·6
Succinic acid	3 2	0·2	3·9	0·24	3·3	0·2
Acetic acid	20·95	1·26	24·5	1·47	14·1	0·85
Alcohol	14·33	1·12	20·1	1·57	4·6	0·36
Formic acid	1·4	0·06	3·4	0·13	9·0	0·35
Carbon dioxide ...	14·6	0·60	18·5	0·76	none	none
Per cent. of active *l*-acid	64·4		80		32	

It appears from this that organisms A and B do not differ essentially from the normal *B. coli communis* in their action on glucose, although they yield rather less lactic acid and more alcohol and acetic acid (see p. 618).

The organism C, on the other hand, differs distinctly in its mode of action, both from *B. coli communis* and *B. typhosus*. The amount of lactic acid is greater, and the proportion of active acid less, whilst the alcohol and acetic acid are produced in different proportions. Like *B. typhosus*, this organism produces formic acid, which it appears to be unable to decompose.

This result shows that an organism which appears by its cultural tests to be closely related to the colon bacillus and to *B. typhosus*, may nevertheless differ from both of these in its action on glucose.

Action of B. coli communis *on Lævulose* (d-*Fructose*).

Lævulose is readily fermented by this organism, both in peptone water and in peptone beef broth. The products are identical in character with those produced from glucose. The following results were obtained in peptone water; 20 grams of the sugar containing 0·8 gram of water, and 19·2 grams of lævulose were employed:

	I.		II.	
	Per cent.	Carbon atoms.	Per cent.	Carbon atoms.
Lactic acid	43	2·58	48·2	2·89
Succinic acid	not determined		6·1	0·37
Alcohol	,,	,,	11·5	0·90
Acetic acid	13	0·78	16·1	0·97
Carbon dioxide ...	15·5	0·63	13·07	0·53
Formic acid	—	—	trace	

The lactic acid produced in both cases was lævorotatory. That derived from experiment I was not further examined. That obtained from experiment II was converted into the zinc salt. As this was found to contain succinic acid, the lactic acid was re-extracted, converted into the calcium salt, and the latter purified and examined as usual. It was found to contain 90 per cent. of the active *l*-lactic acid.

An experiment which was carried out in peptone beef broth yielded a very similar result to experiments with glucose under the same conditions. The lactic acid obtained was greater in quantity (65 per cent.), and contained only 51 per cent. of active acid. As is the case with glucose in this medium, the sum of the products exceeded the weight of the sugar taken.

The difference of constitution between glucose and lævulose, therefore, does not appear to produce any difference in the character of the products of fermentation by this organism. The lactic acid in particular is of the same character, and this shows that the configuration of the lactic acid produced does not depend on that of the carbon atom situated in the α-position to the aldehyde group of glucose, since in lævulose this atom is not asymmetric.

Action of B. coli communis *on Mannitol.*

Although the substitution of lævulose for glucose produces no recognisable difference in the products of fermentation, a very marked change is observed when the sugar is replaced by a hexatomic alcohol.

Dulcitol is not fermented by this organism. The attempt was made in peptone water, and although the organism developed well, no sign of fermentation (gas evolution, or production of acid) could be observed. A similar power of discriminating between mannitol and dulcitol has previously been observed in other organisms.

Sorbitol appears to undergo only a very slight fermentation when submitted to this organism, and the products have not yet been examined.

Mannitol, on the other hand, is very vigorously and completely acted on, the products being qualitatively the same as those produced from glucose and lævulose.

The following are the complete results of two experiments, I in peptone water, II in beef broth:

	I.		II.	
	Per cent.	Carbon atoms.	Per cent.	Carbon atoms.
Lactic acid	18·6	1·13	24·9	1·51
Succinic acid[1]	8·9	0·55	9·4	0·58
Acetic acid	9·5	0·58	8·2	0·50
Alcohol	28·06	2·22	33·5	2·65
Formic acid	3·0	0·12	0·6	0·02
Carbon dioxide	28·44	1·16	30·84	1·26
Per cent. of active *l*-acid	79		29	
Hydrogen (atoms)	2·7		2·7	

The lactic acid from experiment I was first converted into the zinc salt, and this was then treated with pure lime, and the resulting calcium salt purified and examined in the usual manner.

The experiment in beef broth, as usual, yields high results, whilst the lactic acid is less active.

The most noticeable point about this result is the large yield of alcohol, which considerably exceeds 25 per cent. of the weight of the mannitol fermented. It is always produced in this large proportion, two other experiments giving 28·2 per cent. (peptone water) and 26·9 per cent. (sodium asparaginate). When separated by means of potassium carbonate, dried over the same salt and distilled, it boiled constantly at 78°, and was therefore pure ethyl alcohol. The succinic acid is much larger in amount than that obtained from glucose, whilst the lactic and acetic acids are much smaller. The amount of hydrogen evolved is also greater, the total amount, including that of the undecomposed formate, being nearly three atoms of that contained in the mannitol. The fact that the amount of alcohol corresponds with two of the carbon atoms of the mannitol, whereas glucose only yields one of its carbon atoms as alcohol, seems to be definitely connected with the fact that in mannitol there are two groups of the formula $CH_2(OH)\cdot CH(OH)\cdot$, whilst in glucose only one such group occurs.

The view that the alcohol produced originates from this particular group is strikingly confirmed by the action of this organism on glycerol, $CH_2(OH)\cdot CH(OH)\cdot CH_2\cdot OH$. Quite recently, I have found that the chief products of this action are alcohol and formic acid, or its decomposition products, carbon dioxide and hydrogen, only a very small quantity of other acids being produced.

$$C_3H_8O_3 = C_2H_6O + CH_2O_2.$$

5·2 grams of glycerol yielded 2·4 grams of alcohol, 1·8 grams of formic acid (partially as carbon dioxide and hydrogen), together with small amounts of two other acids, probably acetic acid and succinic acid.

Reduction of Asparaginic Acid by B. coli communis *in presence of Glucose and Mannitol.*

When the sole nitrogenous nourishment supplied to this organism consists of sodium asparaginate, glucose and mannitol are fermented in the usual manner, with the single exception that the amount of hydrogen evolved is much less, a large proportion of it serving to reduce the asparaginic acid to ammonium succinate.

$$CO_2H\cdot CH_2\cdot CH(NH_2)\cdot CO_2H + 2H = CO_2H\cdot CH_2\cdot CH_2\cdot CO_2H + NH_3.$$

This is shown in the following table of results, in which col. I gives (*a*) the volume of hydrogen actually evolved; (*b*) the volume of hydrogen taken up by the asparaginic acid, calculated from the ammonia found in the solution, and (*c*) the volume of hydrogen evolved in another experiment in presence of peptone from the same weight of glucose. Col. II gives the same information for mannitol. The solutions contained 20 grams of sodium asparaginate and 20 grams of glucose and mannitol respectively per litre.

	I. Glucose. c.c.		II. Mannitol. c.c.	
a.	663	1884	887	2925
b.	1221		2038	
c.	1776		3274	

The occurrence of this reduction is of some interest, since asparaginic acid is not very easily reduced by ordinary reagents, but requires to be heated with hydriodic acid. Whether the reduction is due to a purely chemical reaction or whether the protoplasm of the organism plays a direct part in the reduction must remain at present undecided.

The reduction is also interesting from another point of view, as showing that an organism like *B. coli communis*, which has very little direct action on proteids, may, nevertheless, under appropriate circumstances, assist in the process of still further decomposing the products formed from them by putrefaction.

Experiments made with aminoacetic and aminopropionic acids to ascertain whether a similar reduction occurred gave negative results, the growth of the organism being very feeble and no ammonia being formed.

Action of B. coli communis *on* d-*Galactose and* l-*Arabinose.*

Both of these sugars are fermented by this organism, with formation of the same products as are derived from glucose. The complete course of the action is still under investigation, but it may be stated that both yield a mixture of *l*-lactic acid with the inactive acid. *l*-Arabinose, moreover, yields alcohol and acetic acid in about the same molecular proportion as glucose, whilst the amount of lactic acid is only half as great.

Action of B. coli communis *on Formates and Lactates.*

The fact that *B. coli communis* differs in its action on glucose from *B. typhosus*, mainly by the production of carbon dioxide and hydrogen instead of formic acid, led to experiments on the direct action of the former organism on formates. It was found that this organism effects the decomposition of these salts according to the equation

$$H{\cdot}CO_2Na + H_2O = NaHCO_3 + H_2.$$

Hoppe-Seyler, in 1876, described a decomposition of this kind by the action of the bacteria of sewage, and quite recently Messrs. Pakes and Jollyman (this vol., p. 386) have drawn attention to this function of *B. coli communis* and other organisms.

The particular organism employed in these experiments exerted comparatively little action on a solution containing 2 per cent. of sodium formate and 1 per cent. of Witte's peptone. In presence of a small amount of sugar, however, a much larger amount of the formate was decomposed, as is shown in the following table, the hydrogen due to the sugar being deducted:

Sodium formate	2 per cent.	2 per cent.	2 per cent.
Glucose	0 ,,	0·2 ,,	0·4 ,,
Hydrogen evolved ...	105 c.c.	1122 c.c.	1481

Increase of pressure appears to favour the formation of formic acid instead of its decomposition products. Thus, under an additional pressure of 0·5 atmosphere, the amount of formic acid produced was increased in one case 3 times, and in a second experiment 10 times.

Pressure.	C.c. of *N*-formic acid from 1 gram of glucose.
1 atmosphere.	0·15
1·5 ,,	0·42
1·5 ,,	1·32

This organism has practically no action on inactive lactic acid in the form of ammonium, calcium, or sodium salt, and even when a small amount of glucose is added, the amount of active lactic acid produced is not greater than is produced from the sugar alone. Thus, a 2 per cent. peptone solution, containing 9 grams of lactic acid as sodium salt, together with 1·8 grams of glucose, only yielded 0·4 gram of active lactic acid, the whole of which is derived from the sugar.

This fact is important as showing that the active lactic acid which results from the action of the organism on so many sugars is almost certainly not formed by the selective decomposition of previously formed inactive acid.

Theoretical Considerations.

The results communicated above seem to show that in the fermentation of *d*-glucose and *d*-fructose, at least two molecules are concerned, and that the chief reaction may be best represented by the equation

$$2C_6H_{12}O_6 + H_2O = 2C_3H_6O_3 + C_2H_4O_2 + C_2H_6O + 2CO_2 + 2H_2,$$

which finds a simple diagrammatic expression as follows:

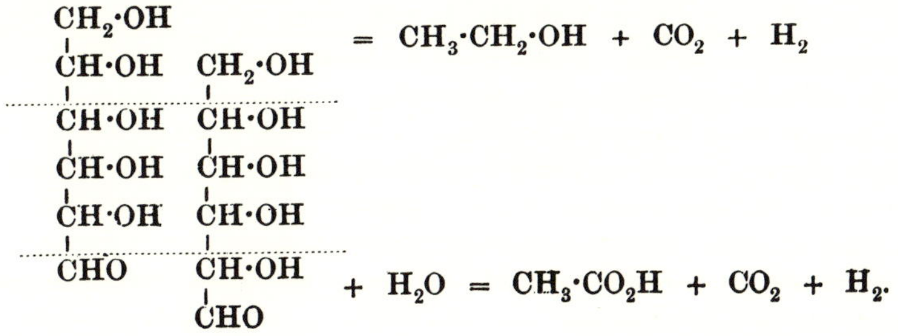

```
CH2·OH
|                           = CH3·CH2·OH + CO2 + H2
CH·OH     CH2·OH
|         |
.................................
CH·OH     CH·OH
|         |
CH·OH     CH·OH
|         |
CH·OH     CH·OH
|         |
.................................
CHO       CH·OH
          |                 + H2O = CH3·CO2H + CO2 + H2.
          CHO
```

According to this scheme, two groups, $\cdot CH(OH)\cdot CH(OH)\cdot CH(OH)\cdot$, are left, which may then be converted into lactic acid or partly break down into other compounds, among which alcohol, acetic acid, hydrogen, and succinic acid are probably to be numbered. This secondary change probably accounts for the fact that in some cases (experiments A and B, p. 620), whilst the lactic acid represents considerably less than 3 atoms of carbon, the alcohol and acetic acid are correspondingly larger in amount.

In the case of mannitol, the fact that the presence of two $\cdot CH_2(OH)\cdot CH(OH)\cdot$ groups gives rise to a largely increased yield of alcohol and a decreased yield of acetic acid, must certainly be considered significant. The chief reaction in this case may also be most easily imagined as embracing two molecules :

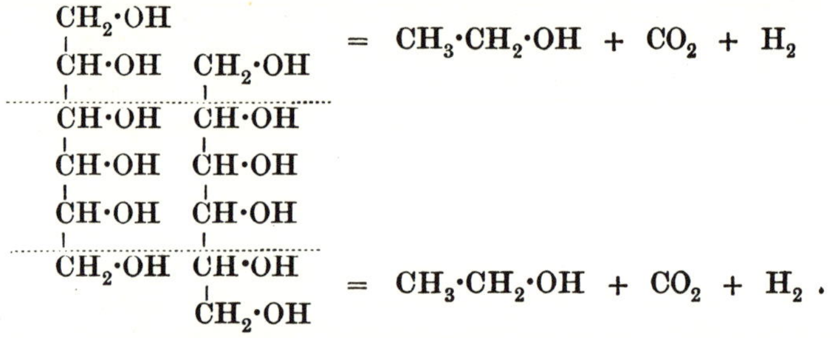

$$
\begin{array}{ll|l}
CH_2\cdot OH & & \\
| & & \\
CH\cdot OH & CH_2\cdot OH & = CH_3\cdot CH_2\cdot OH + CO_2 + H_2 \\
| & | & \\
\hdashline
CH\cdot OH & CH\cdot OH & \\
| & | & \\
CH\cdot OH & CH\cdot OH & \\
| & | & \\
CH\cdot OH & CH\cdot OH & \\
| & | & \\
\hdashline
CH_2\cdot OH & CH\cdot OH & \\
 & | & = CH_3\cdot CH_2\cdot OH + CO_2 + H_2\ . \\
 & CH_2\cdot OH &
\end{array}
$$

The two residual groups, $\cdot CH(OH)\cdot CH(OH)\cdot CH(OH)\cdot$, then behave in an analogous manner to those derived from the sugars, but yield in this case less lactic acid and more of the other products.

From this point of view the action of the organism on a pentose like *l*-arabinose is of great interest. If the action is of the same type as that on glucose, fructose, and mannitol, the maximum yield of lactic acid should only be 30 per cent. of the weight of the sugar, only three carbon atoms from two molecules being available for the production of the acid, whereas in the case of glucose six atoms are available :

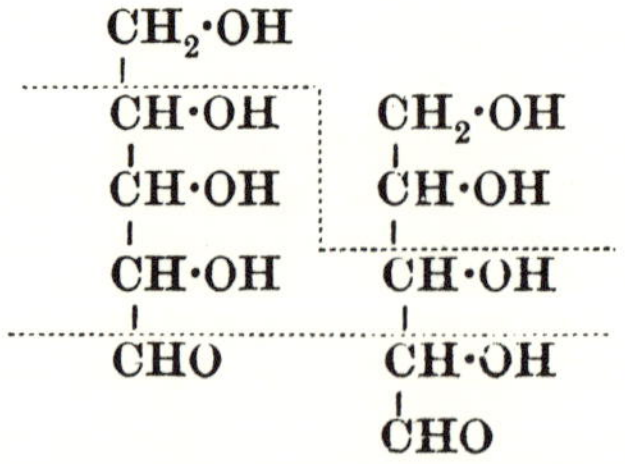

$$
\begin{array}{ll}
CH_2\cdot OH & \\
| & \\
\hdashline
CH\cdot OH & CH_2\cdot OH \\
| & | \\
CH\cdot OH & CH\cdot OH \\
| & | \\
\hdashline
CH\cdot OH & CH\cdot OH \\
| & | \\
\hdashline
CHO & CH\cdot OH \\
 & | \\
 & CHO
\end{array}
$$

The acetic acid and alcohol, on the other hand, should represent each one carbon atom per molecule of the sugar. The single experiment so far completed is in good agreement with this scheme, the amounts of alcohol and acetic acid formed bearing nearly the required ratio to the lactic acid, but the experiment must be multiplied before any definite conclusion can be drawn.

The group ·CH(OH)·CH(OH)·CH(OH)·, from which it is suggested that the lactic acid is formed, has the same empirical composition as lactic acid, and would yield this acid by the interchange of a hydrogen and oxygen atom between the two terminal groups, or between the terminal groups of a second similar chain derived from another molecule of the sugar.

The unaltered asymmetric group, ·CH(OH)·, would then retain its asymmetry in the new molecule of lactic acid thus produced. Such a rearrangement, however, if it could be effected by means of reagents containing only symmetric molecules, would probably result in the production of an inactive acid, the oxidation and reduction taking place to an equal extent at either end of the chain.

$$
\begin{array}{c}
CH_3 \\
| \\
H\text{Ⓒ}OH \\
| \\
CO_2H
\end{array}
\qquad\qquad
\begin{array}{c}
\dot{C}H\cdot OH \\
| \\
H\text{Ⓒ}OH \\
| \\
\dot{C}H\cdot OH
\end{array}
\qquad\qquad
\begin{array}{c}
CO_2H \\
| \\
H\text{Ⓒ}OH \\
| \\
CH_3
\end{array}
$$

The change being, however, brought about in the presence of the asymmetric molecules of the protoplasm of the organism—or possibly of an enzyme—would most probably be influenced so as to proceed entirely in one direction, or more rapidly in one direction than in the other, and thus give rise to an active acid, the activity of which would be independent of the previous configuration of the group and depend only on the nature of the organism. This view is therefore not inconsistent with the fact observed by Péré that the same sugar when submitted to the same organism, but under different conditions of nourishment, may yield lactic acids of opposite signs. Before admitting this explanation it would, however, be necessary to show that the type of reaction had remained unaltered.

The question of the origin of the lactic acid within the molecule of the sugar may also be attacked by examining the products formed from sugars of different configuration, and the work is being proceeded with in this direction ; it is also being extended to embrace a study of the various other types of formation of lactic acid from the carbohydrates, and the action of these organisms on compounds of different constitution.

Summary of Results.

1. In the fermentation of glucose by *B. coli communis*, the lactic acid produced never exceeds in amount one half of the sugar fermented, but may be less. It is optically active, but is probably not derived from previously formed inactive acid. Alcohol and acetic acid are produced in approximately equivalent amounts.
2. The products of fermentation of glucose by *B. typhosus* are similar to those produced by *B. coli communis*, with the exception that formic acid is produced instead of a mixture of carbon dioxide and hydrogen.
3. Certain bacteria which appear by the ordinary tests to be intermediate between *B. coli communis* and *B. typhosus* decompose glucose according to a different type of reaction.
4. *d*-Fructose, when fermented by *B. coli communis*, yields the same products as glucose, and in the same proportion.
5. Mannitol, when fermented by *B. coli communis*, yields about 28 per cent. of alcohol, corresponding to two of its carbon atoms, together with less lactic acid and acetic acid than are formed from glucose. The alcohol appears to be produced from the group $CH_2(OH)\cdot CH\cdot OH$.
6. Asparaginic acid is reduced to ammonium succinate by *B. coli communis* in presence of glucose or mannitol, the volume of hydrogen evolved being correspondingly diminished.

Chemical and Water Laboratory,
Jenner Institute of Preventive Medicine.

3

THE THIRD FORM OF SUGAR FERMENTATION

Carl Neuberg and Julius Hirsch

This article was translated expressly for this Benchmark volume by D. W. Thayer and D. Dolata from "Die dritte Vergärungsform des Zuckers," Biochem. Z., **100,** *304–322 (1919)*

A. DEVELOPMENT AND PROBLEM

We have been able to explain the facts announced by Neuberg and Färber[1] in 1916 that, aside from the sulfurous acid salts, a number of other substances of alkaline reaction produce an essential transformation in the course of alcoholic fermentation.[2] This transformation consists of an increased formation of acetaldehyde, acetic acid, and glycerin, as well as a corresponding reduction of the common fermentation products ethyl alcohol and carbonic acid.

The mechanism of the sulfite action was elucidated by Neuberg and Reinfurth.[3] They showed that it was a question of an active participation of the salts of the sulfurous acid. These react with an intermediate compound of sugar decomposition, acetaldehyde, which is blocked from further reaction. The consequence of this fixation of a stage of oxidation is the correlative formation of a reduction product, glycerin; acetaldehyde and glycerin are present in molecular proportions at any time[4] during the fermentation. The immediate preliminary stage of acetaldehyde is pyroracemic acid (pyruvic acid), which, in the process of decomposition by carboxylase, yields, besides acetaldehyde, carbon dioxide of fermentation. Based on these discoveries, Neuberg and Reinfurth were able to add to the classical alcoholic cleaving of sugar a new and second form of fermentation, the decomposition by the interception method. It occurs as follows:

$$C_6H_{12}O_6 = CH_3\cdot CHO + CO_2 + C_3H_8O_3$$

The reaction mechanism of the other alkaline salts appeared complex at first. We have found that acetaldehyde also plays a determinative role and that here also the transformation of the aldehyde results in the deviation of the course of the entire fermentation process. While under the influence of the sulfurous salt (as mentioned, fixation of the acetal-

[1] C. Neuberg and E. Färber, *Biochem. Z., 78,* 238 (1916).
[2] C. Neuberg and J. Hirsch, *Biochem. Z., 96,* 175 (1919).
[3] C Neuberg and E. Reinfurth, *Biochem. Z., 89,* 365; *92,* 234 (1918).
[4] C. Neuberg and J. Hirsch, *Biochem. Z., 98,* 141 (1919).

dehyde in the form of acetaldehyde-sulfurous salt—which is not accessible to the biological agents any longer—takes place), an intervention of a different kind comes about at the level of acetaldehyde under the influence of alkali-reacting salts, such as carbonates, bicarbonates, phosphates, and so on. These salts, which do not have specific affinity for the acetaldehyde, are not able to unite with it and secure it, but they cause a new reaction in the acetaldehyde, the enzymatic rearrangement of two molecules under absorption of 1 mole of water to ethyl alcohol and acetic acid. These are then metabolic end products. The remaining transformations caused by the alkaline addition are the same as those caused by sulfite. On account of the use of the acetaldehyde, the hydrogen, which normally is used in the final hydrogenation to ethyl alcohol, cannot work in its usual way and instead reacts with—because it is not in the unreactive molecular state—another part of the sugar and reduces it to glycerin. Because the formation of 1 mole of acetic acid is due to the "Cannizzaro reaction" of two molecules of acetaldehyde, each one of these originally equivalent to 1 mole of glycerin, there is a correlation here of 1 mole of acetic acid to 2 moles of glycerin.

Thus we have established the third form of sugar cleavage by yeast.[5] The decomposition of sugar into acetic acid, ethanol, carbon dioxide, and glycerin takes place according to the equation

$$2C_6H_{12}O_6 + H_2O = CH_3 \cdot COOH + C_2H_5OH + 2CO_2 + 2C_3H_8O_3$$

In view of the importance of a new form of fermentation, it seemed necessary for us to prove that under the stated fermentation conditions the above equation generally proves to be correct. We had earlier demonstrated the existence of the mentioned relations for the fermentation of sugar in the presence of sodium hydrogen carbonate. Their general validity is shown by carrying out experiments with the different salts already used by Neuberg and Färber and with other related combinations, as well as by the exact analysis of the reaction products.

We have again undertaken fermentations in the presence of dipotassium carbonate, dipotassium phosphate, trisodium phosphate, disodium phosphate followed by others with mixtures of acidic and secondary phosphates of sodium, and furthermore with magnesium oxide and zinc hydroxide.

In all cases, without exception, we found the equation for the third form of fermentation confirmed. Exactly as we had shown with the sodium hydrogen carbonate, acetic acid and glycerin are formed—fully in compliance with the theory—in the ratio of 1 : 2 moles. With some examples we have likewise determined, as was done before, that initially acetaldehydes are accumulated, which obviously disappear again owing to the occurrence of the rearrangement. The disproportionation is supported, it appears, by the alkaline medium, as the experiment[6] in this

[5] C. Neuberg and J. Hirsch, *Biochem. Z., 96,* 175 (1919); *98,* 141 (1919).
[6] C. Neuberg and J. Hirsch, *Biochem. Z., 96,* 192 (1919).

case likewise shows specifically where added acetaldehyde—not produced in the act of fermentation through breakdown of sugar—comes in contact with yeast.

It was possible in all the experiments to use sucrose, because the invertase stays active in the presence of all earlier-mentioned additions (Neuberg and Färber, footnote 1).

With regard to the extent to which the third form of fermentation of sugar can be obtained, we note the following: with sodium hydrogen carbonate, we have reached—surpassing the former maximum value of 29.8 percent—up to 35.4 percent of the theoretical amount of acetic acid and glycerin, so that more than one third of the sugar broke down.[7] The salts examined here gave 27 percent of the total decomposition. These values may be further increased. The sugar not included in this type of cleavage is subject to the usual alcoholic fermentation in the sense of the Gay–Lussac equation, that is, the decomposition to carbonic acid and ethanol. In other words, under the present conditions the maximum concentrations found feasible by us represent the tolerance limit of yeast for these alkalizers; because of both the limited solubility of the additives and their poisonousness, it is simply not possible to strengthen their content as far as necessary to force the total fermentation into the new path.

The chemical character of the examined additives is different; among them are easily soluble ones as well as completely insoluble compounds. All appear to be of the same kind, just as we established earlier with the soluble and insoluble sulfurous acid salts. The similarity of the final effect allows the conclusion that a common reaction mechanism is being dealt with, which probably consists of a predisposed but normally limited fermentation pathway, that is, the catalysis of the dislocation[8] of the intermediate acetaldehyde. Removal of a significant quantity of acetaldehyde from the normal chain of oxidation steps must then be accompanied by a compulsory correlative increase in glycerin as a result of reduction. The details are shown in the following experiments.

B. FERMENTATION IN THE PRESENCE OF POTASSIUM CARBONATE

First, in preliminary experiments the maximum concentration of potassium carbonate that could be present in the sugar solutions without causing damage to the yeast, consequently leaving the fermentation incomplete, was established. In five tests the concentration of potassium carbonate was adjusted from 0.1 to 0.5 *M* in a fermentation broth containing 10 percent sucrose and an amount of yeast equal to the weight of sugar. At concentrations of 0.5 and 0.4 *M* K_2CO_3, no fermentation started at all; at 0.3 and 0.2 *M* K_2CO_3, the conversion was not

[7] The corresponding evidence will be given in another communication.
[8] Compare C. Oppehheimer, *Fermente,* II, 684 (1913).

complete; only at 0.1 *M* K_2CO_3 could the process be followed through to the end. The borderline was approximately at the alkalinity, which Neuberg and Färber[9] had earlier established with the cell-free fermentation.

Experiment 1. The mash consisted of the following: 47.5 g of sucrose (equivalent of 50 g of hexose), 50 g of yeast M, 7 g of K_2CO_3, and water to a volume of 500 cm^3; concentration of K_2CO_3 = 0.1 *M*. After 48 hours all sugar was gone. The quantitative estimations of alcohol, glycerin, and acetic acid were accomplished by the same methods as previously described.[10] We emphasize that the exact determination of acetic acid requires a completely clear filtrate of the fermented medium; otherwise volatile acid values that are too high are obtained from the yeast cells.

In 100 cm^3 of fermentation broth (containing the decomposition products of 10 g of 6-carbon sugar) the following were found: alcohol, 4.17 g; glycerin, 0.67 g; and acetic acid, 0.23 g.

Experiment 2. The starting mixture for fermentation was the same as in Experiment 1 (0.1 *M* K_2CO_3). For 100 cm^3 the gains in fermentation products amounted to the following: alcohol, 4.16 g; glycerin, 0.56 g; and acetic acid, 0.18 g.

These two experiments show that the formation of glycerin as well as acetic acid is clearly increased in the presence of potassium carbonate. The yield of alcohol is thereby decreased in accordance with fermentation proceeding exclusively by the Gay–Lussac equation.

C. FERMENTATION IN THE PRESENCE OF DIPOTASSIUM PHOSPHATE

The examination of the tolerance limit for dipotassium phosphate showed in preliminary experiments that no fermentation at all occurred with a concentration of 3.0 *M* K_2HPO_4; with concentrations of 2.0 and 1.5 *M* K_2HPO_4, the change could not be completed with the available yeast; while from 1.0 *M* down a sugar solution of 10 percent was completely decomposed in 24 hours by an amount of yeast equal to the quantity of sugar.

Experiment 3. 500 cm^3 of fermentable liquid consisted of the following: 47.5 g of sucrose, 50 g of yeast, and 87 g of dipotassium phosphate, K_2HOP_4 (corresponding to a 1.0 *M* solution). The fermentation expired after 48 hours; for 10 g of converted hexose the following was determined: alcohol, 3.46 g; glycerin, 1.37 g; and acetic acid, 0.46 g.

Experiment 4. A similar formulation as in Experiment 3 resulted (for 100 cm^3) in the following: alcohol, 3.71 g; glycerin, 1.25 g; and acetic acid, 0.40 g. In a 1.0 *M* solution of dipotassium phosphate 4 to 4.6 percent and 12.5 to 13.7 percent of the sugar by weight was cleaved into

[9] C. Neuberg and E. Färber, *Biochem. Z., 78,* 238 (1916).
[10] C. Neuberg and J. Hirsch, *Biochem. Z., 96,* 175 (1919).

acetic acid and glycerin, respectively. These yields reach 26.8 percent of the quantity that may appear according to the equation for the third form of fermentation. They scarcely fall short of the yield accomplished by sodium hydrogen carbonate.

D. FERMENTATION IN THE PRESENCE OF MAGNESIUM OXIDE

The preliminary experiments showed that in a solution of 10 percent sugar complete fermentation was accomplished only in a 0.125 *M* suspension of magnesium oxide (MgO). At 0.25 and 0.5 *M* the fermentation occurred after considerable delay, but came to a stop within 24 hours.

Experiment 5. The mash had the following ingredients: 47.5 g of sucrose, 50.5 g of yeast M, and 2.5 g of MgO (= 0.125 *M*); volume = 500 cm^3. After 48 hours the fermentation was terminated. For 100 cm^3 fermentation, the analysis showed the following: alcohol, 4.21 g; glycerin, 0.73 g; and acetic acid, 0.24 g. To identify acetic acid, the silver salt was prepared as described previously from the distillate of the volatile acids. The silver determination gave the following: 0.1554 g of substance, 0.0987 g of Ag; CH_3;COOAg: calculated 64.66 percent Ag, found 63.51 percent Ag.

Experiment 6. With an equal mixture of yeast and sucrose, but without MgO and under otherwise similar conditions, the following was obtained per 100 cm^3: alcohol, 4.72 g; and glycerin, 0.31 g. The figure ascertained here for the titration of the volatile acids was 1.0 cm^3 of *n*-sodium hydroxide for 100 cm^3 of converted fermentation; neither it nor the quantity of glycerin exceeded the values obtained with the common form of fermentation.

The amount of acetic acid found until now with fermentation under different conditions does not fluctuate significantly. Bechamp,[11] who identified the constant appearance of acetic acid with the alcoholic cleavage of sugar, has determined 0.05 to 0.40 percent from the weight of the applied carbohydrate. Higher acidity, from 0.40 to 2.96 percent, was found by Fernbach[12] under many varied conditions. He detected that here mainly acetic acid existed and that it is found in even greater quantities if the acidity at the very beginning is reduced. He emphasized the significance of the existence of the yeast with these words:

> Ainsi nous arrivons à la conclusion que le fait d'une augmentation d'acidité d'autant plus faible que le liquide est primitivement plus acide est la règle générale dans la vie de la levure. Elle se retrouve, quelles que soient les conditions dans lesquelles la cellule de levure est placée.

This capacity for the formation of considerable quantities of acetic acid is also shown by the yeast in presence of phosphates and carbonate, as

[11] A. Bechamp, *Compt. Rend.*, *56,* 1231, *57,* 520 (1863); *75,* 1036 (1872).

[12] A. Fernbach, *Ann. Brasserie Destill,* 1913.

Euler and Svanberg[13] and Kerb[14] recently were able to establish. The observations by Reisch[15] concerning the roll of the fermentation of acetic acid (which we explain below) are in agreement. According to them the strongest types of fermentation yeast are also the strongest producers of acetic acid. The production of acetic acid is maximal at the beginning of the fermentation and stops completely soon thereafter. The course of development shows a certain analogy with the normal development of glycerin studied by W. Seiffert and R. Reisch in as much as the latter also begins with the actual cleavage of the sugar; but at a time when the acetic quantity has already become constant, the glycerin yield increased still further. The quantitative relation between acetic acid and glycerin produced by fermentation and the internal dependency of both products, however, have apparently not been studied.

Even the insoluble magnesia causes a portion of the sugar to decompose in agreement with the equation of the third form of fermentation.

E. FERMENTATION IN THE PRESENCE OF THE VARIOUS FORMS OF SODIUM PHOSPHATE (TERTIARY, SECONDARY, AND PRIMARY + SECONDARY PHOSPHATE)

Experiment 7 (with trisodium phosphate). Qualitative experiments with Na_3PO_4 showed complete inhibition of fermentation by 0.5 *M*, incomplete decomposition by 0.25 *M*, and total transformation by 0.125 *M* Na_3PO_4 with 48 hours.

The limiting alkalinity was approximately the same as that found applicable, years ago, in the cell-free system by Neuberg and Färber.[16] Exactly as in this case, a distinct evolution of carbon dioxide takes place very early. The statement by Wilenko[17] that sugar could become fermented in the presence of phosphates without the production of CO_2 was not compatible with the results that Euler and Tholin[18] had obtained earlier in experiments with the interesting accelerating and delaying effect of the phosphate ion. Oelsner and Koch[19] independently refuted the former opinion concerning live yeast and convincingly confirmed the statement by Neuberg and Färber in regard to the changes in the fermentation yields.

We mention the following experiment with trisodium phosphate. In 1000 cm^3 of the used mash the following were present: 95 g of sucrose (equivalent to 100 g of hexose), 100 g of yeast M, and 48 g of Na_3PO_4 (+ 12 H_2O = 0.125 *M*). After 48 hours the following were present per 100

[13] H. Euler and O. Svanberg, *Z. Physiol. Chem., 105,* 187 (1919).
[14] Johannes Kerb, *Ber., 52,* 1795 (1919).
[15] R. Reisch, *Zentral. Bakteriol. Parasitenk.,* II, *14,* 572 (1905).
[16] C. Neuberg and E. Färber, *Biochem. Z., 78,* 258 (1916).
[17] G. G. Wilenko, *Z. Physiol. Chem., 98,* 255 (1917).
[18] H. Euler and T. Tholin, *Z. Physiol. Chem., 97,* 274 (1916).
[19] A. Oelsner and A. Koch, *Z. Physiol. Chem., 104,* 175 (1919).

cm^3 of fermentation: alcohol, 3.90 g; glycerin, 1.03 g; and acetic acid, 0.34 g. Accordingly, the yield of glycerin and acetic acid was 20.1 percent with the use of trisodium phosphate following the method of the third form of fermentation.

Experiment 8 (with disodium phosphate). Orientating experiments showed that 10 g of sucrose began to ferment by means of 10 g of compressed yeast at 34°C in 100 cm^3 of liquor containing 0.25 to 1.5 *M* Na_2HPO_4. At incubator temperature the disodium phosphate remains in solution up to a content of 1 *M*, which is also the limit for the complete decomposition of sugar.

Next an experiment with a weaker Na_2HPO_4 concentration is given; 700 cc of the starting mixture for fermentation contained the following: 66.5 g of sucrose (= 70 g of hexose), 56.0 g of disodium phosphate, $Na_2HPO_4 + 12H_2O$ (= 0.22 *M*), and 70.0 g of yeast M. In the fermented mash the following were detected: alcohol, 41.98 percent (of sugar used); glycerin, 8.10 percent (of sugar used); and acetic acid, 2.68 percent (of sugar used).

That the increase in the amount of phosphate—as we had established earlier with the bicarbonate experiments—likewise causes an increased incidence of the third form of fermentation is pointed out next.

Experiment 9 (disodium phosphate in m-concentration was used). 500 cm^3 of liquid contained the following: 47.5 g of sucrose, 50.0 g of yeast M, and 180.0 g of $Na_2HPO_4 + 12H_2O$. After 60 hours all sugar was used up. To avoid crystallization of the disodium phosphate in cold temperatures the volume was doubled before processing by addition of water. In 100 cm^3 of the original starting mixture for fermentation (with 10 g of invert sugar) the following were retained: alcohol, 4.02 g; glycerin, 0.97 g; and acetic acid, 0.38 g.

Experiment 10 (with a mixture of disodium and monosodium phosphate). In 700 cm^3 of the starting mixture for fermentation the following were found: 66.5 g of sucrose (= 70 g of invert sugar), 70.0 g of yeast M, and a mixture of 23 g of $Na_2HPO_4 + 12H_2O$ + 12 g of $NaH_2PO_4 + H_2O$. After 48 hours the analysis for 100 cm^3 fermentation was as follows: alcohol, 4.05 g; glycerin, 1.13 g; and acetic acid, 0.40 g. The well-known buffering mixture of both phosphates had a considerable influence on the fermentation progress: it caused a decomposition of 22 percent of the sugar according to the third form of fermentation.

F. FERMENTATION IN THE PRESENCE OF ZINC HYDROXIDE

In each of four starting mixtures for fermentation, which contained 10 percent sucrose and an amount of yeast M equal in weight to the amount of sugar, were suspended 0.125, 0.25, 0.5, and 1 mole of carefully washed $Zn(OH)_2$ freshly prepared from zinc nitrate. The mixture with 0.5 mole or lesser amounts was completely fermented after 48 hours, whereas in a mixture containing 1 mole, sugar was still present.

Experiment 11. Into 240 cm^3 of water containing 57 g of sucrose and 60 g of yeast M were mixed 360 cm^3 of an aqueous suspension of zinc hydroxide that contained 99 g of $Zn(OH)_2$ per 1200 cm^3. Thus 30 g of $Zn(OH)_2$, that is, 0.5 mole in a total volume of 600 cm^3, were used. After 48 hours the action stopped and the following determination of the fermentation products was established: alcohol, 4.22 g; glycerin, 0.69 g; and acetic acid, 0.24 g. Therefore, under the influence of zinc hydroxide, more than twice the amount of glycerin and a corresponding gain in acetic acid were obtained than in ordinary fermentation.

G. FERMENTATION IN THE PRESENCE OF ALUMINUM HYDROXIDE

Preliminary experiments showed that a 10 percent sugar solution was completely fermented if the content of suspended, well-washed aluminum hydroxide—made from nitrous alumina—was not higher than $M/2$-Al(OH).

Experiment 12. Sixty grams of yeast was added to 57 g of sucrose dissolved in water and the total volume then adjusted to 200 cm^3. To this was added 400 cm^3 of an aluminum hydroxide suspension that contained 78 g of freshly precipitated substance per 1333 cm^3 of water. Thus 23.4 g of $Al(OH)_3$—corresponding to 0.5 mole in a total volume of 600 cm^3—were present. The appearance of the fermentation was that of a highly viscous paste; the activity of the yeast, which was very lively at first, stopped after 48 hours.

The alcohol analysis gave 4.70 g in 100 cm^3 mash for 10 g of fermented sugar. Thus practically all sugar was converted according to the normal equation. The yield of glycerin (0.27 g) was also the usual yield. An influence of aluminum hydroxide on the nature of the fermentation products was not noticeable under the chosen conditions of the experiments.

H. FERMENTATION IN THE PRESENCE OF COLLOIDAL IRON HYDROXIDE

Experiment 13. Sucrose to the amount of 23.75 g was diluted in 125 cm^3 of water and mixed with 25 g of yeast. To this mixture 150 cm^3 of a 5 percent solution of colloidal iron hydroxide was added. Total volume was 300 cm^3. After 48 hours all sugar had disappeared; in 100 cm^3 of fermentation broth (one third of the original solution) with a sugar content of 8.33 g of hexose the following was present: 4.12 g of alcohol. The fermentation had therefore practically occurred according to the Gay–Lussac equation and an influence of the iron hydroxide was not noticeable.

J. DETERMINATION OF THE ACETALDEHYDE APPEARING DURING THE COURSE OF FERMENTATION

Earlier we demonstrated in accord with the observations by Neuberg and Färber that fermentations in the presence of sodium carbonate in the first 6 hours show an increased formation of acetaldehyde, which in the course of further decomposition again disappears except for those traces encountered also in the common alcoholic sugar cleavage. This indicated to us that the acetaldehyde was the original oxidation equivalent, which—according to the theory—must be formed as the correlative counterpart to the glycerin produced by reduction. On the consumption of acetaldehyde, we were able to base the massive formation of acetate during fermentation in a sodium bicarbonate alkaline solution, in that the acetaldehyde is converted through disproportionation to acetic acid and ethanol.

On account of similar behavior during phosphate fermentation we were able—as shown in the following experiments—to find at the beginning of the conversion an increased formation of acetaldehyde and to establish at the same time the fact that it will go back after a few hours to the normal quantity found during the common form of fermentation.

Experiment 14. In 261 g of dipotassium phosphate, 142.5 g of sucrose was dissolved and mixed with 150 g of yeast. Then the volume was adjusted to 1500 cm^3 (content of $K_2HPO_4 = 1\ M$). Carbon dioxide evolution was clearly evident after a 3-hour incubation at 31°C.

The aldehyde analyses were accomplished in the manner described earlier. The amounts of acetaldehyde, based on the amount of sugar used, were as follows:

Hours elapsed	%
4	0.2
$5\frac{1}{2}$	0.7
7	1.3
$8\frac{1}{2}$	1.3
10	1.3
$14\frac{1}{2}$	0.3
24	0.3

Experiment 15. For comparison, a series of continued aldehyde estimations in a normal fermentation mixture are cited, in the presence of 10 percent sugar and an equal quantity of yeast M. The content of acetaldehyde, based on the amount of sugar used, was as follows:

Hours elapsed	%
1	0.2
2	0.2
4	0.25
5	0.2
24	0.2

The last two experiments show that in the fermentation pathway in *M* dipotassium phosphate solution, the amount of aldehyde can increase at the beginning approximately up to six times the normal value, and then decrease again to the usual amount during the further course of fermentation.

Table 1 is a summary of the current information concerning the extent of the third form of fermentation of sugar under various conditions

Table 1 Data on third form of sugar fermentation from preceding and earlier experiments

In the presence of:	*Conversion of sugar according to the third form of fermentation in an amount of (%):*
1.5 *M* sodium carbonate	26.8
1.0 *M* dipotassium phosphate	26.8
0.09 *M* Na_2HPO_4 + buffering mixture 0.123 *M* NaH_2PO_4	22.1
0.125 *M* trisodium phosphate	20.1
1.0 *M* disodium phosphate	19.0
0.22 *M* disodium phosphate	15.8
0.125 *M* magnesium oxide	14.3
0.5 *M* zinc hydroxide	13.5
0.1 *M* potassium carbonate	13.1

Without trying to draw conclusions, one gets the impression after looking at Table 1 that in general the activity of the third form of fermentation can be connected with the changing molarity of the additives in the mashes. Naturally, these figures are not directly comparable with one another; the condition of the yeast used was not uniform because some of the experiments were carried out at widely separated times.

In all cases the ratio of 1 mole of acetic acid : 2 moles of glycerin established in the theory has been proved experimentally. With sufficient accuracy all these experiments show that the acetic acid number is one third the size of the glycerin, according to a molecular weight of 60 for $1Ch_3{\cdot}COOH$ and 184 for $2C_3H_8O_3$. Only the glycerin data were used as a basis for the calculation because the quantitative determination of the glycerin is analytically the more precise operation. In the process of distillation of any kind of fermentation product a certain quantity of "nonspecific" volatile acid flows over and becomes mixed—as we pointed out earlier—with the acetic acid, which is forming according to the third form of fermentation. The nonspecific volatile acids, besides acetic acid, which usually predominates, are acids formed through side reactions and partially also from other sources than sugar, such as yeast protein.

Notable and worthy of further experimentation is the circumstance that many, but by no means all, compounds with basic characteristics cause the third form of sugar fermentation to begin. In view of other aspects of the fermentation problems, we would like to point out that the phosphates of all neutralization levels have proved to be effective in this respect. Since they are among the best exciters of the third form of fermentation, they prevent the normal progress of the common alcoholic sugar cleavage. This characteristic of the salts combined with phosphoric acid supports our earlier opinion. We shall come back to the proportions during the treatment of other problems.

K. RELATIONSHIP AND MEANING

Some brief observations probably merit consideration concerning the relations of the third form of fermentation to the common alcoholic sugar cleavage. One probably may assume that both types of decomposition in the norm proceed side by side. Glycerin is formed identically by every yeast fermentation. If the usual yield of 3 percent is taken as a base, the entrance of the other form of fermentation at 5.8 percent can be calculated from this. However, acetaldehyde or acetic acid is not present in the amount equivalent to the value of glycerin mentioned above; the reason for this could be, as mentioned earlier,[20] that the acetaldehyde or its preliminary step, pyroracemic acid, undergoes a multiple consumption during the fermentation process. Pyroracemic acid and acetaldehyde are, for example, the very substances of which synthetic usage is best clarified through the organisms in their chemical action.[21] It can also be assumed that with the use of sugar as building material for the different cell constituents the same fractional parts are utilized which are formed in the physiological sugar cleavage; un-

[20] C. Neuberg and E. Reinfurth, *loc. cit.*

[21] Reference is made to the importance of the acetaldehyde for the biochemical synthesis of butyl derivatives (E. Friedman, *Contrib. Chem. Physiol. Pathol., 11,* 202 1908); A. Harden and D. Norris, *Contrib. Chem. Physiol. Pathol., 12,* I. 1045, 2051); to the transition of the pyroracemic acid to alanine (G. Emden and E. Schmitz, *Biochem. Z., 38,* 393; H. Fellner, *Biochem. Z., 38,* 414, 1911); also to the experiments by P. Maze (*Compt. Rend., 134,* 240, 1902) and A. Perrier (*Compt. Rend., 151,* 163, 1910) on the utilization of acetaldehyde through microorganisms; even for technical purposes, this utilization has recently received attention (F. Ehrlich and Cons. F. Electr. Industry, *Contrib. Chem. Physiol. Pathol., 19,* III/IV, t. T. 462).

changed sugar is not used in any case, except in the deposition of certain polysaccharides. This view of the function of the acetaldehyde and the pyroracemic acid has been supported substantially in the meantime through the experiments by C. Neuberg and F. F. Nord[22] who were able—with help of the "interception process"—to establish the role of acetaldehyde in metabolism in a number of other microorganisms.

In any case, one arrives through these deliberations to the result that it is a normal function of the yeast, which is stimulated to an unusually strong development by the additions. The third form of fermentation is an expression of this intensified development.

It is not impossible to picture the activity of the different ferments. We know that common alcoholic sugar cleavage requires a certain acidity.[23] The optimum H-ion concentration lies at pH 2.71, according to measurements by Luers[24]; Boas and Leberle[25] allow the margin pH 2.94 to 3.80. In any case, the yeast strives for a maximal value. If, at the beginning, the right H-ion concentration is not present, the yeast itself produces acid. This acid is volatile and consists completely or mainly of acetic acid, and already Fernbach has shown that with acid deprivation the yeast reacts with an increased production of volatile acids from sugar. Because the enzyme of the disproportionation (Batelli and Stern, Parnass, Neuberg) unfolds the enzyme's optimal activity with bicarbonate alkaline reaction, it is understandable that the enzyme does not start action after the formation of a certain acidity. Because we know today of the absolute dependency of the activity of many enzymes on a closely bounded H-ion concentration, an agreement of this kind may be the reason why, after the formation of the acid reaction favorable for the zymase (cell-free sap of yeast), no aldehyde is disproportioned to acetic acid (and alcohol). In complete accord with oxidation–reduction processes, when the zymatic cleavage of the sugar occurs, the resulting acetaldehyde will always be reduced to ethanol as long as no consumption of acetaldehyde or pyroracemic acid, thus no interception of any kind, occurs. The internal usage of acetaldehyde and pyroracemic acid in transformations and metabolic processes represents a physiological interception. Through it "hydrogen" becomes available exactly as in the fixation of the aldehyde through purely chemical means; because it cannot develop freely, it is absorbed by a part of the sugar and glycerin is formed. Therefore, all glycerin occurring in any kind of fermentation is the visible expression of the intermediate consumption of acetaldehyde or pyroracemic acid.

This concept explains best why the fermentation quotient of alcohol

[22] C. Neuberg and F. F. Nord, *Biochem. Z., 96,* 133, 158 (1919).

[23] That CO_2 development in cell-free fermentation can at first be stimulated through alkalinity, but will be reduced in the long run, has been explained sufficiently by Buchner.

[24] H. Luers, *Contrib. Chem. Physiol. Pathol., 14,* I. 1101.

[25] F. Boas and H. Leberle, *Biochem. Z., 90,* 78 (1918).

to carbonic acid, which is theoretically 46 : 44 = 1.045, is often found to be smaller than 1 in sugar cleavage by yeast. In these cases probably more CO_2 than alcohol is released because the synthetic use of the acetaldehyde causes the entrance of the second or third fermentation, which still produces 1 mole of CO_2 to 1 mole of bound acetaldehyde; but at most only 0.5 mole of alcohol appeared through the Cannizzaro reaction. Where formerly it was maintained that the carbonic acid of fermentation would get mixed with carbon dioxide formed by other metabolic processes (respiration, basal metabolism), a part of these processes now seems to be expressed in the formation of glycerin, which constantly has to be verified.

If the OH-ion concentration is raised through the addition of mineral substances, causing the reaction to come closer to the optimum of the Cannizzaro fermentation and start the disproportionation, the zymatic process is going to be changed through inevitable removal of the acetaldehyde only so far that now another "hydrogen" acceptor (the glycerin-forming one) steps into its place. Whether one should see in the increased formation of acetic acid an effort by the yeast to create the most suitable medium for its characteristic work, the forming of alcohol, is uncertain. It must also be the object of further research to determine if differences in the speed of the individual fermentation reactions caused by the additions would be noticeable and therefore would influence the entrance of one of the other forms of fermentation.

The end products of the three forms of fermentation are distinctly divided chemically, yet the three kinds of sugar cleavage are closely connected to each other. Based on the above statements, the condition of the medium—H-ion concentration and internal need for oxidation steps (pyroracemic acid, acetaldehyde)—determines what final shape the intermediate formation will take. Thereby, it has no direct or side path nor main or side products. None of the three fermentation equations can be obtained in pure form, not even the Gay–Lussac ideal form, because the second or third form of fermentation occurs on account of different uses of acetaldehyde or pyroracemic acid at the same time.

4

Reprinted from *J. Bacteriol.*, **36**(1), 67–76 (1938)

INFLUENCE OF pH ON THE DISSIMILATION OF GLUCOSE BY AEROBACTER INDOLOGENES[1]

MILO MICKELSON AND C. H. WERKMAN

Department of Bacteriology, Iowa State College, Ames, Iowa

Received for publication January 31, 1938

Osburn, Brown and Werkman (1937) observed a critical pH level in the dissimilation of glucose by *Clostridium butylicum* near 6.3. Above this level butyl alcohol and isopropyl alcohol were not formed. The significance of this critical pH was attributed to the fact that in mixtures of acetic (or butyric) acid and its sodium salt below 6.3 there will be "free" acid, i.e., acid which can be distilled from solution and can be converted into alcohols by *C. butylicum.*

This concept of critical pH is applied to the dissimilation of glucose by *Aerobacter indologenes* in the present investigation, particularly with reference to the findings of Reynolds and Werkman (1937), that when sufficient acetic acid is present in a glucose dissimilation by *A. indologenes,* no gaseous hydrogen is evolved and the acetic acid disappearing in the medium is accounted for as 2,3-butylene glycol + acetylmethylcarbinol. Their conclusions were to the effect that acetic acid occurring in the fermentation was transformed into 2,3-butylene glycol + acetylmethylcarbinol.

The principle of critical pH probably has a general application to cellular metabolism. Although pH 6.3 was found to be the critical level in the case of *C. butylicum* in converting butyric and acetic acids into corresponding alcohols, it is likely that each organism will show several such levels which will be determined by the dissociation of the substance subject to change. Furthermore, as pointed out by Osburn and Werkman (1937), the change

[1] Supported in part by Industrial Science Research Funds of Iowa State College.

in metabolism does not occur sharply at pH 6.3, but the effect is gradual and extends over a relatively narrow range.

The dissimilation of glucose by *A. indologenes* at pH levels below and above 6.3 has been studied in the present work. The results are of particular significance for an understanding of bacterial dissimilation as well as having certain practical implications in procedures involving the presence of acetylmethylcarbinol. In the latter respect, reference may be made to the Voges-Proskauer test and the formation of acetylmethylcarbinol in butter cultures.

In the work of Reynolds and Werkman (1937) no mention of a critical pH level at which the mechanism of dissimilation changes its nature, is made. In the present investigation the significance of levels above and below pH 6.3 is shown in the dissimilation of glucose by *A. indologenes*. The results indicate that the gas ratio of carbon dioxide to hydrogen, the occurrence of acetylmethylcarbinol and 2,3-butylene glycol and the amounts of acetic and formic acids are dependent on the pH of the fermentation and that the range near pH 6.3 is a critical level; below this range, the course of dissimilation is markedly different from that above.

METHODS

Glucose plus acetate fermentations were conducted in 4-liter Erlenmeyer flasks fitted with proper openings for removal of gas, taking of samples, and addition of alkali. Brom-thymol-blue was used as an indicator and the pH maintained by adding normal sodium hydroxide.

The medium consisted of glucose 2 per cent, ammonium sulfate 0.3 per cent, peptone 0.5 per cent, sodium acetate 0.6 per cent and 0.1 M dipotassium phosphate. It is of interest that distilled water was not suitable for making up the medium. Tap water provided excellent growth. When Speakman's salts in distilled water replaced the tap water, growth was not nearly as good. When tap water was evaporated and the residue ashed, addition of the latter provided growth equal to that in the tap water medium. Apparently the essential constituent is inorganic.

The pH was adjusted with sulfuric acid and the medium sterilized at 20 pounds pressure for 30 minutes. Glucose was sterilized separately. Incubation was at 30°C.

Carbon dioxide was collected in Bowen potash bulbs in a drying train. Residual carbon dioxide was determined on a sample collected in alkali which was acidified and refluxed in a stream of CO_2-free air. The carbon dioxide was collected and weighed in a Bowen bulb.

Hydrogen was determined by continuous combustion over heated copper oxide and weighing the water formed. Oxygen-free nitrogen was continuously led through the flasks to remove gases.

An aliquot part of the fermented medium was made acid to congo red and distilled to half volume. The distillate was neutralized to phenolphthalein, again distilled and the alcohol determined on this distillate. The two residues were combined and steam distilled to recover the volatile acids. Total volatile acidity was determined on this distillate by titration. The formic acid was determined according to Auerbach and Zeglin (1922), and the acetic acid obtained by difference. In the presence of acetylmethylcarbinol it was necessary to neutralize the volatile acid distillate, evaporate to small volume and remove at least six volumes by alkaline steam distillation to avoid interference with the determination of formic acid.

Ethyl alcohol was determined on an aliquot part of the neutral volatile fraction by oxidation with a sulphuric acid-potassium dichromate mixture in a closed flask on the steam bath. The mixture was then steam distilled and the acetic acid determined. The alcohol determination was corrected for the acetic acid originating from the acetylmethylcarbinol.

Acetylmethylcarbinol was determined on an aliquot part of the fermented medium according to Stahly and Werkman (1935).

2,3-Butylene glycol was steam distilled from a sample of the medium from which the sugar had been removed by the copper-lime technique of Hewitt (1931); 13 volumes or more were collected from a constant volume of 20 ml. which contained 25 grams of $MgSO_4 \cdot 7H_2O$. Butylene glycol was determined on an aliquot part of the distillate by oxidation with periodic acid and directly distilling the acetaldehyde into a solution of sodium bisulphite. The aldehyde was determined by titration with iodine.

The lactic acid was determined on the residue according to Friedemann and Graesser (1933).

The experiments with cell suspensions were conducted in Erlenmeyer flasks containing 300 ml. of 0.1 M phosphate with substrate, adjusted to the desired pH. Carbon dioxide and hydrogen were collected in bottles containing alkali; the former was absorbed and determined on an aliquot part, the latter was determined by volume displacement. The hydrogen evolved was in contact with the medium and was available for reduction of the acetic acid. The inoculum consisted of about 5 grams of cell paste per 300 ml. of medium. Air was displaced from the flasks with nitrogen when the fermentations were started. After four days at 30°C. the flasks were removed and contents analyzed according to the procedure used in the glucose ferementations.

EXPERIMENTAL

In tables 1 and 2 are given results of typical experiments in which the influence of pH on the mechanism of dissimilation of glucose by *A. indologenes* is shown. In the fermentation kept above pH 6.3 (near 7.0) the yields of carbon dioxide and 2,3-butylene glycol are greatly suppressed as compared with an acid fermentation. Acetic and formic acids accumulate and added acetic acid is not attacked. The quantitative relationships of the products are quite different from those in a dissimilation occurring below pH 6.3. Here, acetic acid, even that added, disappears rapidly with the simultaneous occurrence of the 4-carbon compounds, acetylmethylcarbinol and 2,3-butylene glycol (fig. 1); also the production of carbon dioxide is large. Whether the failure of acetic acid to act as an intermediate in alkaline fermentation is due only to a lack of free acid or a dearth of available hydrogen or both, is not clear. Should the presence of formic acid be the result of a synthesis from carbon dioxide and hydrogen, the failure of acetic acid reduction probably results from the utilization of hydrogen in the reduction of carbon

TABLE 1

Fermentation of glucose plus added acetic acid by Aerobacter indologenes. pH maintained above 6.3

TIME	SUGAR FERMENTED	CO_2	H_2	FORMIC ACID	ACETIC ACID	LACTIC ACID	ACETYLMETHYLCARBINOL	2,3-BUTYLENEGLYCOL	ETHYL ALCOHOL	2,3-BUTYLENEGLYCOL + ACETYLMETHYLCARBINOL	CARBON RECOVERY	O/R INDEX
hours											*per cent*	
0	0				25.77							
16	19.91	15.03		19.15	37.65	2.72	0.24	9.35	15.20	9.59	108	0.89
20	50.90	33.90		49.90	47.70	2.76	1.27	12.50	36.70	13.77	89	1.06
23½	89.30	51.00		74.00	96.05	7.10	1.50	15.60	64.60	17.10	92	0.97
31½	113.10	58.5	2.26	95.00	109.00	5.85	0	22.40	74.80	22.40	86.3	0.97
75	113.10	60.24	9.26	95.00	109.10	6.35	0	22.30	75.60	22.30	87	0.95

Products in millimoles per liter.

TABLE 2

Fermentation of glucose plus added acetic acid by Aerobacter indologenes at pH 6.3 or below

TIME	SUGAR FERMENTED	CO_2	H_2	FORMIC ACID	ACETIC ACID	LACTIC ACID	ACETYLMETHYLCARBINOL	2,3-BUTYLENEGLYCOL	ETHYL ALCOHOL	2,3-BUTYLENEGLYCOL + ACETYLMETHYLCARBINOL	CARBON RECOVERY	O/R INDEX
hours											*per cent*	
0	0				24.66							
24	35.93	59.59		10.40	11.84		3.64	31.40	20.8	35.04	99	0.92
29	48.73	74.57		10.20	7.00	4.76	4.30	42.00	27.16	46.30	102	0.89
33	59.13	86.52		10.08	4.80	2.28	4.64	45.50	31.46	50.14	95	0.88
48	96.98	132.69	1.24	19.72	5.31	2.20	2.20	70.00	51.54	72.20	90	0.87
96	111.3	215.66	17.69	12.20	3.35	2.46	1.05	82.00	58.40	83.05	96	1.12
168	111.3	225.66	29.07	1.35	4.95	1.78	2.36	82.20	55.80	84.56	96.5	1.15
264	111.3	230.74	30.83	1.26	6.00	2.70	0.72	87.20	60.02	87.92	100	1.14

Products in millimoles per liter.

dioxide rather than acetic acid. It is to be noted that there was some carbon dioxide and glycol formed in the alkaline fermentation, probably owing to the difficulty of holding the medium alkaline as at times it became sufficiently acid to allow the formation of these products. In subsequent experiments conducted at pH 7.0 or above, the formation of 2,3-butylene glycol was diminished to a still greater extent—from a normal of approximately 70 mM to 2 mM or less. The accumulation of acetic

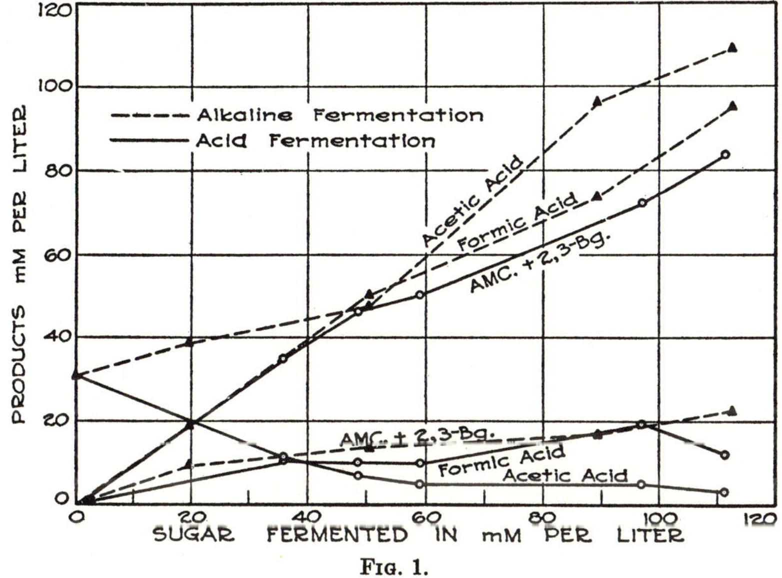

Fig. 1.

acid demonstrates its necessity in an available form for the formation of acetylmethylcarbinol and 2,3-butylene glycol. Incomplete carbon recoveries with the alkaline fermentations suggest that some undetermined product is formed.

It is interesting to note the influence of pH on the gas ratio which varies within large limits. When the pH is maintained below 6.3 and sufficient acetate added, the production of hydrogen is completely suppressed and the carbon dioxide formation is large, whereas, in an alkaline medium the production of both carbon dioxide and hydrogen can be practically prevented. Under the usual conditions of glucose dissimilation the pH drops within the vicinity of 6.0–5.5 but insufficient acetic acid is present to prevent the liberation of hydrogen and a $H_2:CO_2$ ratio of 0.3–0.5 is obtained.

As the fermentation nears completion, the pH shows a reversion owing to the conversion of the acetate ion into neutral compounds.

In regard to the source of hydrogen used in the reduction of acetic acid, it is possible that this is furnished by the dehydrogenation of formic acid. However, using heavy cell suspensions, it was impossible to demonstrate the conversion of acetic acid to acetylmethylcarbinol or 2,3-butylene glycol in the presence of formic acid as H_2-donator (table 3). The fermentations were

TABLE 3

Fermentation of acetic and formic acids by cell suspensions of Aerobacter indologenes in acid and alkaline medium

REACTION OF MEDIUM	ACETIC ACID ADDED	FORMIC ACID ADDED	ACETIC ACID	FORMIC ACID	CARBON DIOXIDE	HYDROGEN	FORMIC ACID FERMENTED
pH 6.0–6.3	112.25	137.76	112.35	72.85	62.20	60.56	64.91
pH 7.0–7.3	111.53	136.90	113.30	121.16	12.40	13.38	15.74

Voges-Proskauer test negative. Products in millimoles per liter.

conducted at pH levels of substantially 6.0–6.2 and at 7.0. It will be observed that the acetic acid was recovered quantitatively in each case and that the formic acid broken down was accounted for quantitatively by equimolar amounts of carbon dioxide and hydrogen. It is also of significance that the formic acid decomposed was about four times greater in an acid than alkaline medium. This finding supports in part, the results from a glucose fermentation under the same conditions. The hydrogen from formic acid did not reduce acetic acid under these conditions. However, in the presence of glucose plus acetic acid the yield of hydrogen is greatly suppressed. If formic acid is the only hydrogen-yielding intermediate it is difficult to account for this behavior which suggests that hydrogen from some other source is necessary for the conversion of acetic acid to acetylmethylcarbinol. Formic acid hydrogen is able to reduce acetylmethylcarbinol to 2,3-butylene glycol by *Escherichia coli* (table 4).

The effect of pH on the action of the formic hydrogenlyase is important. Krebs (1937) found that *E. coli* did not form lactic acid from pyruvic acid anaerobically when the pH was between 6.6–8.0. However, if the pH was lowered to 5.4, considerable lactic acid and carbon dioxide were produced. In an alkaline medium formic and acetic acids were the only final products of pyruvic breakdown. Woods (1936) found in the case of *E. coli* that an alkaline reaction is necessary for a synthesis of formic acid from carbon dioxide and hydrogen. We have

TABLE 4

Reduction of acetylmethylcarbinol by cell suspensions of Escherichia coli and formic acid

	ACETYLMETHYLCARBINOL	CARBON DIOXIDE	HYDROGEN	FORMIC ACID	ACETIC ACID	2,3-BUTYLENEGLYCOL	CARBON RECOVERY
							per cent
Initial	54.25			112.77			
Final	0	76.31	30.10	39.69	25.73	41.91	102

Products in millimoles per liter.

found that formate in an acid medium is decomposed vigorously. However, at alkaline levels the conditions are relatively more favorable for the synthesis of formic acid than its breakdown. Since formic hydrogenlyase is a reversible system, the retention of carbon dioxide in the medium may tend to drive the reaction toward the synthesis of formate.

SUMMARY AND CONCLUSIONS

A critical pH level has been shown to exist in the dissimilation of glucose by *A. indologenes* in the region of 6.3. Fermentation carried out above this level results in an accumulation of acetic and formic acids. The production of hydrogen and carbon dioxide is greatly suppressed and the formation of acetylmethylcarbinol and 2,3-butylene glycol may be prevented if care is taken not to allow the pH to drop within the vicinity of 6.3. When the fermentation occurs below pH 6.3, the acetic acid is converted into acetylmethylcarbinol and 2,3-butylene glycol. If sufficient acetic acid is added to an acid fermentation, the production of gaseous hydrogen is prevented.

It is suggested that the acidity and alkalinity of the medium expresses itself by determining the relative hydrogen accepting ability of acetic acid and carbon dioxide. When the medium is alkaline, carbon dioxide is the better hydrogen acceptor resulting in the formation of formic acid; under acid conditions (approximately 6.3 or less) carbon dioxide cannot compete with acetic acid and acetylmethylcarbinol and 2,3-butylene glycol are formed.

For the conversion of acetic acid into neutral compounds, it is important to have both free acetic acid and available hydrogen. Evidence suggests that the general occurrence of formic acid among the final products of bacterial dissimilation of glucose may be the result of a synthesis from carbon dioxide and hydrogen, a fact which would have important implications in formulating schemes of dissimilation particularly with reference to the breakdown of pyruvic acid into formic and acetic acids or acetaldehyde and carbon dioxide.

The results point clearly to the fact that the gas ratio of (H_2:CO_2) 0.5 is fortuitous. This generally accepted ratio of 0.5 is the result of the conditions under which the determination is made and may vary within wide limits. Review of the literature dealing with quantitative experiments shows that the ratio is usually near 0.3, indicating a marked utilization of hydrogen for the reduction of acetic acid to 2,3-butylene glycol.

Appreciation is expressed to Dr. A. R. Stanley for assistance in preliminary experiments.

REFERENCES

AUERBACH, F., AND ZEGLIN, H. 1922 Beitrage zur Kenntnis der Ameisensaure. I. Mitteilung: Zur gravimetrischen Bestimmung der Ameisensaure. Z. physik. Chemie, **103**, 161–177.

FREIDEMANN, T. E., AND GRAESSER, J. B. 1933 The determination of lactic acid. J. Biol. Chem., **100**, 291–308.

HEWITT, L. F. 1931 Bacterial metabolism. I. Lactic acid production by hemolytic streptococci. Biochem. J., **26**, 208–217.

KREBS, H. A. 1937 The rôle of fumarate in the respiration of *Bacterium coli commune*. Biochem. J., **31**, 2095–2124.

OSBURN, O. L., BROWN, R. W., AND WERKMAN, C. H. 1937 The butyl alcohol-isopropyl alcohol fermentation. J. Biol. Chem., **121**, 685–695.

REYNOLDS, H., AND WERKMAN, C. H. 1936 The intermediate dissimilation of glucose by *Aerobacter indologenes*. J. Bact. **33**, 603–614.

STAHLY, G. L., AND WERKMAN, C. H. 1935 The determination of acetylmethylcarbinol. Ia. State Coll. Jour. Sci. **10**, 205–211.

WOODS, D. D. 1936 Hydrogenlyases. IV. The synthesis of formic acid by bacteria. Biochem. J., **30**, 515–527.

5

Reprinted from *J. Bacteriol.*, 57(2), 147–158 (1949)

FERMENTATION OF GLUCOSE BY SUSPENSIONS OF ESCHERICHIA COLI

J. L. STOKES

Hopkins Marine Station, Pacific Grove, California

Received for publication October 29, 1948

In 1926 Rona and Nicolai (1926) reported that manometric measurements of the fermentation of glucose by *Escherichia coli* showed this decomposition to be a typical lactic acid fermentation with the formation of two moles of lactic acid per mole of glucose. This conclusion is in contrast with the one reached by Harden (1901), who, by direct chemical analyses, found that the fermentation can be represented approximately by the equation:

$$\underset{\text{glucose}}{2C_6H_{12}O_6} + H_2O = \underset{\text{lactic acid}}{2CH_3CHOHCOOH} + \underset{\text{acetic acid}}{CH_3COOH} + \underset{\text{ethanol}}{C_2H_5OH} + 2CO_2 + 2H_2$$

Harden's work received support from later investigations (Grey, 1914, 1918; Kay, 1926; Scheffer, 1928; Tasman, 1935; Tikka, 1935) also conducted on a scale large enough to permit chemical analyses. Although Harden's studies were made with growing cultures of the bacteria, essentially similar results were later obtained by the use of nonproliferating cell suspensions (Grey, 1918; Tasman, 1935; Tikka, 1935). Even with cell-free extracts of *E. coli* a decomposition more complicated than a lactic acid fermentation has been demonstrated (Kalnitsky and Werkman, 1943).

The discrepancy between the results of Rona and Nicolai and of Harden, Grey, and others has never been explained or reinvestigated, although it has caused much perplexity. The experimental support for Rona and Nicolai's conclusion is rather meager; apart from a qualitative determination of lactic acid and the establishment that no gaseous products are formed directly from the sugar, their case rests upon manometric data that purportedly show the formation of two moles of acid per mole of sugar.

That no gaseous fermentation products were found by Rona and Nicolai is understandable enough today; the subsequent investigations of Stephenson, Stickland, and Yudkin (Stephenson, 1937) have clearly shown that cell suspensions of *E. coli* do not liberate gas from glucose under anaerobic conditions if the organisms are grown in the presence of an abundant supply of oxygen. Thus, the method of preparing cell suspensions used by Rona and Nicolai—growing the organisms on the surface of agar media—would have precluded the appearance of gas in their experiments, other than through the decomposition of bicarbonate by acidic products. But it is also known that such cell suspensions yield, instead of the normally found mixture of CO_2 and H_2, an equivalent amount of formic acid.

From the evidence submitted in their publication, it is not clear how Rona and Nicolai could clain the production of 2 moles of acid per mole of glucose; recalculation of their data pertaining to the two fermentations that were allowed to go to completion indicates that much less acid was formed:

GLUCOSE CONSUMED		ACID FORMED	MOLES OF ACID PER MOLE OF GLUCOSE
mg	*mm³*	*mm³*	
0.15	18.7	11.5	0.61
0.30	37.4	44.5	1.19

One might, perhaps, consider the experiments of Cattaneo and Neuberg (1934) as providing support for Rona and Nicolai's contentions. By the use of dried cells of *E. coli* in the presence of toluene and glutathione, a quantitative conversion of hexose diphosphate to lactic acid was accomplished. Such a system, however, is far removed from that operating in a normal fermentation.

In view of the existing difficulty of accounting satisfactorily for the claims put forward by Rona and Nicolai, it seemed advisable to repeat their manometric experiments and to supplement and support the manometric data by chemical analyses of the fermentations.

EXPERIMENTS

Experiments were made with three strains of *Escherichia coli*, obtained from different sources to ensure that the results would have general significance. Strain PA4.1 was obtained from the culture collection of the Hopkins Marine Station; strain E was isolated from raw sewage by Dr. S. Elsden, to whom we are indebted for this culture; and strain S was isolated from a fecal suspension streaked on eosin methylene blue agar plates.

All three strains consisted of short gram-negative rods. They fermented glucose and lactose in broth with the formation of acid and gas, formed indole in tryptone medium, did not grow with citrate as the sole source of energy, produced sufficient acid in glucose broth to change the color of methyl red indicator, and did not produce acetylmethylcarbinol. The strains were therefore typical *E. coli*. Stock cultures were maintained on yeast extract, 2 per cent glucose (YED) agar slants in the refrigerator and subcultured at bimonthly intervals.

Manometric experiments. The Barcroft-Warburg apparatus was used. Cell suspensions were prepared from YED agar plate cultures incubated at 35 C for approximately 18 hours. The cells were washed from the plates with 0.01 M $NaHCO_3$, centrifuged, and resuspended in sufficient 0.01 M $NaHCO_3$ to give a concentration of 5 mm^3 of cells per ml. Two ml of suspension were used in each Warburg vessel. Two-tenths ml of M/30 glucose were placed in one side cup; the other cup received 0.2 ml of 2 N H_2SO_4 for the determination of initial or residual bicarbonate. The atmosphere was either CO_2, or N_2 containing 1 or 5 per cent CO_2. The gases were freed of oxygen by passage over heated copper gauze. The bath temperature was 30.4 C, ± 0.1 C. Fermentations were allowed to continue until all the glucose had been utilized as indicated by the cessation of acid formation. This usually required 2 to 3 hours. The endogenous fermentation was negligible. Acid formation was measured as CO_2 liberated from the interaction of the fermentation acids with the bicarbonate of the cell suspensions, and the total amount of acid formed was calculated from the difference between initial and residual bicarbonate. The course of a typical fermentation is shown in figure 1.

The results of a series of experiments to determine the number of moles of acid produced by *E. coli* per mole of glucose fermented are given in table 1. Without exception and regardless of the gas phase, the three strains produced an average of about 2.5 moles of acid per mole of glucose. All strains formed approximately the same amount of acid. These values are greater than the 2 moles of acid claimed by Rona and Nicolai and immediately rule out the possibility that each mole of sugar is converted into 2 moles of lactic acid. They indicate, on the contrary, a decomposition more in line with the results of Harden, Grey, and others.

That the CO_2 produced in the vessels is due entirely to the reaction of the fermentation acids with the bicarbonate and does not arise, in part, metabolically from the glucose is shown by the close agreement between initial HCO_3^--CO_2 and the sum of the residual HCO_3^--CO_2 and the CO_2 output (table 2). If any metabolic CO_2 had been formed, the sum of CO_2 liberated to the gas phase and residual bicarbonate would have exceeded the initial HCO_3^--CO_2 by an amount equal to that of the metabolic CO_2.

It was to be expected that the pressure changes would be due entirely to CO_2 and not in part to hydrogen, since the cells were grown aerobically and would therefore not contain hydrogenlyase, the enzyme that splits formic acid to CO_2 and H_2 (Stephenson, 1937). This hypothesis was further supported by the absence of gas formation in fermentations in phosphate buffer. Finally, the absence of hydrogen was firmly established by gas analyses at the termination of several of the manometric experiments. This was necessary in view of the following considerations:

The close agreement between initial bicarbonate-CO_2 and final bicarbonate CO_2 plus liberated gaseous CO_2 does not rigorously exclude the possibility that hydrogen might have been formed. Since *E. coli*, in common with other heterotrophic bacteria, can fix CO_2 (Elsden, 1938; Wood *et al.*, 1941), it is at least theoretically possible that H_2 production might have occurred, counterbalanced by an exactly equivalent CO_2 assimilation. The gas analyses made by a procedure similar to that of the "second method" of Dickens and Simer (Dixon, 1943) showed convincingly that neither H_2 nor any gas other than CO_2 and the initially introduced nitrogen were present.

The three strains of *E. coli* will form both acid and gas from glucose, providing the cell suspensions are prepared from YED broth cultures (100 ml of medium per 150-ml Florence flask) instead of from YED agar plates. Such cells, having been cultivated under reduced oxygen tension, contain hydrogenlyase and thus produce H_2 and CO_2 in place of formic acid. Data for two of the strains grown in broth are given in table 3.

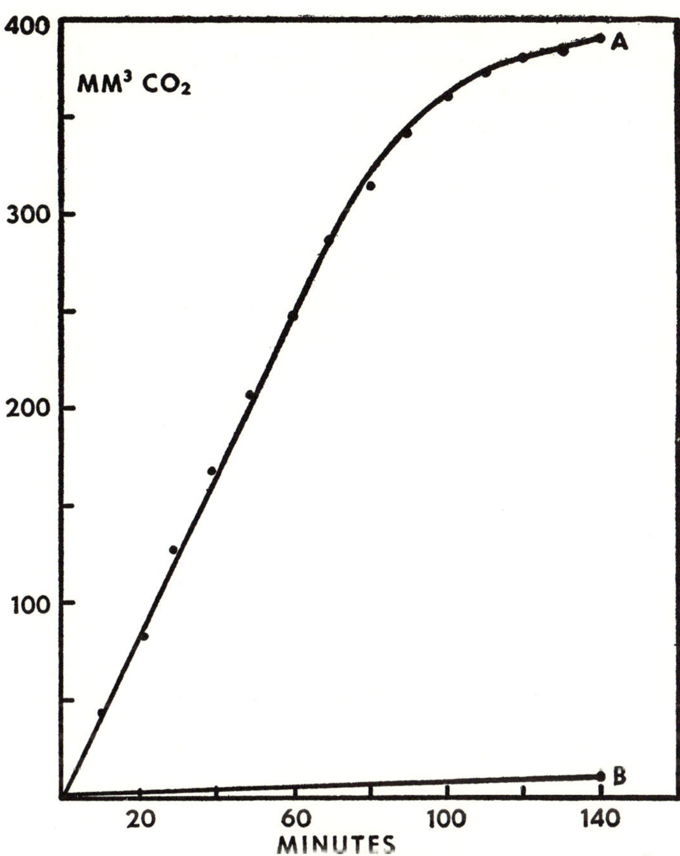

Figure 1. Fermentation of 6.66 μM of glucose by 10 mm³ of *E. coli* cells. A. CO_2 production with glucose. B. CO_2 production without glucose (endogenous fermentation).

TABLE 1

Formation of acid from glucose by suspensions of E. coli

STRAIN	MM³ GLUCOSE	MM³ ACID: Percentage of CO_2 in atmosphere 1	5	100	MOLES ACID PER MOLE OF GLUCOSE
PA4.1	149.3	384			2.57
	149.3	372			2.49
S	149.3	355			2.38
	149.3		386		2.58
	149.3		389		2.60
	149.3			355	2.38
E	149.3	369			2.47
	149.3		380		2.54

The excess of 300 mm[3] of gas, calculated as CO_2, over the total initial HCO_3^--CO_2 is that produced metabolically and not from the interaction of the fermentation acids with bicarbonate. From the ΔHCO_3^-, acid production is computed as 1.3 moles per mole of glucose, i.e., 1.2 moles less than in fermentations with aerobically grown cells. This difference could readily be accounted for as due to the decomposition of formic acid. As will be shown, this is exactly the amount of formic acid produced by cells grown under aerobic conditions. Also, such

TABLE 2

Carbon dioxide balances in the fermentation of glucose by E. coli suspensions

	STRAIN		
	PA4.1	S	E
	mm³	*mm³*	*mm³*
Initial HCO_3^--CO_2	451	420	428
CO_2-liberated	386	389	380
Residual HCO_3^--CO_2	61	33	62
Total	447	422	442

TABLE 3

Formation of acid and gas from 0.2 ml of M/30 glucose by E. coli suspensions from broth cultures

	E. COLI (S)	E. COLI (E)
	mm³	*mm³*
1. Initial HCO_3^--CO_2	413	414
2. Gas liberated (CO_2 + H_2)	492	503
3. Residual HCO_3^--CO_2	205	220
4. Total gas (2 + 3)	697	723
5. Metabolic gas (4 minus 1)	284	309
6. CO_2 due to acid formation (1 minus 3)	208	194
7. Moles of acid per mole of glucose (6 ÷ 149.3)	1.37	1.30

decomposition should give rise to a 1:1 mixture of H_2 and CO_2 and that is what was found.

Chemical analyses. Manometry shows that 2.5 moles of acid are produced per mole of glucose. This is in line with the amounts to be expected from the investigations of Harden, Grey, and others. One might, however, expect that acidic products other than lactic acid had been produced, notably, formic, acetic, and succinic acids. It was therefore desirable to support this view further by chemical analyses.

In preparing material for analysis, a pilot manometric experiment was always made in order to be certain that the particular batch of cells used behaved typically and also to indicate when the sugar was completely utilized. Alongside the Warburg vessels was placed a 500-ml Erlenmeyer flask, fitted with an inlet and outlet tube for aeration, which received exactly 100 times the quantity of cell suspension and glucose that went into each Warburg vessel. The flask thus received 200 ml of cells suspended in 0.01 M $NaHCO_3$ and 20 ml of M/30 glucose (120 mg). The gas phase in the flask was either nitrogen with 5 per cent CO_2 or CO_2 alone, the same as in the Warburg vessel. The flask culture was shaken by hand at frequent intervals during incubation. At the conclusion of the manometric experiment, the flask culture was removed from the bath and acidified with 3 ml of 10 N H_2SO_4 to stop the fermentation and to preserve the fermented liquid from microbial contamination during storage. The cells were removed by centrifugation, washed once with about 50 ml of water, and the total supernatant liquids were adjusted to exactly 300 ml. Aliquots were removed for chemical analysis. The remainder of the fermentation liquor was stored in the refrigerator.

Some difficulties were encountered in the analyses because of the small amounts of metabolic products. About 150 ml of liquor, representing only 60 mg of the initial glucose, was used for a complete analysis of ethanol, the various organic acids, and residual glucose. Semimicro methods, adequate for quantitatively measuring 0.5 mg to 1.0 mg of each of the expected end products, were required. The methods finally adopted were first tested on known solutions of each of the fermentation products and then on an artificial mixture of all of them, mixed in the proportion normally found in the fermentations. Recoveries were of the order of 90 per cent or better.

Residual glucose was tested for by the modified method of Luff-Schoorl (Browne and Zerban, 1941). Even with considerable amounts of fermented solutions, concentrated by evaporation, no trace of unfermented sugar was ever detected. Ethanol was determined by the dichromate oxidation method of Northrup *et al.* (1919). The total volatile acids were obtained by steam distillation. Formic and acetic acids were determined by Duclaux distillation of an aliquot of the volatile acid fraction. Because of the small quantities of acids involved, it was necessary to use special precautions and very carefully to standardize the distillation procedure in order to obtain accurate and reproducible results. The distillation flask was lagged with asbestos and covered with a tin can to reduce condensation. Also, it was essential to use a flame strong enough to distill 25-ml fractions in $5\frac{1}{2}$ minutes with a reproducibility of ± 5 seconds; slower distillation gave erratic results. Care was taken to avoid any change in apparatus or procedure after establishment of the proper conditions for attainment of reproducible constants with known solutions of formic and acetic acids. Formic acid was also determined directly by the $HgCl_2$ method of Fincke (1913). This was done, usually, on an aliquot of the volatile acid fraction but occasionally also directly on the fermented liquor; the same results were obtained in both cases. Since the quantities of formic and acetic acids determined by the two procedures were in good agreement, it is obvious that volatile fatty acids other than these were not present. Lactic acid was determined by the acid $KMnO_4$ method of Friedemann and Graeser (1933), either directly on the fermented liquor or on an aqueous solution of an ether extract of the nonvolatile acid fraction. Succinic acid was measured in the Warburg apparatus with a succinic dehydrogenase preparation made from cormorant breast muscle. The procedure of Cohen (Umbreit *et al.*, 1947) was modified to include disintegration of the muscle tissue in a Waring blender for about 20 seconds in between the first and second washing of the minced tissue. The final preparation had relatively small particles because of the blender treatment, was easy to pipette, and had a high activity, probably due to the finer dispersion of the tissue, so that determinations were completed in 20 minutes as compared to the 60 to 90 minutes normally required. Difficulty was encountered in the ether extraction of the nonvolatile acids after mixing the acidified, concentrated alcohol- and volatile-acid-free fermentation liquor with anhydrous Na_2SO_4. This was traced to the simultaneous extraction of small amounts of the H_2SO_4, initially used in excess for the purpose of acidifying the solution. Although the amount of H_2SO_4 extracted, about 1.0 ml of 0.01 N acid, would be insignificant in the determination of macro quantities of nonvolatile acids, it was equal to about 25 per cent of the total nonvolatile acids being measured. This difficulty was eliminated by acidifying the nonvolatile acids with only sufficient H_2SO_4 to give a pH of 2 prior to mixing with Na_2SO_4. This procedure eliminated excess H_2SO_4 and therefore the extraction of the latter by ether.

The results of the chemical analyses have been combined in table 4. Each mole of glucose fermented gave rise to 0.8 moles of ethanol and of acetic acid, 1.2 moles of formic acid, 0.1 to 0.2 moles of lactic acid, and 0.3 to 0.4 moles of succinic acid. All the carbon, hydrogen, and oxygen of the fermented glucose is accounted for by these end products. The recoveries, although somewhat high, are reasonably good in view of the small quantities of metabolic products involved. The chemically recovered equivalents of acid are substantially in agreement with the quantities determined manometrically. There were no significant differences between strains E and S.

Lactic acid, which is stated by Rona and Nicolai to be the only end product of the fermentation, actually is produced in only relatively small amounts, 0.1 to 0.2 moles per mole of sugar fermented, which is less than that of any of the other organic acids. These quantities are also considerably less than the approximately 1 mole of lactic acid found by Harden (1901), Scheffer (1928), and Tasman (1935) in their experiments. At first, it was suspected that the low yield of lactic acid might be due to the relatively high pH (7.1) at which the fermentations proceeded, since Tikka (1935) has shown such a correlation in his experiments. For example, he found 0.9 moles of lactic acid formed per mole of glucose fermented at pH 6.4, but only 0.4 moles at pH 7.1, and 0.05 moles at pH 7.6. In one experiment, conducted in bicarbonate buffer at pH 7.6, Tikka obtained 0.2 moles of lactic acid, a value which corresponds exactly with our own results. The profound effect of pH on the outcome of microbial fermentations has frequently been observed (see, e.g., Mickelson and Werkman, 1938; Gunsalus and Niven, 1942).

In order to test this possibility, a fermentation was conducted with strain S, suspended in 0.01 M $NaHCO_3$, and equilibrated with a gas phase of pure CO_2, resulting in a pH of 5.8. This experiment yielded essentially the same small amount of lactic acid as fermentations at the higher pH; also the quantities of the other fermentation products were in agreement with those obtained at pH

TABLE 4

Metabolic products formed in the dissimilation of glucose by suspensions of E. coli

	E. COLI (E)	E. COLI (S)	
Gas phase	5% CO_2 in N_2	5% CO_2 in N_2	CO_2
Initial pH	7.1	7.1	5.8
	moles per mole of glucose dissimilated		
End products			
Ethanol	0.84	0.77	0.82
Formic acid	1.16	1.21	1.34
Acetic acid	0.81	0.78	0.70
Lactic acid	0.10	0.20	0.21
Succinic acid	0.34	0.39	0.27
	per cent		
Recoveries			
Carbon	102	108	102
Hydrogen	110	114	111
Oxygen	107	115	110
Redox index	0.90	1.04	0.98
Equivalents of acid			
Manometrically	2.54	2.60	2.38
Chemically	2.75	2.97	2.79

7.1 (table 4). Obviously the hydrogen ion concentration is not the only factor which influences lactic acid formation.

The high yields of lactic acid have invariably been associated with fermentations in media containing phosphate. Hence a number of experiments were conducted in M/15 phosphate buffers at different levels of pH. The results (table 5) leave no doubt that it is the combined effect of phosphate and low pH that gives rise to yields of lactic acid of the order of magnitude of 1 mole per mole of sugar fermented—yields which have so often been encountered ever since Harden's investigations. The spectacular decrease in lactic acid formation at high pH levels is accompanied by an increase in ethanol and volatile acids; analysis of the latter has proved that it is chiefly formic acid production that is affected. This phenomenon is directly comparable with the findings of Gunsalus and Niven

(1942), who showed that homofermentative lactic acid bacteria produce considerable amounts of ethanol and fatty acids at high pH. It would be important to establish whether the behavior of the lactic acid bacteria, too, depends upon the presence of phosphate.

From the foregoing results it follows, therefore, that Harden's equation can be considered to paraphrase the coli fermentation only under a restricted set of environmental conditions. It is possible to view the coli fermentation, especially that with hydrogenlyase-deficient cells, as one in which the preliminary or glycolytic phase gives rise to the key intermediate, pyruvate. The subsequent fate of pyruvate will largely determine the nature of the fermentation. The

TABLE 5

Effect of pH on the quantities of metabolic products formed in the fermentation of glucose by E. coli (S)*

pH	LACTIC ACID	SUCCINIC ACID	VOLATILE ACID	ETHANOL
	moles per mole of glucose fermented			
5.62	0.95	0.14	1.05	0.48
6.00	0.74	0.19	0.75	0.50
6.50	0.32	0.31	0.79	0.78
7.00	0.10	0.26	1.51	0.81
7.46	0.07	0.26	1.39	0.82
7.96	0.05	0.24	1.44	0.83

* Fermentations were conducted in M/15 phosphate buffer at the indicated pH's and under an atmosphere of nitrogen.

pertinent reactions of pyruvate in coli probably are reduction, phosphoroclastic splitting, and dismutation:

$$(1)\ \underset{\text{pyruvic acid}}{CH_3COCOOH} + 2H = \underset{\text{lactic acid}}{CH_3CHOHCOOH}$$

$$(2)\ CH_3COCOOH + H_2O \overset{H_3PO_4}{=} \underset{\text{acetic acid}}{CH_3COOH} + \underset{\text{formic acid}}{HCOOH}$$

(Utter and Werkman, 1943; Utter, Lipmann, and Werkman, 1945)

$$(3)\ 2CH_3COCOOH + H_2O = CH_3CHOHCOOH + CH_3COOH + CO_2$$

(Krebs, 1937)

These are competitive reactions which are regulated by the conditions of the fermentation and which, in turn, determine the quantities of end products formed. Below pH 7 reduction of pyruvate to lactic acid occurs to a considerable extent, whereas above pH 7 this reduction is limited in favor of a phosphoroclastic split of pyruvate to acetic and formic acids. This view is strengthened by the observation that in the absence of phosphate the amount of lactic acid formed is not influenced by pH—and even more so by the lack of CO_2 production in the experiments with aerobically grown cells, which rules out the occurrence of the Krebs dismutation reaction.

The three reaction equations for the decomposition of pyruvic acid do not account for the formation of ethanol, nor of succinic acid. In connection with the latter, the present investigation poses a new problem. Since 1938 succinic acid production in sugar fermentations has been generally considered as resulting from the addition of CO_2 to pyruvic acid and the subsequent reduction of the condensation product, oxaloacetic acid (Wood and Werkman, 1938; Krebs and Eggleston, 1940; Wood, 1946). The experiments here reported render this interpretation doubtful for the fermentation by formic hydrogenlyase-deficient *E. coli*. The CO_2 analyses at the termination of the manometric experiments, as well as the close agreement between initial bicarbonate-CO_2 and the sum of the residual HCO_3^--CO_2 and CO_2 production resulting from acid formation, show conclusively that CO_2 was not assimilated during the fermentation. Nevertheless, appreciable amounts of succinic acid were formed. It would appear, therefore, that this substance must have arisen by a mechanism other than the "Wood-Werkman reaction," and it seems possible that a condensation of a 3-carbon compound with formic acid rather than with CO_2 may have occurred. The possibility is, of course, not excluded that other mechanisms may be involved, such as the postulated condensation of acetate with a C_2 or C_3 compound (Slade and Werkman, 1943; Kalnitsky, Wood, and Werkman, 1943). However, some implications of the present data seem to favor the former supposition.

According to the three equations representing the modes of decomposition of pyruvic acid, the formation of acetic acid should be accompanied by the liberation of an equimolar quantity of a 1-carbon compound. It is reasonable to assume that the other 2-carbon product, ethanol, is formed by a similar mechanism. But in that event the molecular ratio of ethanol and acetic acid, on the one hand, and of the 1-carbon compound, here restricted to formic acid, on the other, should be unity. And this is evidently not the case; in the three fermentations listed in table 4 the quantities of the 2-carbon products are far in excess of the formic acid. By postulating a condensation of a 3-carbon intermediate substance with formic acid as one of the steps leading to succinic acid production, the total extent of formic acid formation can be computed by adding the presumable assimilated quantity, 1 mole for each mole of succinic acid, to that finally observed as remaining in the fermented liquid. In doing so, the following figures result:

	QUANTITY IN MOLES OF			
	C_2-products	Formic acid	Succinic acid	"Corrected" formic acid (formic + succinic acids)
Fermentation 1......	1.65	1.16	0.34	1.50
Fermentation 2......	1.55	1.21	0.39	1.60
Fermentation 3......	1.52	1.34	0.27	1.61
Average............	1.57			1.57

This reveals that the "corrected" values for formic acid production closely approximate those of the C_2 compounds, so that the expected 1:1 ratio of C_2 and C_1 products is actually realized.

Alternatively, a mechanism of succinic acid formation involving condensations of 2-carbon compounds would fit the data only if the latter substances could originate from the breakdown of sugar without being accompanied by 1-carbon products. No evidence exists at present for such a degradation.

As for ethanol, the most probable mode of its formation now appears to consist in a reduction of acetic acid. These problems are being investigated further.

ACKNOWLEDGMENTS

These investigations would not have been possible without the constant advice and help of Professor C. B. van Niel. I am also indebted to him for assistance in the preparation of the manuscript. The capable assistance of Mr. William Byrne with the experiments on the effect of pH on the coli fermentations in phosphate buffer is gratefully acknowledged.

SUMMARY

The claim of Rona and Nicolai, based on manometric experiments, that aerobically grown cells of *Escherichia coli* ferment glucose with the production of two moles of lactic acid per mole of sugar could not be confirmed. Instead, our manometric experiments show that 2.5 moles of acid are formed. Chemical analyses demonstrate that 0.8 moles each of ethanol and of acetic acid, 1.2 moles of formic acid, 0.2 moles of lactic acid, and 0.4 moles of succinic acid are produced in the fermentations. These end products, in the amounts formed, account completely for all the glucose dissimilated.

The pH greatly influences the yields of metabolic products in fermentations conducted in phosphate, but not in those in bicarbonate buffer.

The quantitative data strongly suggest that succinic acid originated by a condensation of a 3-carbon compound with formic acid.

REFERENCES

Browne, C. A., and Zerban, F. W. 1941 Sugar analysis. 3d ed. John Wiley and Sons, New York.

Cattaneo, C., and Neuberg, C. 1934 Umstellung der Coli-Gärung auf reine Milchsäure-Gärung. Biochem. Z., **272**, 441–444.

Dixon, M. 1943 Manometric methods. 2d ed. Cambridge University Press, Cambridge, Eng.

Elsden, S. R. 1938 The effect of CO_2 on the production of succinic acid by *Bact. coli commune*. Biochem. J., **32**, 187–193.

Fincke, H. 1913 Nachweis und Bestimmung der Ameisensäure. Biochem. Z., **51**, 253–287.

Friedemann, T. E., and Graeser, J. B. 1933 The determination of lactic acid. J. Biol. Chem., **100**, 291–308.

Grey, E. C. 1914 The enzymes which are concerned in the decomposition of glucose and mannitol by *Bacillus coli communis*. Proc. Roy. Soc. (London), B, **87**, 472–484.

Grey, E. C. 1918 The enzymes concerned in the decomposition of glucose and mannitol by *Bacillus coli communis*. Part II. Experiments of short duration with an emulsion of the organisms. Proc. Roy. Soc. (London), B, **90**, 75–92.

Gunsalus, I. C., and Niven, C. F., Jr. 1942 The effect of pH on the lactic acid fermentation. J. Biol. Chem., **145**, 131–136.

Harden, A. 1901 The chemical action of *B. coli communis* and similar organisms on carbohydrates and allied compounds. J. Chem. Soc., **79**, 610–628.

KALNITSKY, G., AND WERKMAN, C. H. 1943 The anaerobic dissimilation of pyruvate by a cell-free extract of *Escherichia coli*. Arch. Biochem., **2**, 113–124.

KALNITSKY, G., WOOD, H. G., AND WERKMAN, C. H. 1943 CO_2-fixation and succinic acid formation by a cell-free enzyme preparation of *Escherichia coli*. Arch. Biochem., **2**, 269–281.

KAY, H. D. 1926 Note on the variation in the end-products of bacterial fermentation from increased combined oxygen in the substrate. Biochem. J., **20**, 321–329.

KREBS, H. A. 1937 Dismutation of pyruvic acid. Biochem. J., **31**, 661–671.

KREBS, H. A., AND EGGLESTON, L. V. 1940 Biological synthesis of oxalacetic acid from pyruvic acid and CO_2. Biochem. J., **34**, 1383–1395.

MICKELSON, M., AND WERKMAN, C. H. 1938 Influence of pH on the dissimilation of glucose by *Aerobacter indologenes*. J. Bact., **36**, 67–76.

NORTHRUP, J. H., ASHE, L. H., AND SENIOR, J. K. 1919 Biochemistry of *Bacillus acetoethylicum* with reference to the formation of acetone. J. Biol. Chem., **39**, 1–21.

RONA, P., AND NICOLAI, H. W. 1926 Über den Fermentsstoffwechsel der Bacterien. I. Mitteilung. Atmung und Glykolyse bei *Bakterium coli*. Biochem. Z., **172**, 82–104.

SCHEFFER, M. A. 1928 De Suikervergistung door Bakterien der Coli-Groep. Dissertation, Delft.

SLADE, H. D., AND WERKMAN, C. H. 1943 Assimilation of acetic and succinic acids containing heavy carbon by *Aerobacter indologenes*. Arch. Biochem., **2**, 97–111.

STEPHENSON, M. 1937 Formic hydrogenylase. Ergeb. Enzymforsch., **6**, 139–156.

TASMAN, A. 1935 The formation of hydrogen from glucose and formic acid by the so-called "resting" *B. coli*. II. Biochem. J., **29**, 2446–2457.

TIKKA, J. 1935 Über den Mechanismus der Glucosevergärung durch *B. coli*. Biochem. Z., **279**, 264–288.

UMBREIT, W. W., BURRIS, R. H., AND STAUFFER, J. B. 1947 Manometric techniques and related methods for the study of tissue metabolism. Burgess Publ. Co., Minneapolis.

UTTER, M. F., LIPMANN, F., AND WERKMAN, C. H. 1945 Reversibility of the phosphoroclastic split of pyruvate. J. Biol. Chem., **158**, 521–531.

UTTER, M. F., AND WERKMAN, C. H. 1943 Role of phosphate in the anaerobic dissimilation of pyruvic acid. Arch. Biochem., **2**, 491–492.

WOOD, H. G. 1946 The fixation of carbon dioxide and the interrelationships of the tricarboxylic acid cycle. Physiol. Revs., **26**, 198–246.

WOOD, H. G., AND WERKMAN, C. H. 1938 The utilization of CO_2 by the propionic acid bacteria. Biochem. J., **32**, 1262–1271.

WOOD, H. G., WERKMAN, C. H., HEMMINGWAY, A., AND NIER, A. O. 1941 Heavy carbon as a tracer in heterotrophic carbon dioxide assimilation. J. Biol. Chem., **139**, 365–376.

6

Reprinted from *Can. J. Tech.*, **29**(2), 123–129 (1951)

DISSIMILATION OF GLUCOSE BY YEAST AT POISED HYDROGEN ION CONCENTRATIONS[1]

BY A. C. NEISH AND A. C. BLACKWOOD

Abstract

A distiller's yeast was grown anaerobically in media containing yeast extract (0.5%) and glucose (5%), the pH being controlled within ± 0.05 pH units by automatic addition of ammonium hydroxide. The sugar was almost completely fermented in 14–47 hours over the range pH 2.4 to 7.4 but very little at pH 2.0 or 8.0. When sodium hydroxide is used in place of ammonia the fermentation is completed at pH 8.0 but not at pH 8.2. If the initial glucose concentration is increased to 25% about 80–85% of the sugar is fermented in five to seven days in the range pH 4.0–pH 6.4 but at pH 7.0 only half of it is utilized. The yield of glycerol, based on the sugar fermented, is increased by increasing the initial glucose concentration or the pH or by using ammonia in place of sodium hydroxide. The highest yield of glycerol obtained was 29% by weight of the sugar fermented. Aeration increases the rate of fermentation but causes lower yields of alcohol and glycerol.

Introduction

Numerous papers have been published on the growth and metabolism of yeast. Some of these deal with the effect of pH on the fermentation of glucose, using buffers to obtain control. Some workers have used large quantities of yeast, hence they have been observing primarily the effect of pH on fermentation while others using relatively small inocula have been measuring the effect on growth as well.

When comparatively large amounts of yeast are used it has been found (4) that fermentations require about the same time to reach completion at any pH in the range 4.0–8.5. In strongly alkaline media fermentation is inhibited more than respiration (16).

When small inocula are used, and the yeast grown on synthetic media, the optimum growth occurs at pH 3.4 to 3.9 although there is an initial lag period which is much less pronounced at a higher pH (15). This optimum varies with different strains and may be as high as pH 6.0 for some yeasts (7). If yeast is grown in a medium containing large amounts of glucose and no attempt is made to control the pH it may fall as low as pH 2.2 before growth ceases (5). Additions of organic acids to a medium containing sucrose has shown that the hydrogen ion concentration must be increased to pH 2.3–2.7 before growth of yeast is prevented (17). Growth in alkaline media is very poor above pH 7.7–8.0 although tolerance to alkali is increased if large amounts of inoculum are used (3).

It is well known that the amount of glycerol produced by yeast is considerably increased if alkaline fermentations are carried out (14). With some

[1] *Manuscript received May 29, 1950.*
Contribution from the National Research Council of Canada, Prairie Regional Laboratory, Saskatoon, Saskatchewan. Issued as paper No. 102 on the Industrial Utilization of Wastes and Surpluses and as N.R.C. No. 2315.

strains of yeast, glycerol may be obtained in yields up to 25% of the sugar fermented (2). It is possible that even better yields might be obtained under more closely controlled conditions.

The present paper reports an investigation on the effect of pH on the yeast fermentation. It differs from previous work in that a monitor capable of giving control within ± 0.05 pH units was employed. Comparatively small inocula were used and the effects of varying the sugar concentration, the oxygen tension, and the neutralizing agent were studied at a number of closely controlled hydrogen ion concentrations.

Experimental

Preparation of Inoculum and Media

The same strain of distiller's yeast was used in all experiments. It was obtained from Dr. Elizabeth McCoy, University of Wisconsin, and numbered Y-2 in our collection. The inoculum was grown at 30°C. in a medium containing 1.0% glucose and 0.5% Difco yeast extract. Ten milliliters of a 20 hr. old inoculum was used for 250 ml. of medium in each fermentation. The medium was prepared by mixing three volumes of a glucose solution with one volume of 2.5% yeast extract and one volume of a salts solution. Each of these solutions was sterilized separately and cooled before mixing. The salts solution contained KH_2PO_4(0.25%), K_2HPO_4 (0.20%), $MgSO_4.7H_2O$ (0.10%), $FeSO_4$ (0.025%), $CaCl_2$ (0.05%), and NaCl (0.10%). It is a suspension which must be shaken well before aliquoting. The concentration of glucose was chosen to give a final concentration of 5–25% as desired.

Control of Fermentations

The pH was controlled by automatic addition of sodium hydroxide or ammonium hydroxide solutions using the same type of flask and apparatus as previously described (13). The glass electrodes were sterilized by soaking overnight in 0.1% mercuric chloride, rinsing with sterile water, and then exposing to ultraviolet light (13). Anaerobic conditions were obtained by continuously bubbling purified nitrogen (9) through the fermenting solution while carbon dioxide–free air was used to obtain aerobic conditions. The temperature was maintained at 30°C.± 0.25°.

Analysis of Fermentation Solutions

When the fermentations were finished, as shown by cessation of glucose utilization or alkali addition, the solutions were acidified with hydrochloric acid and swept out with the gas stream in order to remove all of the carbon dioxide. The carbon dioxide was determined gravimetrically, by absorption in Caroxite. The solution was cleared with zinc hydroxide as described previously (9) but some changes were made in the analytical methods. The ethanol was determined as before but the glycerol was estimated by a colorimetric method based on measurement of the formaldehyde formed on periodate oxidation (8). The organic acids were determined by partition chromatography on silica (10) after extraction from the acidified fermentation solution by ether. A considerable amount of water is extracted from alcoholic solutions, hence the extract was made to a definite volume with water, after evaporation of the ether, rather than attempting to add enough *tert*-amyl alcohol to the ether extract to give a solution of the acids in an organic solvent, as before (10). This allows the acids to be concentrated 10-fold during the extraction. An aliquot of the aqueous solution of the extracted acids (0.5 ml.) was analyzed on a silica-water column (10) packed with the upper 15% of it left dry to absorb the water in the sample (12). 2, 3-Butanediol and acetoin were determined by procedures found useful with bacterial fermentations (9), at first, but these were changed to more sensitive and specific methods during the investigations. Acetoin was finally estimated in distillates by the reaction with alkaline creatine and alpha-naphthol (18) while 2, 3-butanediol was determined by measurement of the acetaldehyde formed on periodate oxidation. The acetaldehyde was separated from the fermentation solution during periodate oxidation by microdiffusion into bisulphite (19), and estimated by measurement of the color produced on reaction with piperazine and sodium nitroprusside (1). Methods based on oxidation to diacetyl, such as that of Hooreman (6), are probably more specific but attempts to use them were abandoned when it was found that the yield of diacetyl was different for different isomers of 2, 3-butanediol. For example D-(*levo*)-2, 3-butanediol gave a yield of 72 + 3% while the (*meso-dextro*)-2, 3-butanediol produced by *Aerobacter aerogenes* gave a yield of 81 ± 2% under the same conditions. Since the 2, 3-butanediol produced by yeast is predominately D-(*levo*)-2, 3-butanediol (11) results obtained by these methods are too low when *meso*-2, 3-butanediol is used as the standard. The specificity of methods based on periodate oxidation can be greatly increased by partition chromatography (12).

Results and Discussions

The results of a fairly extensive investigation into the effect of pH on the anaerobic dissimilation of dilute glucose solutions (5%) are shown in Table I. Ammonium hydroxide was used as the neutralizing agent. This fermentation was rather insensitive to the hydrogen ion concentration in the range pH 2.4 to 7.4, considering that there are 100,000 times as many hydrogen ions at the lower pH. However, there was only a small amount of sugar fermented at pH 2.0 (8%) or 8.0 (19.6%), most of this being fermented in the first 10 hr. after inoculation. Compared to the bacteria which have been studied in this way (13) yeast is characterized by a somewhat greater sensitivity to alkali and a tolerance for at least 1000 times as great a hydrogen ion concentration. The rate of fermentation shows a rather broad optimum from pH 4.0 to 6.6. As expected from previous work the yield of glycerol and acetic acid increased with increase of the pH. The highest yield of glycerol obtained was 17.9% of the weight of the sugar fermented (pH 7.4).

TABLE I

ANAEROBIC DISSIMILATION OF DILUTE GLUCOSE SOLUTIONS BY YEAST WITH AUTOMATIC pH CONTROL USING AMMONIUM HYDROXIDE

The fermentations were run in a medium containing 5% glucose, the pH being controlled by automatic addition of ammonium hydroxide

Product	Millimoles of product per 100 millimoles of glucose fermented										
	pH 2.4	pH 3.0	pH 3.4	pH 4.0	pH 5.0	pH 5.6	pH 6.0	pH 6.6	pH 7.0	pH 7.4	pH 7.6
2, 3-Butanediol	0.76	0.75	0.43	0.48	0.46	0.45	0.53	0.45	0.45	0.51	0.68
Acetoin	Nil	Nil	0.02	Nil	Nil	Nil	Nil	Nil	0.07	0.07	0.19
Ethanol	169.7	171.5	171.8	177.0	172.6	164.0	160.5	149.2	149.5	136.9	129.9
Glycerol	8.13	6.16	6.56	6.60	7.82	13.0	16.2	19.2	22.2	35.0	32.3
Butyric acid	0.16	0.13	0.18	0.32	0.25	0.16	0.36	0.44	0.25	0.45	0.21
Acetic acid	0.98	0.52	0.46	0.69	0.84	1.95	4.03	6.14	8.68	15.1	15.1
Formic acid	0.27	0.36	0.26	0.42	0.63	0.09	0.82	0.14	0.35	0.75	0.49
Succinic acid	0.62	0.53	0.48	0.26	0.32	0.52	0.49	0.52	0.23	0.51	0.68
Lactic acid	0.59	0.82	0.77	0.38	0.47	1.29	1.63	1.44	1.93	1.26	1.37
Carbon dioxide	181.2	180.8	176.8	189.8	187.6	174.8	177.0	164.8	161.0	160.0	148.5
Glucose carbon assimilated	15.3	12.4	–	16.1	14.0	–	12.4	16.5	–	–	–
Fermentation time, hr.	41	29	25	14½	17½	14½	15½	14½	35	47	25
% Glucose fermented	98.4	98.5	98.1	97.0	96.5	98.2	98.0	98.5	98.3	96.9	60.3
% Carbon recovered	95.1	93.8	91.2	98.0	96.3	92.5	96.4	93.1	92.5	97.3	91.3
O/R Index	1.04	1.03	1.01	1.05	1.06	1.02	1.05	1.03	1.00	1.03	1.01

An investigation of the effect of using sodium hydroxide as the neutralizing agent has shown that the fermentation can be carried to completion in somewhat more alkaline media (see Table II). However this does not result in any better yields of glycerol. Ammonia was slightly better than sodium hydroxide for production of glycerol, in these experiments.

TABLE II

ANAEROBIC DISSIMILATION OF DILUTE GLUCOSE SOLUTIONS BY YEAST WITH AUTOMATIC pH CONTROL USING SODIUM HYDROXIDE

Conditions as stated for Table I except that sodium hydroxide was used in place of ammonium hydroxide

Product	Millimoles of product per 100 millimoles of glucose fermented					
	pH 5.0	pH 6.0	pH 7.0	pH 7.6	pH 8.0	pH 8.2
2, 3-Butanediol	0.40	0.39	0.38	0.33	0.36	0.33
Acetoin	nil	nil	0.01	0.01	0.01	0.03
Ethanol	177.0	165.9	168.8	148.0	142.5	136.2
Glycerol	6.10	10.4	11.3	25.1	29.4	31.3
Butyric acid	0.27	0.39	0.04	0.35	0.26	0.43
Acetic acid	1.20	4.27	4.30	9.16	9.31	10.5
Formic acid	0.14	0.46	0.41	0.43	0.52	0.25
Succinic acid	1.16	1.14	0.51	0.43	0.45	0.43
Lactic acid	1.53	1.73	0.94	0.87	0.94	0.92
Carbon dioxide	187.0	178.0	170.5	167.8	160.2	160.1
Fermentation time, hr.	14	16	15½	32	46	48
% Glucose fermented	99.1	98.5	99.9	98.1	98.0	55.7
% Carbon recovered	96.0	94.0	91.2	94.1	93.1	94.5
O/R index	1.04	1.03	1.01	1.04	1.02	1.05

The difference in the yields of 2, 3-butanediol between Tables I and II is due to a change in the analytical methods. The results in Table I, obtained by the older methods (9), are undoubtedly high but are reported since they at least set a maximum value. The figures in Table II for 2, 3-butanediol and acetoin were obtained by the more sensitive and specific methods described above. It is of interest that the yields of these compounds are insensitive to variation of the hydrogen ion concentration over a range which markedly affects their production by bacteria (13).

The fermentation of more concentrated glucose solutions was investigated in order to see how much glucose could be fermented at a favorable pH. It was found that about 22–23 gm. of glucose per 100 ml. of medium could be utilized (see Table III), in the range pH 4.0–6.4, both aerobically and anaerobically.

TABLE III

DISSIMILATION OF CONCENTRATED GLUCOSE SOLUTION BY YEAST WITH AUTOMATIC pH CONTROL USING AMMONIUM HYDROXIDE

The fermentations were run in a medium containing 23–25% glucose. Anaerobic conditions were maintained by bubbling a stream of purified nitrogen through the medium. Aerobic fermentations were given about 13 volumes of air per hour

	Gm. per 100 ml. of medium								
	Anaerobic fermentation						Aerobic fermentation		
	pH 4.0	pH 5.0	pH 5.6	pH 6.0	pH 6.4	pH 7.0	pH 6.6*	pH 7.0	pH 7.4
Glucose at start	25.3	24.4	26.6	23.8	25.8	25.2	23.7	22.6	22.9
Glucose at end	3.7	3.3	4.3	2.6	4.0	13.1	0.7	0.6	12.9
Glucose fermented	21.6	21.1	22.3	21.2	21.8	12.1	23.0	22.0	10.0
Glycerol formed	1.11	1.34	2.51	2.46	3.05	3.50	1.35	1.37	2.08
Ethanol formed	10.1	10.2	9.90	8.89	9.59	4.45	8.10	7.84	3.33
Fermentation time, days†	4	5	7	7	7	5	2	2	2
Time actually run, days	6	6	9	8	9	9	3	2	2

* *Contained 1.09% acetic acid—not determined in the others.*
† *Determined by daily analysis for glucose and from the rate of alkali addition.*

The fermentations are more rapid under aerobic conditions but the yields of glycerol and ethanol are not as good, possibly because of respiration. The most interesting result of using a high initial glucose concentration is the increased yield of glycerol relative to the ethanol. This can best be seen by reference to figures in Table IV, which have been calculated from the data in Tables I and III. It is obvious that the yield of glycerol is increased by raising either the initial concentration of glucose or the pH. By using concentrated solutions, yields of glycerol were obtained up to 29% of the sugar fermented (see Table

III, anaerobic fermentation at pH 7.0). This is a good yield considering the yeast used has not been selected for, or adapted to, glycerol production.

Whenever high yields of glycerol are obtained the fermentations are slow and incomplete. It seems to be necessary to inhibit the ethanol-producing mechanism of yeast before glycerol can be obtained in good quantities. This may be done by raising the pH, as has been known for some time, or by in-

TABLE IV

EFFECT OF THE INITIAL GLUCOSE CONCENTRATION ON THE RATIO OF GLYCEROL TO ETHANOL FORMED UNDER ANAEROBIC CONDITIONS

pH of medium	Ratio of glycerol/ethanol by weight	
	5% Glucose	23–25% Glucose
4.0	0.067	0.110
5.0	0.091	0.131
5.6	0.158	0.254
6.0	0.202	0.277
6.4	.....	0.318
6.6	0.257	.....
7.0	0.297	0.786
7.4	0.512	.....

creasing the concentration of glucose. It can be seen from Table III that fairly close control of the pH is important when concentrated glucose solutions are fermented. The result obtained at pH 6.4 is interesting since most of the sugar was fermented to give a solution containing over 9.5% ethanol and 3.0% glycerol. The difficulties of recovering glycerol from such a solution on a commercial scale may not be as great as in some of the patented fermentation processes. However ethanol is still the major product, the yield of glycerol being only 14% of the weight of the sugar fermented. A small amount of aeration might increase the rate of fermentation without much decrease in the yield of glycerol and ethanol. Good yields of products might also be obtained by fermenting a concentrated glucose solution at pH 7.0 for two days and then allowing the pH to fall to 6.4 so the rest of the sugar may ferment.

References

1. DESNUELLE, P. and NAUDET, M. Bull. soc. chim. France, 12: 871. 1945.
2. EOFF, J. R., LINDEN, W. V., and BEYER, G. F. J. Ind. Eng. Chem. 11: 842. 1919.
3. EULER, H. V. and SVANBERG, O. Arkiv Kemi Mineral. Geol. 7 (11). 1919. (Chem. Abstracts, 14: 1128. 1920.)
4. HAGGLÜND, E., SODERBLÖM, A., and TROBERG, B. Biochem. Z. 169: 200. 1926.
5. HARTELIUS, V. Compt. rend. trav. lab. Carlsberg. 20: 44. 1933.
6. HOOREMAN, M. Thesis. Paris. 1949.
7. LAER, M. H. VAN. Wochschr. Brau. 39: 226. 1922.
8. LAMBERT, MARGUERITE and NEISH, A. C. Can. J. Research, B, 28: 83. 1950.
9. NEISH, A. C. Report 46-8-3, National Research Council of Canada. 1946. Free copies sent by author on request.

10. Neish, A. C. Can. J. Research, B, 27: 6. 1949.
11. Neish, A. C. Can. J. Research, B, 28: 660. 1950.
12. Neish, A. C. Can. J. Research, B, 28: 535. 1950.
13. Neish, A. C. and Ledingham, G. A. Can. J. Research, B, 27: 694. 1949.
14. Neuberg, C. and Hirsch, J. Biochem. Z. 96: 175. 1919.
15. Taxner, C. J. Inst. Brewing, 41: 27. 1935.
16. Trautwein, K. and Wasserman, J. Biochem. Z. 236: 35. 1931.
17. Weldin, J. Proc. Iowa Acad. Sci. 22: 95. 1945.
18. Westerfeld, W. W. J. Biol. Chem. 161: 495. 1945.
19. Winnick, T. H. J. Biol. Chem. 142: 461. 1942.

7

Reprinted with permission from *Biotech. Bioeng.*, **15**(2), 239–255 (1973)

The Influence of Environmental Conditions on the Macromolecular Composition of *Candida utilis*

Y. ALROY and S. R. TANNENBAUM, *Department of Nutrition and Food Science, Massachusetts Institute of Technology, Cambridge, Massachusetts 02139*

Summary

Glucose-limited chemostat cultures of *Candida utilis* were cultivated at various pH levels (3.0–7.5), temperatures (15–37.5°C), dilution rates (0.06–0.42 hr^{-1}), and with different nitrogen sources (NH_4^+ and NO_3^-). The ratio of total nucleic acid to protein increased with increase in dilution rate at constant temperature and decreased with increase in temperature at constant dilution rate. The pattern of these variations is consistent with the hypothesis that the nucleic acid to protein ratio is a function of the ratio of the actual dilution rate to the critical dilution rate corresponding to each one of the cultivation temperatures. This ratio is called "reduced dilution rate." A basis is proposed on which various microorganisms may be compared with respect to the ratios of cell protein to nucleic acid, RNA, ribosomal RNA, and polysomes.

The rate of balanced growth of microorganisms in a batch culture can be varied either by adjusting the medium composition or by changing the incubation temperature. With the first approach, the amount of RNA per unit biomass increases with growth rate whereas protein remains relatively constant.[1–6] Based on such observations, Schaechter et al.[7] and Ecker and Schaechter[1] have calculated that in growing organisms the rate of protein synthesis per ribosome particle is constant and independent of growth rate. With more refined methods, Rosset et al.[4] and Sykes and Young[8] have shown that the efficiency of ribosomes in protein synthesis increases with an increase in growth rate.

The effect of cultivation temperature on RNA and protein contents of microorganisms has been studied by several authors. Schaechter et al.[7] demonstrated that although the growth rate of *Salmonella typhimurium*, cultivated batchwise in five different media, varies markedly with temperature; the amount of RNA, mass, DNA, and number of nuclei per cell remain virtually unchanged. Similarly, with *Candida utilis*, Brown and Rose[9] found that RNA and protein vary only slightly per unit of biomass for batch cultures cultivated between 15 and 30°C. Thus Ecker and Schaechter[10] have suggested that temperature alters the rate constant for growth but not the amount or nature of the reactants that control growth. In the chemostat, growth rate is controlled directly by dilution rate. Therefore, it is possible to vary the temperature of cultivation without simultaneously altering the growth rate of organisms in the culture. Tempest and Hunter[11] and Brown and Rose[9] have shown that at a constant dilution rate, the content of RNA in *Aerobacter aerogenes* and *C. utilis* increases with a decrease in cultivation temperature. Under the same conditions, variations in protein content are smaller than those in RNA content and their direction and extent seem to depend on the organism and the growth-limiting substrate. Tempest and Hunter[11] assumed that the rate of protein synthesis, per unit of ribosome, is constant at a particular temperature, and that the concentration of ribosomes within an organism cannot exceed a certain fixed amount. Lowering the temperature decreases ribosomal activity so that, in order to maintain a constant growth rate in the chemostat, the organisms compensate by synthesizing additional ribsomal RNA.

The present paper reports the changes in total nucleic acid and protein contents of chemostat-grown glucose-limited *C. utilis* when the cultivation temperature, the dilution rate, the pH of the medium or the nitrogen source were altered serially. A unified approach is proposed to describe the effect of growth rate and temperature on RNA and protein contents of microorganisms.

MATERIALS AND METHODS

Organism

The strain of *C. utilis* NRRL-Y900 used in this study was maintained by monthly transfers on yeast extract (0.5%)—tryptic soy agar (4.0%) (Difco Laboratories, Detroit, Mich.). Purity of growing cultures was checked periodically by microscopic examination and by examining the morphology of colonies on agar plates.

Media

The cultivation medium had the following composition per liter: glucose, 2.0 to 10.0 g; KH_2PO_4, 5.0 g; $MgSO_4 \cdot 7H_2O$, 0.3 g (0.5 g at 6 and 10 g glucose per liter); $CaCl_2$, 0.2 g (0.35 g at 10 g glucose per liter); $CuSO_4 \cdot 5H_2O$, 0.08 μg; $FeSO_4 \cdot 7H_2O$, 0.7 μg; $MnSO_4 \cdot H_2O$, 0.8 μg; $Na_2MoO_4 \cdot 2H_2O$, 0.4 μg; $ZnSO_4 \cdot 7H_2O$, 0.8 μg: D-biotin, 5.4 μg. The content of trace elements (Cu, Fe, Mn, Mo, and Zn) and D-biotin was doubled when the concentration of glucose was 10 g per liter. This medium was supplemented with nitrogen in the form of either $(NH_4)_2SO_4$ or a mixture of KNO_3 and HNO_3. The concentration of the nitrogen source was 4 mmoles per g of glucose. When $(NH_4)_2SO_4$ was used, the final pH of the medium was brought to 7.2 by substituting a mixture of KH_2PO_4 and K_2HPO_4 (1.75 and 3.6 g per liter, respectively) for KH_2PO_4. The adjustment of the final pH of media in which nitrate was present is discussed below under pH Control. The nature of the nitrogen source and the concentration of glucose used are indicated under Results.

Chemostat

A glass chemostat vessel of 500 ml working volume was used. The apparatus was similar in design to that described by Cooney and Mateles.[12] The capability of the chemostat to transfer excess oxygen at high productivity rates was checked by showing that a batch culture cultivated under similar conditions could be maintained in the logarithmic phase past the maximum cell density (5 g/liter) attained at steady state.

pH Control

For the experiments in which nitrate was used as nitrogen source (except for that in which the pH was maintained at 7.5), the initial pH of the medium was lowered with HNO_3 to give a desired steady-state pH, taking into consideration the number of hydrogen ions which disappeared due to nitrate (the sole nitrogen source) consumed by the growing culture. For each nitrate ion consumed, one hydrogen ion disappeared. Nitric acid was ordinarily added to the medium after sterilization to avoid browning. Additional adjustments during the course of the experiment were made by adding equivalent amounts of either concentrated KOH or H_2SO_4, as required, to both the medium reservoir and the fermentor. The attained steady-state pH was usually constant within less than ±0.1 unit.

Harvesting and Processing of Cells

A culture was assumed to be in a steady state when the cell concentration and pH remained constant for five doubling times. Harvest of cells in steady state was performed by rapidly siphoning cells and medium through ¼ in. polycarbonate tubing connected to a stainless steel coil of the same inner diameter immersed in a stirred saltwater-ice mixture at about −10°C. It took less than 1 min to collect most of the broth (approximately 450 ml) at a final temperature of 10°C or less in a glass container which was kept cold in an ice bucket. The broth was centrifuged in the cold at 10,000 x *g* for 15 min and the supernatant fluid was frozen and saved for analyses. The cell pellet was washed twice with distilled water and frozen. Significant loss of cellular material during growth and washings was ruled out by checking the absorbance of the supernatant fluids at 260 nm. Later, the frozen pellet was thawed and immediately transferred for freeze-drying. The freeze-dried material was placed in air-tight containers and kept frozen until analyzed.

Analyses

Moisture content of freeze-dried cells was determined by drying in a vacuum oven at 70°C for 6 hr. Cell protein was measured with a modification of the biuret reaction of Stickland[13] using bovine serum albumin (Pentex, Kankakee, Ill.) as a standard. The alkaline cell suspensions were not heated, but instead, stored for 48 hr in the refrigerator, and yielded up to 7% more protein than with the original procedure. Cell nucleic acid was extracted by the Schmidt-Tannhauser procedure as described by Munro and Fleck.[14] The alkaline hydrolysis step was not performed so that only total nucleic acid was determined. The absorbance of the extract was measured at 260 nm and, neglecting the difference in absorbtivity between DNA and RNA, the nucleic acid was quantitated using an absorbtivity of 32 liter/g-cm which had been determined previously for this organism,.[15] Glucose in the cell-free broth was measured using the Glucostat procedure (Worthington Biochemical Corp., Freehold, N.J.). The above analyses were performed in triplicate. The coefficients of variations for protein, nucleic acid, and glucose did not exceed, respectively, 1.5%, 1.2%, and 1.0% of the mean.

Critical Dilution Rate

The critical dilution rate (D_c) at a given combination of medium composition, pH and temperature was determined by the "wash-out" technique described by Herbert et al.[16]. It was, however, necessary to modify the technique to compensate for the lack of automatic pH control. The initial pH of the medium was set so that, at the maximum cell density (which was determined by the concentration of glucose in the feed), the pH of the broth would be well above the one at which the maximum dilution rate was to be determined. The dilution rate of a continuous culture at steady state was then increased to about 10% over the expected maximum dilution rate so that wash-out occurred until the desired pH was reached. Then the dilution rate was adjusted to the expected maximum dilution rate and the cell density and pH were measured. At this point, the latter two variables varied only very slowly with time so that over several turnover times the culture could be considered virtually at steady state. If the variations with time of cell density and pH were larger than desirable, the dilution rate was readjusted as needed.

RESULTS

The results of three experiments are summarized in Table I. In each experiment, one or two of the cultivation variables were altered: dilution rate in experiment 1, pH and dilution rate in experiment 2, and dilution rate and temperature in experiment 3. In addition, experiment 1 differs from experiments 2 and 3 in the nitrogen source (NH_4^+ and NO_3^-) employed. Each line in Table I represents one run. In some cases the same cultivation conditions have been used in two separate runs to demonstrate the extent of biological variation.

Increasing the dilution rate at 30°C from 0.06 to 0.35 hr^{-1} (experiment 1) increases the nucleic acid content and decreases the protein content. Varying the growth pH from 3.05 to 7.5 (experiment 2) causes no significant change in the levels of these two macromolecular fractions. On the other hand, by increasing the cultivation temperature from 15 to 37.5°C, keeping all other conditions constant (experiment 3), the nucleic acid content decreases markedly (from 7.74% to

TABLE I

Effect of Growth Variables on Nucleic Acid and Protein Contents

	D, hr^{-1}	D/D_c	Nitrogen source	Temperature, °C	pH	Nucleic acid, % dry wt	Protein, % dry wt	Nucleic acid / protein
Exp. 1	0.06	0.13	NH_4^+	30.0	5.9	5.51	40.5	0.136
	0.23	0.51	same	same	same	6.07	37.5	0.162
	0.35	0.78	same	same	same	7.17	36.4	0.197
Exp. 2	0.12	0.28	NO_3^-	30.0	3.05	6.69	44.1	0.152
	0.13	0.30	same	same	3.05	7.19	45.6	0.158
	0.12	0.27	same	same	3.60	6.05	41.2	0.147
	0.31	0.69	same	same	4.35	7.57	40.6	0.186
	0.12	0.27	same	same	4.65	6.43	45.7	0.141
	0.12	0.27	same	same	5.30	6.30	44.6	0.141
	0.30	0.67	same	same	5.45	6.99	38.8	0.180
	0.12	0.27	same	same	5.75	6.15	45.4	0.135
	0.13	0.29	same	same	6.50	6.61	47.0	0.141
	0.30	0.67	same	same	6.50	7.70	40.6	0.190
	0.30	0.67	same	same	7.50	7.08	40.1	0.177
Exp. 3	0.03	0.23	NO_3^-	15.0	3.3–3.5	—	—	0.157
	0.09	0.69	same	15.0	same	—	—	0.182
	0.12	0.92	same	15.0	same	7.74	35.6	0.217
	0.12	0.48	same	22.5	same	6.74	40.8	0.165
	0.12	0.27	same	30.0	same	6.36	39.8	0.160
	0.12	0.27	same	30.0	same	—	—	0.145
	0.42	0.93	same	30.0	same	—	—	0.200
	0.21	0.29	same	37.5	same	4.36	34.7	0.126

4.36%) and the protein content alternates moderately (range of 40.8% to 34.7%). Changing the nitrogen source from ammonium to nitrate (compare experiment 1 with experiment 2 at pH range of 5.3 to 6.5) raises the levels of both protein and nucleic acid, but leaves unchanged the ratio of nucleic acid to protein. Figure 1 shows that the latter ratio increases linearly with dilution rate at a constant temperature. At a constant dilution rate, the nucleic acid to protein ratio increases as the cultivation temperature decreases. The latter effect can be mathematically reduced, if the term D_c is introduced. D_c is the "critical" dilution rate obtainable at a given temperature and medium composition before wash-out of the culture. It is

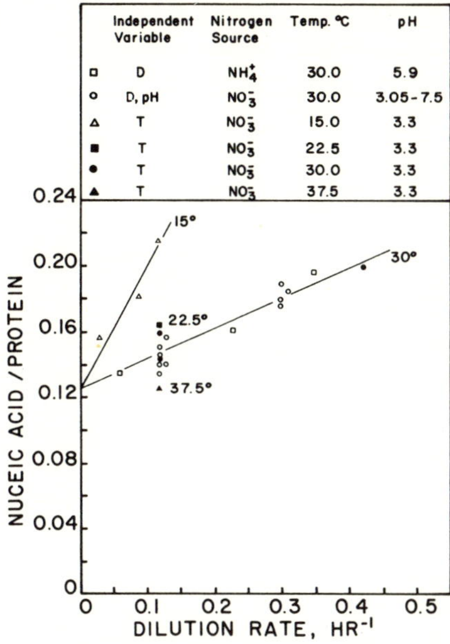

Fig. 1. Nucleic acid to protein ratio vs. dilution rate.

quite often found to be equal to the maximum specific growth rate, μ_m, obtained with a batch culture under similar environmental conditions.[16] At a given medium composition D_c or μ_m is an exponential function of the temperature over a considerable range (see Fig. 2; also ref. 17). If instead of the actual dilution rate (D), its ratio to the critical dilution rate (D/D_c, "reduced dilution rate") is used, a dimensionless scale of dilution rate is obtained which is applicable at all temperatures employed. Figure 3 demonstrates that the nucleic acid to protein ratios at all temperatures and dilution rates employed fall on one line when plotted against the reduced dilution rate.

Another way to express the relationship between the growth rate and the nucleic acid and protein is to use the term (protein/nucleic acid) × dilution rate, or $P/NA \times D$. The latter is equal to the rate of protein synthesis per unit mass of nucleic acid, and is called the "efficiency of nucleic acid in protein synthesis."

The efficiency of nucleic acid in protein synthesis increases with temperature and dilution rate (Fig. 4). These effects of growth rate

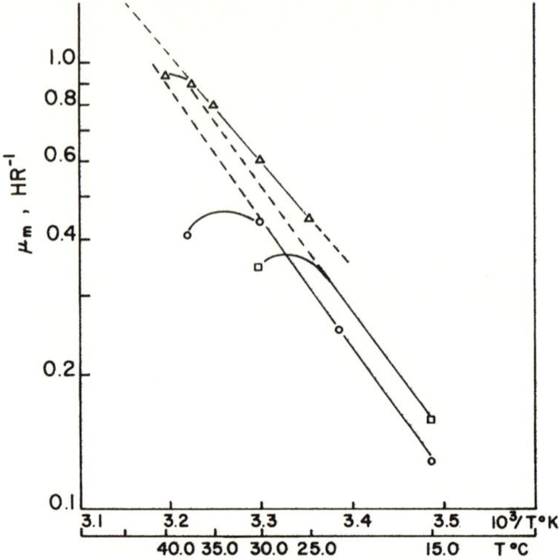

Fig. 2. Maximum specific growth rate vs. reciprocal of temperature: (○) *C. utilis* (NRRL-Y900); (□) *C. utilus* (NCTC-321; ref. 9); (△) *A. aerogenes* (NCTC-418; refs. 4, 11, 22, 23).

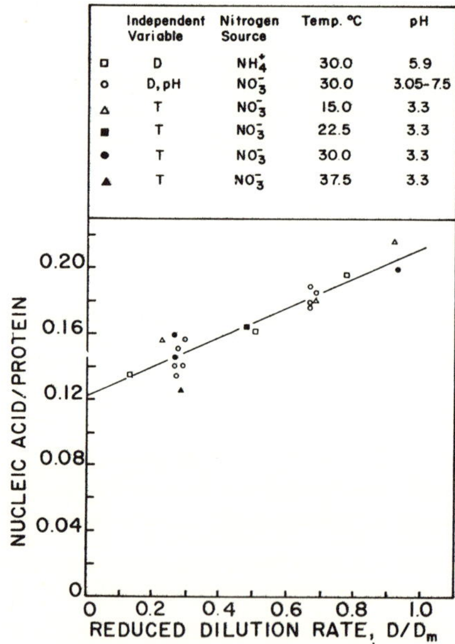

Fig. 3. Nucleic acid to protein ratio of *C. utilis* NRRL-Y900 as a function of the reduced dilution rate.

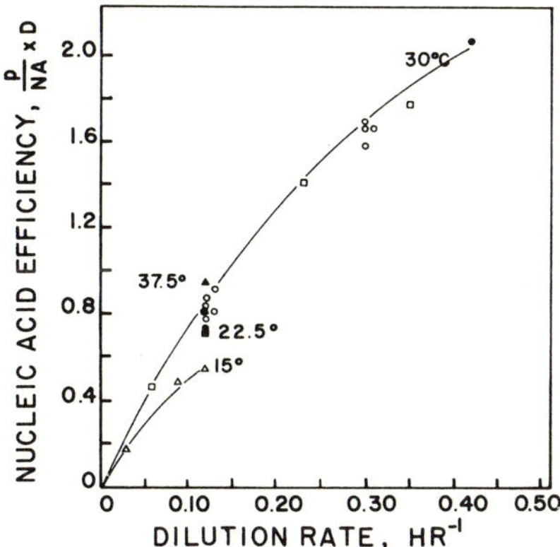

Fig. 4. Efficiency of nucleic acid in protein synthesis as a function of dilution rate (*C. utilis* NRRL-Y900).

and temperature can be treated in a way similar to the one described above, namely by substituting D/D_c for D. In addition, the actual efficiency of nucleic acid in protein synthesis can be divided by the one at the corresponding critical dilution rate. The resulting plot (Fig. 5) thus provides an arithmetic method for reducing protein synthesis efficiency data taken at various temperatures. Note, however, that in the absence of measurements of nucleic acid and protein contents at the critical dilution rates the nucleic acid efficiency at these points must be obtained by extrapolation.

DISCUSSION

The term "ribosome efficiency in protein synthesis" will be defined as the rate of synthesis of protein per unit of ribosomes. As such, it has been first used by Schaechter et al.[7] Other authors studied the rate of protein synthesis per units of total RNA (e.g., ref. 18), ribosomal RNA (rRNA) (e.g., ref. 8) and, in this study, total nucleic acid. Consequently, and as a matter of convenience, reference will be made in this discussion to the following corresponding terms: "ribosomal efficiency," "RNA efficiency," "rRNA efficiency," and "nucleic acid efficiency." The use of such terminology should not be taken to mean that nucleic acid efficiency or RNA efficiency necessarily has biological significance. These are merely mathematical terms which express the rate of protein synthesis per unit weight of nucleic acid or RNA, respectively. On the other hand,

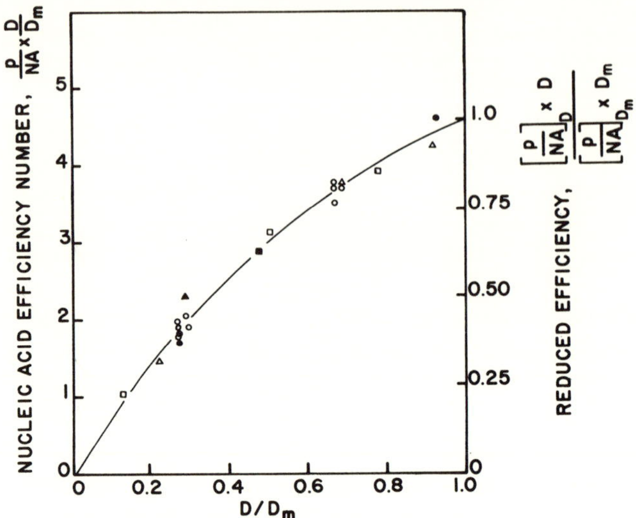

Fig. 5. Reduced efficiency of nucleic acid in protein synthesis as a function of reduced dilution rate (*C. utilis* NRRL-Y900).

these terms may bear quantitative relationships to biologically significant terms such as ribosomal efficiency or rRNA efficiency. The ribosomal efficiency (E_{Rb}) is defined as $dP/(Rb \times dt)$, P and Rb being the protein and ribosome contents, respectively. However, at steady state, $dP/dt = \mu P$ (μ is the specific growth rate). Hence,

$$E_{Rb} = \frac{dP}{Rb \times dt} = \frac{P}{Rb} \times \mu$$

Therefore, the efficiency of the ribosomes in protein synthesis at any given growth rate determines the ribosome to protein ratio and vice versa. Similar relationships obviously hold also for the efficiencies and corresponding ratios of cell protein to rRNA, total RNA, and total nucleic acid.

The ratio of nucleic acid to protein is independent of the levels of other bulky components in the biomass and is predictable under a wide variety of cultivation conditions (Table I, Fig. 1, also see ref. 3). This is illustrated by comparing ammonia and nitrate cultures which have been cultivated under otherwise similar conditions. Although the nucleic acid to protein ratio in these cultures is approximately the same, the contents of both protein and nucleic acid are higher in cells fed nitrate than in cells grown on ammonia. The difference is probably due to a higher content of carbohydrate in the cultures incubated with ammonia.

Our data (Fig. 1) further indicate that under carbon limitation, the ratio of nucleic acid to protein increases linearly with the growth rate at a constant temperature and is not influenced to any significant extent by the pH. Since RNA makes up more than 92% of the nucleic acid in *C. utilis*,[1] it is not surprising that this observation is in agreement with reports on RNA and protein contents by Tempest and Hunter[11] and Tempest et al.[18] The data previously presented by these authors can be recalculated to show that the RNA to protein ratio similarly remains invariable with the pH of the medium and increases linearly with the dilution rate in a (glycerol) carbon-limiting chemostat. It should also be added that the growth-limiting substrate and the method of cultivation (batch or chemostat) should not influence these relationships, as long as the limitation remains in the same class, which in this case is carbon and energy.[10]

Raising the temperature of cultivation at a constant dilution rate causes an increase in efficiency of nucleic acid in protein synthesis (Fig. 4) and, consequently, a corresponding decrease in the ratio of nucleic acid to protein. This effect can be observed also with recalculated data on RNA to protein ratios in carbon-limited cultures of *A. aerogenes*[11] and *C. utilis*.[9] Since in batch culture, the contents of RNA and protein are determined by the medium and are independent of the cultivation temperature, the term (protein/RNA $\times$ μ_m) increases with temperature in exactly the same way as μ_m. In the chemostat, where the dilution rate can be varied from zero to D_c, the efficiency of RNA in protein synthesis changes from an apparent zero to a value which is proportional to the critical dilution rate. If both dilution rate and RNA efficiency are divided by D_c to yield "reduced dilution rate" and "reduced RNA efficiency", respectively, the resulting boundaries of both scales will be identical regardless of temperature (namely, 0 and 1). It would then be logical to postulate that reduced RNA efficiency is dependent only on reduced dilution rate, irrespective of temperature as shown in Figure 5. Recalculation of previously reported data on *A. aerogenes*[11,18] under carbon or magnesium limitation (Figs. 6, 7) and on *C. utilis* cultivated under carbon or nitrogen limitation[9] (Figs. 8, 9) demonstrates that the observations described in Figure 5 may be a general phenomenon.

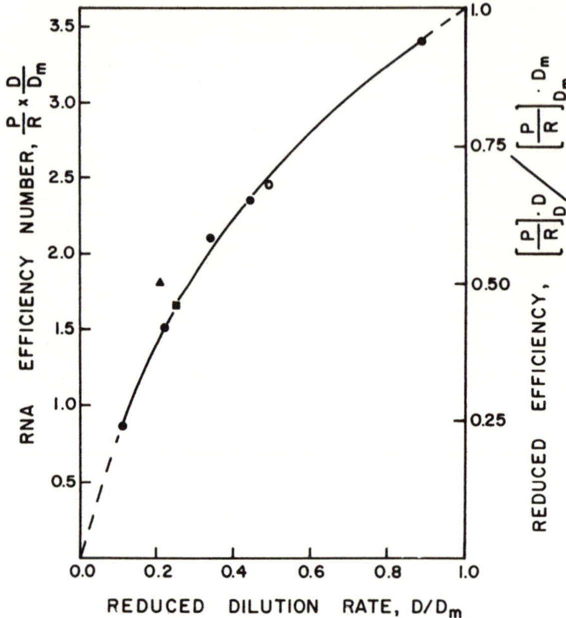

Fig. 6. Reduced efficiency of RNA in protein synthesis as a function of reduced dilution rate (*A. aerogenes* NCTC-418, carbon-limited chemostat). Cultivation temperature: (▲) 40°C; (■) 35°C; (●) 30°C; (○) 25°C. Original data from refs. 11 and 18.

The reduced efficiency of RNA in protein synthesis under nitrogen limitation, as recalculated from Brown and Rose,[9] displays a discontinuous variation with the reduced dilution rate (Fig. 9). On the basis of Figure 8 and the broken line in Figure 9, there is no difference between nitrogen-limited and carbon-limited cultures with respect to reduced efficiency of RNA. A similar conclusion has been presented for *A. aerogenes.*[12]

Theoretical Implications

The approach presented here can be used to predict the ratio of RNA to protein, as well as the efficiency of RNA in protein synthesis, for a given organism at various temperatures and dilution rates, provided the variation in the specific growth rate with temperature and R/P for a number of reduced dilution rates are known. To demonstrate this procedure, the reduced dilution rate for a given

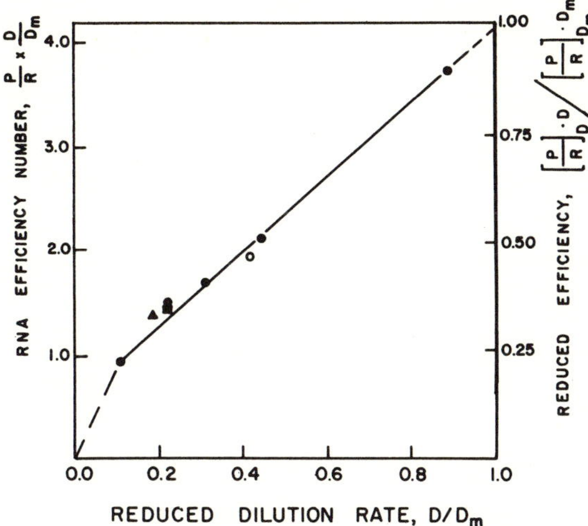

Fig. 7. Reduced efficiency of RNA in protein synthesis as a function of reduced dilution rate (*A. aerogenes* NCTC-418, magnesium-limited chemostat). Symbols as in legend Fig. 6.

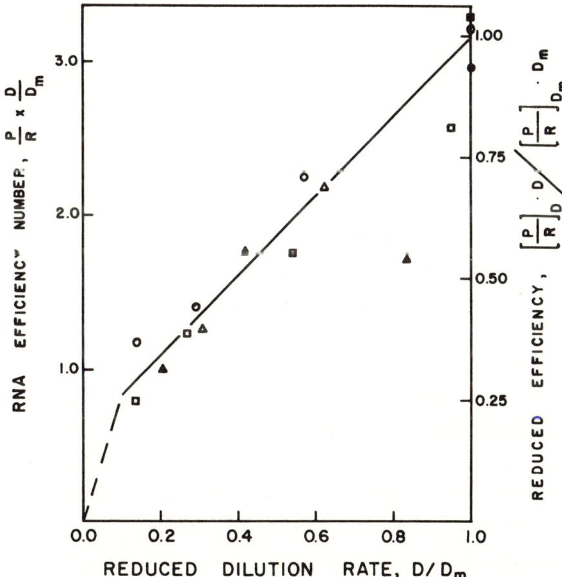

Fig. 8. Reduced efficiency of RNA in protein synthesis as a function of reduced dilution rate (*C. utilis* NCTC-321, carbon-limited chemostat). Cultivation conditions: (○) 30°C chemostat; (□) 25°C chemostat; (▲) 20°C chemostat; (△) 15°C chemostat; (■) 30°C batch; (●) 15°C batch. Original data from ref. 9.

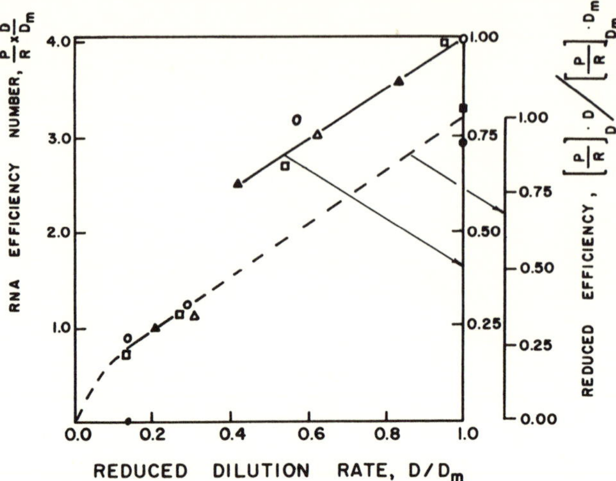

Fig. 9. Reduced efficiency of RNA in protein synthesis as a function of reduced dilution rate (*C. utilis* NCTC-321). Symbols as in Fig. 8.

organism is multiplied by the corresponding critical dilution rate at 30°C (yielding the respective "modified dilution rate" at 30°C) and the RNA or nucleic acid to protein ratio is then plotted against the new product (Fig. 10). The plot makes it evident that performance of a particular organism at temperatures other than 30°C are consistent with and predictable from data taken at 30°C. Figure 10 also suggests that the ratio of RNA to protein is approximately the same for various microorganisms cultivated in the chemostat under carbon limitation. One should, however, realize that the validity of this similarity at other cultivation temperatures is uncertain since the activation energies for the maximum specific growth rates of microorganisms are not identical (e.g., see Fig. 2).

Leick[19] has shown that, at 30°C on rich media, efficiency of RNA in protein synthesis was reasonably close (1.65–2.16 hr^{-1}) for 12 out of 15 organisms studied. If his data are recalculated and presented as RNA to protein ratio versus the specific growth rate, the apparent scatter of the data becomes meaningful, namely, there is a linear increase in the RNA to protein ratio with the specific growth rate attained by *all* the organisms studied (Fig. 11). Note also that Leick's data are consistent with carbon-limiting chemostat data,

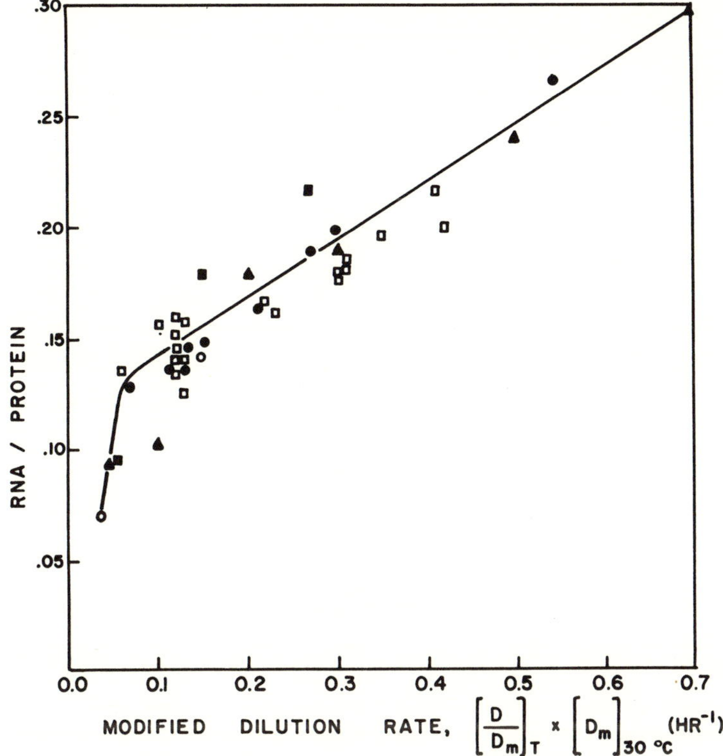

Fig. 10. Universal RNA to protein ratio as a function of modified dilution rate in a carbon-limited chemostat: (●) *A. aerogenes*; [11,18] (□) *C. utilis*; [9] (▲) *B. megaterium*; [10] (○) *Polytomella caeca*; [24] (■) *Azotobacter chroococcum*. [26]

which were presented above in Figure 10 and are represented by a solid line in Figure 11. The somewhat higher values of the RNA to protein ratios derived from Leick's data, and possibly also their scatter, can be explained by the fact that this RNA term included low molecular weight RNA derivatives which are normally extracted from the cells by cold perchloric or trichloroacetic acids. For our cells, this low molecular weight material would represent an additional 20% of total nucleic acids.

Recently, Varricchio and Monier[21] have shown that *E. coli* K-12, which has an excessively high ratio of RNA to DNA, displays,

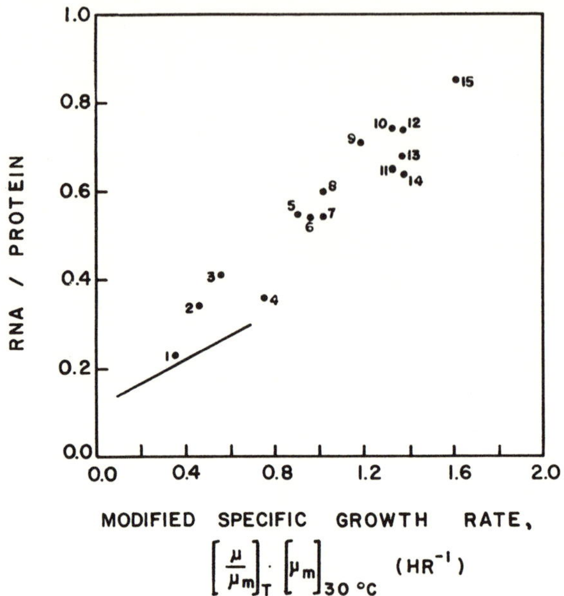

Fig. 11. Universal RNA to protein ratio as a function of modified specific growth rate. Temperature = 30°C. 1. *Tetrahymena pyriformis* GL; 2. *Saccharomyces cerevisiae* (haploid) 1507–7A; 3. *Saccharomyces cerevisiae* (diploid) No. 66; 4. *Lactobacillus bulgaricus*; 5. *Cl. perfringens*; 6. *Micrococcus anhaemolyticus*; 7. *Pseudomonas aeruginosa*; 8. *Erwinia carotovora*; 9. *Bacillus subtilis*; 10. *Serratia marcescens*; 11. *E. coli* B; 12. *Citrobacter freundii*; 13. *E. coli*; 14. *Salmonella typhimurium;* 15. *A. aerogenes*. Solid line as in Fig. 10. Original data from ref. 19.

nevertheless, a ratio of polysomes to DNA which is the same as that of the "normal" strain B at all growth rates studied. In addition, since the protein to DNA ratio is constant at all growth rates, their data indicate that the efficiency of polysomes in protein synthesis is constant and those of RNA and ribosomes increase with the growth rate. Since the polysome is the active particle in protein synthesis, it is postulated that the efficiency of polysomes at a given temperature and at all growth rates is identical or similar in many microorganisms. The reason such a universality also exists with respect to total RNA is probably due to the similarity of control of the levels of polysomes, individual ribosomes, native ribosomal subunits and tRNA in organisms grown under carbon limitation.

The work for this report was supported by the Alfred P. Sloan Foundation, Grant No. 70–6–2. The paper is Contribution No. 1933 from the Department of Nutrition and Food Science, M.I.T. The authors would like to acknowledge the comments and criticism of Professor A. L. Demain in the course of this investigation.

References

1. A. C. R. Dean, *Proc. Roy. Soc.*, **13,**(155), 580 (1962).
2. W. Harder and H. Veldkamp, *Arch. Mikrobiol.*, **59,** 123 (1967).
3. F. C. Neidhardt and B. Magasanik, *Biochim. Biophys. Acta*, **42,** 99 (1960).
4. R. Rosset, J. Julien, and R. Monier, *J. Mol. Biol.*, **18,** 308 (1966).
5. J. Sykes and D. W. Tempest, *Biochim. Biophys. Acta*, **103,** 93 (1965).
6. H. E. Wade and D. M. Morgan, *Biochem. J.*, **65,** 321 (1957).
7. M. Schaechter, O. Maaløe, and N. O. Kjeldgaard, *J. Gen. Microbiol.*, **19,** 592 (1958).
8. J. Sykes and T. W. Young, *Biochim. Biophys. Acta*, **169,** 103 (1968).
9. C. M. Brown and A. H. Rose, *J. Bacteriol.*, **97,** 261 (1969).
10. R. E. Ecker and M. Schaechter, *Biochim. Biophys. Acta*, **76,** 275 (1963).
11. D. W. Tempest and J. R. Hunter, *J. Gen. Microbiol.*, **41,** 267 (1965).
12. C. L. Cooney and R. I. Mateles, *Recent Advan. Microbiol.*, **31,** 441 (1971).
13. L. H. Stickland, *J. Gen. Microbiol.*, **5,** 698 (1951).
14. H. N. Munro and A. Fleck, *Analyst*, **91,** 78 (1966).
15. S. Ohta, S. Maul, A. J. Sinskey, and S. R. Tannenbaum, *Appl. Microbiol.*, **22,** 415 (1971).
16. D. Herbert, R. Ellsworth, and R. C. Telling, *J. Gen. Microbiol.*, **14,** 601 (1956).
17. H. Ng, J. L. Ingraham, and A. G. Marr, *J. Bacteriol.*, **84,** 331 (1962).
18. D. W. Tempest, J. R. Hunter, and J. Sykes, *J. Gen. Microbiol.*, **39,** 355 (1965).
19. V. Leick, *Nature*, 217, 1153 (1968).
20. I. I. Ivanova, Z. V. Sakharova, and N. D. Jerusalimski, *Mikrobiologiya*, **36,** 253 (1967).
21. F. Varricchio and R. Monier, *J. Bacteriol.*, **108,** 105 (1971).
22. J. R. Postgate and J. R. Hunter, *J. Gen. Microbiol.*, **29,** 233 (1962).
23. R. E. Strange and F. A. Dark, *J. Gen. Microbiol.*, **29,** 719 (1962).
24. R. Jeener, *Arch. Biochem. Biophys.*, **43,** 381 (1953).
25. H. Dalton and J. R. Postgate, *J. Gen. Microbiol.*, **56,** 307 (1969).

Accepted for Publication August 24, 1972

Part II

WATER ACTIVITY AND OSMOTIC PRESSURE

Editor's Comments on Papers 8 Through 12

The newcomer to studies of the water relations of microorganisms would be well advised to read the excellent review by Scott (1957) of the water relations of food spoilage microorganisms. Unfortunately, this article (45 pages) is much too long for inclusion in this collection. Scott and his co-workers in the Division of Food Preservation and Transport, C.S.I.R.O. in Ryde, New South Wales, Australia, have published many of the basic studies on water relations of microorganisms. The paper by Scott (1953) on *Staphylococcus aureus* and the paper by Tomkins (1929) on fungal growth should also be consulted.

Christian and Waltho (Paper 8) review the evidence for defining the water requirements of staphylococci and micrococci in terms of water activity. The influence of the water activity of the medium upon cell composition is discussed also. The ability of staphylococci to grow in concentrated environments was dependent on water activity (a_w) not on water content or solute concentration. Christian and Ingram in 1959 (Paper 15) discovered that the a_w of cells was never greater than that of the growth medium. The cell water content decreased from 1.66 g/g dry weight at 0.993 a_w to 0.71 g/g at 0.90a_w. The water contents of staphylococci were not under direct cellular control at low levels of a_w. This conclusion was reached by comparison of the equilibrium water content of heat-killed cells to that of cells grown at various a_w values.

Baird-Parker and Freame (Paper 9) studied the combined effects of water activity, pH, and temperature on the growth of *Clostridium botulinum* from spore and vegetative cell inocula. Spores germinated at a_w values as low as 0.93, but vegetative growth was inhibited between a_w of 0.96 to 0.98.

Kuo and Lampen (Paper 10) studied the osmotic regulation of invertase formation and secretion by protoplasts of *Saccharomyces.* Invertase synthesis was reversibly inhibited at high osmolarity. A KCl concentration of 0.4 *M* (equivalent to 0.75 osmolal) was most effective for synthesis and secretion of invertase; synthesis was inhibited at concentrations greater than 0.6 *M*. Similar results were obtained with NaCl, $MgSO_4$, and sorbitol. Formation and secretion of invertase were inhibited by cycloheximide. Uptake of ^{14}C-fructose or ^{14}C-threonine by the protoplasts decreased significantly when the concentration of the osmotic stabilizer, sorbitol, was increased. The results suggest that the release of invertase is enzymatically controlled and very specific.

Scheie (Paper 11) reported the development of a spectrophotometric technique for the determination of cellular osmotic pressure. One of the most interesting findings was the wide range of osmotic pressures exhibited by *E. coli,* which depended on the medium in which the cells had been grown and the phase of growth. *Escherichia coli* B/r cells in the exponential phase (grown in nutrient broth) had a differential pressure of 6.9 atm but late log phase cells registered 35 atm, a fivefold variation. The technique allows detection of damage not visible with a light microscope and offers a new and sensitive tool for evaluating osmotic effects.

Charlang and Horowitz (Paper 12) describe the effects of low a_w on membranes of *Neurospora crassa.* Exposure of *Neurospora* conidia to media of low a_w released a germination essential component regardless of the use of an electrolyte or nonelectrolyte to lower the value. Agents that damaged the cytoplasmic membrane also caused a loss of the germination fator from the conidia. The results are offered as one possible explanation for xerotolerance.

REFERENCES

Scott, W. J. 1953. Water relations of *Staphylococcus aureus* at 30°C. *Aust. J. Biol. Sci.* 6(4): 549–564.

———. 1957. Water relations of food spoilage microorganisms. *Advan. Food Res.* 7: 83–127.

Tomkins, R. G. 1929. Studies of the growth of moulds—I. *Proc. Roy. Soc. (London) B105*(738): 375–401.

SELECTED READINGS

Abrams, A. 1959. Reversible metabolic swelling of bacterial protoplasts. *J. Biol. Chem.* 234(2): 383–388.

Allwood, M. C. 1971. Heat induced sensitivity to osmotic shock in *Escherichia coli* and its relationship to cellular potassium content. *Microbios.* 4: 93–96.

Baird-Parker, A. C., M. Boothroyd, and E. Jones. 1970. The effect of water activity on the heat resistance of heat sensitive and heat resistant strains of salmonellae. *J. Appl. Bacteriol.* 33(3): 515–522.

Bateman, J. B., C. L. Stevens, W. B. Mercer, and E. L. Carstensen. 1962. Relative humidity and the killing of bacteria: the variation of cellular water content with external relative humidity or osmolality. *J. Gen. Microbiol.* 29(2): 207–219.

Bayer, M. E. 1967. Response of cell walls of *Escherichia coli* to a sudden reduction of the environmental osmotic pressure. *J. Bacteriol.* 93(3): 1104–1112.

Beers, R. J. 1956. *Effect of moisture level on germination of bacterial endospores.* Thesis Publ. No. 18, 112. University of Illinois, Urbana, Ill.

Broekman, J. H. F. F., and J. F. Steenbakkers. 1973. Growth in high osmotic medium of an unsaturated fatty acid auxotroph of *Escherichia coli* K-12. *J. Bacteriol.* 116(1): 285–289.

Bullock, K., and A. Tallentire. 1952. Bacterial survival in systems of low moisture content. Part IV. The effect of increasing moisture content on heat resistance, viability and growth of spores of *Bacillus subtilis*. *J. Pharm. London* 4: 917–931.

Christian, J. H. B., and J. A. Waltho. 1964. The composition of *Staphylococcus aureus* in relation to the water activity of the growth medium. *J. Gen. Microbiol.* 35(2): 205–213.

Corner, T. R., and R. E. Marquis. 1969. Why do bacterial protoplasts burst in hypotonic solutions? *Biochim. Biophys. Acta* 183(3): 544–558.

Cota-Robles, E. H. 1963. Electron microscopy of plasmolysis in *Escherichia coli*. *J. Bacteriol.* 85(3): 499–503.

Diamond, R. J., and A. H. Rose. 1970. Osmotic properties of spheroplasts *Saccharomyces cerevisiae* grown at different temperatures. *J. Bacteriol.* 102(2): 311–319.

Eisenberg, A. D., and T. R. Corner. 1973. Osmotic behavior of bacterial protoplasts: temperature effects. *J. Bacteriol.* 114(3): 1177–1183.

Gale, E. F., and J. M. Llewellin. 1970. Release of lipids from, and their effect on aspartate transport in osmotically shocked *Staphylococcus aureus*. *Biochim. Biophys. Acta 222*(2): 546–549.

Gilby, A. R., and A. V. Few. 1959. Osmotic properties of protoplasts of *Micrococcus lysodeikticus*. *J. Gen. Microbiol. 20*(2): 321–327.

Grodzinski, B., and B. Colman. 1973. Loss of photosynthetic activity in two blue-green algae as a result of osmotic stress. *J. Bacteriol. 115*(1): 456–458.

Hale, C. M. F. 1957. A note on the relationship between the gram reaction and plasmolytic effects in bacteria. *Exptl. Cell. Res. 12*(3): 657–659.

Hawthorne, D. C., and J. Friis. 1964. Osmotic-remedial mutants. A new classification for nutritional mutants in yeast. *Genetics 50*(5): 829–839.

Indge, K. J. 1968. The effects of various anions and cations on the lysis of yeast protoplasts by osmotic shock. *J. Gen. Microbiol. 51*(3): 425–432.

Ingram, M. 1957. Micro-organisms resisting high concentrations of sugars or salts. *Symp. Soc. Gen. Microbiol. 7*: 90–133.

Landman, O. E., A. Ryter, and C. Frehel. 1968. Gelatin-induced reversion of protoplasts of *Bacillus subtilis* to the bacillary form: electron-microscopic and physical study. *J. Bacteriol. 96*(6): 2154–2170.

Livingston, L. R. 1969. Locus-specific changes in cell wall composition characteristic of osmotic mutants of *Neurospora crassa*. *J. Bacteriol. 99*(1): 85–90.

Lovett, S. 1965. Rapid changes in bacteria following introduction into hypertonic media. *Proc. Soc. Exptl. Biol. Med. 120*(2): 565–569.

Lucke, B., and M. McCutcheon. 1932. The living cell as an osmotic system and its permeability to water. *Physiol. Rev. 12*(1): 68–139.

Mager, J. 1955. Influence of osmotic pressure on the polyamine requirement of *Neisseria perflava* and *Pasteurella tularensis* for growth in defined media. *Nature (London) 176*(4488): 933–934.

Marquis, R. E. 1967. Osmotic sensitivity of bacterial protoplasts and the response of their limiting membrane to stretching. *Arch. Biochem. Biophys. 118*(2): 323–331.

———, and T. R. Corner. 1967. Permeability changes associated with osmotic swelling of bacterial protoplasts. *J. Bacteriol. 93*(3): 1177–1178.

Mitchell, P., and J. Moyle. 1956. Osmotic function and structure in bacteria. *Symp. Soc. Gen. Microbiol. 6*: 150–180.

Munro, G. F., and C. A. Bell. 1973. Effects of external osmolarity on phospholipid metabolism in *Escherichia coli* B. *J. Bacteriol. 116*(1): 257–262.

———, K. Hercules, J. Morgan, and W. Sauerbier. 1972. Dependence of the putrescine content of *Escherichia coli* on the osmotic strength of the medium. *J. Biol. Chem. 247*(4): 1272–1280.

Neelon, V. J., and F. Bernheim. 1973. Effect of hypertonicity and sugars on potassium uptake in a strain of *Enterobacter*. *Microbios 7*: 137–148.

Neu, H. C., and L. A. Heppel. 1965. The release of enzymes from *Es-*

cherichia coli by osmotic shock and during the formation of spheroplasts. *J. Biol. Chem. 240*(9): 3685–3692.

Okrend, A. G., and R. N. Doetsch. 1969. Plasmolysis and bacterial motility: a method for the study of membrane function. *Arch. Microbiol. 69*(1): 69–78.

Pethica, B. A. 1958. Lysis by physical and chemical methods. *J. Gen. Microbiol. 18*(2): 473–480.

Riemann, H. 1968. Effect of water activity on the heat resistance of *Salmonella* in "dry" materials. *Appl. Microbiol. 16*(10): 1621–1622.

Scheie, P., and H. Dalen. 1968. Spatial anistotropy in *Escherichia coli. J. Bacteriol. 96*(4): 1413–1414.

———, and R. Rehberg. 1972. Response of *Escherichia coli* B/r to high concentrations of sucrose in a nutrient medium. *J. Bacteriol. 109*(1): 229–235.

Skujins, J. J., and A. D. McLaren. 1967. Enzyme reaction rates at limited water activities. *Science 158*(3807): 1569–1570.

Smith, L. 1962. Structure of the bacterial respiratory-chain system. Respiration of *Bacillus subtilis* spheroplasts as a function of the osmotic pressure of the medium. *Biochim. Biophys. Acta 62*(1): 145–152.

Wodzinski, R. J., and W. C. Frazier. 1960. Moisture requirements of bacteria. I. Influence of temperature and pH on requirements of *Pseudomonas fluorescens. J. Bacteriol. 79*(4): 572–578.

Wolf, A. V., and M. G. Brown. 1971. Concentrative properties of aqueous solutions: conversion tables, D181-226. *In* R. C. Weast (ed.), *Handbook of Chemistry and Physics,* 52nd ed. The Chemical Rubber Co., Cleveland, Ohio.

8

Reprinted from *J. Appl. Bacteriol.*, **25**(3), 369–377 (1962)

THE WATER RELATIONS OF STAPHYLOCOCCI AND MICROCOCCI

By J. H. B. CHRISTIAN and JUDITH A. WALTHO

Division of Food Preservation, C.S.I.R.O., Ryde, N.S.W., Australia

CONTENTS

1. INTRODUCTION

THE WATER RELATIONS of micro-organisms have received increasing attention in recent years and the available data, spread rather thinly over many species of bacteria, yeasts and moulds, have been reviewed in detail by Scott (1957). Interest in the water relations of staphylococci and micrococci stems from the importance of these bacteria as agents of food poisoning and food spoilage, and their ability to grow in relatively dry or concentrated environments is well known. Three traditional methods of food preservation—salting, syruping and dehydration—achieve a similar end, namely a substantial reduction in the activity of the water in the food. Thus accurate information about the water requirements of micro-organisms such as staphylococci and micrococci allows useful predictions to be made concerning the safety and stability of various foods.

It is clear that the water requirements of at least some bacteria are best defined in terms of water activity of the substrate (Scott, 1953; Christian & Scott, 1953; Wodzinski & Frazier, 1960). In this paper we review the evidence for defining the water requirements for the growth of the staphylococci and micrococci in this way. The influence of the water activity of the medium upon the composition of the cell, and the relationship of cell composition to water requirement, are also discussed.

2. THE CONCEPT OF WATER ACTIVITY

Bacteria, in common with all forms of life, require water for growth. The object in examining their water relations is to determine their minimum water requirements and their responses when water is available in excess of this minimum. Experience and nutritional considerations show that bacteria will not proliferate in pure water. It follows that any substrate, be it liquid or solid, which supports growth is, in essence, an aqueous solution. When the water content of such a solution is reduced its vapour pressure and freezing point decrease and its boiling point increases. These changes in colligative properties result from a decrease in activity of the solvent water. Clearly the same effect on the water activity may be obtained by adding solutes as by removing the solvent.

By definition, the water activity (a_w) of a solution is the ratio of its water vapour pressure to the vapour pressure of pure water at the same temperature, so that $a_w = p/p_0$. Biologists have frequently discussed solute concentrations in terms of equilibrium relative humidity (E.R.H.) or of osmotic pressure (O.P.), and the relationships between a_w and these expressions are as follows: E.R.H. (%) $= a_w \times 100$ and O.P. (atmospheres) $= (-RT \log_e a_w)/\bar{V}$, where $\bar{V}$ is the partial molal volume of water. In this discussion a_w is used to describe the status of water in the environment, since relative humidity refers more strictly to the surrounding atmosphere and osmotic pressure presupposes a membrane of specific permeability properties.

When the a_w of a solution is altered the relative concentrations of water and of solutes also change. Thus before asserting that a microbial response to a concentrated environment is an effect of a_w it is necessary to ensure that one of these accompanying changes is not, in fact, responsible. An illustration of how this has been achieved with staphylococci is given below.

3. Water Activity and Growth

The unusual tolerance of staphylococci for high concentrations of sodium chloride is well known and forms the basis of the selective media devised for these bacteria by Chapman (1946) and by Maitland & Martyn (1948). In experiments with 14 food poisoning strains of *Staphylococcus aureus*, Scott (1953) showed that this high salt tolerance was a manifestation of a very low requirement for water. He determined the lowest a_w at which growth would occur in several nutrient media, controlling

Table 1. *The minimum water activities* (a_w) *and water contents for the growth of* Staphylococcus aureus *in various media at* 30°

(after Scott, 1953)

Nutrient substrate	Solutes added to control a_w	Minimum values for growth	
		a_w	Water content (% dry wt)
Casamino Acids, yeast extract, Casitone	Salts mixture*	0·88	375
Nutrient broth	Salts mixture*	0·86	315
Nutrient broth	Sucrose	0·88	60
Nutrient broth	Sucrose 3·44 m + salts mixture*	0·86	75
Milk	None	0·86	16
Meat	None	0·88	23
Soup	None	0·86	63

* $NaCl:KCl:Na_2SO_4::5:3:2$ moles.

water activity by adding salts or sugars or mixtures of both. Similar experiments were performed in three dried foods, reconstituted with water to give the required range of a_w. The data in Table 1 show that growth occurred at levels of a_w down to 0·88 or 0·86 in every case. In contrast, the lowest water contents at which growth was observed in the various substrates ranged from 16% to 375% of the dry weight. Growth at these a_w levels was not affected appreciably by the nature of the major solutes present, whether these were salts, sugars, or constituents of the foods themselves. Clearly the ability of these bacteria to grow in a concentrated environment is determined not by the water content or the solute concentration but by the a_w. The water requirement of such bacteria is therefore best defined in terms of the minimum a_w that will support growth.

Using a thin layer equilibration technique, Clayson & Blood (1957) obtained growth of *Staph. aureus* at a_w levels above 0·85. Earlier reports, reviewed by Scott (1953), also support his conclusion that the food poisoning staphylococci are a homogeneous group in respect of their water requirements. The few data available suggest that other species of staphylococci also exhibit a low water requirement. The minimum levels of a_w for the growth of *Staph. aureus*, *Staph. albus* and *Staph. citreus* have been recorded as 0·88, 0·89 and 0·89 respectively (Christian & Waltho, 1961).

Information about the water relations of other micrococci is fragmentary. This is a larger group than the staphylococci and it is not surprising that greater variations in water requirements have been reported. Of 17 species of bacteria examined by Burcik (1950), the only micrococcus tested, *Micrococcus roseus*, had the lowest water requirement, growing at levels of a_w down to 0·905. In a survey which included 32 strains of bacteria (Christian & Waltho, 1961), an unspecified *Micrococcus*, which grew at 0·84 a_w, had the lowest water requirement. Much more exacting were *M. roseus* and *M. lysodeikticus*, with minimum levels of a_w for growth of 0·925 and 0·94 respectively. Solar salts are a rich source of micrococci, most of which will grow in very concentrated environments. Bain, Hodgkiss & Shewan (1957), who reviewed the bacteriology of salt, quoted Petrova's report that of 10 *Micrococcus* spp. isolated from solar salt and capable of growth in salt free medium, 6 grew in a medium saturated with NaCl, 2 in 20% and 2 in 10% NaCl. The equivalent a_w levels are approximately 0·75; 0·86 and 0·93.

In spite of the variations found in the minimum levels of a_w for growth of various strains, none of the micrococci examined had a very high requirement for water. In their summary, Mossel & van Kuijk (1955) observed that micrococci (and staphylococci) frequently grow at lower a_w levels than can be tolerated by most other bacteria.

Thus far only the lowest a_w levels which will permit growth have been discussed. Also important are the quantitative effects of higher a_w levels upon the rates of growth of bacteria. Fig. 1 (from Scott, 1953) shows the relationship between a_w and rate of growth for 14 food poisoning strains of *Staph. aureus*. The a_w of the basal medium was 0·999 and lower a_w levels were obtained by adding NaCl, KCl and Na_2SO_4 in the mole ratio 5:3:2. The distinct optimum between 0·995 and 0·990 a_w is of interest. Every nonhalophilic bacterium that has been examined with sufficient care exhibits an optimum a_w, which generally lies between 0·98 and 0·995.

Since most common bacteriological media have a_w levels above 0·995 it is likely that they provide suboptimal conditions for many bacteria. Scott (1957) has suggested that at a_w levels above the optimum cells may experience difficulty in maintaining an adequate intracellular osmotic pressure. This possibility has not been examined.

Reduction of the a_w from 0·97 to 0·90 resulted in an approximately linear reduction

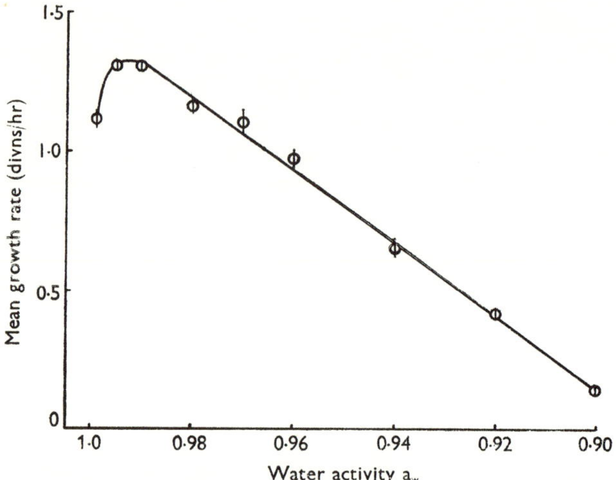

Fig. 1. The relationship between mean growth rate and water activity (a_w) for 14 strains of *Staph aureus*. Standard errors of means are shown as vertical lines. (After Scott, 1953).

in growth rate. The small standard errors of the mean rates, shown as vertical lines in Fig. 1, indicate that the 14 strains were homogeneous in respect of rates of growth at all a_w levels as well as in minimum water requirement. Other effects of reducing the a_w were an increase in the length of the lag phase and a reduction in the màximum cell population.

To our knowledge, no quantitative data relating rates of growth to a_w have been reported for other staphylococci or micrococci.

4. Factors Affecting Water Relations

Several environmental factors have been shown to influence the water relations of staphylococci. Scott (1953) found that growth was slower at all levels of a_w under anaerobic conditions than when shaken in air. Aerobically, the minimum a_w permitting growth was 0·86; anaerobically it was above 0·90. The effects of intermediate oxygen tensions have not been studied.

All the results reviewed so far were obtained at temperatures and pH values close to the optimum for growth of the various bacteria. The relationships between these variables and salt tolerance (i.e. water requirement) of *Staph. aureus* have been reported recently (Scott, 1961). The results were only semiquantitative, being given as presence or absence of growth after incubation for 14 days. They showed that the water requirement was least at temperatures and pH values close to the normal optimum and that any substantial departure from these optima resulted in an increased requirement for water.

The data of Table 1 show no marked effects of nutrients upon the water relations of staphylococci. However, all the substrates employed were complex and unlikely to be nutritionally deficient. It has been observed that the water requirement of *Salmonella oranienburg* was much greater in a minimal medium than in rich substrates (Christian, 1955). Similar investigations with staphylococci would be of interest.

The water relations of the halophilic bacteria, a group which contains many micrococci, are complicated by their requirement for NaCl. Extremely halophilic bacteria require about 3 M NaCl and will grow in media saturated with NaCl. The moderate halophile, *M. halodenitrificans*, grows at NaCl concentrations between about 0·55 M and 4·0 M. In both cases part but not all of the NaCl requirement may be replaced by some other salts, but not by nonelectrolytes. The high minimum salt requirements result from a need for specific ions and for an adequate ionic strength, and are not primarily requirements for a low a_w level. At the other extreme, growth in the presence of 4 M or saturated NaCl certainly demonstrates a low requirement for water among the halophilic bacteria. The a_w of a saturated solution of NaCl is 0·75, and is the lowest a_w at which bacterial growth has been reported.

5. Water Activity and Cell Composition

The data discussed so far afford no explanation for the low water requirement of these groups of bacteria. It is probable that an almost immediate effect of reducing the a_w of the environment will be a reduction of the a_w within the cell, and hence an increase in the concentration of intracellular solutes. Measurements of the freezing points of bacteria by Christian & Ingram (1959) showed that the a_w of cells was never greater than the a_w of the growth medium. Thus the red halophilic bacteria, which grow only in media of low a_w, should possess very concentrated internal solutions. In fact, these bacteria contain sodium, potassium and chloride ions at extremely high concentrations (Christian & Waltho, 1962). The enzymes extracted from such cells are themselves halophilic (Gibbons, 1957), requiring high concentrations of salts for optimum activity. Thus the low water requirement of red halophiles is probably explained in part by the high salt tolerance of their enzymes. Similar information concerning the composition and enzyme properties of non-halophilic bacteria might assist in determining the basis of the water requirement of this group. Unfortunately little is known of the effect of a_w or concentration upon the activities of the enzymes of staphylococci, but an impression of the composition and a_w of these cells is emerging from recent work and will be reviewed briefly.

The cell of *Staph. aureus* contains a remarkably concentrated solution. Mitchell & Moyle (1956) estimated it from equilibration experiments to be osmotically equivalent to about 1 molal sucrose, or 0·98 a_w. A freezing point of $-1{\cdot}7°$ was obtained for a cell pellet containing an undefined amount of interstitial medium freezing at $-0{\cdot}1°$ (Scott, 1953). From these values it is likely that the cell a_w was slightly less than

0·98. In both cases cells were harvested from media of about 0·999 a_w. The turgor pressure was thus close to 25 atmospheres.

Partial analyses of staphylococci after growth at a_w levels from 0·993 to 0·90 were made by Christian & Waltho (paper in publication). Fig. 2 shows that the cell water content decreased from 1·66 g/g dry weight at 0·993 a_w to 0·71 g/g at 0·90 a_w. Essentially the same curve was obtained for cells grown at 0·993 a_w, washed in water and transferred to NaCl solutions of the various a_w levels prior to analysis. Fig. 2 shows the equilibrium water contents of heat killed cells at four levels of a_w. At high a_w levels these preparations took up more water than whole cells, whose expansion was limited by intact cell boundaries. At low levels of a_w the water contents were similar for each type of cell preparation, and the authors concluded that in this region of a_w cells have little direct control over their water content.

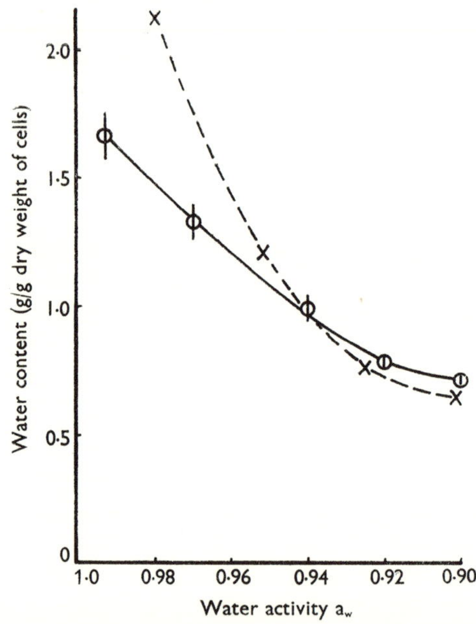

Fig. 2. The effect of water activity (a_w) on the water content of *Staph. aureus* cells. Open circles, cells grown at levels of a_w from 0·993 to 0·90. Standard errors of means are shown as vertical lines. Crosses, cells grown at 0·993 a_w, heat killed and equilibrated to the various a_w levels.

The concentrations of sodium, potassium and amino acids found in staphylococci after growth in media adjusted by the addition of NaCl to a_w levels from 0·993 to 0·90 are shown in Fig. 3. The increased sodium concentration at low a_w was partly a result of the higher sodium content of the medium. The increased concentrations of potassium and amino acids, however, were caused by the decrease in cell water content. The reciprocal changes in potassium and amino acid concentrations at low a_w are of interest. Reduction of the a_w from 0·92 to 0·90 also produced sharp decreases in the protein and inorganic phosphate contents. The authors suggest that further investigations of cell composition in this region may reveal which steps in metabolism are primarily affected by restriction of the a_w.

6. Prediction of the Water Requirement

In a study of 32 nonhalophilic bacteria a large negative correlation was found between the potassium content of cells grown in basal medium (0·993 a_w) and the minimum a_w permitting growth (Christian & Waltho, 1961). The authors suggested that high potassium content might indicate a low intracellular a_w, which would protect the cell from dehydration or plasmolysis in a concentrated environment. The relationship between potassium content and cell a_w has been examined in the following experiments (unpublished data). Four strains, *M. lysodeikticus*, *M. roseus*, *Staph. aureus* and an unspecified *Micrococcus* sp. (designated MRI), were grown in basal brain-heart infusion broth (0·993 a_w) and analyzed for intracellular sodium, potassium, amino acids and water. The freezing points of heat killed cells were also determined. In Fig. 4 the values obtained are plotted against the minimum a_w at

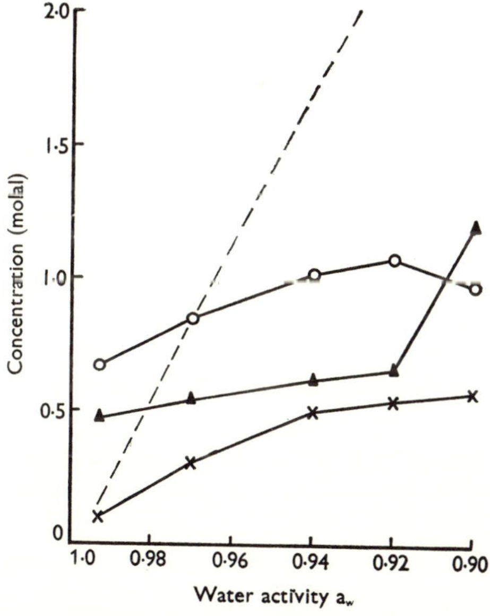

Fig. 3. The concentrations of sodium (crosses), potassium (open circles) and amino acids (full triangles) within cells of *Staph. aureus* grown in brain-heart infusion broth adjusted to levels of a_w from 0·993 to 0·90 by the addition of NaCl. The NaCl content of the medium is shown by the broken line.

which growth of the respective strains occurred. The postulated relationship between potassium content and water requirement does not hold for strain MRI, which grew at a_w levels down to 0·84. However, in this organism the low potassium content appears to be compensated by an enhanced amino acid pool. It can be seen that as the minimum a_w decreased from 0·94 to 0·84 the sum of the intracellular sodium, potassium and amino acid concentrations increased. This suggests a concomitant decrease in intracellular a_w, which is confirmed by the freezing point depressions also shown in Fig. 4.

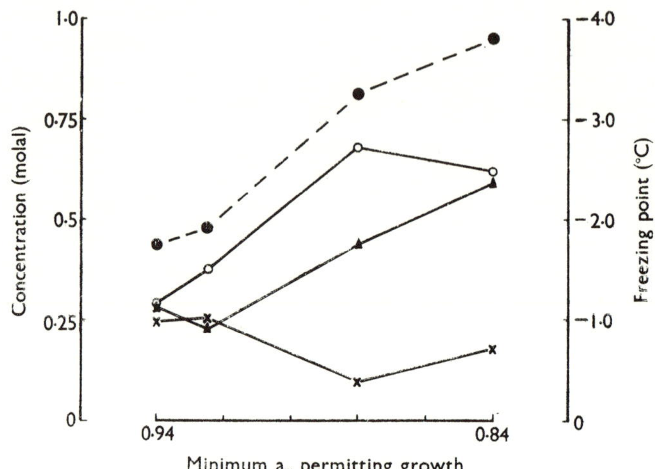

Fig. 4. The relationship between the sodium (crosses), potassium (open circles), and amino acid (full triangles) contents, and the freezing points (full circles), of four micrococci to the minimum water activity (a_w) which permitted growth. The analyses and freezing points were obtained with cells grown at 0·993 a_w.

It may be concluded that, for some micrococci at least, the minimum water requirement is related to the a_w of the cell after growth in basal medium, and that potassium content is often a reasonable guide to cellular a_w. Much more information is needed, particularly for bacteria of very high and very low water requirements, to test the validity of these conclusions. If the ability to accumulate low molecular weight solutes determines the water requirements of micrococci, the primary effects of low a_w may be upon active transport or upon the systems providing energy for transport. It is of interest that, for the four micrococci discussed here, the water requirements for respiration were similar to the water requirements for growth (unpublished data).

7. References

Bain, N., Hodgkiss, W. & Shewan, J. M. (1957). The bacteriology of salt used in fish curing. *2nd Int. Symp. Fd Microbiol., Cambridge*, p. 1. London: H.M.S.O.

Burcik, E. (1950). Über die Beziehungen zwischen Hydratur und Wachstum bei Bakterien und Hefen. *Arch. Mikrobiol.* **15**, 203.

Chapman, G. H. (1946). A single culture medium for selective isolation of plasma coagulating staphylococci and for improved testing of chromogenesis, plasma coagulation, mannitol fermentation and the Stone reaction. *J. Bact.* **51**, 409.

Christian, J. H. B. (1955). The influence of nutrition on the water relations of *Salmonella oranienburg*. *Aust. J. biol. Sci.* **8**, 75.

Christian, J. H. B. & Ingram, M. (1959). The freezing points of bacterial cells in relation to halophilism. *J. gen. Microbiol.* **20**, 27.

CHRISTIAN, J. H. B. & SCOTT, W. J. (1953). Water relations of salmonellae at 30°C. *Aust. J. biol. Sci.* **6**, 565.

CHRISTIAN, J. H. B. & WALTHO, J. A. (1961). The sodium and potassium content of non-halophilic bacteria in relation to salt tolerance. *J. gen. Microbiol.* **25**, 97.

CHRISTIAN, J. H. B. & WALTHO, J. A. (1962). Solute concentrations within cells of halophilic and non-halophilic bacteria. *Biochim. biophys. Acta* **65**, (in press.)

CLAYSON, D. H. F. & BLOOD, R. M. (1957). Food perishability: the determination of the vulnerability of food surfaces to infection. *J. Sci. Fd Agric.* **8**, 404.

GIBBONS, N. E. (1957). The effect of salt on the metabolism of halophilic bacteria. *2nd Int. Symp. Fd Microbiol., Cambridge*, p. 67. London: H.M.S.O.

MAITLAND, H. B. & MARTYN, G. (1948). A selective medium for isolating *Staphylococcus* based on the differential inhibiting effect of increased concentrations of sodium chloride. *J. Path. Bact.* **60**, 553.

MITCHELL, P. & MOYLE, J. (1956). Osmotic structure and function in bacteria. In *Bacterial Anatomy. Symp. Soc. gen. Microbiol.* no. 6, p. 150. Cambridge: University Press.

MOSSEL, D. A. A. & VAN KUIJK, H. J. L. (1955). A new and simple technique for the direct determination of the equilibrium relative humidity of foods. *Food Res.* **20**, 415.

SCOTT, W. J. (1953). Water relations of *Staphylococcus aureus* at 30°C. *Aust. J. biol. Sci.* **6**, 549.

SCOTT, W. J. (1957). Water relations of food spoilage microorganisms. *Advanc. Fd Res.* **7**, 83.

SCOTT, W. J. (1961). Available water and microbial growth. *Symp. Low Temp. Microbiol., Camden, N.J.*, p. 89. Camden, New Jersey: Campbell Soup Coy.

WODZINSKI, R. J. & FRAZIER, W. C. (1960). Moisture requirements of bacteria. 1. Influence of temperature and pH on requirements of *Pseudomonas flucrescens*. *J. Bact.* **79**, 572.

(*Received* 19 *July*, 1962)

9

Reprinted from *J. Appl. Bacteriol.*, **30**(3), 420–429 (1967)

Combined Effect of Water Activity, pH and Temperature on the Growth of *Clostridium botulinum* from Spore and Vegetative Cell Inocula

A. C. Baird-Parker and Barbara Freame

Unilever Research Laboratory, Colworth House, Sharnbrook, Bedford, England

(*Received* 28 *March* 1967)

Summary. The initiation of vegetative growth by spores in Reinforced Clostridial Medium (RCM) adjusted to different water activities (a_w) by the addition of NaCl or glycerol does not depend solely on the a_w of the medium; minimum a_w at which spores of *Cl. botulinum* types A, B and E could initiate growth were higher in the media containing NaCl. The minimum a_w at which vegetative cell inocula could grow were similar to those at which spore inocula could initiate growth. The combined effect of lowering the pH and a_w of the medium was to increase the minimum at which spores of types A and E could germinate and also increase the minimum at which spore and vegetative cell inocula of *Cl. botulinum* types A, B and E could initiate growth.

The separate effects of water activity (a_w), pH and temperature on the growth of vegetative cells from bacterial spores are well known (Murrell, 1961; Riemann, 1962; Segner, Schmidt & Boltz, 1966). However, few investigations have been done to determine whether these factors interact to prevent the germination of spores and their further development into vegetative cells. Two such studies have been published, one by Ohye & Christian (1966) on *Clostridium botulinum* types A, B and E, and one by Mundt, Mayhew & Stewart (1954) on *Cl. sporogenes*. These and other studies were made to find if a_w and pH are factors controlling the growth of spore formers in foods. We are interested in such studies as part of an investigation into factors controlling the growth of *Cl. botulinum* in foods that have not been sufficiently heated, or otherwise processed, to destroy spores of this organism. Our studies have been done using spores of single strains of *Cl. botulinum* types A, B and E inoculated into Reinforced Clostridial Medium (RCM) adjusted to pH values between 5 and 7 and with NaCl or glycerol added to give a_w values in the range 0·997–0·890. The development of spores in such media when incubated at 20 or 30° has been followed by examining preparations under phase contrast. The stages in the development of spores into vegetative cells were divided into germination, outgrowth and vegetative cell division using the criteria suggested by Pulvertaft & Haynes (1951), Campbell (1957) and Gould (1964).

Materials and Methods

Preparation of spore suspensions

Spores of *Cl. botulinum* type A were prepared by inoculating 1 ml of an overnight RCM culture of strain ZK3 (a laboratory isolate from groundnuts) into a freshly steamed bottle of a medium containing (% w/v): Trypticase (BBL) (Baltimore Biological Laboratories, Baltimore, U.S.A.), 10; sodium thioglycollate, 0·1; pH, 7·1

(Brown, 1956). The medium was dispensed in 23 ml amounts into McCartney bottles and sterilized at 121° for 15 min. One ml amounts of a 6 h culture of ZK3 in this medium were used to inoculate further bottles of the medium. The inoculated bottles were incubated with loose caps at 30° in anaerobic jars and examined for spores daily. Good spore yields were usually obtained after 2–3 days but they were mostly still in the vegetative cells and required incubation for a further 4 days or longer before they became free. At intervals during the incubation period the bottles of culture were removed from the anaerobic jars and after tightening their caps gently shaken to resuspend the sedimented organisms; the caps were again loosened before reincubating.

Spores of *Cl. botulinum* type B were prepared in a medium containing (% w/v): Trypticase (BBL), 5; ammonium sulphate, 1; pH, 7·0; sterilized at 121° for 20 min (Tsuji & Perkins, 1962). One ml amounts of an overnight RCM culture of *Cl. botulinum* type B (strain ATCC 438) were used as inocula and bottles were incubated aerobically in tightly closed screwcapped McCartney bottles for 22 days at 30°. Although a good spore yield was apparent early in the incubation period the majority of the spores were not free from the vegetative cells. Despite the prolonged incubation period many were still not free, but they were harvested at this time as they were germinating and forming new rods.

Spores of *Cl. botulinum* type E were prepared by using Roberts' A1 medium (Roberts, 1965). A heavily grown RCM culture of the Beluga strain of *Cl. botulinum* type E was used as inoculum. One ml amounts of this culture were used to inoculate freshly steamed 20 ml amounts of Roberts' medium which were then incubated aerobically in tightly closed screwcapped McCartney bottles for 20–30 h at 30°. Spores were harvested as soon as they were present in large numbers as they germinated on further incubation.

It should be noted that not all batches of BBL Trypticase are suitable for obtaining spore crops of *Cl. botulinum* (Perkins, 1965) and that several batches were tested to find the most suitable.

Harvesting and testing spore crops

The spores were harvested after their various incubation periods by centrifuging and washed 3 times in cold (5°) sterile distilled water. The final sediments were bulked to give suspensions containing 10^8–10^9 spores/ml. Samples of the spore suspensions of types A and B were tested to check that they were fully heat resistant, and then, together with type E, they were inoculated into RCM and incubated for 4 days at 30°. Cultures were plated aerobically and anaerobically for purity and 0·25 ml amounts of the culture supernatant fluids were injected intraperitoneally (IP) into pairs of 25–30 g mice. Specific neutralization tests were made by mixing the culture supernatant fluids with equal volumes of 1:12 dilutions of antisera to type A, B and E toxins (Pasteur Institute, Paris) and after standing for 30 min at room temperature injecting 0·5 ml IP into pairs of mice. After confirming the purity of the spore suspensions, and that they were able to give rise to toxic cultures, they were sealed in 2 ml ampoules and stored at −20°. The spore suspensions were finally checked under phase contrast to see if they were fully phase-bright (ungerminated) and free from vegetative cell debris. Type A and E spore suspensions were free from vegetative cells and contained *c.* 98% of ungerminated spores. Type B spore suspensions contained *c.* 60% of ungerminated spores, 30% of germinated spores and 10% of rods.

Preparation of media of different a_w and pH

A modified RCM was used for all experiments. It contained: peptone (Evans), 1·0 g; Lab-lemco (Oxoid), 1·0 g; hydrated sodium acetate, 0·5 g; yeast extract (Difco), 0·15 g; soluble starch, 0·1 g; glucose, 0·1 g; L-cysteine, 0·05 g; ascorbic acid, 1·0 g; distilled water, 100 ml; pH 7·1–7·2. It was prepared double strength, dispensed in 100 ml amounts in 4 oz screwcapped bottles, sterilized at 121° for 15 min and stored at 5° until required. The medium was adjusted to the required a_w by the addition of glycerol or NaCl of analytical reagent grade. The amounts of glycerol or NaCl required were calculated from data listed by Scott (1957) and Robinson & Stokes (1955). Media were prepared with a_w in steps of 0·01 between 0·89 and 0·99. The glycerol or sodium chloride equivalent of the medium was calculated from the a_w of the modified RCM, which was 0·997. The glycerol was rapidly weighed into 4 oz screwcapped bottles and sterilized in a hot air oven at 160°. The volumes of glycerol were calculated, and sterile, freshly steamed distilled water added to give an estimated volume of 50 ml. To this volume was added 50 ml of freshly steamed double strength modified RCM. The final medium was stored overnight at 5°. For the media containing NaCl, the salt was dried in an oven (100°) overnight, weighed into 4 oz screwcapped bottles and sterilized at 160°. Fifty ml of distilled water and 48·6 ml of the modified double strength RCM was added to give the required weight of salt in 100 g of modified single strength RCM. Where media of different pH were required, appropriate amounts of N HCl were added; the pH values selected were 5·0, 5·3, 5·5, 6·0 and 7·0. The volumes of HCl required to give these pH was taken into account when diluting the media to their final volumes.

Water activity measurements

The a_w of all media was confirmed experimentally after preparation. An electrical dewpoint apparatus with an accuracy of $\pm 0{\cdot}002$ a_w was built for the purpose. Before determining the a_w of the media, 1·0 ml of distilled water was added to 24 ml of each medium to allow for the water added with the inoculum.

Setting up and inoculating tubes for experiments

The media were distributed in 2·4 ml amounts into 100×6 mm tubes. The spore suspensions were thawed and the spores resuspended by mixing on a Whirlimix (Fisons Scientific Ltd., Loughborough). Type E spores were extremely difficult to resuspend and up to 10 min mixing was required to disintegrate the clumps of spores formed on freezing. Amounts of 0·1 ml of the spore suspension were inoculated into each tube with 1 ml disposable tuberculin syringes. The tubes were stoppered with rubber bungs and their contents thoroughly mixed by inverting several times. The type A and B spores were then immediately heat shocked by placing them in a water-bath at 82–84° for 12 min in a metal rack fitted with a lid to hold the bungs in position.

The tubes were cooled in cold water and incubated in stirred waterbaths at 20 or 30° (accuracy ±0·25°). Tubes containing the type E spores were incubated immediately after mixing their contents. In some experiments, type A and B spores were heat shocked in distilled water and then added to the unheated medium.

For experiments using vegetative cells, overnight cultures in RCM were used to inoculate media dispensed in bijou bottles; sufficient media was added to almost fill the bottles. The cultures used for inoculation were judged to be free from spores by microscopical examination.

Reading and recording of results

At intervals during the incubation period the amount of spore germination, outgrowth and vegetative cell growth was determined by phase contrast observations. Initial observations of type A and B spores were made immediately after inoculation and heat shocking. The tubes of media were removed one at a time from the waterbath and after inverting to mix their contents the bungs were removed and the liquid further mixed by gently pipetting up and down a Pasteur pipette. A single drop of medium was placed on a glass slide, 0·8–1·0 mm thick, and covered with a 22 mm^2 '0' thickness coverslip (Chance Ltd., Birmingham). The bungs were replaced and the tubes put back in their waterbaths with a minimum delay. The slide preparations were then examined under phase contrast using a Reichert Neopan microscope (Shandon Scientific Co.) at a magnification of ×1000. Random fields were counted until *c.* 50 spores, germinated spores or vegetative cells had been counted. Phase bright spores were counted as ungerminated spores and phase dark spores as germinated spores (Pulvertaft & Haynes, 1951). In some preparations from media of low a_w many spores were not completely phase-dark. Viable counts from one of these media, before and after pasteurization, were done and it was established that the spores had lost their heat resistance; they were therefore counted as germinated spores. Cells emerging from the germinated spore and still attached to the spore coat were scored as 'outgrowths' (Campbell, 1957). Free cells undergoing division were recorded as vegetative cells.

During the first 5–6 h of incubation the tubes were examined at $\frac{1}{2}$–1 h intervals. Subsequent readings were made daily for the first 2–3 days, and then during the following week, they were examined daily or every second day. After this time they were examined at least once a week. Percentage germination was estimated by expressing the number of germinated spores, together with any outgrowths and vegetative cells that were present, as a percentage of the total spores, outgrowths and vegetative cells counted. When vegetative cell division occurred, as indicated by an increase in the total spore and vegetative cell count/field, we estimated the approximate amount of cell division by the increase in total count/field and allowed for this in estimating the approximate percentage germination. The vegetative cells content was expressed as a percentage of the total count.

Growth from vegetative cell inocula was determined by examining bottles at frequent intervals for turbidity. Media from bottles showing growth was plated on yeast-glucose agar and incubated aerobically to check absence of contamination.

Results

To reduce the amount of information to reasonable proportions much of our data has been condensed (Tables 1–5). ± in the Tables means that the amount of germination or vegetative cells was 1–10% of the total count; + is used for percentages of 11–50% and ++ for those >50%. Outgrowths and vegetative cells occurring together have been included in the percentages of vegetative cells. Times at which the observations were made are listed in the Table as follows: $\overset{+}{1}$h–$5\overset{+}{d}$ means that germinated spores or vegetative cells were first observed after 1 h and that germination or vegetative cells exceeded 50% after 5 days. It was frequently observed that with continued incubation, vegetative cell lysis occurred and that the vegetative cell count decreased.

In preliminary experiments the same results were obtained when spores of *Cl. botulinum* were heat shocked in the glycerol or salt-containing media as when they were heat shocked in distilled water and then added to the unheated media. Therefore in subsequent experiments the spores of type A and B were heat shocked in the media under test; spores of types E were not heat shocked.

The effect of water activity and temperature on the growth of Cl. botulinum *from spore and vegetative cell inocula*

In modified RCM at pH 7 and incubated at 20 or 30°, spores of *Cl. botulinum* types A, B and E were able to germinate at a_w down to 0·89 in RCM adjusted with glycerol (R.Gl) and to 0·93 in RCM adjusted with NaCl (R.NaCl) (Tables 1–3). Germination rate was dependent on the organism, incubation temperature and medium. All spores germinated faster at 30° than at 20° and spores of all strains germinated faster and to a greater extent at the lower a_w in R.Gl than in R.NaCl.

Type A and B spores when incubated at 30° in R.NaCl were able to develop into vegetative cells at a minimum a_w of 0·96, whereas in R.Gl vegetative cells developed at much the lower a_w of 0·93; the B spores outgrew but did not form vegetative cells in R.Gl at a_w of 0·91 and 0·90, respectively. Type E spores initiated vegetative growth at a minimum a_w of 0·94 in R.Gl and 0·97 in R.NaCl. At 20°, the minimum a_w for vegetative cell formation from spores of types A and B was 0·97 in R.NaCl and 0·93 in R.Gl. Type E spores formed vegetative cells at a minimum a_w of 0·96 in R.Gl and outgrew at 0·94; in R.NaCl, vegetative cells were formed at an a_w of 0·98, with possibly some formation at 0·97. The limiting a_w for the growth of *Cl. botulinum* types A, B and E from vegetative cell inocula are given in Table 4.

The effect of pH and temperature on the growth of Cl. botulinum *from spore and vegetative cell inocula*

Spores of *Cl. botulinum* types A, B and E when incubated at 20 or 30° in modified RCM without added glycerol or NaCl were able to germinate and initiate vegetative cell growth at pH values between 7·0 and 5·3 (Tables 1–3). Type B spores were also able to germinate when incubated at 20 or 30° in RCM at pH 5, but vegetative cell growth occurred only at 30°. Type A spores were unable to germinate at pH 5; some type E spores germinated very slowly at pH 5 but no further development was observed. The rates of germination of type A and E spores were similar at pH 7 and

6 but were slower at pH 5·5 and 5·3; the effect of temperature on the germination rates of type A at pH 5·5 and 5·3 was marked. The germination rates of type B spores were less affected by pH.

Vegetative cell inocula of types A, B and E grew at pH 5·3. Types A and B were also able to grow at pH 5·0 when incubated at 30° but not at 20° (Table 4).

The combined effect of pH, water activity and temperature on the growth of Cl. botulinum *from spore and vegetative cell inocula*

The minimum a_w in R.NaCl and R.Gl at which spores of *Cl. botulinum* types A and E germinated, increased markedly with a decrease in the pH of the media (Tables 1 & 3). The minimum a_w at all pH levels were lower in R.Gl than R.NaCl, and for types A and B were also lower in media incubated at 30° than at 20°; there was little or no temperature effect on the germination of type E spores. Type B spores, unlike those of types A and E, germinated at all a_w between 7 and 5 (Table 2). The development of germinated spores of types A, B and E into vegetative cells was influenced by pH, a_w and incubation temperature. The minimum a_w at which vegetative cell inocula could grow was generally identical, or within 0·01, of the minimum level at which germinated spores initiated vegetative cell growth. The differences between the minimum a_w permitting vegetative cell development from spore inocula in R.NaCl and R.Gl were also similar to those permitting vegetative cell inocula to grow in these media (Table 5).

Discussion

There is little information on the effect of sodium chloride and other salts on the germination of *Cl. botulinum* spores; by germination we mean the change of the phase-light spore to phase-dark, with loss of heat resistance. Halvorson (1955) found that the germination of spores of a number of *Bacillus* and *Clostridium* spp. was almost completely inhibited by 9% of NaCl, KCl or $MgCl_2$ but up to 15% of Na_2SO_4 and $MgSO_4$ did not interfere with germination. It should be noted, however, that Halvorson was comparing germination rates over a 15 min period, in which time his control suspension containing no added salts had almost completely germinated. Gould (1964) showed that at least 15% of NaCl was required to inhibit the germination of *Bacillus* spores for 24 h and that between 4 and 7% was required to inhibit the development of germinated spores into vegetative cells. Mundt *et al.* (1954) observed 90% germination of *Cl. sporogenes* in the presence of 8% of NaCl and at pH values down to 5·3, but that germinated spores were unable to outgrow under these conditions. It would thus appear to be generally agreed that spore germination can occur in the presence of higher concentrations of NaCl than can vegetative growth. This is supported by our observations on spores of *Cl. botulinum* types A, B and E. Spores of these organisms will germinate in the presence of at least 10·3% w/w of NaCl but are unable to initiate vegetative growth; the rate and amount of germination was reduced, however, at this NaCl concentration. The difference between the NaCl concentrations inhibiting germination and vegetative cell development and the similarity between those inhibiting growth from spore and vegetative cell inocula indicates that NaCl mainly affects the stage between germination and vegetative cell development. As outgrowths were not detected at NaCl concentrations that were inhibitory to vegetative cell development, but not to spore germination, it would appear that NaCl must affect some stage in the germination sequence between germination and outgrowth. There is some indication from our results that glycerol may affect vegetative cell development from spores at the outgrowth stage as outgrowths were observed in media containing glycerol concentrations that were inhibitory to vegetative cell division.

The initiation of growth by spore and vegetative cell inocula of *Cl. botulinum* types A and B is prevented by adding 6–10% w/w of NaCl (Tanner & Evans, 1933; Yesair & Cameron, 1942; Greenberg, Silliker & Fatta, 1959; Pederson, 1957; Ohye & Christian, 1966). The precise limiting concentration of NaCl depends on the germination medium, incubation temperature, strain tested and the number of spores used as inoculum. Segner *et al.* (1966), studying in detail the effect of NaCl and temperature on the growth of four strains of *Cl. botulinum* type E from spore inocula, found that in a Trypticase peptone-glucose medium at temperatures between 16 and 30° the development of vegetative cells from spores of three of the strains was inhibited by 5·0% w/w, and the fourth (the Beluga strain) by 4·5% w/w, of NaCl; at 8 and 10° growth of all four strains was inhibited by 4·5%, but not by 4%, w/w of NaCl. Although 4% of NaCl had little effect on the development of vegetative cells at 30° it gave rise to a 3–4 fold increase in time to produce visible growth at 8°.

The growth of vegetative cells of type E at different temperatures has been studied recently by Ohye, Christian & Scott (1966). Their most salt tolerant strain grew in the presence of 5·8% w/w of NaCl at 30°, 5·1% at 20° and 4·3% at 15°. Our results for type A, B and E spores are mainly in agreement with these data (Tables 1–5).

The limiting pH for the growth of *Cl. botulinum* from spore inocula depends on a variety of factors. It is generally agreed that the limiting pH for the growth of *Cl. botulinum* type A and B spores is pH 4·8–5·0 (Ingram & Robinson, 1951; Ingram & Handford, 1957; Wagenaar & Dack, 1954) and that *Cl. botulinum* type E spores are inhibited at a slightly higher pH. However, the limiting pH for type E spores is controversial. The problem is reviewed by Segner *et al.* (1966), who made a detailed study of the effect of pH on the germination of type E spores in laboratory media and found that the limiting pH depended on such factors as the number of spores used as inoculum, the reducing agent in the medium and the incubation temperature. At 30°, the limiting pH for growth in the best medium and with an inoculum of 2×10^7 heat shocked spores/tube was pH 5·03; at 8° the limiting pH was 5·9; for unheated spores it was higher. In liver broth the limiting pH for vegetative cell growth for spore inocula was pH 5·22, but germination occurred at pH 5·01. Our results for *Cl. botulinum* type E are in agreement with those of Segner *et al.* (1966). The Beluga strain we studied was able to germinate at pH 5·0 but was unable to develop further; the strain grew from both vegetative and spore inocula at pH 5·3. Spores of the *Cl. botulinum* type B strain studied were able to grow at pH 5·0 when incubated at 30° but not at 20°; the type B spores germinated at 20° but did not develop further. Type A spores were unable to germinate at pH 5·0. Type A vegetative cells were able to grow at pH 5·0 when incubated at 30° but not at 20°.

Scott (1953) and Christian & Scott (1953) showed that the inhibitory effect of different carbohydrates and salts on certain bacteria can be explained in terms of their effect on a_w. Scott (1955) extended this concept to canned ham and showed that the a_w of canned ham corresponds closely to the brine concentration of the ham. Ham would appear to be exceptional in this respect as the relationship between salt concentration and a_w does not hold for all salt meat products (Anderton, 1963). The a_w of canned ham varies between 0·95 and 0·97 (Scott, 1955; Pearson, pers. comm.), and it is interesting to relate these figures to water activities required to inhibit the growth of clostridial spores in laboratory media. Bever & Halvorson (1948) found that *Cl. botulinum* type B would grow in the presence of 50% of sucrose (a_w = 0·935) but not in 55% of sucrose (a_w = 0·917). However, Beers (1957) reports that the germination of *Cl. botulinum* type B was prevented by 31% of sucrose (a_w = 0·975) and growth by 30% of sucrose (a_w = 0·976). Williams & Purnell (1953) used liver powder adjusted to different moisture levels to obtain a range of water activities for a study of the effect of a_w on the germination of *Cl. botulinum* spores, and observed germination at all moisture concentrations tested (the minimum was <0·90) but vegetative cell development was reduced at 0·94–0·96 a_w and generally prevented at 0·95–0·90 a_w. Scott (1955) pointed out that the accuracy of these a_w measurements was poor, and that his unpublished observations indicated that the minimum a_w allowing development of spores into vegetative cells in media containing various sugars and electrolytes was *c.* 0·95 a_w.

The most recent results of Ohye and colleagues (Ohye *et al.*, 1966; Ohye & Christian, 1966) are the most exact results so far obtained on the effect of a_w on the growth of *Cl. botulinum* from spore and vegetative cell inocula. Ohye *et al.* (1966) studied the effect of a_w on the growth of *Cl. botulinum* type E from vegetative cell inocula and the effect of temperature on the limiting a_w. They found that the growth of *Cl. botulinum* type E from vegetative cell inocula was limited at almost the same a_w when a mixture of three salts (NaCl, KCl, Na_2SO_4) or NaCl alone was used to adjust the a_w of the growth medium; NaCl was slightly more inhibitory than the three salts. The most salt tolerant strain they studied grew at a minimum a_w of 0·965 when incubated at 30 or 25°; of 0·97 when incubated at 20°, and 0·975 when incubated at 15°. They showed that the growth medium affected the minimum a_w for growth and that any reduction of a_w below that of the optimum (0·995) affected the growth rate of the strains at all temperatures. Ohye & Christian (1966) studied the effects of water activity and pH on the ability of spores of *Cl. botulinum* types A, B and E to initiate growth at temperatures between 0 and 50°; a mixture of salts was used to obtain media of different a_w. At the optimum growth temperatures (30–40°) and pH (7·0), the minimum a_w limiting growth were 0·95 for type A, 0·94 for type B and 0·97 for type E. Reduction of the incubation temperature from 40° to 20° increased the limiting a_w for growth from spores of *Cl. botulinum* types A and B to 0·97. The minimum a_w for type E at 20° was still 0·97, but at 10° it was increased to 0·99. The effect of increasing or decreasing the pH was to increase the minimum a_w at which vegetative cell growth was prevented. An interaction between pH and temperature was obtained for *Cl. botulinum* type E spores but was less apparent with type A and B spores.

Although the Australian workers have shown that limiting a_w for the growth of

Cl. botulinum type A, B and E spores from spore inocula, and the growth of type E spores from vegetative inocula, is unaffected by whether NaCl or a mixture of salts is used to adjust a_w, our results show that media adjusted to the same water activities using glycerol or NaCl give different results. To introduce a discussion of our findings it is pertinent to mention the results obtained by Kim (1965) for the growth of vegetative cell inocula of *Cl. welchii* in media adjusted to different a_w by the addition of glycerol, sucrose or NaCl. He found that when sucrose or NaCl was used, the lowest a_w supporting the growth of *Cl. welchii* was 0·97–0·95, but in media adjusted with glycerol the minimum a_w were 0·95–0·93. We obtained similar differences between minimum a_w with NaCl and glycerol for the growth of *Cl. botulinum* types A, B and E from spore and vegetative cell inocula. We cannot therefore support the general view of the Australian work that it is possible to explain the effect of NaCl on the initiation of growth by bacterial spores or vegetative cells solely in terms of a_w; another factor is also involved. At the minimum inhibitory concentration and the optimum pH for growth, NaCl does not appear to have a marked effect on the germination of spores of *Cl. botulinum*, but it affects their outgrowth and further development into vegetative cells. Gould (1964) found similarly that NaCl did not markedly affect germination of spores of *Bacillus* spp. and suggested that the effect was mainly on the development of outgrowth stage into vegetative cells; however his results show that this effect on the outgrowth stage was not true for all *Bacillus* spp. and some were inhibited after the germination stage. We agree with the finding of Ohye & Christian (1966) that decreasing the pH of the growth medium increases the minimum a_w at which spores will initiate growth.

A further effect of pH is to reduce the rate of germination, and in the case of type A spores to inhibit germination completely; this is apparent in media containing either glycerol or NaCl. With decrease in pH, the minimum a_w permitting germination increases and at pH 5·5 and below the minimum a_w preventing spore germination of spores of *Cl. botulinum* type A is almost identical with the minimum preventing vegetative cell development. Similar results were obtained with the type E strain studied, but pH had little or no effect on the germination of type B spores other than to slow the rate of germination. Further strains of *Cl. botulinum* types B and A are being studied to see whether this is a true difference between type A and B spores.

The combined effects of pH and salt concentration on the initiation of growth by *Cl. botulinum* spores may prove useful in reducing the botulism hazard of foods. As will be seen from Table 5 at pH levels that could be achieved with a variety of foods quite low concentrations of salt will prevent the development of spores into vegetative cells and hence toxin production.

References

Anderton, J. I. (1963). Pathogenic organisms in relation to pasteurized cured meats. B.F.M.I.R.A. Scientific and Technical Survey No. 40. Leatherhead, British Food Manufacturing Industries Research Association.

Beers, R. J. (1957). Effect of moisture activity on germination. In *Spores*, p. 45. Ed. H. O. Halvorson. Washington: American Institute of Biological Sciences. Publication No. 5.

Beyer, J. S. & Halvorson, H. O. (1948). *Rep. Hormel, Inst. Univ. Minn* 1947–8, in Scott (1955).

Brown, W. L. (1956). The production and germination requirements of putrefactive anaerobe 3679. PhD thesis, University of Illinois.

TABLE 1. *Effect of carbon source on the formation of invertase by Saccharomyces strains 303-67 and 1016*

Carbon source	Concn (mM)	Strain 303-67		Strain 1016	
		Turbidity[a] (Klett units)	Invertase[b] (units/ 5 × 10^7 cells)	Turbidity[a] (Klett units)	Invertase (units/ 5 × 10^7 cells)
Glucose	10	70	15.0	59	50.5
	100	105	4.5	127	7.6
Fructose	10	73	18.2	58	72.5
	100	90	4.7	105	8.5
Mannose	10	65	14.6	63	67.0
	100	88	7.9	100	25.8
Maltose	5	35	3.0	25	10.0
	50	30	5.0	120	32.8

[a] Vogel's medium N (reference 26, modified as in Methods) was inoculated with 10^7 cells per ml (Initial reading was 35 Klett units for strain 303-67 and 31 for strain 1016). Cultures were incubated for 4 hr.

[b] The one-step assay of Neumann and Lampen (19) was used; units are micromoles of sucrose hydrolyzed in 30 min.

Stock cultures of *Saccharomyces* 1016 were maintained on 0.5% peptone - 0.3% yeast extract - 1% glucose medium containing 1.5% agar. The cells from one slope were transferred into 10 ml of Vogel's medium N (26) supplemented with (per liter): 2.0 mg of inositol, 0.2 mg of calcium pantothenate, 0.2 mg of pyridoxine-hydrochloride, 0.2 mg of thiamine-hydrochloride, and containing 0.05 M maltose (nonrepressive conditions) or 0.2 M glucose (repressive conditions). The cultures were incubated for 16 hr at 28 C on a reciprocating shaker. For preparing exponential-phase cells, 10 ml of a stationary-phase culture was added to 100 ml of the same medium, the mixture was incubated as above for 4 to 5 hr until an optical density corresponding to a Klett reading (filter no. 66) of 130 to 140 (equivalent to about 4×10^7 cells per ml) was reached. The cells were washed twice with distilled water and suspended at a concentration of 4×10^8 cells per ml [equivalent to 8 mg (dry weight)].

Preparation of protoplasts. Protoplasts were obtained from exponential-phase cells by a slight modification of the procedure described by Gascon and Lampen (7). To 5 ml of yeast suspension containing 4×10^8 cells per ml, the following additions were made: 0.05 M tris(hydroxymethyl)aminomethane (Tris)-hydrochloride buffer (*p*H 7.5), 1 ml; 1.2 M KCl containing 0.02 M $MgSO_4$, 6 ml; snail enzyme, 0.5 ml; and 1 M 2-mercaptoethanol, 0.2 ml. The mixtures were incubated at 30 C on a reciprocating shaker (75 cycles/min with a 2.5-cm stroke); after 1 hr, more than 95% of the cells were converted into protoplasts. The protoplasts were centrifuged at $1,000 \times g$ for 3 min and washed twice with 5 mM $MgSO_4$ and 0.02 M phosphate buffer (*p*H 6.0) containing 0.6 M KCl or 0.8 M sorbitol.

Measurement of invertase formation and secretion by protoplasts. Washed protoplasts were suspended at a concentration of 5×10^7 per ml in the desired osmotic stabilizer with 10 mM fructose as the energy source unless otherwise stated. The suspensions were incubated at 30 C in a reciprocating shaker as described for the preparation of protoplasts. For measurement of invertase that is released into the medium, 1 ml of the incubation mixture was withdrawn at intervals and centrifuged at $1,000 \times g$ for 3 min; 0.2 ml of the supernatant was pipetted into 0.8 ml of ice-cold water and assayed as described below. For determination of the total amount of invertase synthesized, 0.2-ml samples were periodically removed from the incubation mixture and transferred to chilled tubes containing 0.8 ml of ice-cold water to lyse the protoplasts. The resulting suspensions were assayed.

Separation of the two forms of invertase. A Sephadex G-200 column (2 by 47 cm) was used for separation of large and small invertase as described by Gascon and Lampen (7). Supernatant fluids ($1,000 \times g$, 3 min) from the incubation mixtures (5×10^7 protoplasts per ml) and protoplast pellets lysed by resuspension to the original volume in distilled water were filtered through membrane filters (0.45 μm porosity, Millipore Corp.) before gel filtration. Less than 5% of the invertase activity of either supernatant fluids or lysates was retained by the filters. One milliliter of the filtrate was applied to the column bed, and 3-ml fractions were eluted with 0.05 M Tris-hydrochloride buffer (*p*H 7.5).

Invertase assay. The assay was essentially that described by Gascon and Lampen (7). One unit of invertase is the amount of enzyme which hydrolyzed 1 μmole of sucrose in 30 min at 30 C in 0.05 M sodium acetate buffer (*p*H 5.0) containing 0.125 M sucrose.

Assay of α-glucosidase and UV-absorbing materials. For determination of materials released into the medium, 1 ml of the incubation mixture was taken at zero time and after 2 hr and centrifuged at $1,000 \times g$ for 3 min; 0.2 ml of the supernatant was pipetted into 0.8 ml of ice-cold water and assayed as below.

Alpha glucosidase activity was estimated by the procedure of Halvorson and Ellias (9). Samples (0.1 ml) were mixed with 2.7 ml of 0.065 M potassium phosphate buffer (*p*H 7.0) and 1.0 ml of reduced glutathione solution (1 mg/ml). The reaction was started by adding 0.1 ml of substrate (3 mg of *p*-nitrophenyl-α-D-glucoside). The production of *p*-nitrophenol was measured continuously with a Beckman DB spectrophotometer at 400 nm with the cuvette compartment maintained at 30 C. A unit of enzyme is defined as the quantity releasing 1 nmole of *p*-nitrophenol per min at 30 C. Ultraviolet (UV) absorption was measured at 260 nm.

Estimation of amino acid incorporation into protein fraction and fructose uptake. Protoplasts (5×10^7/ml) were incubated at 30 C in sorbitol-stabilized medium with 10 mM unlabeled fructose and 0.1 mM L-threonine-*U*-^{14}C (0.05 μCi/ml of suspension). At intervals, protoplast suspension (0.8 ml) was added to 1 ml of 10% trichloroacetic acid and heated at 95 C for 30 min. The residues (protein fraction) were collected on membrane filters (0.45-μm porosity) and washed twice with 5% cold trichloroacetic acid. For measuring fructose uptake, only 10 mM fructose-*U*-^{14}C (0.1 μCi/ml of suspension) was included in the incubation mixtures. At various intervals, the protoplasts from 0.8-ml samples of the incubation mixtures were collected on membrane filters (Millipore Corp.) and washed twice with 4 ml of ice-cold 0.8 M KCl solution. Filters were transferred directly to scintillation vials containing 10 ml of Bray's solution (1). The radioactivity was measured in a Packard Tri-Carb liquid scintillation spectrometer.

Chemicals. All chemicals were reagent quality. Radioisotopes were purchased from New England Nuclear, Boston, Mass. Snail enzyme (Glusulase) was obtained from Endo Lab Inc., New York; cycloheximide from Upjohn Co., Kalamazoo, Mich., and D-sorbitol from Nutritional Biochemical Corp., Cleveland, Ohio

RESULTS

Effect of osmotic support on the synthesis and secretion of invertase and leakage of intracellular materials. In the previous study of the distribution of invertase isoenzymes in intact cells and protoplasts of *Saccharomyces*, the protoplasts were prepared and maintained in 0.6 M KCl to prevent osmotic lysis (7, 8). However, we discovered that under these conditions some of the protoplasts lysed during the incubation period for formation and secretion of invertase. This led us to examine the effect of the concentration of KCl, sorbitol, and other osmotic supports on the stability of the protoplasts and on their formation and secretion of invertase.

Stability of the protoplasts was determined by measuring the release of an intracellular enzyme, α-glucosidase (12) and of UV-absorbing substances. The results in Fig. 1 show the effect of various KCl concentrations on invertase formation and on release of intracellular materials. At 0.2 M KCl the protoplasts were fragile and immediately released large amounts of invertase, α-glucosidase, and UV-absorbing materials (Fig. 1A and B). There was no net synthesis of invertase under these conditions (Fig. 1A). Protoplast stability was excellent in KCl concentrations higher than 0.6 M; however, invertase synthesis appeared to be severely reduced. A KCl concentration of 0.4 M (equivalent to 0.75 osmolal) was most effective for the synthesis and secretion of invertase, even though lysis of some protoplasts with release of intracellular materials occurred (Fig. 1B). Similar data, which will not be presented in detail, were obtained when NaCl was

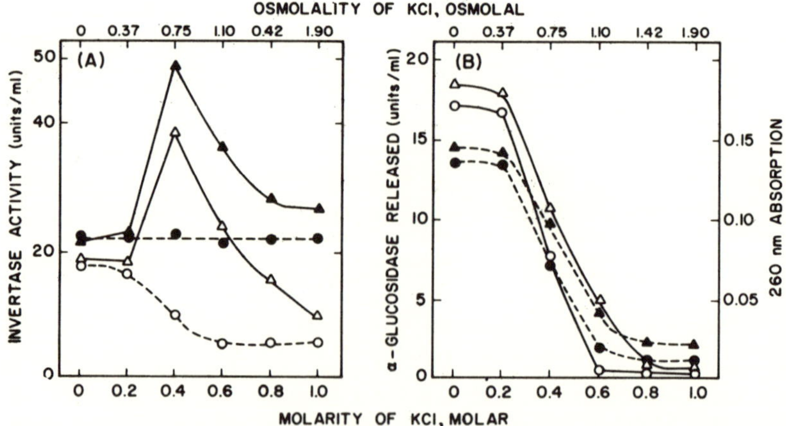

FIG. 1. *Formation and secretion of invertase, and release of α-glucosidase and UV-absorbing materials from protoplasts suspended in various concentrations of KCl. Protoplasts from cells grown in 0.05 M maltose were suspended at 5×10^7 per ml in the indicated concentrations of KCl and 10 mM fructose as energy source, and were incubated at 30 C. At zero time and after 2 hr, 0.2-ml samples of the incubation mixtures or of the supernatant fluids were transferred into 0.8 ml of ice-cold water and assayed. Enzyme activity is expressed as units (micromoles of disaccharide hydrolyzed in 30 min) per milliliter of the incubation mixture or of the supernatant fluid. (A) Invertase activity: (●) 0 time, total activity; (▲) 2 hr, total activity; (○), 0 time, supernatant fluid; (△) 2 hr, supernatant fluid. (B) α-Glucosidase and UV-absorbing materials in the supernatant fluid: (○) 0 time, α-glucosidase; (△) 2 hr α-glucosidase; (●) 0 time, absorbancy at 260 nm; (▲) 2 hr, absorbancy at 260 nm. Osmolality of KCl was calculated from freezing-point depression of the same concentration of KCl (27).*

used in place of KCl. These results were surprising, since the concentrations of KCl and NaCl tested were not extreme and were used in many investigations for stabilization of yeast protoplasts (25).

An attempt was made to see whether the reduction of invertase synthesis and secretion in high concentrations of KCl was due to specific solute (or ionic) effects or to osmotic effects. The same type of experiment was therefore performed with protoplasts suspended in various concentrations of sorbitol or $MgSO_4$ (Fig. 2 and 3, respectively). Invertase was most actively synthesized and secreted when the protoplasts were suspended in 0.6 M sorbitol or $MgSO_4$ (equivalent to about 0.65 osmolal), and the synthesis of invertase was reduced if the concentration of these osmotic stabilizers was increased. Thus, for the four osmotic supports tested, the optimal osmotic range for invertase formation and secretion by the protoplasts was about 0.65 to 0.75 osmolal. It is important to note that the effectiveness of invertase synthesis by protoplasts suspended in 0.8 M sorbitol was only 75 to 80% of that of protoplasts in 0.6 M sorbitol, but 0.8 M sorbitol gave a greater protection of the protoplasts with no leakage of intracellular materials into the medium during the 2-hr incubation (Fig. 2). For this reason, protoplasts suspended in 0.8 M sorbitol solution were considered to be the most suitable for the study of invertase secretion from the cell membrane and were used in all subsequent experiments.

De novo synthesis of invertase and its secretion. Cycloheximide is known to be active against yeast cells, and its primary action is to inhibit cytoplasmic protein synthesis at the ribosomal level (15, 20). To determine whether the secretion of invertase is dependent on de novo protein synthesis, protoplasts from cells grown in nonrepressive conditions (0.05 M maltose) were incubated with 10 mM fructose and various concentrations of sorbitol as the osmotic stabilizer and in the absence or presence of 5 μg of cycloheximide per ml. During a 2-hr incubation with fructose and 0.6 M sorbitol but without cycloheximide, the level of invertase increased about fourfold (Fig. 4A). Both synthesis and secretion of the enzyme were gradually reduced by increasing sorbitol concentrations, but in 1.0 M sorbitol, most of the invertase originally present was released into the suspending medium without a significant increase in total activity. The release of invertase was also prevented in 1.2 M sorbitol. In the presence of cycloheximide, invertase synthesis and secretion were almost completely inhibited, including the release of the original activity (Fig. 4B).

Form of invertase secreted by protoplasts. It was demonstrated that *Saccharomyces* cells contain two forms of invertase (7, 8). The large form has a molecular weight of 270,000, approximately 50% mannan, and the small form has a molecular weight of 135,000 and is devoid of carbohydrate. The large form of invertase (in the periplasm and cell wall) is released from the cells during the formation of protoplasts.

It was of interest to examine which type of enzyme was released by protoplasts when the enzyme was being actively synthesized and secreted. Protoplasts from cells grown in 0.05 M maltose (nonrepressive) or 0.2 M glucose (repressive) medium were incubated for 2 hr in 0.8 M sorbitol and 10 mM fructose. The invertase present on the protoplasts and that secreted into the suspending medium was fractionated by gel filtration on a Sephadex G-200 column (Fig. 5). Protoplasts from cells grown on nonrepressive

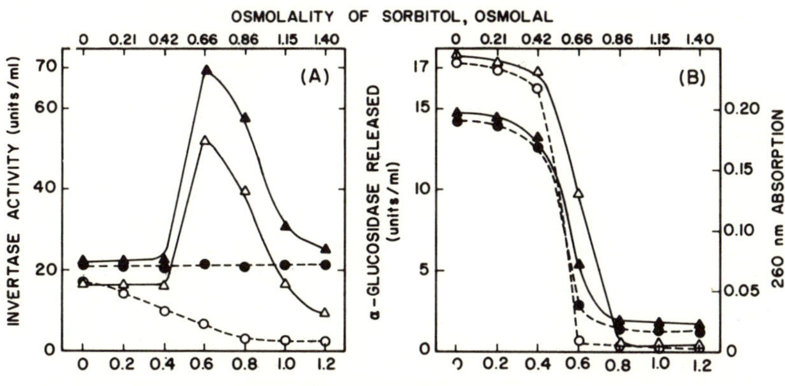

FIG. 2. *Formation and secretion of invertase and release of α-glucosidase and UV-absorbing materials from protoplasts suspended in various concentrations of sorbitol. All conditions were as described in the legend to Fig. 1, except that the osmotic support was sorbitol. Osmolality of sorbitol was measured and calculated from freezing-point depression of the same concentrations of sorbitol (27).*

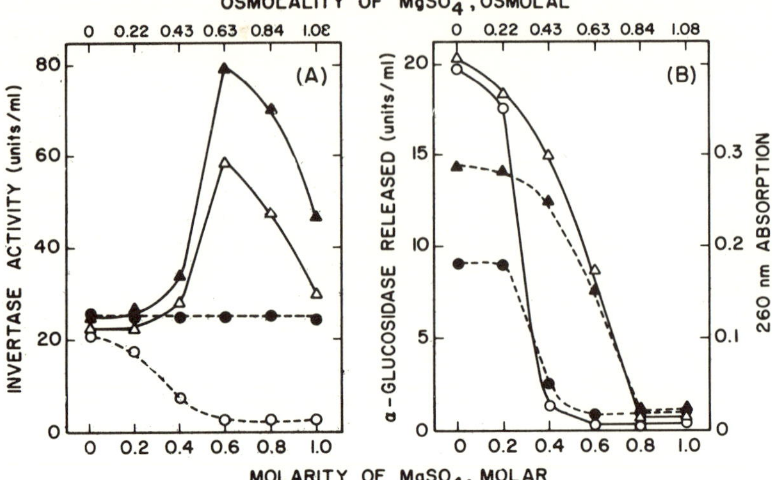

FIG. 3. *Formation and secretion of invertase and release of α-glucosidase and UV-absorbing materials from protoplasts suspended in various concentrations of $MgSO_4$. All conditions were as described in the legend to Fig. 1, except that the osmotic support was $MgSO_4$. Osmolality of $MgSO_4$ was calculated from freezing-point depression of the same concentration of $MgSO_4$ (27).*

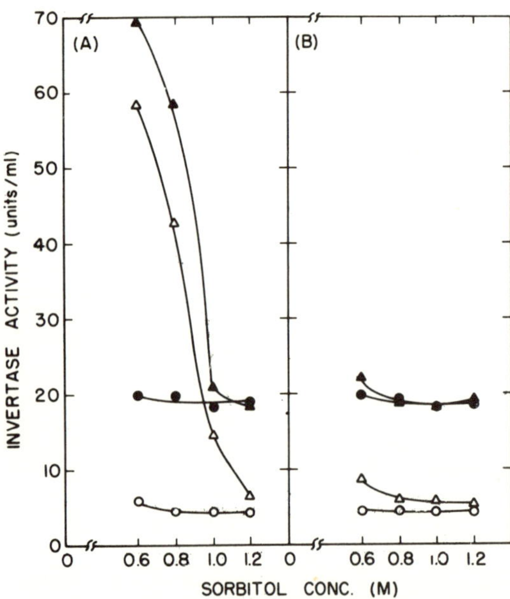

FIG. 4. *Effect of cycloheximide on the synthesis and secretion of invertase by protoplasts. (A) No cycloheximide added (experimental conditions as for Fig. 2); (B) cycloheximide (5 μg/ml) added to incubation mixture at zero time. Invertase activity: (●) 0 time, total; (○) 0 time, supernatant fluid; (▲) 2 hr total; (△) 2 hr, supernatant fluid.*

medium initially contained mainly the large form (Fig. 5A); those from repressive medium contained less total activity but most of this was the small form (Fig. 5B). Essentially identical curves were obtained for the extracts from protoplasts incubated for 2 hr and are not repeated in Fig. 5. Both types of protoplasts secreted only the large form of invertase, with no detectable release of the small form initially predominant in the protoplasts from cells grown in repressive medium. The results strongly support the concept that invertase is secreted through the plasma membrane by a specific process and not as a result of membrane damage.

Reversal of osmotic effect on invertase synthesis in protoplasts. It was important to learn whether the inhibitory effect of high osmotic pressure on invertase synthesis is reversible. For this purpose, a suspension of protoplasts in 1.2 M sorbitol with 10 mM fructose was incubated as previously described. Under these conditions, there was no net synthesis of invertase (Fig. 6). After 60 and 120 min of incubation, portions of the incubation mixture were transferred to 0.8 M sorbitol with fructose. Invertase synthesis commenced promptly, and the total activity increased nearly 10-fold in 2 hr. Surprisingly, there was no significant difference in the time course of invertase synthesis by the protoplasts whether or not they were preincubated in 1.2 M sorbitol. This indicates that the inhibitory effect of high osmolarity on enzyme synthesis is not due to irreversible alteration of the metabolic functions of the protoplast.

Sugar uptake and amino acid incorporation by protoplasts in various sorbitol concentrations. Since synthesis of invertase by protoplasts was severely reduced by high osmolarity, the osmotic

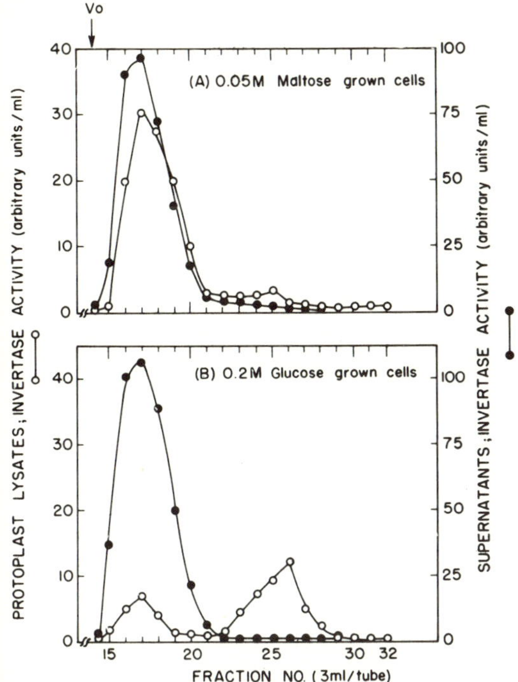

FIG. 5. *Form of invertase produced by protoplasts from cells grown in repressive (0.2 M glucose) and nonrepressive (0.05 M maltose) media. The protoplasts were suspended at 5 × 10[7] per ml in 0.8 M sorbitol and 10 mM fructose and incubated at 30 C. Sephadex G-200 column (2 by 47 cm) was equilibrated with 0.05 M Tris-hydrochloride (pH 7.5); samples equivalent to 1 ml of incubation mixture were added and eluted with the same buffer. Vo, void volume of column. (○) Zero time, protoplast lysate; (●) 2 hr, supernatant fluid.*

effect on sugar uptake and protein synthesis was determined. Both the rate of ^{14}C-fructose uptake and of incorporation of ^{14}C-threonine into the protein fraction decreased significantly as the concentration of sorbitol increased (Fig. 7). Protoplasts in 1.2 M sorbitol incorporated less than 20% of the radioactivity taken up by protoplasts in 0.6 M sorbitol. Similar results were obtained when the incorporation of ^{14}C-fructose into macromolecules such as mannan and glucan was measured (*unpublished data*).

If the inhibitory effect of high osmolarity on invertase synthesis in protoplasts is the result of a decreased penetration of sugar into the cells, synthesis of the enzyme should be restored if a high level of extracellular fructose is present. To test this hypothesis, protoplasts in 0.8 M or 1.2 M sorbitol were incubated with concentrations of fructose ranging from 10 to 150 mM. Protoplasts in 0.8 M sorbitol actively synthesized enzyme when the external fructose concentration was 10 mM (Fig. 8A). Formation of the enzyme was gradually repressed, however, as the fructose concentration in the suspending medium was increased, even though the total level of osmotic support (fructose and sorbitol) remained constant. These results are consistent with previous reports of catabolic repression of invertase synthesis in yeast cells or protoplasts by high concentration of sugar (2, 6, 7, 13, 17). Protoplasts suspended in 1.2 M sorbitol formed little, if any, enzyme when 10 mM fructose was included in the incubation mixture (Fig. 8B). Synthesis of the enzyme was significantly stimulated as the concentration of fructose was increased (at constant osmolarity) and with prolonged incubation. Such results would be expected if the permeability of the plasma membrane had decreased in high concentrations of sorbitol, as the experiments of Fig. 7 had indicated. However, repression of invertase synthesis occurred at higher levels of external fructose. The results were rather surprising, since the amounts of fructose incorporated by these

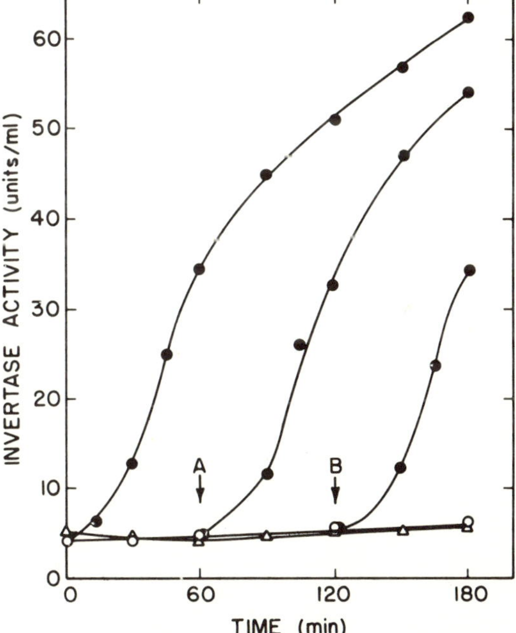

FIG. 6. *Reversibility of the osmotic inhibition of invertase synthesis. Protoplasts from cells grown in 0.2 M glucose medium were suspended in 0.8 or 1.2 M sorbitol in the presence or absence of 10 mM fructose and incubated at 30 C in a reciprocating shaker. At various intervals, 0.2 ml of incubation mixture from each flask was lysed in 0.8 ml of ice-cold water, and total invertase activity was measured. Symbols: (●) 0.8 M sorbitol, 10 mM fructose; (○) 1.2 M sorbitol, 10 mM fructose. At 60 min (arrow A) and 120 min (arrow B), samples were centrifuged at 1,000 × g for 3 min and suspended in 0.8 M sorbitol with 10 mM fructose; (△) 0.8 M sorbitol, fructose omitted.*

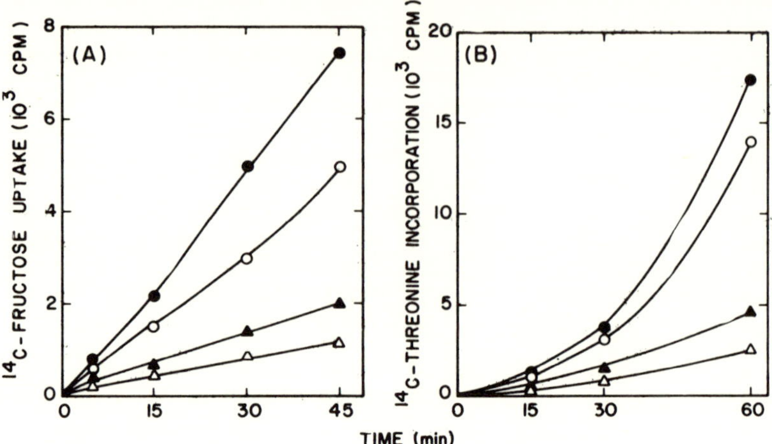

FIG. 7. *Effect of various concentrations of osmotic stabilizer on fructose uptake and threonine incorporation into protein fraction. Protoplasts were suspended at 5 × 10^7 per ml in various concentrations of sorbitol containing 10 μmoles (0.1 μCi) of ^{14}C-fructose per ml, or 10 μmoles of unlabeled fructose plus 0.1 μmole (0.05 μCi) of ^{14}C-threonine per ml at 30 C. Samples (0.8 ml) were removed at the times indicated and analyzed for the amount of radioactivity incorporated. Symbols:* (●) *0.6 M sorbitol;* (○) *0.8 M sorbitol;* (▲) *1.0 M sorbitol and* (△) *1.2 M sorbitol.*

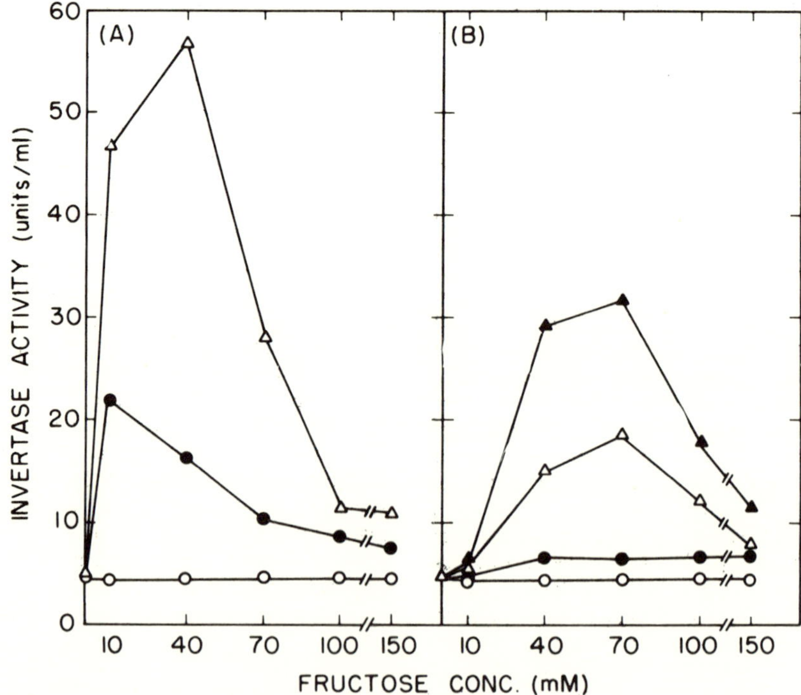

FIG. 8. *Invertase formation by protoplasts in low and high osmolarity as a function of fructose concentration. Experimental conditions are those described in the legend of Fig. 6, except that the protoplasts were incubated at 30 C with the indicated concentrations of fructose and with sorbitol added to yield a total concentration of 0.8 M (A) or 1.2 M (B). Incubation time:* (○) *0 time,* (●) *1 hr,* (△) *2 hr,* (▲) *3 hr.*

protoplasts were less than 30% of those taken up in 0.8 M sorbitol. This suggests that the repression of invertase formation may be initiated by an effect of the hexose at an external site on the protoplast membrane. This phenomenon is being investigated further.

DISCUSSION

The synthesis and secretion of invertase by yeast protoplasts is remarkably sensitive to the concentration of the agent used for osmotic stabilization. The inhibitory effect of several agents, such as KCl, NaCl, $MgSO_4$, and sorbitol, was correlated with the osmolarity of the supporting medium rather than the absolute concentration of the supporting species, indicating that osmolarity is the factor responsible for the changes observed. Although there are a number of reports on protein (enzyme) and nucleic acid synthesis by protoplasts (3, 4, 5, 10, 11, 16, 23), the conditions essential for maximum synthetic capacity have seldom been established or stated. For example, in studies on protein and ribonucleic acid synthesis by protoplasts of a strain of *S. cerevisiae*, Hutchison and Hartwell (10) employed 1.0 M sorbitol to stabilize the protoplasts; total protein increased about sixfold during incubation for as long as 15 hr. From the data presented here, one would expect that the synthetic activity of the protoplasts would have been greatly enhanced if a lower osmolarity had been used, although other factors might also be involved.

One of the few studies concerned with the effect of osmotic pressure on *Saccharomyces* has shown that both intact cells and protoplasts have mechanisms for adaptation to an increase in osmolarity of the medium. Lillehoj and Ottolenghi (14) found that cells or protoplasts of *Saccharomyces* suspended in KCl or sorbitol at concentrations as high as 2 to 3 M initially shrank and then slowly returned to normal volume. Return to normal volume was not accompanied by complete return to normal metabolism, however, since respiratory activity, reduced in the high osmolarity medium, did not return to a normal rate. They speculated that the adaption occurs by an increase in the internal pressure of the cells. Recently, Ikeda and Ottolenghi (10th Int. Congr. Microbiol., Abstr., p. 2, 1970) have shown that the materials responsible for the adaption of the cells to growth in media of high osmotic pressure are trehalose and glycerol, as these two intracellular compounds were significantly accumulated under these conditions. The reduction of invertase formation that we observed with media of high osmolarity was eliminated when the protoplasts were transferred into a medium of lower osmolarity. There was no evidence of irreversible alteration of metabolic functions; rather, the time course of invertase formation by the protoplasts was unchanged after they had been preincubated in a high osmotic medium. These observations, coupled with the fact that the concentrations used for intact cells by Ottolenghi et al. (14) were much higher than those used in the present study, make correlation of the results difficult.

Smith (21) found that the respiratory activity of spheroplasts of *Bacillus subtilis* was lower in 0.8 M sucrose medium than in 0.3 to 0.4 M sucrose, either with endogenous substrates or with added amino acids as substrates.

The observations that the rate of sugar uptake and amino acid incorporation into protein by the yeast protoplasts was severely reduced in a medium of high osmolarity and that inhibition of enzyme formation could be partially reversed by increasing the level of external sugar and incubating for a longer period of time indicate that permeability changes are involved. It seems possible that the primary effect is a structural change in the plasma membrane due to the high osmolarity. This concept is supported by the fact that invertase formation was most active under osmotic conditions in which some release of intracellular materials occurred, with the additional implication that metabolic functions are at a maximum for the protoplasts when the plasma membrane is in a loosened condition.

Under the optimal conditions established in this study, the total invertase activity of a protoplast suspension increased 5- to 10-fold during 2 hr at 30 C. Formation of invertase was inhibited in the presence of cycloheximide, thus supporting previous evidence that de novo synthesis of protein is required (12, 22). The fact that cycloheximide or a very high sorbitol level (1.2 M) completely inhibited release of the invertase originally present suggests that the process of liberation is also enzymatically mediated. The present study has also established that the secretion of invertase is very specific; only the large form was released into the suspending medium. The ability to alter the stability and functions of protoplasts by changes in osmolarity should prove useful for study of the biosynthetic relationship between the two forms of invertase and of the secretion of the enzyme through the plasma membrane into the suspending medium.

ACKNOWLEDGMENTS

We thank Jane Sharkey and Jan Tkacz for their stimulating discussions during the course of this investigation.

This research was supported by Public Health Service grant AI-04572 from the National Institute of Allergy and Infectious Diseases.

LITERATURE CITED

1. Bray, G. A. 1960. A simple efficient scintillator for counting aqueous solutions in a liquid scintillation counter. Anal. Biochem. **1**:279-285.
2. Davies, A. 1956. Invertase formation in *Saccharomyces fragilis*. J. Gen. Microbiol. **14**:109-121.
3. De Kloet, S. R. 1965. Accumulation of RNA with a DNA like base composition in *Saccharomyces carlsbergensis* in the presence of cycloheximide. Biochem. Biophys. Res. Commun. **19**:582-586.
4. De Kloet, S. R. 1966. Ribonucleic acid synthesis in yeast. The effect of cycloheximide on the synthesis of ribonucleic acid in *Saccharomyces carlsbergensis*. Biochem. J. **99**:566-581.
5. De Kloet, S. R., R. K. A. Van Wermeskerken, and V. V. Koningsberger. 1961. Studies on protein synthesis by protoplasts of *Saccharomyces carlsbergensis*. I. The effect of ribonuclease on protein synthesis. Biochim. Biophys. Acta **47**:138-143.
6. Dodyk, F., and A. Rothstein. 1964. Factors influencing the appearance of invertase in *Saccharomyces cerevisiae*. Arch. Biochem. Biophys **104**:478-486.
7. Gascon, S., and J. O. Lampen. 1968. Purification of the internal invertase of yeast. J. Biol. Chem. **243**:1567-1572.
8. Gascon, S., and P. Ottolenghi. 1967. Invertase isozymes and their localization in yeast. Compt. Rend. Trav. Lab. Carlsberg **36**:85-93.
9. Halvorson, H., and L. Ellias. 1958. The purification and properties of an alpha-glucosidase of *Saccharomyces italicus* Y1225. Biochim. Biophys. Acta **30**:28-40.
10. Hutchison, H. T., and L. H. Hartwell. 1967. Macromolecule synthesis in yeast spheroplasts. J. Bacteriol. **94**:1697-1705.
11. Islam, M. F., and J. O. Lampen. 1962. Invertase secretion and sucrose fermentation by *Saccharomyces cerevisiae* protoplasts. Biochim. Biophys. Acta **58**:294-302.
12. Lampen, J. O. 1968. External enzymes of yeast: their nature and formation. Antonie van Leeuwenhoek J. Microbiol. Serol. **34**:1-18.
13. Lampen, J. O., N. P. Neumann, S. Gascon, and B. S. Montenecourt. 1967. Invertase biosynthesis and the yeast cell membrane, p. 363-372. *In* H. J. Vogel, V. Bryson, and J. O. Lampen (ed.), Organizational biosynthesis. Academic Press Inc., New York.
14. Lillehoj, E. B., and P. Ottolenghi. 1967. Osmotic effect on yeast cells and protoplasts, p. 145-152. *In* R. Muller (ed.), Symposium on yeast protoplasts, Jena, 1965. Akademie-Verlag, Berlin.
15. McKeehan, W., and B. Hartesty. 1969. The mechanism of cycloheximide inhibition of protein synthesis. Biochem. Biophys. Res. Commun. **36**:625-630.
16. McLellan, W. L., Jr., and J. O. Lampen. 1963. The acid phosphatase of yeast. Localization and secretion by protoplasts. Biochim. Biophys. Acta **67**:324-326.
17. McMurrough, I., and A. H. Rose. 1967. Effect of growth rate and substrate limitation on the composition and structure of cell wall of *Saccharomyces cerevisiae*. Biochem. J. **105**:189-203.
18. Millbank, J. W. 1964. DNA and protein synthesis in the growing yeast protoplast. Exp. Cell Res. **35**:77-83.
19. Neumann, N. P., and J. O. Lampen. 1967. Purification and properties of yeast invertase. Biochemistry **6**:468-475.
20. Siegel, M. R., and H. G. Sisler. 1964. Site of action of cycloheximide in cells of *Saccharomyces pastorianus*. II. The nature of inhibition of protein synthesis in a cell-free system. Biochim. Biophys. Acta **87**:83-89.
21. Smith, L. 1962. Structure of the bacterial respiratory-chain system. Respiration of *Bacillus subtilis* spheroplasts as a function of the osmotic pressure of the medium. Biochim. Biophys. Acta **62**:145-152.
22. Trevithick, J. R., and R. L. Metzenberg. 1964. Invertase isoenzyme formed by *Neurospora* protoplasts. Biochem. Biophys. Res. Commun. **16**:319-325.
23. Van Wijk, R. 1968. Alpha-glucosidase synthesis, respiratory enzymes and catabolite respression in yeast. I. The effects of glucose and maltose on inducible alpha-glucosidase synthesis in protoplasts of *S. carlsbergensis*. Proc. Koninkl. Nederl. Akad. Wetenschap. Ser. C **71**:60-71.
24. Van Wijk, R., J. Ouwehand, T. Van Den Bos, and V. V. Koningsberger. 1969. Induction and catabolite repression of α-glucosidase synthesis in protoplasts of *Saccharomyces carlsbergensis*. Biochim. Biophys. Acta **186**:178-191.
25. Villanueva, J. R. 1966. Protoplasts of fungi. p. 3-66. *In* G. C. Ainsworth and A. S. Sussman (ed.), The fungi, vol. 2. Academic Press Inc., N.Y.
26. Vogel, H. J. 1956. A convenient growth medium for Neurospora (medium N). Microb. Genet. Bull. **13**:42-43.
27. Wolf, A. V., and M. G. Brown. 1969. Concentrative properties of aqueous solutions: conversion tables, p. D-171-D-210. *In* R. C. Weast (ed.), Handbook of chemistry and physics, 50th ed. The Chemical Rubber Co., Cleveland, Ohio.

11

Reprinted from *J. Bacteriol.*, **114**(2), 549–555 (1973)

Osmotic Pressure in *Escherichia coli* as Rendered Detectable by Lysozyme Attack

PAUL SCHEIE

Department of Biophysics, The Pennsylvania State University, University Park, Pennsylvania 16802

Received for publication 8 January 1973

The enhanced susceptibility of plasmolyzed *Escherichia coli* to lysozyme attack was used to estimate the internal osmotic pressure of these cells under various conditions. Differences were detected between strains, culture media, stages in the growth cycle, and the osmotically active material used to produce plasmolysis. Lysozyme also was found to attack unplasmolyzed cells at 0 C and between 50 and 70 C.

The osmotic pressure of bacteria is believed to afford a measure of the concentration of small, free solute molecules contained within the semipermeable plasma membrane. This concentration helps establish the internal environment in which the chemistry of the cell takes place. Moreover, the differential osmotic pressure, that is, the difference between the osmotic pressure in the cell and that of the medium, will determine how firmly the plasma membrane is pressed against the rigid peptidoglycan layer. This differential osmotic pressure thereby may affect one or more of the several membrane-related activities.

Studies on the osmotic properties of *Escherichia coli* generally have involved varying the osmotic pressure of the medium and observing the response of the cells. Envelopes of *E. coli* cells are subject to alteration by osmotic stress. Hypotonicity leads to the release of small metabolites (11) and several envelope-bound enzymes (8) as well as to the formation of finger-like protrusions (2) and, perhaps, lysis (12). Hypertonicity leads to plasmolysis during which the plasma membrane shrinks away from the rigid peptidoglycan layer (5, 13). The osmotic pressure of the medium at which plasmolysis first becomes microscopically evident generally is cited as the internal osmotic pressure of the cell. It is tacitly assumed that the plasma membrane is free to shrink away from the cell envelope, at least in some regions. Values so obtained range from 2 atm for resting cells to 15 atm for rapidly growing cells (10). It is, of course, difficult and tedious to quantitate the prevalence of plasmolysis for large numbers of cells by microscope examination. Furthermore, several of the studies on plasmolysis have involved washed cells from which many osmotically active particles probably had been removed. Hence, the osmotic pressure of *E. coli* should not be considered a well-determined parameter.

Osmotic injury in the form of plasmolysis renders *E. coli* susceptible to lysozyme attack (3). Subsequent dilution of such cells in a hypotonic medium can result in the formation of spheroplasts or cell lysis. Lysis, in turn, is easily monitored by turbidity measurements. While this method is commonly utilized in the preparation of protoplasts or gently lysed cells, little attention has been given to the possibility of using the same technique to measure the internal osmotic pressure of *E. coli* or as an assay for certain injuries, osmotic or other, to the cell envelope. Thermal damage, for instance, would seem to be a likely candidate for detection. Thermal inactivation of cells is believed by some to involve a leaky plasma membrane (12), and microscopic observations have shown heat-induced alterations in the appearance of the outer membrane (6). Moreover, one might anticipate a role for internal osmotic pressure in thermal inactivation to the extent that outward pressure on a thermally weakened envelope effects irreversible changes

Osmotic pressure in *E. coli* cells subjected to various conditions and assayed by susceptibility to lysozyme attack are reported here. Differences were found between strains, culture media, stages in the growth cycle, and osmotically active materials used in the assay. Investigation of the effect of lysozyme on heated cells indicated an increased susceptibility to attack between 50 and 70 C, in addition to which evidence was found for damage to the envelopes of some cells when placed at 0 C.

MATERIALS AND METHODS

E. coli strains B/r, B_{s-1}, and W3110 obtained from Stanley Person were grown at 37 C in 10, 15, or 25 ml of medium bubbled with air. Growth media used were nutrient broth (Difco) at 8 g/liter and Roberts C-minimal medium with 5% glucose (11).

Microscope observations were carried out with a Bausch and Lomb bright-field phase-contrast microscope at a magnification of ×900.

Osmotic effects. Samples (5 ml) of cultures in exponential phase were placed in 12-ml heavy-wall centrifuge tubes (Sorvall). The initial absorbance (~0.2) of the suspensions in these tubes was determined at 425 nm with a Bausch and Lomb spectrophotometer, model 20, after which the cells were centrifuged at 8,000 × *g* at room temperature for 3 min. The supernatant fluid was decanted and the inside walls of the tubes were blotted to remove most of the remaining liquid. The tubes were than placed in a bath at the appropriate temperature. Room temperature was approximately 25 C; a stable 11 C bath was obtained with running cold tap water; and the 0 C bath consisted of an ice slurry.

Two to three minutes were allowed for the pellet to reach the desired temperature. The desired sucrose or salt solution (0.1 ml), maintained at the same temperature as the pellet, was added to the tubes, and the pellets were suspended by a few seconds of shaking on a Vortex mixer. A 0.1-ml sample of egg white lysozyme [56 μg/ml in 0.0005 M tris (hydroxymethyl)aminomethane (Tris) at pH 7.2] was added and the tube was reshaken. After an additional 30 to 60 s in the bath the tubes were removed and the cells were diluted. This dilution was accomplished with 5 ml of nutrient broth at 25 C for cells grown in nutrient broth and with 5 ml of double-distilled water for cells grown in the C-minimal medium. The absorbance was measured immediately following dilution and again after 5 to 10 min. It was found that little change occurred after 5 min. Results were expressed in terms of the ratio of final absorbance/initial absorbance which, multiplied by 100, is per cent initial absorbance. The mean values of 3 to 8 separate measurements were plotted as a function of solute concentration along with error bars indicating estimates of the standard deviation of the mean (16).

One departure from this procedure began with cells grown in nutrient broth to an initial absorbance of ~0.65 when the osmotic pressure of cells from cultures after the exponential phase was measured. For another set of experiments cells grown in C-minimal medium were washed twice with double-distilled water before being subjected to sucrose and lysozyme. The initial absorbance was measured prior to pelleting the cells after the second suspension in water. The lysozyme applied to these washed cells was dissolved in 0.01 M Tris at pH 7.2.

High-temperature effects. Cells to be subjected to high temperature were grown in C-minimal salts to a concentration of 1.5 to 2 × 10^8 cells/ml (absorbance ≃0.4). A 2-ml amount of this suspension was added to a preincubated test tube containing 8 ml of C-minimal medium lacking the $MgCl_2$ and Na_2SO_4 but with lysozyme and ethylenediaminetetracetic acid (EDTA) added to give final concentrations of 56 μg/ml and 4%, respectively. The tubes were removed at intervals from the bath and the absorbance was measured at 425 nm. The first readings were made 20 s after adding the cells to the hot medium and these served as the values for initial absorbance. The measurement process required removing the tubes from the bath for about 15 s per reading.

RESULTS

Osmotic pressure and differential pressure. The decrease in absorbance of lysozyme-treated *E. coli* B/r after subjection to various concentrations of sucrose at room temperature (~25 C) is indicated in Fig. 1. Figure 2 shows the two positions in the growth cycle from which samples were taken. Plasmolysis and lysis first became evident in exponential phase cells at about 0.2 M sucrose and by 0.5 M sucrose the maximum amount of lysis had been attained. These results agreed with those reported by Birdsell and Cota-Robles (3), who used a similar technique, as well as with those of Scheie (13) who noted the percentage of cells plasmolyzed by counting under a microscope. The residual absorbance, even at high concentrations of sucrose, might be accounted for in one or more ways. Ghosts of either cells or spheroplasts contributed a small amount to the absorbance. On the other hand, up to 20% of the cells may have been resistant to plasmolysis; or, it is possible that some of the cells were permeable to the sucrose after lysozyme attack and did not lyse; or perhaps the outer portions of the cell envelope on some of the cells, even if plasmolyzed, were impermeable to lysozyme and, consequently, the lysozyme still could not reach its substrate. Microscope observation of exponential-phase cells treated with 0.8 M sucrose and lysozyme indicated the presence of both rods and spheres after dilution into nutrient broth; hence, more than one of the above must have been involved.

E. coli B/r cells from cultures after exponential phase were less susceptible to plasmolysis and lysozyme attack for a given concentration of sucrose. Fifty percent lysis was not exhibited until 1.0 M sucrose was used, at which point the maximum amount of lysis probably had not been reached.

Results such as those in Fig. 1 make it clear that all cells in a culture did not have the same osmotic pressure. Nevertheless, there is some

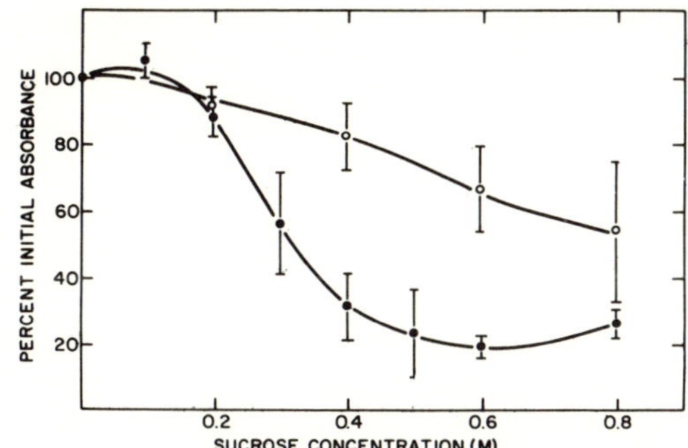

FIG. 1. *Lysozyme-produced change in absorbance for cells of E. coli B/r grown in nutrient broth and treated at 25 C with various concentrations of sucrose. Initial absorbance of 0.2 (●) and 0.65 (○).*

advantage to selecting some single value for comparative purposes. Traditionally, the sucrose concentration at which plasmolysis is observed was used but may not be the most appropriate value. A more suitable choice would be the osmotic pressure of the sucrose solution at which there was a 50% decrease in final absorbance. From Fig. 1 this would correspond to 0.31 M sucrose, or 7.6 atm (1) for exponential phase *E. coli* B/r. Considering that the medium in which these cells were grown had an osmotic pressure of about 30 mosmol, or 0.7 atm (13), it follows that the differential pressure (the difference between internal and external osmotic pressure) in these cells grown in nutrient broth was approximately 6.9 atm.

By the same criterion, cells with an initial absorbance of 0.65 had an osmotic pressure of almost 35 atm. Microscope observation of these cells after subjection to 1 M sucrose confirmed that many cells were not plasmolyzed and that the plasmolysis in evidence was not as severe as that produced in exponential-phase cells with 1 M sucrose (14). Rods, spheres, and ghosts of both were visible after dilution.

Similar data for strains B_{s-1} and W3110 in exponential phase are presented in Fig. 3. The osmotic pressure, using the above criterion, in B_{s-1} was 0.21 M sucrose or 5.0 atm and that in W3110 was 0.41 M sucrose or 10.5 atm. It may be significant that strain W3110 cells normally appeared to have a smaller cross section than either B/r or B_{s-1} cells; thus their cytoplasm may have been more concentrated.

In Fig. 4 data are shown for strain B/r grown in Roberts C-minimal medium and for strain B/r grown in C-minimal medium but washed twice before being subjected to sucrose and lysozyme. The cells grown in C-minimal medium had an osmotic pressure of 17 atm (0.6 M sucrose), more than twice that of cells grown in nutrient broth. Since C-minimal medium has an osmotic pressure of about 7.2 atm (13), the differential pressure was almost 10 atm. This is 40% higher than the differential pressure for the same cells grown in nutrient broth. Washed cells, on the other hand, showed a markedly lower osmotic pressure, which illustrates one of the hazards of obtaining information of an osmotic nature from washed cells.

A decrease in the final absorbance of cells grown in C-minimal medium was observed even with the sucrose treatment deleted. This is believed to have been due to a hypotonic osmotic shock, brought about when the cells were resuspended in the lysozyme medium,

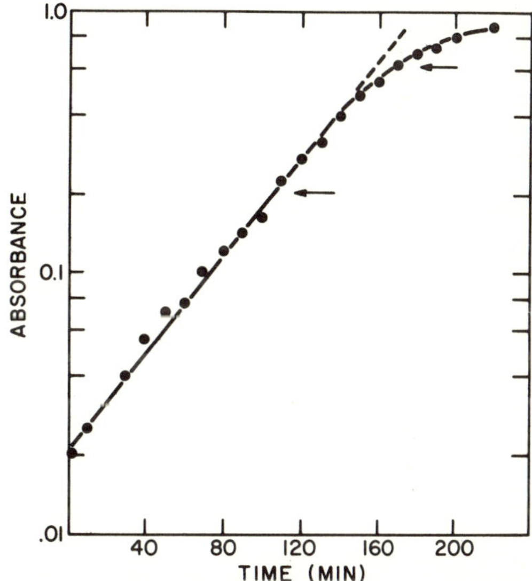

FIG. 2. *Growth of E. coli B/r in nutrient broth at 37 C. Arrows indicate times at which cells were removed for treatment with sucrose and lysozyme.*

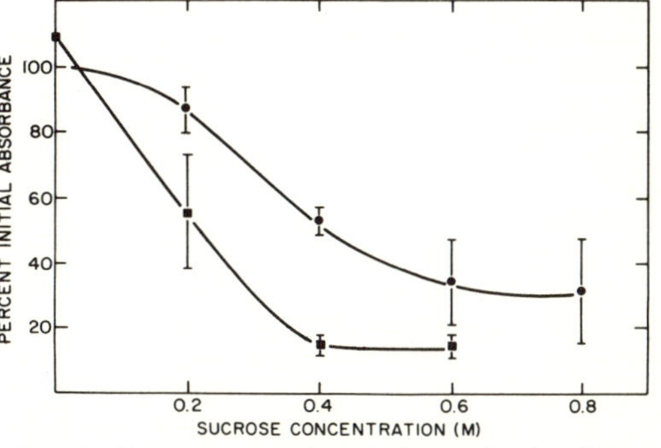

FIG. 3. *Lysozyme-produced change in absorbance for cells of E. coli W3110 (●) and E. coli B_{s-1} (■) grown in nutrient broth and treated at 25 C.*

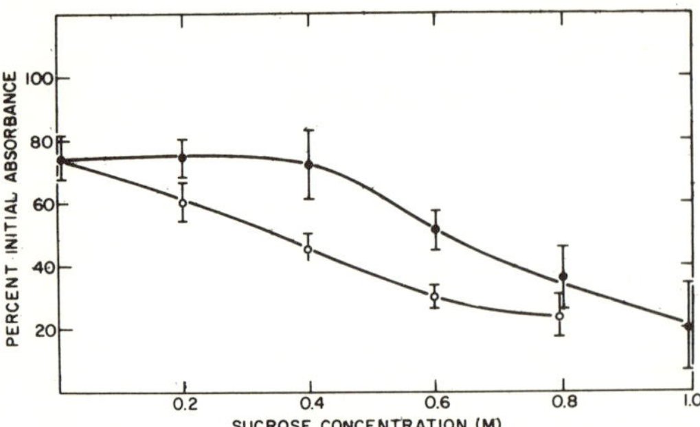

FIG. 4. *Lysozyme-produced change in absorbance for cells of E. coli B/r grown in Roberts C-minimal medium (●) and E. coli B/r grown in C-minimal medium but washed twice with water before treatment* (O).

that injured the envelope and left it susceptible to lysozyme attack.

Salts also are osmotically active and one expects them to produce plasmolytic effects similar to those caused by sucrose. Figure 5a shows the decrease in absorbance produced by resuspending *E. coli* B/r cells (grown in nutrient broth) in various concentrations of NaCl or $MgCl_2$ prior to adding lysozyme. Since dissociation of salts produces a greater effect than does sucrose for a given molar concentration, Fig. 5b presents the same results but with salt concentrations expressed in terms of the concentration of sucrose that would give the same freezing point depression (1). There was little difference between effects produced by NaCl and sucrose. $MgCl_2$ was not as effective in potentiating cell lysis although microscope examination of cells suspended in 0.6 M $MgCl_2$ showed them to be extensively plasmolyzed. Thus, lack of lysis was not due to absence of plasmolysis. Moreover, the $MgCl_2$-treated cells were predominantly rod shaped after the final dilution which suggests that lysozyme did not reach its substrate. Mg^{2+} is believed to be important to the integrity of the outer membrane (9), and it is conceivable that the excess Mg^{2+} in the medium caused the outer membrane to become less permeable to the lysozyme.

Deplasmolysis. Plasmolysis is known to be a transient phenomenon, lasting only a few minutes in 0.4 M sucrose (13). Lysozyme attack was used in an attempt to follow deplasmolysis. Pellets of *E. coli* B/r grown in nutrient broth were suspended in 0.2 or 0.4 M sucrose and then incubated at 37 C for various intervals before lysozyme was added. No increase in final absorbance could be detected for periods in sucrose of up to 30 min, although cells observed under phase optics had lost all visible signs of plasmolysis within 10 min. Presumably, the restoration of the original cytoplasmic space did not entail a return to normal contact between the plasma membrane and the remainder of the cell envelope.

Deplasmolysis takes place even faster if nutrients are present when the cells are subjected to sucrose (15). Figure 6 indicates that the injury to B/r cells grown in nutrient broth and suspended in sucrose plus nutrient broth was slightly greater than what would have been expected on the basis of the additional osmotic pressure due to the nutrient broth. Microscope observation showed no signs of plasmolysis so, again, lysozyme attack indicated the existence of an injury in the cell envelope persisting beyond the time at which recovery was assumed to have taken place.

Thermal effects. The lysis produced by treating *E. coli* B/r with sucrose and lysozyme at different temperatures is indicated in Fig. 7. Data not shown for 37 C closely resembled

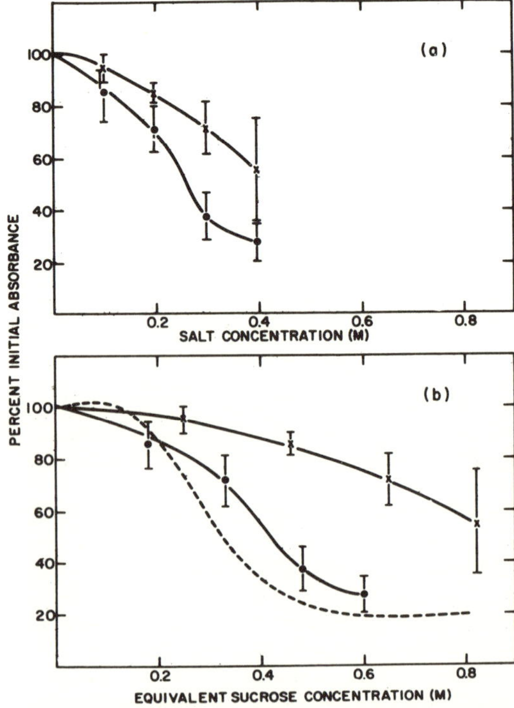

FIG. 5. *Lysozyme-produced change in absorbance for cells of E. coli B/r grown in nutrient broth and treated at 25 C with NaCl (●) and with $MgCl_2$ (X). (a) Salt concentration expressed in molarity; (b) salt concentration expressed in terms of molarity of sucrose producing the same freezing point depression. Dotted line represents data from Fig. 1 for cells treated with sucrose.*

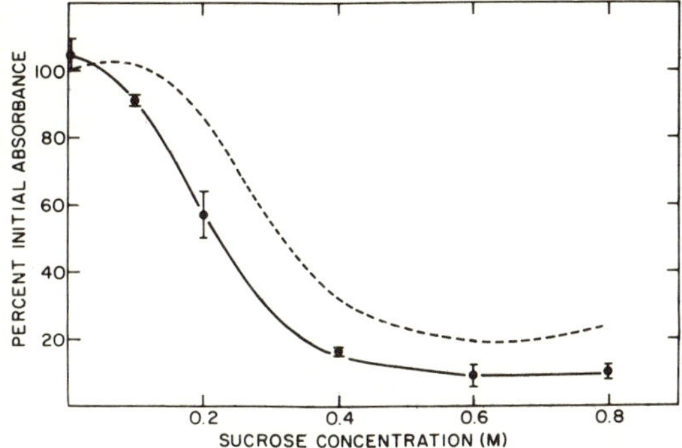

FIG. 6. *Lysozyme-produced change in absorbance of cells of E. coli B/r grown in nutrient broth and treated at 25 C with sucrose in the presence of nutrient broth. Dotted line shows the data from Fig. 1 for cells treated with sucrose.*

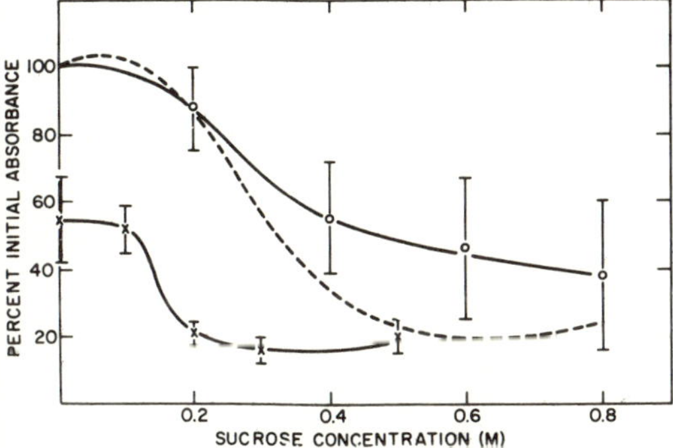

FIG. 7. *Lysozyme-produced change in absorbance for cells of E. coli B/grown in nutrient broth and treated with sucrose at 11 C* (O) *and at 0 C* (×). *Dotted line represents data from Fig. 1 in which cells were treated at 25 C.*

those for 25 C. The most striking feature, one exhibited by all three strains, was that at 0 C no sucrose was needed for lysozyme to cause lysis. Resuspension of the cooled cells in nutrient broth with no lysozyme present did not produce any decrease in absorbance, nor was any noted for similarly treated cells grown in C-minimal medium. Between 11 and 0 C the envelope of *E. coli* cells grown in nutrient broth underwent some alteration that increased susceptibility to lysozyme attack. This effect is currently under investigation.

Implication by other investigators (6, 12) for participation of the cell envelope in thermal inactivation at temperatures greater than 50 C suggested that an increase in susceptibility to lysozyme might be found at similar temperatures. Cell suspensions heated to 55 C for 10 min, cooled to room temperature, and then treated as above did not respond with a decrease in absorbance. Although this was consistent with a similar finding by Harries and Russell (7), it merely indicated that after cooling no increased susceptibility could be detected. Cells grown in C-minimal medium and then heated in C-minimal medium lacking Mg^{2+} but containing lysozyme did suffer a decrease in absorbance. Figure 8 shows the enhancement of this effect with EDTA present at a concentration of 0.4%. Some loss in absorbance was noted, even without lysozyme or EDTA, and probably resulted from thermally induced lysis, which is consistent with other suggestions of such lysis (12). Results of heating and treating the three strains, B/r, B_{s-1}, and W3110 are shown in Fig. 9. Strain B_{s-1}, reportedly more thermally sensitive than B/r (4), showed less sensitivity to lysozyme attack. Perhaps the cause of such increased thermal sensitivity was a more easily disrupted envelope which became leaky and therefore not subject to hypotonic lysis after lysozyme attack. No lysis was detected in cells heated to 70 C. This may have been due to heat fixation of the cells after which they could no longer respond osmotically. Measurements of similar effects in nutrient broth were not possible as the lysozyme formed a precipitate in the broth that prevented accurate measurement of the absorbance of the cells.

DISCUSSION

Determination of osmotic pressure by lysozyme attack has distinct advantages over techniques utilizing microscopy. A quantitation is possible that would require considerable time

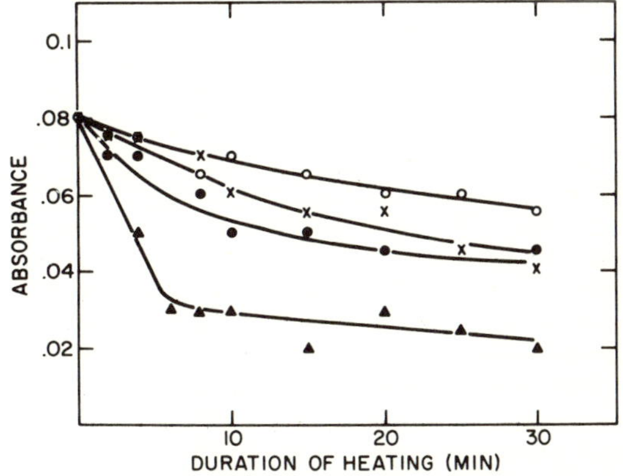

FIG. 8. *Absorbance of E. coli B/r grown in C-minimal medium and heated at 55 C in the presence of C-minimal medium* (O), *C-minimal medium minus the $MgCl_2$ and with EDTA added* (●), *C-minimal medium plus lysozyme* (×), *and C-minimal medium minus $MgCl_2$ plus EDTA and lysozyme* (▲).

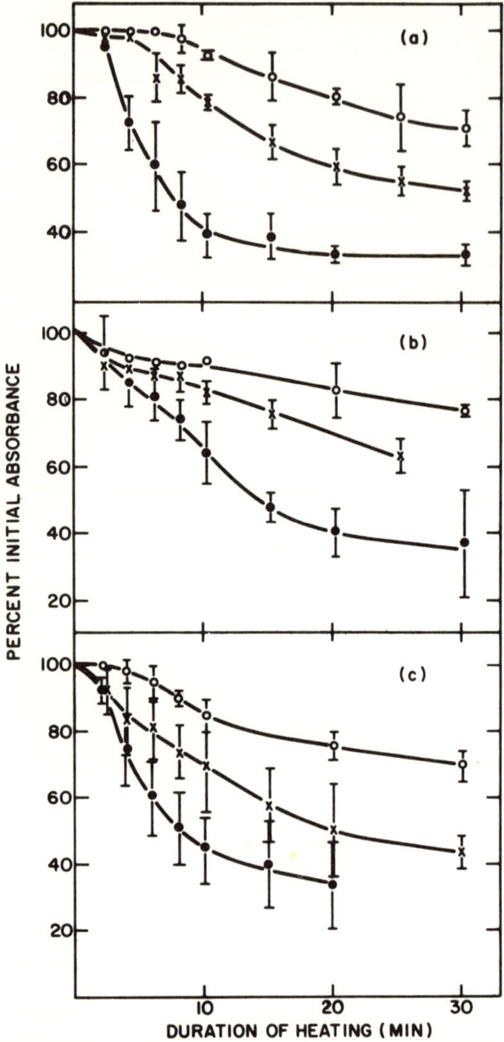

FIG. 9. *Absorbance decrease for (a) E. coli B/r, (b) E. coli* B_{s-1}*, and (c) E. coli W3110 grown in C-minimal medium and heated in C-minimal medium minus the* $MgCl_2$ *and* Na_2SO_4 *but in the presence of lysozyme and EDTA at 51 C* (O), *53 C* (×), *and 55 C* (●).

and effort with a microscope. In addition, lysozyme apparently can be used to detect osmotic damage not visible with a light microscope.

The osmotic pressure of bacteria often has been used by investigators to produce or prevent major changes in the cell. This includes most studies of plasmolysis, hypotonic shock, and the conversion of cells to an osmotically sensitive form. Values such as those reported here should assist future pursuit of similar studies. At the same time, one might inquire as to how osmotic pressure is used by the cell. Does it have a causative or regulatory role, or is it merely the result of other processes or properties more fundamental? Differences have been pointed out here between strains and growth media that, while they do not provide an answer, do suggest that such a line of investigation may now be worth pursuing.

One of the most interesting results is the apparent range of osmotic pressures exhibited in every situation presented. Some cells growing in nutrient broth plasmolyzed in the presence of just over 0.1 M sucrose while others required 0.5 M sucrose. This corresponds to a five-fold difference either in the internal concentration of osmotically active particles as assumed here, in the permeability of *E. coli* to sucrose, or in the degree to which the plasma membrane is fastened to the cell wall. Moreover, there were alsays some cells that did not lyse even after being subjected to 1.0 M sucrose. No other physical parameter comes to mind that exhibits such a variation in an exponential-phase culture. A study of the osmotic pressure in synchronized cells might prove instructive.

The data also suggest that osmotic pressure plays a role in thermal inactivation. Cells with a high internal osmotic pressure (grown in C-minimal medium) lysed in the presence of lysozyme and EDTA at temperatures similar to those associated with ribonucleic acid (RNA) degradation, leakage of metabolites (12), and deoxyribonucleic acid (DNA) strand breaks (18). Some lysis occurred even without lysozyme or EDTA, which reinforces the suggestion made elsewhere that up to 20% of those cells maintained in a salt medium may lyse at these temperatures while washed cells heated in water do not (12). Washing cells in water leads to the release of metabolic pools (11) and a decrease in osmotic pressure as reported here. It follows that the differential osmotic pressure would likewise be reduced.

A plausible hypothesis for thermal death might include some critical temperature at which a portion of the rigid cell wall is weakened so that it can no longer restrain the plasma membrane which ruptures, whereupon enough cytoplasmic material is released to equalize the internal and external pressures. The point of rupture would provide the entry point for lysozyme. The blebs which de Petris (6) found in the outer membrane of some heated cells may have been formed at the site of such a rupture and one could acount for EDTA-enhanced lysis by noting that, if the EDTA removes portions of the outer membrane as believed (9), the point of rupture would become more rapidly accessible to attack by lysozyme. Furthermore, it also has been reported that *Aerobacter aerogenes* cells heated in water are more resistant to inactivation than are those heated in a growth-supporting medium (17).

One might reasonably expect a lesion in the membrane to heal once the pressure differential is removed. However, subsequent degradation of RNA may again increase the internal osmotic pressure, to the point of producing lysis, providing the integrity of the peptidoglycan layer had been destroyed. According to such an hypothesis one would expect lysozyme to be ineffective when added after cells had been cooled. What remains, of course, is the question as to which of the various heat-induced phenomena is the primary cause of thermal inactivation.

LITERATURE CITED

1. American Optical Co. 1964. Tables of properties of aqueous solutions related to index of refraction. American Optical Co., Buffalo, N.Y.
2. Bayer, M. E. 1967. Response of cell walls of *Escherichia coli* to a sudden reduction of the environmental osmotic pressure. J. Bacteriol. **93:**1104–1112.
3. Birdsell, D. C., and E. H. Cota-Robles. 1967. Production and ultrastructure of lysozyme and ethyleneaminetetraacetate-lysozyme spheroplasts of *Escherichia coli*. J. Bacteriol. **93:**427–437.
4. Bridges, B. A., M. J. Ashwood-Smith, and R. J. Munson. 1969. Correlation of bacterial sensitivities to ionizing radiation and mild heating. J. Gen. Microbiol. **58:**115–124.
5. Cota-Robles, E. H. 1963. Electron microscopy of plasmolysis in *Escherichia coli*. J. Bacteriol. **85:**499–503.
6. de Petris, S. 1967. Ultrastructure of the cell wall of *Escherichia coli* and chemical nature of its constituent layers. J. Ultrastruc. Res. **19:**45–83.
7. Harries, D., and A. D. Russell. 1967. A note on some changes in the physical properties of heated suspensions of *Escherichia coli* J. Pharm. Pharmacol. **19:**740–743.
8. Heppel, L. A. 1967. Selective release of enzymes from bacteria. Science **156:**1451–1455.
9. Leive, L., U. K. Shoulin, and S. E. Mergenhagen. 1968. Physical, chemical, and immunological properties of lipopolysaccharide released from *Escherichia coli* by ethylenediaminetetraacetate. J. Biol. Chem. **243:**6384–6391.
10. Mitchell, P., and J. Moyle. 1956. Osmotic function and structure in bacteria. Symp. Soc. Gen Microbiol. **6:**150–180.
11. Roberts, R. B., P. H. Abelson, D. B. Cowie, E. T. Bolton, and R. J. Britten. 1957. Studies of biosynthesis in *Escherichia coli*. Carnegie Inst. Wash. Publ. 607.
12. Russell, A. D., and D. Harries. 1968. Damage to *Escherichia coli* on exposure to moist heat. Appl. Microbiol. **16:**1394–1399.
13. Scheie, P. O. 1969. Plasmolysis of *Escherichia coli* B/r with sucrose. J. Bacteriol. **98:**335–340.
14. Scheie, P., and H. Dalen. 1968. Spatial anisotropy in *Escherichia coli*. J. Bacteriol. **96:**1413–1414.
15. Scheie, P., and R. Rehberg. 1972. Response of *Escherichia coli* B/r to high concentrations of sucrose in a nutrient medium. J. Bacteriol. **109:**229–235.
16. Snedecor, G. W. 1956. Statistical methods, p. 38. The Iowa State University Press, Ames, Iowa.
17. Strange, R. E., and M. Shon. 1964. Effects of thermal stress on viability and RNA of *Aerobacter aerogenes* in aqueous suspension. J. Gen. Microbiol. **34:**99–114.
18. Woodcock, E., and G. W. Grigg. 1972. Repair of thermally induced DNA breakage in *Escherichia coli*. Nature N. Biol. **237:**76 79.

12

Reprinted from *J. Bacteriol.*, **117**(1), 261–264 (1974)

Membrane Permeability and the Loss of Germination Factor from *Neurospora crassa* at Low Water Activities

GISELA CHARLANG AND N. H. HOROWITZ

Division of Biology, California Institute of Technology, Pasadena, California 91109

Neurospora crassa conidia incubating in buffer at low water activities (a_w) release a germination-essential component as well as 260-nm absorbing and ninhydrin-positive materials, regardless of whether an electrolyte or non-electrolyte is used to reduce a_w. Chloroform and antibiotics known to increase cell-membrane permeability have a similar effect. This suggests that membrane damage occurs in media of low a_w and that an increase in permeability is responsible for the release of cellular components. The damage caused in media of low a_w is nonlethal in most cases, and the conidia recover when transferred to nutrient medium.

Living organisms have a high and, within narrow limits, apparently irreducible requirement for water. Most microorganisms are unable to grow in media whose water activity (a_w) is below 0.9. A few unusual bacteria, the extreme halophiles (7, 18), grow at a_w 0.75. Growth of some fungi has been reported at a_w 0.62 (18), but the growth is so slow even under ideal conditions of temperature and nutrition that its biological significance is doubtful. Some very dry natural environments, such as locales in the dry valleys of Antarctica where the prevailing a_w is 0.45 or less at 0 C, are abiotic (6). In many other regions of the world, also, water is the life-limiting factor. Yet little is known about what determines the water requirements of microorganisms, or what cellular responses result when these requirements are not met.

Charlang and Horowitz (3) have presented evidence for the existence of a substance in conidia of *Neurospora crassa* which is essential for their germination and which is lost from the cells in media of low a_w. The substance, called "germination factor" (GF), can be extracted from conidia or mycelium. When added to media of low a_w, it shortens the long germination lag that is observed in these media. GF activity is found in extracts of other fungi, as well—e.g., *Aspergillus nidulans* and *Penicillium chrysogenum*—but not yeast; *Escherichia coli* contains a trace (unpublished data from this laboratory). GF is not present in complete medium (see composition under Materials and Methods).

GF loss occurs in media of low a_w regardless of whether an electrolyte or non-electrolyte is used to reduce a_w. In this paper, we present evidence that a nonlethal alteration of the permeability of the cell membrane in such media is responsible for the loss of GF, along with other molecules. A similar, but usually lethal, effect is obtained when antibiotics known to damage the membrane specifically are used.

MATERIALS AND METHODS

Wild-type *N. crassa*, strain 74A, was grown on slants or in 125-ml flasks of complete medium [Vogel minimal medium N (19) supplemented with 0.5% yeast extract, 0.25% casein hydrolysate, and 1.5% agar] in the manner previously described (3).

Experimental procedures. Conidia from 4-day-old cultures were harvested by suspension in distilled water and filtered through four layers of cheesecloth to remove mycelial debris. They were washed once and suspended in 0.067 M sodium phosphate buffer, pH 6.0, before being counted in a haemocytometer. A total of 10^8 conidia/ml were incubated in buffer at low a_w or in buffer with added antibiotics or chloroform, at 30 C for up to 48 h with shaking. At various time intervals, samples were taken, and the conidia were removed by centrifugation and Millipore filtration. The cell-free solutions were tested for GF activity, 260-nm absorbance, and ninhydrin reaction.

The a_w of the buffer (initially 0.998) was lowered to 0.938, 0.908, or 0.843 by adding NaCl or glycerol. The required amounts of solutes were calculated from the tables of Robinson and Stokes (17), Scott (18), and Ingram (7).

Certain antibiotics have been shown to damage the cell membrane with a resulting increase in permeability (5). Of these, we used nystatin (8, 9, 10) at 5 μg/ml, polymyxin B (5, 8, 15) at 50 μg/ml, and tyrocidine (12) at 20 μg/ml. Nystatin was decomposed photochemically (10) before GF assays were carried out. Nystatin and polymyxin were obtained from Sigma Chemical

Co., and tyrocidine was obtained from Nutritional Biochemicals Corp.

Assays. The GF assay was essentially the one previously described (3). Samples to be tested were added to 50 ml of complete medium plus NaCl at a_w 0.934. Inoculation was with 10^2 conidia/ml, and the flasks were incubated with shaking at 30 C for 90 h.

The maximum absorbance in the 250- to 260-nm range was read from spectral curves obtained on a recording Cary 15 spectrophotometer. The method of Moore and Stein (13) was used for determinations of ninhydrin-positive material.

RESULTS

Experiments at low a_w. Loss of GF from conidia is correlated with loss of 260-nm absorbing and ninhydrin-positive materials. This is shown in Table 1 for solutions of lowered a_w and in Table 2 for substances known to damage the cell membrane. Germination does not occur during 48-h exposure to any of the solutions listed in these tables, except for the tyrocidine case discussed below.

When conidia are incubated in buffer at a_w 0.998, there is some loss of 260-nm absorbing and ninhydrin-positive materials into the medium, but there is no detectable loss of GF. At low a_w, however, GF is released along with larger amounts of 260-nm absorbing and ninhydrin-positive substances (Table 1). It should be noted that the loss of materials from the cells is both a_w- and time-dependent.

All of the solutions (except buffer) listed in

TABLE 1. *Amounts of GF activity, 260-nm absorbing and ninhydrin-positive materials released from conidia at different levels of a_w*

Suspending medium	Molal concentration of solute (NaCl or glycerol)	a_w of suspending medium	GF activity[a]			OD at 260 nm[b]			Ninhydrin-positive material[c]		
			6 h[d]	24 h	48 h	6 h	24 h	48 h	6 h	24 h	48 h
Buffer		0.998	0.0	0.0	0.0	0.13	0.23	0.33	0.014	0.014	0.025
Buffer plus NaCl	1.78	0.938	1.5	5.0		0.20	0.24		0.068	0.12	
	2.57	0.908	6.3	10.5	18.3	0.195	0.30	0.74	0.075	0.164	0.242
	4.15	0.843	6.3	9.4	15.4	0.190	0.33	0.84	0.06	0.151	0.259
Buffer plus glycerol	3.32	0.938	0.0	2.7	2.2	0.12	0.275	0.68	0.039	0.045	0.05
	5.0	0.908	1.3	5.0	10.8	0.17	0.33	1.06	0.05	0.078	0.156
	8.75	0.843	1.9	8.6	47.3	0.16	0.77	3.30	0.092	0.28	0.99

[a] GF bioassay of 0.25 ml of incubation mixture. Amounts are expressed in mg of mycelial dry weight, average of two flasks.
[b] Peak absorbance in 250 to 260 nm range.
[c] Expressed in mM amino acid equivalents based on a leucine standard.
[d] Time of exposure.

TABLE 2. *Amounts of GF activity, 260-nm absorbing and ninhydrin-positive materials released from conidia by antibiotics or chloroform*

Suspending medium	Concentration of antibiotic (μg/ml)	GF activity[a]			OD at 260 nm[b]			Ninhydrin-positive material[c]		
		3 h[d]	6 h	24 h	3 h	6 h	24 h	3 h	6 h	24 h
Buffer		0.0	0.0	0.0	0.13	0.13	0.23	0.012	0.014	0.014
Buffer plus nystatin	5	trace	4.0	57.7	1.43	2.2	4.9	2.26	2.28	3.21
Buffer plus polymyxin B	50		45.0	69.5		4.3	11.1		2.9	6.5
Buffer plus tyrocidine	20	46.4	1.8	1.2	2.83	1.43	0.58	1.35	0.36	0.067
Buffer plus chloroform	saturated	72.2[e]			6.5[e]			3.05[e]		

[a] GF bioassay of 0.25 ml of incubation mixture. Amounts expressed in mg dry weight (average of 2 flasks).
[b] Peak absorbance in 250 to 260 nm range.
[c] Expressed in mM amino acid equivalents based on a leucine standard.
[d] Time of exposure.
[e] Exposure was for 1 h.

Table 1 cause extensive plasmolysis of conidia. No, or at best only partial, recovery from plasmolysis occurs during a 48-h exposure to the NaCl solutions. It thus appears that the alteration in permeability that is induced by these solutions is not so extensive as to permit free entry of NaCl into the cells. Although plasmolyzed, the cells are nevertheless fully viable, as shown by subsequent plate counts on sorbose medium (2). In the case of glycerol solutions, there is complete recovery from plasmolysis within 48 h at a_w 0.938 and a_w 0.908, and the cells show full viability when plated out. In glycerol at a_w 0.843, however, viability is reduced by about 40%, and recovery from plasmolysis is abnormal. The cells are enlarged and have a transparent center with the protoplasm distributed around the periphery. In contrast, normal conidia tend to be uniformly dense.

The data of Table 1 parallel these observations in that they show that glycerol is more damaging than NaCl at high concentrations. At low concentrations (high a_w), the reverse is true. Our previous growth studies showed that glycerol is less toxic than NaCl at water activities in the neighborhood of 0.9 or higher (3). For example, *N. crassa* 74A will grow very slowly in complete medium plus glycerol at a_w 0.898, but it does not grow at all at the same water activity with NaCl.

Experiments with antibiotics and chloroform. The experiments summarized in Table 2 were performed to test the hypothesis that loss of GF from conidia in media of low a_w results from altered permeability of the cell membrane. A number of antibiotics are known to be specifically active against membranes. Nystatin and tyrocidine have been shown to inhibit germination and growth of *N. crassa* by irreversibly damaging the cell membrane (8, 10, 12). Polymyxin B is known to breach the osmotic barrier, especially in gram-negative bacteria (15), probably by inducing a reorientation of membrane lipids (5). The effect of polymyxin B on *N. crassa* cell membranes has not been studied in detail, but Kinsky (8) has shown that it does inhibit both germination and growth. Chloroform causes extensive damage to plasma membranes (5). It was used in our experiments to induce maximum leakage.

No conidia survive after a 1-h exposure to chloroform-saturated buffer, and values for released GF, 250-nm absorbing, and ninhydrin-positive material are high (Table 2). In nystatin-supplemented buffer, viability of conidia decreases gradually to nearly zero in 24 h. The opposite trend is shown in amounts of material released. Polymyxin B has a similar effect, but there were some survivors after 24 h, and they had germinated in the incubation mixture. It appears that the damage caused to the sensitive fraction of the population was extensive, judging from the large quantities of material released.

The reaction of tyrocidine is unusual. At a concentration of 20 μg and 10^7 conidia/ml, the antibiotic is lethal, viability being reduced to less than 1% after a 2-h exposure. With 10^8 conidia/ml, the same tyrocidine concentration causes an immediate and rapid loss of 260-nm absorbing and ninhydrin-positive material, as well as GF. But after 1 h, the reaction is reversed, and the released material begins to disappear from the medium. After 3 h, the conidia germinate. At 6 h, only traces of GF activity, about half of the 260-nm absorbing and 20% of the ninhydrin-positive materials, remain in the medium; at the same time, germination is essentially complete. After 24 h, no further growth has occurred, but the quantities of intracellular material in the medium have been further reduced. Because germination does not occur in unsupplemented buffer, it appears that tyrocidine—or, more likely, an impurity in the commercial preparation—provides nutrient to the conidia.

Dialysis results. Mach and Slayman (12) found that 44% of the 260-nm absorbing material released by *Neurospora* mycelium during a

TABLE 3. *Amounts of nondialyzable 260-nm absorbing material released from conidia at different levels of a_w or by chloroform*

Suspending medium	Time (h)	OD[a] before dialysis	OD[a] after dialysis	Nondialyzable material (% of total)
Buffer only at a_w 0.998	6	0.192	0.036	18.2
	24	0.194	0.032	16.5
	48	0.230	0.043	18.7
Buffer plus NaCl at a_w 0.843	6	0.159	0.023	14.5
	24	0.250	0.035	14.2
	48	0.415	0.051	12.3
Buffer plus glycerol at a_w 0.843	6	0.338	0.076	22.5
	24	0.791	0.135	17.1
	48	4.69	0.350	7.47
Buffer plus excess chloroform	1	7.86	2.49	31.8

[a] Peak absorbance in 250 to 260 nm range.

2-min exposure to an inactivating concentration of tyrocidine was nondialyzable. We dialyzed 260-nm absorbing material released from conidia exposed to low a_w or chloroform for 24 h at 4 C against four changes of water. The results, presented in Table 3, show that increasing absolute amounts of nondialyzable material are correlated with increased length of exposure to low a_w. The fraction of nondialyzable material in the total 260-nm absorbing material decreases with time. It thus appears that, as an index of damage, the absolute amount rather than the percentage of nondialyzable material is more revealing. On the other hand, with a short exposure to a highly damaging compound like chloroform, one observes the release of a large percentage of nondialyzable material, as Mach and Slayman did with tyrocidine.

DISCUSSION

Exposure of *Neurospora* conidia to media of low a_w results in the release of a germination-essential component, together with 260-nm absorbing and ninhydrin-positive materials. Similar results are obtained regardless of whether an electrolyte or a non-electrolyte is used to reduce a_w. We conclude that reduction of a_w, and not specific solute toxicity, is responsible. Because membrane-specific antibiotics also induce a release of GF and other materials from conidia, it is likely that the loss of intracellular substances at low a_w results from alteration of the cell membrane.

Membrane damage at low a_w may explain the water requirements of many microorganisms, and resistance to such damage may explain xerotolerance. Hypertonic solutions of NaCl appear to cause membrane damage in blue-green algae (1), and high sucrose concentrations make *E. coli* permeable to nucleoside triphosphates (4). The loss of putrescine from *E. coli* growing at moderately elevated osmolarities, however, is not the result of membrane damage (14). Preliminary experiments with *A. nidulans* and *P. chrysogenum* in our laboratory have shown that these species are much more xerotolerant than *N. crassa*: their conidia retain GF and germinate at $a_w < 0.9$. Since their GF is readily diffusible—as shown by its rapid loss from conidia in chloroform-saturated buffer—there is an evident correlation between ability to germinate at low a_w and resistance to permeability changes.

Specific toxicities may also be manifested at high solute concentrations, superimposed on the effects of low a_w. The results presented in this paper show that at the lowest water activities tested, glycerol begins to be lethal, but NaCl does not. In our earlier experiments on the effects of reduced a_w on germination and growth, sucrose was more inhibitory than NaCl, glycerol, or glucose (3). Specific solute toxicities have also been observed in yeast (11, 16).

ACKNOWLEDGMENTS

This work was supported by a grant from the National Aeronautics and Space Administration (NGR 05-002-121).

LITERATURE CITED

1. Batterton, J. C., Jr., and C. Van Baalen. 1971. Growth responses of blue-green algae to sodium chloride concentration. Arch. Mikrobiol. Z. **76**:151–165.
2. Brockman, H. E., and F. J. deSerres. 1964. "Sorbose toxicity" in *Neurospora*. Amer. J. Bot. **50**:709–714.
3. Charlang, G. W., and N. H. Horowitz. 1971. Germination and growth of *Neurospora* at low water activities. Proc. Nat. Acad. Sci. U.S.A. **68**:260–262.
4. Gros, F., J. Gallant, R. Weisberg, and M. Cashel. 1967. Decryptification of RNA polymerase in whole cells of *Escherichia coli*. J. Mol. Biol. **25**:555–557.
5. Harold, M. F. 1970. Antimicrobial agents and membrane functions. Advan. Microbial Physiol. **4**:45–104.
6. Horowitz, N. H., R. E. Cameron, and J. S. Hubbard. 1972. Microbiology of the dry valleys of Antarctica. Science **176**:242–245.
7. Ingram, M. 1957. Micro-organisms resisting high concentrations of sugars or salts. Symp. Soc. Gen. Microbiol. **7**:90–133.
8. Kinsky, S. C. 1961. Alterations in the permeability of *Neurospora crassa* due to polyene antibiotics. J. Bacteriol. **82**:889–897.
9. Kinsky, S. C. 1962. Effect of polyene antibiotics on protoplasts of *Neurospora crassa*. J. Bacteriol. **83**:351–358.
10. Kinsky, S. C. 1962. Nystatin binding by protoplasts and a particulate fraction of *Neurospora crassa*, and a basis for the selective toxicity of polyene antifungal antibiotics. Proc. Nat. Acad. Sci. U.S.A. **48**:1049–1056.
11. Koppensteiner, G., and S. Windisch. 1971. Der osmotische Wert als begrenzender Faktor für das Wachstum and die Gärung von Hefen. Arch. Mikrobiol. Z. **80**:300–314.
12. Mach, B., and C. W. Slayman. 1966. Mode of action of tyrocidine on *Neurospora*. Biochim. Biophys. Acta **124**:351–361.
13. Moore, S., and W. H. Stein. 1954. A modified ninhydrin reagent for the photometric determination of amino acids and related compounds. J. Biol. Chem. **211**:907–913.
14. Munro, G. F., K. Hercules, J. Morgan, and W. Sauerbier. 1972. Dependence of the putrescine content of *Escherichia coli* on the osmotic strength of the medium. J. Biol. Chem. **247**:1272–1280.
15. Newton, B. A. 1956. The properties and mode of action of the polymyxins. Bacteriol. Rev. **20**:14–27.
16. Norkrans, B. 1968. Studies on marine occurring yeasts: respiration, fermentation, and salt tolerance. Arch. Mikrobiol. Z. **62**:358–372.
17. Robinson, R. A., and R. H. Stokes. 1968. Electrolyte solutions. Butterworths, London.
18. Scott, W. J. 1957. Water relations of food spoilage microorganisms. Advan. Food Res. **7**:83–127.
19. Vogel, H. J. 1964. Distribution of lysine pathways among fungi: evolutionary implications. Amer. Natur. **98**:435–446.

Part III

HALOPHILIC INTERACTIONS

Editor's Comments on Papers 13 Through 17

Flannery (Paper 13) defines facultative halophiles as microorganisms that will grow in media containing less than 2 percent NaCl, but grow best in media containing greater percentages. Obligate halophiles grow only in media containing more than 2 percent NaCl. These organisms usually require 15 to 25 percent salt. Larsen (1962) breaks the classification down to include: slight, or growing best in media with 2 to 5 percent NaCl; moderate, or growing best in media containing 5 to 20 percent NaCl; and extreme, or growing best in media containing 20 to 30 percent NaCl. The halophiles should not be confused with halotolerant organisms such as staphylococci, which will grow in the presence of 10 percent or greater concentrations of NaCl but which do not require it. There are two distinct groups of halophilic bacteria, the *Halobacteria* and the *Halococci* (previously called the *Sarcina–Micrococcus* group). Both are characteristically red, pink, or orange in color (Larsen, 1962; Lanyi,

1974). Many are extremely pleomorphic and can appear as rods or cocci depending on the cultural conditions.

Much of the earlier history of the halophiles is reviewed by Flannery (Paper 13). Halophiles have been isolated from many different sources and are associated with food spoilage and industrial processes. Bacteria were isolated from both crude and refined NaCl, and the reddening of salted fish was linked to these organisms very early in this century (Beckwith, 1911; Rappin, 1920; Clayton and Gibbs, 1927; Gibbons, 1937). The reddening of salted hides was also linked to the presence of bacteria (Robertson, 1931; Lockhead, 1934). Halophiles have been isolated from both the Great Salt Lake and the ocean by ZoBell (Smith and ZoBell, 1937; ZoBell, 1946). A much more detailed review is given by Flannery.

A regular and measurable growth response by *Vibrio costicolus* to increasing concentrations of NaCl was demonstrated by Flannery, Doetsch, and Hansen in 1952. The effect of substituting different salts for NaCl was investigated. Growth was obtained without adding NaCl, providing a sufficient amount of a number of other salts was present. This does not always occur; Robinson and Gibbons (1952) found that the moderate halophile *Micrococcus halodenitrificans* failed to grow in any of 11 salts other than NaCl. Other reports have indicated that the growth response of nonobligate halophiles at the optimum a_w diminished following the Hofmeister series Na > K > Li > Mg > Ca (Ingram, 1957).

The ability of bacteria to grow in high concentrations of NaCl has been attributed to the resistance of enzymes to salting out (Ingram, 1947) or to mechanisms which maintain the intracellular salt concentration below that of the medium. Flannery, Doetsch, and Hansen (1953) found that the optimum concentration of NaCl for the oxidation of glucose by *Vibrio costicolus* was 1.2 *M*, which was less than the concentration required for growth. Baxter and Gibbons (1954) compared the glycerol dehydrogenases of *Pseudomonas salinaria, Vibrio costicolus,* and *Escherichia coli,* which are extreme, moderate, and nonhalophilic, respectively. The results suggested that the extremely halophilic bacteria contain high concentrations of salt and that their enzymes function best at these concentrations. In contrast, the moderate and nonhalophilic organisms contained relatively little salt. The enzyme from *P. salinaria* was most active in the presence of 1.5 *M* NaCl, which is approximately half that required for growth. Evidence also was presented by Baxter and Gibbons indicating that the enzymes were more active in the presence of potassium chloride. Robinson, Gibbons, and Thatcher (1952) presented evidence that the moderate halophile *Micrococcus halodenitrificans* maintained a low intracellular salt concentration by an energy-dependent mechanism. Apparently, a fundamental physiological difference existed between extreme halophiles and other organisms, but later reports (Yamada, Shiio, and Egami, 1954) described both salt-requiring and non-salt-requiring enzymes from moderate halophiles.

As a result of the above, Baxter and Gibbons (Paper 14) undertook a comparative study of the effects of sodium and potassium chloride on several enzymes of *Micrococcus halodenitrificans* and *Pseudomonas salinaria*. The activities of isocitric, succinic, malic, and lactic dehydrogenases and cytochrome oxidase were determined. Lactic dehydrogenase and cytochrome oxidase were more active in high concentrations of salt in both the moderate and extreme halophiles. The remaining enzymes were most active at high or low salt concentrations, depending on their isolation from the extreme or moderate halophiles, respectively. The authors concluded that the extreme halophiles have a high intracellular salt content. They suggested that the moderate halophile *M. halodenitrificans* may contain an intracellular salt content of 0.5 *M* and that this level is maintained with the aid of cytochrome oxidase.

The concept that halophiles achieve salt tolerance by exclusion of the salt in the medium was tested (Paper 15) by measurements of the freezing points of cells grown in varying concentrations of 1 to 4 *M* salts. The freezing points of *Micrococcus halodenitrificans, Vibrio costicolus, Halobacterium halobium,* and *Sarcina litoralis* were determined. It was found that the freezing points of both the halophilic and nonhalophilic cells were close to those of the media in which they were grown. Thus neither the extreme nor moderate halophiles maintain an intracellular water activity greater than the surrounding medium. Christian and Waltho (1961, 1962) extended this study and found that the K level generally exceeded that of the medium.

Baxter (1959) found that the lactic dehydrogenase of *Halobacterium salinarium* was unstable and inactive at low solute concentrations. The decrease in rate was first order and inversely dependent on solute concentration. From the data obtained from studies of crude sonic extracts, Baxter suggested

> that halophilic enzymes differ from other enzymes in being rather loosely held in their native, enzymatically active confirmations, so that it is only when the intramolecular elec-

> trostatic repulsions are reduced by the presence of salts that they are able to assume the structure in which they are active as catalysts. On lowering the salt concentration they first expand reversibly to an inactive conformation, and with further lowering of the salt concentration irreversible disruption of the molecule may occur.

Holmes and Halvorson (1965) suggested a similar function for salt in the stabilization of purified malic dehydrogenase from *H. salinarium*.

Hubbard and Miller (Paper 16) describe the purification of NADP-isocitric dehydrogenase from *Halobacterium cutirubrum*. The purified enzyme was rapidly inactivated at low NaCl levels, but 75 percent of the initial activity could be restored by dialysis against 4 *M* NaCl. The active and inactive forms of the enzyme could be distinguished by sedimentation rate, electrophoretic mobility, and elution rate from Sephadex. The authors interpret the faster elution rate from Sephadex of the inactive form as indicating an increase in the molecular radius of the protein by aggregation of several molecules or by expansion of a single molecule. The results tend to support the concept of molecular expansion during inactivation rather than disaggregation into inactive subunits. The properties of the purified enzyme were further defined by Hubbard and Miller (1970). The active enzyme had a $s_{20,w}$ of 5.3 and a molecular weight of 75,000. Removal of NaCl caused an unfolding of the protein to a less dense form with a $s_{20,w}$ of 2.0. Inactivation exposed SH groups which had to be protected with dithiothreitol to allow 90 percent reactivation of the enzyme in 4 *M* NaCl. Wulff, Hubbard, and Miller (1972) studied the circular dichroism of active, inactivated, and reactivated isocitric dehydrogenase from *H. cutirubrum*. The spectrum of the inactivated enzyme did not show a significant α-helix band. The α-helix was restored by dialysis of the inactive protein against 5 *M* NaCl. These results support the earlier conclusion that removal of NaCl allows the enzyme to expand to an inactive form.

Lanyi and his co-workers have extensively studied the electron transport system in *H. cutirubrum*. The cytochrome components were identified (Lanyi, 1968) and the salt dependence of reduced nicotin adenine dinucleotide oxidase was established (Lanyi, 1969). The results suggested a main branch of electron transport including a flavoprotein, reduced by NADH, and a second branch consisting of flavoproteins and cytochrome b(559), which are reduced by NADH and succinate. The branches join at the C-type cytochromes. Lanyi and Stevenson (1969) reported a threefold stimulation of catalase activity in extracts from *H. cutirubrum* by increasing the concentration from 0.5 to 1.5 *M* of monovalent salts and by 0.1 *M* divalent salts. Greater concentrations resulted in inhibition of activity. The order of effectiveness was $MgCl_2 > LiCl > NaCl > KCl > NH_4Cl$, and $LiCl > LiNO_3 > Li_2SO_4$. The magnitude of enzyme inhibition was correlated with their molar vapor pressure depression in aqueous solutions. Stimulation of enzyme activity could be observed when one salt was added at its optimum concentration in the presence of inhibitory concentrations of another salt. The results were interpreted to mean that the effect was not due to a change in water activity but was the result of enzyme–salt interaction.

Lieberman and Lanyi (Paper 17) studied the mode of action of salts on cytochrome oxidase isolated from *H. cutirubrum*. Cytochrome oxidase from this strain required up to 5 *M* NaCl for maximal activity and stability. Enzyme activity in the presence of low concentrations of $MgCl_2$ was more dependent on pH but less sensitive to ethanol or *n*-propanol than enzyme activity in the presence of high NaCl concentrations. The results suggested that hydrophobic forces predominated in promoting enzyme activity and stability in the presence of high NaCl concentrations. Charge shielding promoted partial activity in the presence of low concentrations of $MgCl_2$. The extensive review by Lanyi (1974) should be consulted for details on the salt-dependent properties of proteins from the extreme halophiles. The article by Gibbons (1969) should be consulted for methods of isolation, growth, and requirements of the halophiles.

Halobacterium salinarium is now the correct name for species previously designated *Pseudomonas salinaria, Halobacter salinaria,* and *Halobacterium cutirubrum. Micrococcus denitrificans* is now designated *Paracoccus denitrificans,* and *Sarcina litoralis* as *Halococcus morrhuae.* (See R. E. Buchanan and N. E. Gibbons (eds.), 1974. *Bergey's Manual of Determinative Bacteriology,* 8th ed., The Williams & Wilkins Co., Baltimore, 1974.)

REFERENCES

Baxter, R. M. 1959. An interpretation of the effects of salts on the lactic dehydrogenase of *Halobacterium salinarium. Can. J. Microbiol.* 5(1): 47–57.

———, and N. E. Gibbons. 1954. The glycerol dehydrogenases of *Pseudomonas salinaria, Vibrio costicolus* and *Escherichia coli* in rela-

tion to bacterial halophilism. *Can. J. Biochem. Physiol. 32*(3): 206–217.
Beckwich, T. D. 1911. The bacteriological cause of the reddening of cod and other allied fish. *Zentr. Bakteriol. Parasitenk. I. Orig. Abt. 60*(1): 351–354.
Christian, J. H. B., and J. A. Waltho. 1961. The sodium and potassium content of non-halophilic bacteria in relation to salt tolerance. *J. Gen. Microbiol. 25*(1): 97–102.
———, and J. A. Waltho. 1962. Solute concentrations within cells of halophilic and non-halophilic bacteria. *Biochim. Biophys. Acta 65*(3): 506–508.
Clayton, W., and W. E. Gibbs. 1927. Examination of halophilic micro-organisms. *Analyst 52*(610): 395–397.
Flannery, W. L., R. N. Doetsch, and P. A. Hansen. 1952. Salt desideratum of *Vibrio costicolus,* and obligate halophilic bacterium. I. Ionic replacement of sodium chloride requirement. *J. Bacteriol. 64*(5): 713–717.
———, R. N. Doetsch, and P. A. Hansen. 1953. Salt desideratum of *Vibrio costicolus,* and obligate halophilic bacterium. II. Effect of salts on the oxidation of glucose. *J. Bacteriol. 66*(5): 526–530.
Gibbons, N. E. 1937. Studies on salt fish. I. Bacteria associated with the reddening of salt fish. *J. Biol. Board Can. 3*: 70–76.
———. 1969. Isolation, growth and requirements of halophilic bacteria. *In* J. R. Norris and D. W. Ribbons (eds.), *Methods in Microbiology,* Vol. 3B, pp. 169–183. Academic Press, New York.
Holmes, P. K., and H. O. Halvorson. 1965. Properties of a purified halophilic malic dehydrogenase. *J. Bacteriol. 90*(2): 316–326.
Hubbard, J. S., and A. B. Miller. 1970. Nature of the inactivation of the isocitrate dehydrogenase from an obligate halophile. *J. Bacteriol. 102*(3): 677–681.
Ingram, M. 1947. A theory relating the action of salts on bacterial respiration to their influence on the solubility of proteins. *Proc. Roy. Soc. (London) B134*(875): 181–201.
———. 1957. Micro-organisms resisting high concentrations of sugars or salts. *Symp. Soc. Gen. Microbiol. 7*: 90–133.
Lanyi, J. K. 1968. Studies of the electron transport chain of extremely halophilic bacteria. I. Spectrophotometric identification of the cytochromes of *Halobacterium cutirubrum. Arch. Biochem. Biophys. 128*(3): 716–724.
———. 1969. Studies of the electron transport chain of extremely halophilic bacteria. II. Salt dependence of reduced diphosphopyridine nucleotide oxidase. *J. Biol. Chem. 244*(11): 2864–2869.
———. 1974. Salt-dependent properties of proteins from extremely halophilic bacteria. *Bacteriol. Rev. 38*(3): 272–290.
———, and J. Stevenson. 1969. Effect of salts and organic solvents on the activity of *Halobacterium cutirubrum* catalase. *J. Bacteriol. 98*(2): 611–616.
Larsen, H. 1962. Halophilism. *In* I. C. Gunsalus and R. Y. Stanier (eds.), *The Bacteria,* Vol. 4, pp. 297–342. Academic Press, New York.

Lockhead, A. G. 1934. Bacteriological studies on the red discoloration of salted hides. *Can. J. Res. 10*(1): 275–286.
Rappin, D. 1920. The dangers of table salt. *J. Amer. Med. Assoc. 75*: 618–619.
Robertson, M. E. 1931. A note on the cause of certain red colorations on salted hides and a comparison of the growth and survival of halophilic or salt-loving organisms and some ordinary organisms of dirt and putrefacation on media of varying salt concentrations. *J. Hyg. 31*: 84–95.
Robinson, J., and N. E. Gibbons. 1952. The effect of salts on the growth of *Micrococcus halodenitrificans* n. sp. *Can. J. Botany 30*(2): 147–154.
———, N. E. Gibbons, and F. S. Thatcher. 1952. A mechanism of halophilism in *Micrococcus halodenitrificans. J. Bacteriol. 64*(1): 69–77.
Smith, W. W., and C. E. ZoBell. 1937. Direct microscopic evidence of an autochthonous bacterial flora in Great Salt Lake. *Ecology 18*(3): 453–458.
Wulff, K., J. S. Hubbard, and A. B. Miller. 1972. Reversible inactivation of the isocitrate dehydrogenase from an obligate halophile: changes in the secondary structure. *Arch. Biochem. Biophys. 148*(1): 318–319.
Yamada, T., I. Shiio, and F. Egami. 1954. On the halophilic alkaline phosphomono-esterase. *Proc. Japan. Acad. 30*: 113–115.
ZoBell, C. E. 1946. *Marine Microbiology: A Monograph on Hydrobacteriology.* Chronica Botanica Co., Waltham, Mass., 240 pp.

13

Reprinted from *Bacteriol. Rev.*, **20**(2), 49–66 (1956)

CURRENT STATUS OF KNOWLEDGE OF HALOPHILIC BACTERIA

WILLIAM L. FLANNERY

Department of Microbiology, Baylor University College of Medicine, Houston, Texas

I. INTRODUCTION

For many years there has been considerable interest in a group of microorganisms known as halophiles. Halophiles are the indigenous organisms of salt packs, brines, or bodies of salt water, and until recently investigations in this field were supported mainly by industries using salt as a preservative. The word halophile is derived from two Greek words, "halos" and "philus," meaning, respectively, "salt" and "loving." Halophiles are, then, the salt-loving organisms and they thrive in relatively high concentrations of salt.

The world of microorganisms can be considered, according to their response to salt concentration, as composed of two groups, the halophiles and the nonhalophiles. This apparently simple demarcation between those organisms which thrive in salt media and those which do not is rather misleading. Numerous intermediate organisms add to the complexity of classification by salt requirement. If, however, the intermediate factors are disregarded for the purpose of simplicity and clarity, a table may be devised which separates the nonhalophiles and halophiles into four major groups (table 1). A second problem then presents itself, even in such a simple separation as that shown in table 1: What concentration of salt can be labeled the boundary concentration—the critical concentration? With so many conflicting and varying reports in the literature, this is extremely difficult to determine. Incomplete descriptions of a wide variety of organisms, of which some are most certainly the same organisms with different names, and lack of essential details for currect identification make comparisons difficult.

Recently Robinson (1) selected 2.5 per cent salt as the critical concentration for defining obligate halophiles. Reports have been made of obligate halophiles growing in salt concentrations as low as 0.5 to 1.0 per cent (2, 3). It is even possible, when the proper conditions are utilized, to grow halophiles in the absence of salt (4, 5) or to grow nonhalophiles in high concentrations of salt (6, 7, 8). Many materials in the nutritional environment influence the effect of the salt concentration in that environment. With these factors in mind a critical concentration, which is a compromise, is selected for table 1. Perhaps later, when experimental data substantiate it, a change will be desirable. Some

TABLE 1

Classification of microorganisms based on their growth response in media containing salt

Groups	Growth Response in Media Plus Salt
I. Nonhalophiles,	
A. Salt-sensitive...	Grow only in media containing less than 2 per cent salt
B. Salt-tolerant....	Grow best in media containing less than 2 per cent salt, but will grow in media containing more than 2 per cent
II. Halophiles,	
A. Facultative.....	Will grow in media containing less than 2 per cent salt, but will grow best in media containing more than 2 per cent
B. Obligate........	Grow only in media containing more than 2 per cent salt

authors prefer to subdivide further obligate halophiles into moderate halophiles and extreme halophiles, defining the moderate halophiles as those requiring the lower concentrations of salt and the extreme halophiles (usually red pigmented) as those requiring about 15 to 25 per cent salt. Difficult as it is to select a dividing line between halophiles and nonhalophiles it is even more difficult to subdivide the group of obligate halophiles on the basis of salt requirement.

II. EARLY HISTORY AND IMPORTANCE TO INDUSTRY

Many hundred years ago salt was extracted from sea water by building mud embankments to contain the water brought from the sea by ditches. This sea water was left in the enclosure for some time until the water had evaporated, leaving the salt on the ground. Baas-Becking (9) quotes an ancient Chinese treatise describing this technique for obtaining salt. The treatise, written about 2700 B.C., also tells that the water became red when left a long time. This is probably the earliest report of the red-pigment-producing halophilic organisms. It is also interesting that red and purple salts were not uncommon in those early days: Pliny recorded that red salt came from Egypt and purple salt from Sicily (9). In the same paper Baas-Becking reports that Hugh Todd in 1683 boiled water from a salt spring in Durham, England, to obtain salt. Todd found that the stones in this spring were a red color.

Coloration of salted products has been observed for many years (10). Originally thought to be the result of a chemical reaction only, it was later discovered that some discolorations were caused by microbial activity. Many bacteria have been isolated from crude salt, and, even after completion of the salt refining processes, a large number of organisms still remain (11, 12). Stuart, Frey and James (13) and Gibbons (10) reported comparisons of the numbers of halophiles present in various salts refined by different methods. Both reports agreed that fewer halophilic organisms were found in mined salt than in solar or sea salt.

Halophiles have been isolated from many bodies of water of high-salt content. They have been found in the Dead Sea (14); Great Salt Lake (15, 16, 17, 18); the Liman area, Odessa (19, 20, 21, 22, 23), and in the ocean (24, 25). It is likely that they may be found in any body of water in which the salt concentration is great enough to allow their growth (2).

There is much information available on the red pigmented halophilic bacteria as a result of the interest of two great industries, fishing and tanning. These industries use salt-preservation processes, and the appearance of red-colored slime on their products causes concern. Whether the slime materially injures the products is questionable (13), but it has certainly cost these industries considerable amounts of money.

A. Reddening of Fish

Canning and refrigeration have largely supplanted salting as the major method for preservation of fish, but large quantities of fish are still salted in many localities. Gibbons (10) states that reddening of salted fish was known in all parts of Novia Scotia, the southern part of Prince Edward Island, the Magdalen Islands, and the Bay of Fundy shore of New Brunswick. This reddening in salted fish was sometimes called "pink" or "pinkeye." It was not at all unusual to find red brine and red fish by the time the bottom of a barrel of salted fish was reached in any grocery or market handling this product (26).

One of the pioneers in the investigation of the reddening of salted fish was Farlow (27). Although he did not isolate the organisms that caused this reddening, he did describe them, where they are found, and their effect upon salted codfish. Many investigators have contributed and enlarged upon these studies (28–40). Numerous microorganisms thought to cause reddening have been isolated. Hess (36, 37) and Hess and Gibbons (38), after an extensive survey, suggest that sodium dihydrogen phosphate and sodium benzoate are the best agents to prevent the growth of halophiles on fish.

B. Reddening of Hides

Hides, when removed from animals, are covered with bacteria and will deteriorate before tanning if not preserved in some way. Since water makes up more than half the hide by weight it is a common practice to preserve the hide with salt (41). Occasionally hides may appear to spoil in spite of the salt applied; a shiny, slimy, red coat is seen, covering the flesh side of the hide. This reddening also has received considerable attention and has been called in various localities "red-heat," "frigorifico reddening," "strawberry," "heating," and "reddening."

The reddening of salted hides was apparently first reported by Becker (42). Browne (43), Lloyd, Marriot, and Robertson (44), and many others have called attention to the halophiles in the curing salt as the organisms of the "red heat" of salted hides. Stuart (45) implicated common saprophytes of dirt, and Horowitz-Wlassowa (3) isolated an organism from the intestines of cattle which experimentally produced reddening of salted hides. In the past 25 years a large number of investigations on the reddening of salted hides have been initiated (13, 28, 44, 46–54). Although there has been disagreement about the causative agent of reddening of salted hides and confusion in classification, most investigators agree that the organisms are probably introduced with the curing salt. Robertson (49) found that 0.5 per cent of sodium fluoride added to the preserving salt for hides was more effective in preventing growth of halophiles and was less deleterious to the hides than borax, boric acid, sodium bicarbonate, or disodium sulfite.

C. Occurrence in Other Industries

It is not meant to infer that halophilic bacteria are associated only with natural bodies of salt water, fish, or hides. Many industries have been confronted with the problem of halophilic growth. For example, it has hampered the petroleum industry's salt-water-injection system. Organisms have also been isolated from olives, anchovies, bacon, cucumbers, canned meat, and other foods (55–61); these organisms may be only salt-tolerant or facultative halophiles, but from the evidence available now it is usually just a matter of time before obligate halophilic forms appear. The presence of salt-tolerant organisms or of halophiles is not always objectionable since they contribute greatly in the ripening process of certain foods (salted meats, sauerkraut, cheeses). It is reasonable to conclude that halophiles may be isolated from any product on which a salt pack or brine is used for preservation.

III. ENVIRONMENTAL INFLUENCE OF SALT

Any investigation into the influence of the environment on microorganisms is complicated by the interplay of many factors. Studies dealing with the influence of relatively low concentrations of salts are no exception. Whether a concentration of salt is inhibitory or stimulatory is characteristic of and dependent upon the particular salt and the test organism. The influence of the salt may be due to the undissociated molecule, the anion, the cation, or all three (62). This influence is also dependent upon the composition of the medium, since the amount of protein, the pH, and the presence of other salts may greatly influence the action of the salt being studied. There is another factor, aptly pointed out by Holm and Sherman (63), which may cause variations in results—bacteria adjust readily to abnormal conditions.

With halophiles the problems are even more involved because of the introduction of new factors or the exaggeration of previous factors by relatively high concentrations of salt. To study the environmental influence of salt on halophiles and to evaluate competently the results obtained with high concentrations, one must be familiar with the problems introduced by low concentrations of salts.

A. Ionic Relationships

Numerous studies dealing with the toxicity or stimulation of cations and anions on nonhalophilic bacteria have been made (64–67). Attempts have been made to correlate the toxicity of various salts to the lyotropic (68, 69) series and, although there appears to be some relationship, experiments by Shaughnessy and Winslow (70), and Holm and Sherman (63) indicate the complexity and the difficulty in classifying salts by their toxic or stimulative effects. It may also be true that too much importance is given to the lyotropic series, for Loeb (71) pointed out a serious error in the evaluation of data from studies on this series. He said that the series are the result of an error because the influence of the salt upon the pH of the solution was not noted, and the effect of the salt upon the solution was attributed to the particular ions of the salt only, with no consideration of the effect of the pH change.

Antagonism, an example of the influence of salts upon one another, is defined by Winslow and Dolloff (72) as the neutralization of the toxic effect of one cation by another cation. Generally antagonism occurs between a monovalent and divalent cation rather than between two monovalent or two divalent cations. For example, Lipman (73) noted antagonism between calcium and potassium, magnesium and sodium, sodium and potassium, but not between magnesium and calcium. It is also established that a certain concentration ratio of the two cations must be reached before the salts are antagonistic to each other. No satisfactory explanation for antagonism has been given. Many investigators have noted discrepancies in studies of the antagonistic effect of salts. It has been suggested that the effect of monovalent and divalent cations is the result of an establishment of favorable quantitative ionic conditions rather than qualitative antagonism.

TABLE 2

Sample protocol for studying the effect of KCl on the sodium-chloride-response curve of Vibrio costicolus*

Class A, Series No.	Salt Concentrations Added to Each Tube of 20 Different Concentrations of NaCl
1	0.2 M KCl
2	0.4 M KCl
3	0.2 M KCl + 0.2 M MgCl
4	0.2 M KCl + 0.2 M LiCl
5	0.2 M KCl + 0.2 M Na_2SO_4
6	0.2 M KCl + 0.2 M Na_2MoO_4
7	0.2 M KCl + 0.2 M Na_2HPO_4
8	0.2 M KCl + 0.2 M NaBr
9	0.2 M KCl + 0.2 M NaI
10	0.2 M KCl + 0.2 M NaF
11	0.2 M KCl + 0.2 M $NaNO_3$
12	0.2 M KCl + 0.2 M KNO_3

* The remaining 10 salts mentioned on this page were used in the same manner.

Ionic relationships are being actively investigated today. Current workers in the field (74–78) believe that there is a real need to understand the action of ions, not only from the standpoint of nutritional requirements but also their relationship and interaction with one another. While investigating the antagonism of toxic zinc ions by magnesium ions, MacLeod and Snell (77) found that zinc was antagonized by a number of divalent cations. They suggest the possibility of inactivation of certain proteins (apoenzymes) by zinc during the formation of metalloenzymes. It was further suggested by them that the antagonizing ions form active complexes with these proteins. Almost all studies on ionic relationships have been made with nonhalophilic microorganisms.

Although few investigations of this type have been made with halophiles, these organisms are especially well suited for ion-toxicity and membrane-permeability studies. This is particularly true of the nonpigmented types since they generally are much easier to handle. Nonpigmented types have less complex nutritional requirements and grow well in a short time, while the red-pigmented halophiles usually require an enriched medium and 7 to 10 days for good growth.

When it was found that a nonpigmented obligate halophile, *Vibrio costicolus*, showed a regular and measurable growth response to increasing concentrations of sodium chloride (79), studies were made on the depressant or stimulatory effects of certain salts on this response curve. A sample protocol is given in table 2. The same procedure was followed for investigating 11 salts. The results accumulated from these studies are summarized in table 3. Even though relatively greater concentrations of the salts are used with halophiles than with nonhalophiles, equal care must be taken when interpreting data concerned with the stimulatory or inhibitory effect of salts on the former. Often the results obtained from such studies are disappointing or in conflict with one another. Although over 3000 optical-density readings were made to gather data for table 3, only the gross effects of the salts could be taken as significant in changing the sodium-chloride-response curve of the organism. Growth with this organism can also be obtained without adding any sodium chloride to the medium if a sufficient amount of any one of a number of salts (magnesium chloride, potassium chloride, lithium chloride, sodium sulfate, sodium molybdate, sodium bromide, and sodium phosphate) is added to the medium. This does not appear to be true for all of the nonpigmented halophiles. Robinson (1) reports also that other salts can be substituted for sodium chloride with *V. costicolus* but that this is not possible with *Micrococcus halodenitrificans*. Some investigators (2, 21, 23, 55, 80) have found that a large number of salts could be used to replace the sodium chloride thought to be required by halophilic bacteria. Although Schoop (59) and Hess (37) believed that this was successful only with facultative halophiles and that obligate halophiles actually required sodium chloride, Hess also reported that trace amounts of magnesium, calcium, or barium ions stimulated growth.

Studies on the depressant action of ions on the growth of halophiles (21, 55, 79) show that the same general effect is obtained with halophiles as has been reported for nonhalophiles (64, 66). In spite of Loeb's objections (71) to the early investigations of the lyotropic series, the toxic action of ions on nonhalophiles and halophiles does have a relationship to the ion sequence of this series.

TABLE 3

Effect of the substitution salts on the sodium-chloride-response curve of Vibrio costicolus

Class	Salts: 0.2 M, 0.4 M, Mixture	Depressant or Stimulatory Action of Each Salt When Added to 20 Concentrations (0 to 4 M) of Sodium Chloride
A	KCl	Slight stimulation in low NaCl concentrations, slight depression after 1.2 M NaCl
B	$MgCl_2$	Stimulation in low NaCl concentrations, slight depression after 1.2 M NaCl
C	LiCl	Slight stimulation in low NaCl concentrations, depressant in remainder of curve
D	KNO_3	Stimulation in low NaCl concentrations, slight depression after 1.2 M NaCl
E	Na_2SO_4	Stimulation in low NaCl concentrations, slight depression after 1.2 M NaCl
F	Na_2MoO_4	Stimulation in low NaCl concentrations, slight depression after 1.2 M NaCl
G	Na_2HPO_4	Stimulation in low NaCl concentrations, slight depression after 1.2 M NaCl
H	NaBr	Slight stimulation in low NaCl concentrations, slight depression after 1.2 M NaCl
	NaI	Slight stimulation in low NaCl concentrations, depressant in remainder of curve
	NaF	Inhibits growth in all concentrations and with all combinations
K	$NaNO_3$	Stimulation in low NaCl concentrations, slight depression after 1.2 M NaCl

B. Physical Effect of Ions

The physical effect of ions on one another must also be considered when working with concentrated solutions of various salts. Ions do not act independently in solutions. The attraction between ions is increased as the solution becomes concentrated, and diminished as the solution becomes dilute. This interionic attraction accounts for some deviations between calculated results and actual results in many problems dealing with electrolytes. A correction factor has been introduced to obtain an "effective concentration." This effective concentration is called the activity of the ion. The activity of the ion has the following relationship to the actual concentration: $\gamma = \frac{\alpha}{m}$ where α is the activity, m is the molal concentration, and γ is the activity coefficient. Formulas exist for determining the activity coefficient, but unfortunately, these are applicable only to dilute solutions and cannot be utilized here. It is sufficient for our purposes to be aware that deviations may be caused by interionic forces and are accentuated as the concentration of the solution increases.

The ions in a medium also influence certain other physical properties of that medium. One of these is osmotic pressure; the osmotic pressure of an electrolyte solution is dependent upon the number of dissolved particles present in a particular volume. Changes in rate of growth, cell division, and physical structure of plants are influenced by osmotic pressure (81). Numerous investigators have studied the effect of osmotic pressure on cell variation, viability, and spore germination of bacteria. Many students of halophilism in bacteria (2, 14, 55, 58, 80, 82–84) have suggested that a major role of the large concentrations of salt necessary in the medium for optimum growth is to regulate osmotic pressure. Much needs to be learned before we clearly understand the effect of osmotic pressure on halophilic bacteria, and even less is known about the influence of the other physical properties of the medium upon halophiles.

IV. METABOLISM AND PHYSIOLOGY

Halophiles, when supplied with the proper environment, are surprisingly active. They ferment a large number of sugars, have proteolytic activity, produce gases, and oxidize many intermediate compounds (1, 19, 79, 85–94). However, it should be remembered that some strains—especially the red pigmented types—grow very slowly and more time is necessary to bring about a change in the substrate.

How halophilic bacteria resist the usually adverse effect of concentrated salt solutions is not known. Why they thrive in concentrated solutions is even more puzzling. Rockwell and Ebertz (95) listed five factors detrimental to microorganisms, all of which were brought about by increased concentrations of sodium chloride. These factors were (a) dehydration, (b) interference with the rapid action of proteolytic enzymes, (c) direct effect of the chloride ion, (d) sensitization to carbon dioxide, and (e) removal of oxygen. These factors have been of special interest to investigators of halophilism; perhaps one or more is beneficial to or is controlled by halophilic bacteria.

A. Dehydration and Enzyme Activity

Control of the dehydration of cellular protoplasm most certainly must play a part in halophilism. Studies by Robinson and coworkers (1, 91–93) have shown that there is a barrier or energy mechanism of some type, since there is considerably less concentration of salts within the cell than outside. They found that the nitritase and lactic acid dehydrogenase of *M. halodenitrificans* were not resistant to the sodium chloride concentration which was optimal for the growth of the organism. It was concluded from these results that the salt concentration in the cell is less than that of the medium, and that Ingram's (96) theory of halophilism based on the resistance of enzymes to "salting out" was disputable. Flannery (87, 88), while studying oxidation of substrates by resting cells of *V. costicolus*, found that an optimum salt concentration was required for maximum enzyme activity. A similar conclusion was made by Yamada and Shiio (97). Recent studies by Baxter and Gibbons (85) were made to compare the activity of an enzyme preparation in several concentrations of sodium chloride. They found that the glycerol dehydrogenases of *Escherichia coli*, *V. costicolus*, and *Pseudomonas salinaria* required greatly different salt concentrations for maximum activity; the concentrations necessary were 0.25 M, 0.5 M, and 1.5 M respectively. These concentrations are approximately half that required by each of the halophiles for maximum growth, which again suggests that the salt concentration within the cell is less than that in the medium. Baxter and Gibbons also found greater glycerol dehydrogenase activity in the presence of the potassium ion as compared with the sodium ion. Robinson and Katznelson (94), while studying the aspartate-glutamate transaminase of *P. salinaria*, found greatest enzymatic activity in salt concentrations (4–8 per cent) much less than that necessary for optimum growth. Yamada and Asano (98) reported a halophilic glucose dehydrogenase and a nonhalophilic formic dehydrogenase which were obtained from the same organism. A halophilic phosphomonoesterase and a nonhalophilic catalase and proteinase have been described by Yamada, Shiio, and Egami (99). These results indicate that some resistance of enzymes to "salting out" or inactivation in certain salt concentrations must also play a part in halophilism.

Several precautions must be observed when resting-cell or cell-free extract studies with obligate halophiles are contemplated. The experimental cells should be handled very carefully—abrupt changes in salt concentration may destroy the cells or inactivate certain enzyme systems. Egami, Yamada and Shiio (100) suggested using distilled water to lyze the cells as an easy way to obtain active cell-free extracts. However, many halophilic enzymes are irreversibly inactivated in distilled water, and the activity studied in such a preparation might result from the nonhalophilic enzymes. Sonic disintegration of halophilic cells has been used successfully by several investigators (85, 98).

Survival studies on *V. costicolus* have shown that this halophile has a much lower survival rate after freeze-drying than after freezing alone. It can be concluded from all the experimental evidence that dehydration by increased concentrations of sodium chloride has little or no beneficial effect on halophiles and that dehydration of the cell protoplasm is prevented by some intimate mechanism of the cell.

B. Direct Effect of Sodium and Chloride Ions

Since obligate halophiles require relatively large amounts of sodium chloride for maximum growth, it is obvious that there is little, if any detrimental effect of the chloride ion. Interest in both the sodium and the chloride ion has been focused on the beneficial effect. Is there a beneficial effect that is specific for either or both ions? All the investigations (1, 20, 21, 23, 55, 59, 79, 80, 87, 88) into the effect of cations and anions have been made by a substitution method; in this method the chloride ion may be retained by using chloride salts of various cations, or the sodium ion may be retained by using sodium salts of various anions. It must be remembered, however, that this method is susceptible to severe criticism. It is not just a simple replacement of ions. The various substituted cations or anions may have a specific toxic or stimulatory effect on the growth or metabolism of the organism. Flannery (87) overcame this difficulty to some degree while studying the effect of the sodium ions and the chloride ions on the growth and metabolism of *V. costicolus*. He used a number of concentrations of substituting salts to obtain a base-line curve, and then investigated the stimulatory effect of low concentrations of sodium chloride. The conclusion from this study and that of others (23, 80) is that both the sodium ion and the chloride ion contribute to the well-being of the organism, but the sodium ion is the more important of the two.

C. Effect of Gas Solubility

The significance of sensitization to carbon dioxide and removal of oxygen by increased concentrations of salt will be considered together. The influence of gases upon halophilic bacteria was of special interest to Stuart and James (101, 102) and Stuart (103). They noted that increased sodium-chloride concentrations gave lower Eh measurements. Studies were made in which varying amounts of nitrogen, carbon dioxide, or illuminating gas were substituted for air. In the presence of 3 M sodium chloride the best growth was obtained in atmospheres containing 40 to 50 per cent of any one of the three gases, but in the presence of 5 M sodium chloride there was no stimulation of growth in any concentration of gas. There was stimulation of growth, however, when cysteine was added to the medium. From these investigations with red-pigmented halophiles it was concluded that they have a preference for reduced oxygen tension. The limiting factor in these studies was the use of a solid surface which necessitated a gross estimation of the amount of growth obtained. Flannery (87) repeated the work quantitatively with a nonpigmented halophile, using 20 different concentrations of sodium chloride in each of 10 concentrations of nitrogen; identical growth curves were obtained in each concentration of nitrogen except in the 100-per cent nitrogen, in which complete depression of growth occurred. It was concluded that oxygen tension had little influence on the growth response of this nonpigmented halophile. Although Brown and Gibbons (104) believe that the red halophilic bacteria are strict aerobes, a similar quantitative study is necessary before any comparison can be made between the influence of oxygen tension on pigmented halophiles and that on nonpigmented halophiles.

D. Nonspecific Particle Effect

It has been proposed that the action of high concentrations of salt upon microorganisms is due to the molecular concentration of solutions. LeFevre and Round (58) suggested that a certain osmotic pressure may be required by halophilic bacteria. They further suggested that if this be so other salts would act similarly and satisfy the requirement. It has already been pointed out that this occurs with at least some of the halophilic bacteria. Richter (80) proposed that sodium chloride has at least two functions to perform for halophilic bacteria. One function is concerned with nutrition and the other is as a regulator of osmotic pressure. He felt that the latter function could be performed by a large number of salts. This does not occur with all halophiles.

According to several reports (1, 37, 59) the sodium-chloride requirement of some halophiles cannot be replaced by other salts. This particular question has been in dispute for some time. As it is incongruous that sodium chloride can be replaced by certain salts for some halophilic bacteria and not for others, future investigations may reveal the reason for this paradox. It may be that the proper conditions for replacement have not been found. A number of investigators (83, 84, 90, 105–107) found, while working with marine halophiles, that the luminescence of these organisms was influenced by osmotic pressure. Certain large bodies in the cells, which had staining reactions of nuclear material, were also influenced by changes in osmotic pressure (106). A nonspecific particle concentration, contributing to the osmotic environment, was suggested by Flannery *et al.* (79) as being important to certain halophiles. Calculations were made for determining the effective particle concentrations. It is well established that an electrolyte with two ions has twice the effect on the osmotic pressure as that of a nonelectrolyte of the same molar concentration, and an electrolyte with three ions has three times the effect. The necessarily crude application of these facts to the growth response of a halophile to concentrations of salts is shown in table 4 (87). In 24 out of 28 calculations (the three solutions allowing no cell multiplication are omitted) the total effective particle concentration was between 1.6 and 2.4. It is well to remember that ions do not act independently as do molecules, and that interionic forces, inherent with the particular ions and their concentrations, slightly influence their effect in a solution. This fact and the depressant or stimulatory effect on the response by some ions may explain the deviations in the particle concentrations giving maximum growth response.

Since nonelectrolytes (sugars) have been used extensively by investigators of osmotic pressure phenomena (108), two very soluble sugars (arabinose and rhamnose) were selected for additional osmotic pressure experiments. It was disappointing, however, to find that these sugars would not substitute for the salt required by *Vibrio costicolus.* The organism would not grow even with the addition of small amounts of sodium chloride to the sugar solutions. These experiments were extended to include a study of respiration on glucose, and it was found that nonelectrolytes could supply the particle concentration necessary in the oxidation of glucose but that sodium chloride, added in small quantities, greatly stimulated the oxidative reaction (88). Johnson and Gray (106), while studying the influence of osmotic pressure on nuclear bodies of luminous bacteria, found that sugar solutions caused a depression of the luminescence. The importance of osmotic pressure to marine forms has been reported, but sugar solutions isotonic with sea water have generally not been satisfactory; in addition to the proper osmotic pressure there has been an electrolyte requirement. Johnson and Harvey (90) found that luminescence and respiration fell off as the sea water used in their medium was diluted. It was suggested by Flannery *et al.* (88) that at least two systems are involved in halophilism—a cell-reproduction system and a respiratory system. From the results obtained with *V. costicolus*, it is likely that not only is a certain particle concentration required to obtain optimum osmotic pressure but that there is also an electrolyte requirement. The electrolyte concentration was required for cell reproduction but was only stimulatory to cell respiration.

E. Permeability of Cell Membranes

Very little is known about the permeability of the cell membranes of halophiles. However, all observations on permeability indicate that it should be a dominant factor in halophilism. For example, some individuals feel that the osmotic pressure of a solution plays a part in modifying the absorption process through the membrane. Mathews (109) believed that salts influence protoplasm through the effect of their ions and that the physiological action of the protoplasm is dependent upon the available potential energy. Stuhlman (110) declared that surface energy is raised by increasing the concentration of salts, and others have stated that since ions tend to concentrate at surfaces, surface tension is thus lowered and permits adsorption. Reviews and books on the general topic of permeability are given in the references (65, 89, 111–113). Some definitive work with halophiles was completed by Robinson *et al.* (93), who reported a diphosphopyridine or triphosphopyridine nucleotide-linked system as the energy-control mechanism of the cell membranes. This work was done with *Micrococcus halodenitrificans* and should be repeated with other halophilic bacteria.

Riley (82) and Pratt and Riley (114), while studying factors causing lysis of resting-cell preparations of a marine bacterium, found that magnesium and sodium ions had a protective effect

TABLE 4

A comparison of the effective particle concentration of the various salts on the growth response of Vibrio costicolus

Salt	Concentration of Salt Yielding Maximum Optical Density	Particle Concentration		Particle Total × 10
		Salt	*NaCl*	
NaCl	1.2	1.2 × 2		24
$MgCl_2$	0.6	0.6 × 3		18
$MgCl_2$ + 0.2 NaCl	0.4	0.4 × 3	+ 0.2 × 2	16
$MgCl_2$ + 0.4 NaCl	0.4	0.4 × 3	+ 0.4 × 2	20
KCl	0.8	0.8 × 2		16
KCl + 0.2 NaCl	0.6	0.6 × 2	+ 0.2 × 2	16
KCl + 0.4 NaCl	0.6	0.6 × 2	+ 0.4 × 2	20
LiCl	0.8	0.8 × 2		16
LiCl + 0.2 NaCl	0.6	0.6 × 2	+ 0.2 × 2	16
LiCl + 0.4 NaCl	0.6	0.6 × 2	+ 0.4 × 2	20
KNO_3	No growth	0		0
KNO_3 + 0.2 NaCl	0.8	0.8 × 2	+ 0.2 × 2	20
KNO_3 + 0.4 NaCl	0.6	0.6 × 2	+ 0.4 × 2	20
Na_2SO_4	0.6	0.6 × 3		18
Na_2SO_4 + 0.2 NaCl	0.4	0.4 × 3	+ 0.2 × 2	16
Na_2SO_4 + 0.4 NaCl	0.4	0.4 × 3	+ 0.4 × 2	20
Na_2MoO_4	0.8	0.8 × 3		24
Na_2MoO_4 + 0.2 NaCl	0.6	0.6 × 3	+ 0.2 × 2	22
Na_2MoO_4 + 0.4 NaCl	0.4	0.4 × 3	+ 0.4 × 2	20
NaBr	0.8	0.8 × 2		16
NaBr + 0.2 NaCl	0.6	0.6 × 2	+ 0.2 × 2	16
NaBr + 0.4 NaCl	0.4	0.4 × 2	+ 0.4 × 2	16
Na_2HPO_4 (mixture)	0.4	0.4 × 3		12
Na_2HPO_4 + 0.2 NaCl	0.4	0.4 × 3	+ 0.2 × 2	16
Na_2HPO_4 + 0.4 NaCl	0.2	0.2 × 3	+ 0.4 × 2	14
$NaNO_3$	No growth	0		0
$NaNO_3$ + 0.2 NaCl	0.6	0.6 × 2	+ 0.2 × 2	16
$NaNO_3$ + 0.4 NaCl	0.6	0.6 × 2	+ 0.4 × 2	20
NaI	No growth	0		0
NaI + 0.2 NaCl	0.2	0.2 × 2	+ 0.2 × 2	8
NaI + 0.4 NaCl	0.2	0.2 × 2	+ 0.4 × 2	12

whereas potassium and ammonium ions did not protect. Magnesium ions, although preserving the cellular integrity, became toxic as the concentration was increased to 1 M. This toxic effect of the magnesium ions has also been noted by others (87, 104). Lysis by bacteriophage of a vibrio which was isolated from ocean mud has been reported recently (115). Some interesting permeability studies could be made with this system.

With tagged materials now available such obviously critical experiments as the passage of materials into and out of the cell and the influence of different concentrations of sodium chloride and other salts on the movement of these materials can be completed. There is no doubt that this information would contribute much to the understanding of the halophilic cell membrane.

V. MORPHOLOGY, ISOLATION, AND CULTIVATION

A. Morphology

It is well established by the observations of numerous investigators that the size and shape of halophilic bacteria may be influenced by the concentration of the salt present. As the concentration of the salt is varied, rod-like forms become elongated or bulbous, or even appear as cocci. Spruit and Pijper (116) found even flat, curved, ribbon forms. When other salts are substituted for sodium chloride, extreme pleomorphism may result (87). This tendency to show different forms is even observed occasionally in cultures which have grown on a medium containing the concentration of sodium chloride thought to be optimal. The appearance of these forms suggests that the medium is not really optimal but lacks some necessary component. Stuart (54) and Lochhead (46) report that the type of basal medium employed also has some influence on the morphology of the cells. Electron-microscope studies with special stains (1, 117) indicate that the cell wall of halophilic bacteria is no different from that of nonhalophilic.

The staining of halophilic bacteria presents some difficulties. Ever present, when bizarre morphological forms are observed, is the possibility that the staining technique is responsible. Methods to avoid these artifacts have been suggested by several investigators (10, 28, 46, 103, 116, 118). The most successful methods are based on a procedure using methanol or an acid-alcohol as a fixing agent and reducing the amount of salt present before applying the stain. Spruit and Pijper (116) used 4 per cent formalin as a fixing agent. The salt present in the smear may be removed by washing gently with 90 per cent ethanol (103) after fixation. Very good results may be obtained using the simple method of Gibbons (10): air-dry the smear, fix 3 to 5 min in absolute methanol, and add the stain directly to the alcohol on the slide. One per cent of either aqueous basic fuchsin or crystal violet will yield well-stained cells.

Recently some excellent morphological studies were completed by Brown and Gibbons (104). Interested in the effect of specific ions on the growth and morphology of red halophiles, Brown and Gibbons made use of modified wet mounts in conjunction with the phase microscope. Figure 1 illustrates the effect of magnesium concentration on a halophilic rod. Some rod strains, after several transfers, changed irreversibly to coccus forms; the authors suggest that this may have taxonomic significance.

Smithies, Gibbons and Bayley (119) have studied the chemical composition of the cell and cell wall of three halophilic bacteria. They found that the chemical composition of these organisms closely resembles that of nonhalophiles except in carbohydrate content—halophilic cells contain only 3 to 10 per cent of the carbohydrate reported for nonhalophilic cells. The cell walls of the halophiles investigated were composed of lipoprotein. Smithies and Gibbons (120) were particularly interested in the slime layer produced by some of the halophiles. They discovered that this slime layer was made up of deoxyribonucleic acid and that the concentration of the salt and the ions present influenced the amount of deoxyribonucleic acid in the layer. Some other observations of Smithies and Gibbons indicate that the presence of large amounts of deoxyribonucleic acid in the slime layer, though not involved in essential physical or chemical reactions, may be the result of altered permeability of the cell. Several laboratories are continuing these studies.

Colonial morphology is influenced by the concentration of salt, the type of salt, the type basal medium, the pH, and the temperature employed (46, 101, 102, 121–124). The time of appearance or type of growth (19, 22) and the amount of growth (1, 10, 59, 87, 101, 102) are dependent upon the environmental conditions provided. Decrease in color, with pigmented halophilic bacteria, is reported to occur with removal of optimal conditions (46, 101, 102, 121). The chemistry of this red pigment has not been thoroughly investigated, but some of its properties were studied by Petter (125) and reported to be a carotinoid type. More recently it has been suggested that it might be a demethylated rhodoviolascin. Petter also reported gas vacuoles in these organisms but this observation has been questioned by Spruit and Pijper (116), who believe that Petter's gas vacuoles are areas of condensed protein. The importance of these areas to the halophilic cell is unknown.

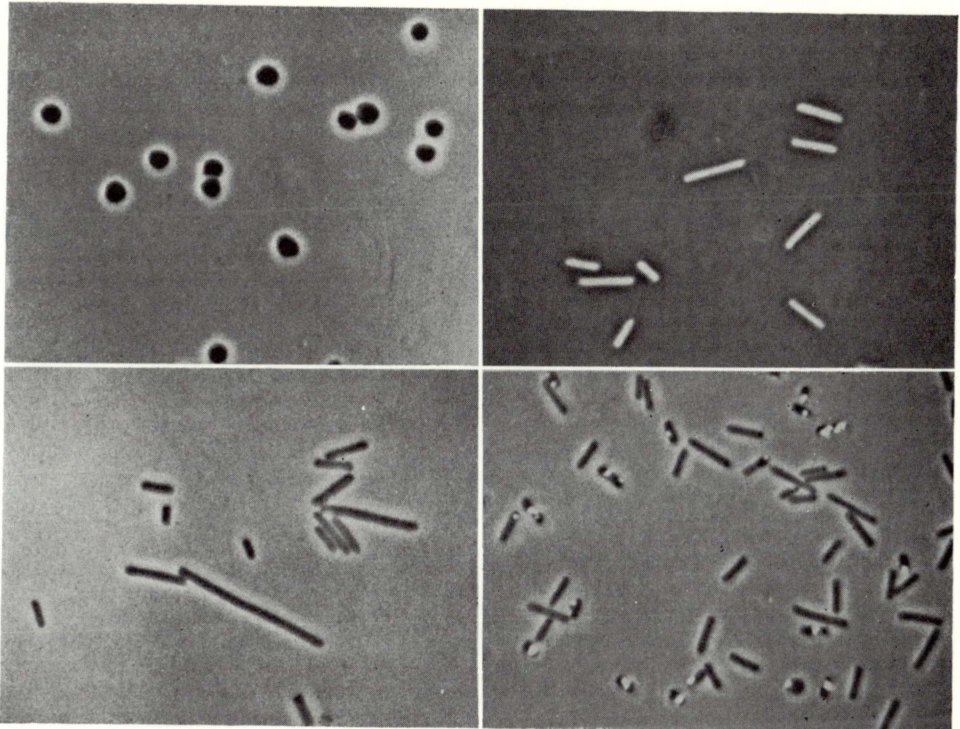

Figure 1. Effect of magnesium concentration on the morphology of *Pseudomonas salinaria.* Phase contrast illumination. 1400 × magnification. *A* (upper left). Cells after one transfer in 0.01 M magnesium. *B* (upper right). Cells in 0.1 M magnesium. *C* (lower left). Cells after one transfer in 0.25 M magnesium. *D* (lower right). Cells after one transfer in 1.0 M magnesium. Refractive bodies are the gas vacuoles of Petter or the condensed protein of Spruit and Pijper. These bodies may be seen in lower concentrations of magnesium ions also.

B. Isolation

Isolation of halophilic bacteria which are already growing on some salted product is not difficult. It is merely a matter of selecting the proper enriched medium (many of which are described under *Cultivation*) and the proper conditions. It is difficult, however, to generalize and describe media and conditions which will be most suitable for all or even a number of halophiles. The selection of these depends a great deal upon the source of organisms. It is fairly well established that most halophilic bacteria grow best on a slightly alkaline medium (13, 52, 53, 58). Usually the optimum salt concentration is between 5 to 15 per cent for nonpigmented halophiles, and 15 to 25 per cent for red-pigmented halophiles. When isolating organisms from salted products, it may take 24 to 72 hr to obtain good growth of the nonpigmented halophiles, and 7 to 14 days for the pigmented types. However, Stuart (45) used incubation periods of 60 to 90 days to isolate halophiles from water, sulfur springs, soil and dung. Pierce (126) also found a long incubation period of 20 to 30 days necessary for isolation of organisms from the Salton Sea. These extremely long incubation periods were probably necessary in selecting strains that had halophilic potentials, or the media utilized were nutritionally poor for the strains isolated. Other investigators have failed in their attempts to repeat this work. The optimum temperature depends upon the source of the organism but usually falls somewhere between 25 and 35 C.

C. Cultivation

The nonpigmented halophiles grow readily in a relatively simple medium. A 1 or 2 per cent peptone or casein digest, with the proper concentration of salt, will usually ensure good growth. In contrast, the red-pigmented halophiles frequently require special media; enrichment techniques are almost always used. Gibbons (10) used a medium consisting of flour, sodium chloride, and fish broth for isolating organisms from salted fish. Clayton and Gibbs (29) utilized a fish-rice medium which later was modified by Stuart, Frey and James (13) for isolation of organisms from different types of refined salt. The modified medium was prepared by steaming freshly flayed calfskin for 1 hr and by adding to the broth so obtained rice, peptone, and sea salt. Spruit and Pijper (116) used salt media with added fish extract or fish. Lochhead (46), while studying the red discoloration of salted hides, used 3 media; one of these, a skim milk-salt medium, was later used by Gibbons (10) and found to be most satisfactory. Dussault and LaChance (121) modified this medium by adding magnesium sulfate, nitrate, ferric chloride, peptone, and glycerol; the incubation time was cut in half and growth was greatly increased. Venkataraman and Screenivasan (127) have reported that a medium consisting of starch, salt, beef infusion, and tryptone was superior to Lochhead's skim milk-salt medium. Many other investigators have used sea or salt-lake water in their media. Lipman (128) and Johnson and Harvey (90) found that dilution of sea water used in media resulted in lower plate counts. Silica gel was employed by Moore (129, 130) as a solidifying agent in media for the cultivation of halophiles. She reported that a medium composed of silica gel, small amounts of five salts, glucose, cystine, peptone, and sodium chloride gave good results. Very little has been done utilizing synthetic media, but these important studies are now in progress in several laboratories. Katznelson and Lochhead (131) found that there was no consistent requirement for growth substances by the six obligate halophiles they studied. They realized that the variation in these requirements was as great as the variation in cellular-morphology and sodium-chloride requirements. A comparison of the nutrition and metabolism of different marine bacteria is being made by MacLeod *et al.* (132). Boring (133) investigated the mineral requirements for respiration of a marine bacterium. He showed, while using ZoBell's synthetic sea water (25) as a base salt solution, that if any one salt was eliminated from the synthetic sea water, respiration of the marine organism continued. Alone, only sodium chloride and magnesium chloride allowed respiration to continue, but the presence of potassium ions was stimulatory. Flannery (87, 88) reported that oxidation of glucose by a halophilic vibrio continued when sodium chloride was replaced by 10 other salts.

In a preliminary report, Flannery (134) describes a synthetic medium which would allow good growth of two nonpigmented halophilic bacteria; it contained 18 amino acids, 10 vitamins, 10 salts, adenine, guanine, and uracil. In experiments to simplify the medium and make it optimal for each organism, he found that it was necessary to include only six of the original 10 salts and none of the bases. None of the vitamins was required by the vibrio on one transfer, and only thiamine was required by the micrococcus. Work is being continued on the amino-acid portion of the medium. There have been no other reports of strictly synthetic media which allow good growth of halophiles. Several laboratories are now attempting to devise synthetic media for the red halophilic bacteria.

Most of the red-pigmented obligate halophiles require a solid surface to give good growth. This complicates matters since high concentrations of sodium chloride, with the usual 1.5 to 2 per cent agar, will raise the temperature of solidification. ZoBell, Anderson and Smith (18) studied this effect of sodium chloride and found that decreasing the concentration of agar lowered the temperature of solidification. It is also difficult to melt agar in high concentrations of sodium chloride. It has been suggested (46) that a separate solution of salt be prepared and not mixed with the other materials until after autoclaving. This technique is successful and aids in preventing precipitation of protein materials.

High concentrations of salt also may influence the pH of the medium. Parsons and Douglas (135), investigating what is usually referred to as "salt error," made determinations of the pH of sodium-chloride solutions of 1, 2 and 3 M, and found colorimetric observations to be about 0.3 pH units higher than electrometric observations. Stuart (103) found that if impure sodium chloride was used, the solutions would become more and more acid as the concentration of the salt increased. He stated that this change in pH could be corrected by using chemically pure grades of sodium chloride.

VI. PROBLEMS OF CLASSIFICATION

The literature on halophilic bacteria is burdened with reports of organisms from various genera which have been isolated from salt, soil, mud, hides, fish and other foods. Halophiles isolated from hides, fish, and salt have been reported in these genera: *Micrococcus* (32, 35, 42, 50, 127, 132, 136–140); *Diplococcus* (26, 34, 35, 140, 141); *Sarcina* (27, 48, 50, 127, 136, 137, 138, 139, 140); and *Bacillus* (32, 35, 52, 136, 140–143). Algae, yeasts, molds, spirochetes, and organisms in the genera *Actinomyces*, *Corynebacterium*, *Proteus*, *Nitrosomonas*, *Vibrio*, *Pseudomonas*, *Flavobacterium* and *Achromobacter* have all been described as halophilic or isolated from salted materials. Some of the organisms are probably identical although they have received different names. Part of the confusion in classification has been the result of changes in morphological and physiological characters of certain organisms in media of different salt concentrations and varied nutrients. Breed (144) and Lochhead (145) have attempted to clarify the classification by suggesting changes for several halophiles.

Some investigators have proposed that differences in sensitivity to sodium chloride might be important in classification. Hill and White (146) suggested using sodium chloride to separate certain gram-negative and gram-positive nonhalophilic forms. Schoop (59) divided bacteria into three groups according to their response to 10 per cent sodium-chloride solutions: (a) nonhalophiles, (b) facultative halophiles, and (c) obligate halophiles. After studying halophilic bacteria Sturges (147) and Sturges and Heidman (148, 149) suggested that a new tribe *Halophileae*, be formed. Spruit and Pijper (116) proposed a genus *Halobacterium*, which would contain all rods with red carotinoid pigment isolated from salt or salted products and growing in not less than a 12 per cent salt media.

Is halophilism of such significance that a new tribe or even a genus should be based on this character? It has been demonstrated many times that some bacteria can become accustomed to high salt concentrations. Stather (51) believed that there is a very close resemblance between some of the organisms of reddening and those of low salt tolerance. Many investigators (13, 45, 58, 150–152) have suggested that obligate halophilic organisms are common saprophytes which develop their halophilic characteristic in a salt environment. Stather (51) found that low salt-tolerance organisms, which resembled those of "red heat," could be cultivated on a 20 per cent salt medium if they were carefully transferred to media of increasing salt concentrations. Kluyver and Baars (4) demonstrated with two vibrios, one halophilic and the other nonhalophilic, that the salt requirements of the two, by gradual alteration of the medium, could be reversed. In contrast to this Hof (153) found that certain organisms carried on a medium of low salt concentration would adjust immediately, obligately and irreversibly, to a medium of high salt concentration. Recently Spruit and Pijper (116) were unable to grow an organism isolated from red salt in media with less than 16 per cent salt.

A doctrine of "physiological artefacts" was advanced by Kluyver and Baars (4). They proposed that a halophilic organism is a physiological artifact selected by the conditions imposed upon the organism by high salt concentration.

From the great variety of genera, the apparent reversibility of salt requirements, and the widespread sources which have been reported for halophilic organisms, it is probable that natural variation accounts for the appearance of nonhalophilic, facultative halophilic, and obligate halophilic forms in the same genus. It is plausible that a process of natural selection, as suggested by Doudoroff (6, 86), occurs when microorganisms are exposed for long periods of time to a high salt environment. The formulation of a new tribe or genus based on halophilism cannot be recommended at this time, for this action would create a repository for a conglomeration of many different organisms whose only relationship would be their halophilic character. When we have sufficient information about the nutrition, metabolism, physiology, and chemical composition of halophilic bacteria, it will be time to decide whether to have a specific tribe for these unique organisms.

ACKNOWLEDGEMENT

Figure 1 was modified from Figures 2, 3, 4, and 5 of Brown and Gibbons (104); originally published in *Canadian Journal of Microbiology*. We thank the copyright owners for permission to reproduce.

REFERENCES

1. Robinson, J. 1950 A possible explanation of microbial halophilism. Dissertation, McGill Univ., Montreal, Can.
2. Golikowa, S. M. 1930 Eine Gruppe von obligat halophilen Bakterien, gezüchtet in Substraten mit hohem NaCl-Gehalt. Zentr. Bakteriol. Parasitenk., Abt. II, **80**, 35–41.
3. Horowitz-Wlassowa, L. M. 1931 Über

die Rotfärbung gesalzener Därme ("der rote Hund"). Zentr. Bakteriol. Parasitenk., Abt. II, **85,** 12–18.

4. KLUYVER, A. J., AND BAARS, J. K. 1932 On some physiological artefacts. Proc. Koninkl. Akad. Wetenschap. Amsterdam, **35,** 370–378.
5. ZOBELL, C. E., AND MICHENER, H. D. 1938 A paradox in the adaptation of marine bacteria to hypotonic solutions. Science, **87,** 328–329.
6. DOUDOROFF, M. 1940 Experiments on the adaptation of *Escherichia coli* to sodium chloride. J. Gen. Physiol., **23,** 585-611.
7. SEVERENS, J. M., AND TANNER, F. W. 1945 The inheritance of environmentally induced characters in bacteria. J. Bacteriol., **49,** 383–393.
8. SHERMAN, J. M., AND CAMERON, G. M. 1934 Lethal environmental factors within the natural range of growth. J. Bacteriol., **27,** 341–348.
9. BAAS-BECKING, L. G. M. 1931 Historical notes on salt and salt manufacture. Sci. Monthly, **32,** 434–446.
10. GIBBONS, N. E. 1936 Bacteria associated with reddening of salt fish. J. Biol. Board Can., **3,** 70–76.
11. RAPPIN, D. 1920 The dangers of table salt. J. Am. Med. Assoc., **75,** 618–619.
12. PETROWA, E. K. 1933 Mikrobiologie des Kochsalzes. Arch. Mikrobiol., **4,** 326–347.
13. STUART, L. S., FREY, R. W., AND JAMES, L. H. 1933 Microbiological studies of salt in relation to the reddening of salted hides. U. S. Dep. Agr. Tech. Bull. 383.
14. ELAZARI-VOLCANI, B. 1940 Studies on the microflora of the Dead Sea. Thesis, Hebrew Univ., Jerusalem, Israel.
15. FREDERICK, E. 1924 On the bacterial flora of the Great Salt Lake and the viability of other microorganisms in Great Salt Lake water. Thesis, Univ. of Utah, Salt Lake City, Utah.
16. KELLERMANN, K. F., AND SMITH, N. R. 1916 Halophytic and lime precipitating bacteria. Zentr. Bakteriol. Parasitenk., Abt. II, **45,** 371.
17. SMITH, W. W. 1936 Evidence of a bacterial flora indigenous to the Great Salt Lake. Thesis, Univ. of Utah, Salt Lake City, Utah.
18. ZOBELL, C. E., ANDERSON, D. Q., AND SMITH, W. W. 1937 The bacteriostatic and bactericidal action of the Great Salt Lake water. J. Bacteriol., **33,** 253–262.
19. BARANIK-PIKOWSKY, M. A. 1927 Über den Einfluss hoher Salzkonzentrationen auf die Limanbakterien. Zentr. Bakt. Parasitenk., Abt. II, **70,** 373–383.
20. RUBENTSCHIK, L. 1925 Über einige neue Urobakterienarten. Zentr. Bakteriol. Parasitenk., Abt. II, **66,** 161–180.
21. RUBENTSCHIK, L. 1929 Zur Nitrifikation beihohen Salzkonzentrationen. Zentr. Bakteriol. Parasitenk., Abt. II, **77,** 1–18.
22. SASLAWSKY, A. S. 1927 Über eine obligat halophile Thionsäurebakterie. Zentr. Bakteriol. Parasitenk., Abt. II, **72,** 236–242.
23. SASLAWSKY, A., AND HARZSTEIN, N. 1930 Über die Einwirkung gewisser Salze auf obligat-halophile Thionsaurebakterien. Zentr. Bakteriol. Parasitenk., Abt. II, **80,** 165–169.
24. HARRISON, F. C., AND KENNEDY, M. E. 1922 The red discoloration of cured codfish. Proc. and Trans. Roy. Soc. Can., Section 5, **16,** 101–152.
25. ZOBELL, C. E. 1946 *Marine microbiology.* Chronica Botanica Co., Waltham, Mass.
26. BECKWITH, T. D. 1911 The bacteriological cause of the reddening of cod and other allied fish. Zentr. Bakteriol. Parasitenk., Orig., Abt. I, **60,** 351–354.
27. FARLOW, W. G. 1880 On the nature of the peculiar reddening of salted codfish during the summer season. U. S. Comm. Fish and Fisheries, Rept. 1878 (pt. 6), 969–973.
28. ANDERSON, H. 1954 The reddening of salted hides and fish. Appl. Microbiol., **2,** 64–69.
29. CLAYTON, W., AND GIBBS, W. E. 1927 Examination for halophilic microorganisms. Analyst, **52,** 395–397.
30. CLOAKE, P. C. 1923 Red discoloration (so-called "pink" or "pink eye") on dried salted fish. Food Invest. Board Spec. Rept. 18.
31. FARLOW, W. G. 1886 Vegetable parasites of codfish. U. S. Fishery Comm. Bull., **6,** 1–4.
32. GAYON, M., AND CARLES, M. 1886 Rouge de la morue. Rev. Sanit. Bordeaux et Provence, **57,** 47–48
33. HANZAWA, H., AND TAKEDA, S. 1931 On the reddening of boned codfish. Arch. Mikrobiol., **2,** 1–22.
34. HØYE, K. 1906 Recherches sur la moisissure de bacalao et quelque autres microorganismes halophiles. Bergens Museums Aarbog, **12,** 3–64.
35. HØYE, K. 1908 Untersuchungen über die Schimmelbildung des Bergfisches. Bergens Museums Aarbog, **4,**, 1–29.

36. HESS, E. 1942a Studies on salt fish. VIII. Effects of various salts on preservation. J. Fisheries Research Board Can., **6**, 1–9.
37. HESS, E. 1942b Studies on salt fish. IX. Effect of environment upon the growth of red halophilic bacteria. J. Fisheries Research Board Can., **6**, 10–16.
38. HESS, E., AND GIBBONS, N. E. 1942 Effect of disinfectants and preservatives on red halophilic bacteria. J. Fisheries Research Board Can., **6**, 17–23.
39. LIEBERT, F. 1930 Über die Ursache des "Rotwerdens" von Pökelhering. Zentr. Bakteriol. Parasitenk., Abt. II, **80**, 33–35.
40. SHEWAN, J. M. 1942 Some bacteriological aspects of fish preservation. Chemistry and Industry, **20**, 312–314.
41. RAY, L. R., AND BEEKMAN, E. 1951 Equipment for tanning of hides and skins. Food and Agr. Organization U. N., Food and Agr. Devel. Paper 13.
42. BECKER, H. 1912 Die Salzflecken. Collegium, **508**, 408–418.
43. BROWNE, W. W. 1922 Halophilic bacteria. Proc. Soc. Exptl. Biol. and Med., **19**, 321–322.
44. LLOYD, D. J., MARRIOT, R. H., AND ROBERTSON, M. E. 1929 "Red heat" in salted hides. J. Intern. Soc. Leather Trades Chemists, **13**, 538–569.
45. STUART, L. S. 1938 Isolation of halophilic bacteria from soil, water, and dung. Food Research, **3**, 417–420.
46. LOCHHEAD, A. G. 1934 Bacteriological studies on the red discoloration of salted hides. Can. J. Research, **10**, 275–286.
47. PAESSLER, J. 1912 Ueber des Salzen von Hauten und Fellen. Collegium, **508**, 379–388.
48. ROBERTSON, M. E. 1931 A note on the cause of certain red colorations on salted hides and a comparison of the growth and survival of halophilic or salt-loving organisms and some ordinary organisms of dirt and putrefaction on media of varying salt concentrations. J. Hyg., **31**, 84–95.
49. ROBERTSON, M. E. 1932 "Red heat"—its causes and prevention. J. Intern. Soc. Leather Trades Chemists, **16**, 564–568.
50. STATHER, F. 1928 Untersuchungen über Salzflecken. Mitteilung über Häuteschälen. Collegium, **703**, 567–599.
51. STATHER, F. 1930 "Rote Verfärbung" und "Rote Erhitzung" auf gesalzenen Häuten. Collegium, **720**, 151–153.
52. STATHER, F., AND LIEBSCHER, E. 1929 Ueber das Rotwerden gesalzener Rohhäute. Collegium, **713**, 427–437.
53. STATHER, F., AND LIEBSCHER, E. 1929 Zur Bakteriologie des Rotwerdens gesalzener Rohhäute. Collegium, **713**, 437–450.
54. STUART, L. S. 1935 The morphology of bacteria causing reddening of salted hides. J. Am. Leather Chemists Assoc., **30**, 226–235.
55. BAUMGARTNER, J. G. 1937 The salt limits and thermal stability of a new species of anaerobic halophile. Food Research, **2**, 321–329.
56. FODA, I. O., AND VAUGHN, R. H. 1950 Salt tolerance in the genus *Aerobacter*. Food Technol., **4**, 182–188.
57. HANKINS, O. G., SULZBACHER, W. L., KAUFFMAN, W. R., AND MAYO, M. E. 1950 Factors affecting the keeping quality of bacon. Food Technol., **4**, 33–38.
58. LEFEVRE, E., AND ROUND, L. A. 1919 A preliminary report upon some halophilic bacteria. J. Bacteriol., **4**, 177–182.
59. SCHOOP, G. 1935 Obligat halophile Mikroben. Zentr. Bakteriol. Parasitenk., Abt. I, **134**, 14–26.
60. SMITH, F. B. 1938 An investigation of a taint in rib bones of bacon. The determination of halophilic vibrios (n. spp.) Proc. Roy. Soc. Queensland, **49**, 29–52.
61. WEST, N. S., GILILLAND, J. R., AND VAUGHN, R. H. 1941 Characteristics of coliform bacteria from olives. J. Bacteriol., **41**, 341–354.
62. MILES, M. A., AND WILSON, G. S. 1946 *Topley and Wilson's Principles of bacteriology and immunity*. 3rd. ed. Vol. I. The Williams & Wilkins Co., Baltimore, Md.
63. HOLM, G. E., AND SHERMAN, J. M. 1921 Salt effects in bacterial growth. I. Preliminary paper. J. Bacteriol., **6**, 511–519.
64. FABIAN, F. W., AND WINSLOW, C.-E. A. 1929 The influence upon bacterial viability of various anions in combination with sodium. J. Bacteriol., **18**, 265–291.
65. FALK, I. S. 1923 The role of certain ions in bacterial physiology. A review. Abstr. Bacteriol. **7**, 33–147.
66. HOTCHKISS, M. 1923 The stimulating and inhibitive effect of certain cations on bacterial growth. J. Bacteriol., **8**, 141–162.
67. WINSLOW, C.-E. A., AND HAYWOOD, E. T. 1931 The specific potency of certain cations with reference to their effect on bacterial viability. J. Bacteriol., **22**, 49–69.
68. FREUNDLICH, H. 1903 Über das Ausfällen

kolloidaler Lösungen durch Electrolyte. Z. physik. Chem., **44,** 129–160.

69. HOFMEISTER, F. 1888 Ueber Regelmässigheiten in der eiweissfällendem Wirkung der Salze und ihre Beziehung zum physiologischen Verhalten derselben. Arch. exp. Pathol. Pharmakol., **24,** 247–260.
70. SHAUGHNESSY, H. J., AND WINSLOW, C.-E. A. 1927 The diffusion products of bacterial cells as influenced by the presence of various electrolytes. J. Bacteriol., **14,** 69–99.
71. LOEB, J. 1920 Ion series and the physical properties of proteins. III. The action of salts in low concentration. J. Gen. Physiol., **3,** 391–414.
72. WINSLOW, C.-E. A., AND DOLLOFF, A. F. 1928 Relative importance of additive and antagonistic effects of cations upon bacterial viability. J. Bacteriol., **15,** 67–92.
73. LIPMAN, C. B. 1909 Toxic and antagonistic effects of salts as related to ammonification by *Bacillus subtilis*. Botan. Gaz., **48,** 105–125.
74. BJERRUM, J. 1950 On the tendency of metal ions toward complex formation. Chem. Revs., **46,** 381–401.
75. LEHNINGER, A. L. 1950 Role of metal ions in enzyme systems. Physiol. Revs., **30,** 393–429.
76. MACLEOD, R. A. 1954 Dependence of the toxicity of cations for lactic acid bacteria on pH and incubation time. J. Bacteriol., **67,** 23–26.
77. MACLEOD, R. A., AND SNELL, E. E. 1950 The relation of ion antagonism to the inorganic nutrition of lactic acid bacteria. J. Bacteriol., **59,** 783–792.
78. TSUYUKI, H., AND MACLEOD, R. A. 1951 Ion antagonisms affecting glycolysis by bacterial suspensions. J. Biol. Chem., **190,** 711–719.
79. FLANNERY, W. L., DOETSCH, R. N., AND HANSEN, P. A. 1952 Salt desideratum of *Vibrio costicolus*, an obligate halophilic bacterium. I. Ionic replacement of sodium chloride requirement. J. Bacteriol., **64,** 713–717.
80. RICHTER, O. 1928 Natrium ein notwendiges Nährelement für eine marine mikroärophile Leuchtbakterie. Akad. Wiss. Wien. Math. Nat. Kl. Denkschr., **101,** 261–264.
81. LIVINGSTON, B. E. 1903 *The role of diffusion and osmotic pressure in plants.* 2nd series, Vol. VIII. The Decennial Publications, Univ. of Chicago Press, Chicago.
82. RILEY, W. H. 1955 A study of factors causing lysis of a marine bacterium. Thesis, Univ. of Florida, Gainesville, Florida.
83. HILL, S. E. 1929 The penetration of luminous bacteria by the ammonium salts of the lower fatty acids. J. Gen. Physiol., **12,** 863–872.
84. JOHNSON, F. H., AND HARVEY, E. N. 1937 The osmotic and surface properties of marine luminous bacteria. J. Cellular Comp. Physiol., **9,** 363–380.
85. BAXTER, R. M., AND GIBBONS, N. E. 1954 The glycerol dehydrogenases of *Pseudomonas salinaria*, *Vibrio costicolus* and *Escherichia coli* in relation to bacterial halophilism. Can. J. Biochem. and Physiol., **32,** 206–217.
86. DOUDOROFF, M. 1942 Studies on the luminous bacteria. I. Nutritional requirements of some species, with special reference to methionine. J. Bacteriol., **44,** 451–459.
87. FLANNERY, W. L. 1953 Salt desideratum of *Vibrio costicolus*, an obligate halophilic bacterium. Dissertation, Univ. of Maryland, College Park, Maryland.
88. FLANNERY, W. L., DOETSCH, R. N., AND HANSEN, P. A. 1953 Salt desideratum of *Vibrio costicolus*, an obligate halophilic bacterium. II. Effect of salts on the oxidation of glucose. J. Bacteriol., **66,** 526–530.
89. JOHNSON, F. H. 1947 Bacterial luminescence. In *Advances in Enzymology*, pp. 215–264. Vol. 7. Edited by F. F. Nord. Interscience Publishers, Inc., New York, N. Y.
90. JOHNSON, F. H., AND HARVEY, E. N. 1938 Bacterial luminescence, respiration and viability in relation to osmotic pressure and specific salts of the sea. J. Cellular Comp. Physiol., **11,** 213–232.
91. ROBINSON, J. 1952 The effects of salts on nitritase and lactic acid dehydrogenase activity of *Micrococcus halodenitrificans*. Can. J. Botany, **30,** 155–163.
92. ROBINSON, J., AND GIBBONS, N. E. 1952 The effect of salts on the growth of *Micrococcus halodenitrificans*. (n. sp.) Can. J. Botany, **30,** 147–154.
93. ROBINSON, J., GIBBONS, N. E., AND THATCHER, F. S. 1952 A mechanism of halophilism in *Micrococcus halodenitrificans*. J. Bacteriol., **64,** 69–77.
94. ROBINSON, J., AND KATZNELSON, H. 1953 Aspartate-glutamate transaminase in a red halophilic bacterium. Nature, **172,** 672–673.
95. ROCKWELL, G. E., AND EBERTZ, E. G. 1924 How salt preserves. J. Infectious Diseases, **35,** 573–575.
96. INGRAM, M. 1947 A theorem relating the

action of salts on bacterial respiration to their influence on the solubility of proteins. Proc. Roy. Soc. (London), B, **134**, 181–201.

97. Yamada, T., and Shiio, I. 1953 Effects of salt concentration on the respiration of a halotolerant bacterium. J. Biochem. (Japan), **40**, 327–337.
98. Yamada, T., and Asano, A. 1954 Oxidation-reduction enzymes of a halophilic bacterium. J. Biochem. (Japan), **41**, 639–645.
99. Yamada, T., Shiio, I., and Egami, F. 1954 On the halophilic alkaline phosphomonoesterase. Proc. Jap. Acad., **30**, 113–115.
100. Egami, F., Yamada, T., and Shiio, I. 1953 Sur une méthode simple pour l'extraction des enzymes bactériens. Compt. rend. soc. biol., **147**, 1531–1533.
101. Stuart, L. S., and James, L. H. 1938 Effect of sodium chloride on the Eh of protogenous media. J. Bacteriol., **35**, 369–380.
102. Stuart, L. S., and James, L. H. 1938 Effect of Eh and sodium chloride concentrations upon the physiology of halophilic bacteria. J. Bacteriol., **35**, 381–396.
103. Stuart, L. S. 1940 The growth of halophilic bacteria in concentrations of sodium chloride above three molar. J. Agr. Research, **61**, 259–265.
104. Brown, H. J., and Gibbons, N. E. 1955 The effect of magnesium, potassium, and iron on the growth and morphology of red halophilic bacteria. Can. J. Microbiol., **1**, 486–494.
105. Farghaly, A. H. 1950 Factors influencing the growth and light production of luminous bacteria. J. Cellular Comp. Physiol., **36**, 165–183.
106. Johnson, F. H. and Gray, D. H. 1949 Nuclei and large bodies of luminous bacteria in relation to salt concentration, osmotic pressure, temperature, and urethane. J. Bacteriol., **58**, 675–688.
107. Schneyer. L. H. 1953 The effects of potassium on bacterial luminescence intensity with reference to temperature and pressure conditions. J. Cellular Comp. Physiol., **42**, 285–293.
108. Morse, H. N. 1914 The osmotic pressure of aqueous solutions. Carnegie Inst. Wash. Publ. 198.
109. Mathews, A. P. 1906 Contribution to general principles of pharmacodynamics of salts and drugs. J. Infectious Diseases, **3**, 572–609.
110. Stuhlman, O. 1943 *An introduction to biophysics*. J. Wiley and Sons, Inc., New York, N. Y.
111. Davson, H., and Danielli, J. F. 1952 *The permeability of natural membranes*. Cambridge Univ. Press, London, England.
112. Johnson, F. H., Eyring, H., and Polissar, M. J. 1954 *The kinetic basis of molecular biology*. John Wiley and Sons, Inc., New York, N. Y.
113. Steinbach, H. B. 1951 Permeability. Ann. Rev. Physiol., **13**, 21–40.
114. Pratt, D., and Riley, W. 1955 Lysis of a marine bacterium in salt solutions. Bacteriol. Proc., **55**, 26.
115. Smith, L. S., and Krueger, A. P. 1955 Characteristics of a new vibrio bacteriophage system. J. Gen. Physiol., **38**, 161–168.
116. Spruit, C. J. P., and Pijper, A. 1952 An obligate halophilic bacterium from solar salt. Antonie van Leeuwenhoek J. Microbiol. Serol., **18**, 190–200.
117. Johnson, F. H., Zworykin, N., and Warren, G. 1943 A study of luminous bacterial cells and cytolysates with the electron microscope. J. Bacteriol., **46**, 167–185.
118. Browne, W. W. 1922 The staining of halophilic bacteria. Abstr. Bacteriol., **6**, 25–26.
119. Smithies, W. R., Gibbons, N. E., and Bayley, S. T. 1955 The chemical composition of the cell and cell wall of some halophilic bacteria. Can. J. Microbiol., **1**, 605–613.
120. Smithies, W. R., and Gibbons, N. E. 1955 The deoxyribose nucleic acid slime layer of some halophilic bacteria. Can. J. Microbiol., **1**, 614–621.
121. Dussault, H. P., and LaChance, R. A. 1952 Improved medium for red halophilic bacteria from salt fish. J. Fisheries Research Board Can., **9**, 157–163.
122. Garrard, E. H., and Lochhead, A. G. 1939 A study of bacteria contaminating sides of Wiltshire bacon, with special consideration of their behavior in concentrated salt solutions. Can. J. Research, D, **17**, 45–58.
123. Stuart, L. S. 1940 Effect of protein concentration and cysteine on the growth of halophilic bacteria. J. Agr. Research, **61**, 267–275.
124. Stuart, L. S., and Swenson, T. L. 1934 Some new morphological and physiological observations on salt-tolerant bacteria. J. Am. Leather Chemists' Assoc., **29**, 142–158.
125. Petter, H. F. M. 1931 On bacteria of salted fish. Koninkl. Akad. Wetenschap. Amsterdam, **34**, 1417–1423.
126. Pierce, George J. 1914 The Salton Sea. Carnegie Inst. Wash. Publ. 193.
127. Venkataraman, R., and Screenivasan, A.

1954 Studies on the red halophilic bacteria from salted fish and salt. Proc. Indian Acad. Sci., **319,** 17–23.

128. LIPMAN, C. B. 1926 The concentration of sea-water as affecting its bacterial population. J. Bacteriol., **12,** 311–313.
129. MOORE, H. N. 1940 The use of silica gels for the cultivation of halophilic organisms. J. Bacteriol., **40,** 409–413.
130. MOORE, H. N. 1941 The use of silica gels for the cultivation of halophilic organisms. II. Quantitative determination. J. Bacteriol., **41,** 317–321.
131. KATZNELSON, H., AND LOCHHEAD, A. G. 1952 Growth factor requirements of halophilic bacteria. J. Bacteriol., **64,** 97–103.
132. MACLEOD, R. A., ONOFREY, E., AND NORRIS, M. E. 1954 Nutrition and metabolism of marine bacteria. I. Survey of nutritional requirements. J. Bacteriol., **68,** 680–686.
133. BORING, J. R. 1955 The mineral requirements for respiration of a marine bacterium. Thesis, Univ. of Florida, Gainesville, Florida.
134. FLANNERY, W. L. 1955 Synthetic media for two nonpigmented obligate halophilic bacteria. Bacteriol. Proc., **55,** 26.
135. PARSONS, L. B., AND DOUGLAS, W. F. 1926 The influence of sodium chloride on the colorimetric determination of pH. J. Bacteriol., **12,** 263–265.
136. BERGMAN, M. 1929 On the red discoloration of salted hides and salt stains. J. Internatl. Soc. Leather Trades' Chemists, **13,** 599–611.
137. LE DANTEC, A. 1891 Étude de la morue rouge. Ann. Inst. Pasteur, **5,** 656–667.
138. KELLERMAN, K. F. 1915 Micrococci causing red deterioration of salted codfish. Zentr. Bakteriol. Parasitenk., Abt. II, **42,** 398–402.
139. POULSEN, V. A. 1880 On Nogle Mikroskopiske Planteorganismer et Morfolgisk Og Kritisk Studie. Vidensk. Meddel. Dansk. Naturhist Forening Kjøbanhavn, **31-32,** 231–254.
140. TATTEVIN, L. 1927 Le sel et les microbes. Thesis, Univ. Nancy, Vannes.
141. BITTING, A. W. 1911 Preparation of the cod and other salt fish for market; including a bacteriological study of the causes of reddening. U. S. Dep. Agr. Bur. Chem. Bull. 133.
142. EDINGTON, A. 1887 An investigation into the nature of the organisms present in "red" cod, and as to the cause of the red coloration. App. Fishery Board, Scotland, Ann. Rpt., **6,** 207–214.
143. HAUSAM, W. 1931 Zur Bakteriologie des Rotwerdens von Salzhäuten. Collegium, **729,** 12–16.
144. BREED, R. S. 1936 The systematic relationship of the red chromogenic organisms. J. Bacteriol., **32,** 357.
145. LOCHHEAD, A. G. 1943 Note on the taxonomic position of the red chromogenic halophilic bacteria. J. Bacteriol., **45,** 574–575.
146. HILL, J. H., AND WHITE, E. C. 1929 Sodium chloride for the separation of certain gram positive cocci from gram negative bacilli. J. Bacteriol., **18,** 43–57.
147. STURGES, W. S. 1923 Studies on halophilic micro-organisms. The flora of meat curing solutions. Abstr. Bacteriol., **7,** 11.
148. STURGES, W. S., AND HEIDMAN, D. 1924 Studies on halophilic microorganisms. II. The flora of meat curing solutions. Abstr. Bacteriol., **8,** 14–15.
149. STURGES, W. S., AND HEIDMAN, D. 1925 Observations on the constancy of the proposed grouping of the halophilic microorganisms. Abstr. Bacteriol., **9,** 2.
150. HAAG, F. E. 1926 *Sphaerotilus natans* (Sack) und *Bac. viridi-glaucescens* (Sack). Zugleich ein Beitrag zur Variabilität des *Bac. megatherium*. Zentr. Bakteriol. Parasitenk., Abt. II., **69,** 4–14.
151. OESTERLE, P. 1929 Der Einfluss von Bodenextrakt, Faulflüssigkeit und Lichtstrahlen. Zentr. Bakteriol. Parasitenk., Abt. II, **79,** 2–16.
152. STAHL, C. A. 1929 Der Einfluss von Kochsalz, Soda, Chloramin-Heyden, und Sublimat. Zentr. Bakteriol. Parasitenk., Abt. II, **79,** 16–25.
153. HOF, T. 1935 Investigations concerning bacterial life in strong brines. Rec. trav. botan. neerl., **32,** 92–173.

14

Reprinted from *Can. J. Microbiol.*, **2**(6), 599–606 (1956)

EFFECTS OF SODIUM AND POTASSIUM CHLORIDE ON CERTAIN ENZYMES OF MICROCOCCUS HALODENITRIFICANS AND PSEUDOMONAS SALINARIA[1]

BY R. M. BAXTER AND N. E. GIBBONS

Abstract

The isocitric, succinic, malic, and lactic dehydrogenases and cytochrome oxidase from *Pseudomonas salinaria*, an extreme halophile, are most active in high concentrations of sodium or potassium chloride. The lactic dehydrogenase and cytochrome oxidase of *Micrococcus halodenitrificans* are also more active at high concentrations of salt, but the isocitric, succinic, malic, α-ketoglutaric, and glutamic dehydrogenases of this moderate halophile are most active at much lower concentrations. It is concluded that the extreme halophiles have a high intracellular salt content. It is suggested that *M. halodenitrificans* maintains an intracellular salt concentration of about 0.5 *M* under optimal conditions for growth and that cytochrome oxidase is important in maintaining this level against a concentration gradient.

Introduction

Most halophilic bacteria fall into one of two groups on the basis of the sodium chloride concentration range in which they will grow. Moderate halophiles grow in concentrations of sodium chloride ranging from 1 or 2% to about 20%. Extreme halophiles do not grow at concentrations below about 15%, and grow even in saturated brines (about 31%, w/v).

In a previous paper (1) we reported that the glycerol dehydrogenase of *Pseudomonas salinaria*, an extreme halophile, required high concentrations of sodium chloride or potassium chloride for maximum activity, while the glycerol dehydrogenases of *Vibrio costicolus*, a moderate halophile, and of *Escherichia coli* were most active at rather low salt concentrations. About the same time Robinson and Katznelson (9) described a transaminase from *P. salinaria* which was most active at high salt concentrations.

Ingram (5) suggested that the proteins of halophiles may be less readily salted out than those of other bacteria. Robinson, Gibbons, and Thatcher (8), on the other hand, presented evidence that the moderate halophile *Micrococcus halodenitrificans* maintains a low intracellular salt concentration by an energy dependent mechanism. Our observations on glycerol dehydrogenases indicated that extreme halophiles may possess a high intracellular salt concentration, to which their enzymes are adapted, as suggested by Ingram, while moderate halophiles may maintain a low intracellular salt concentration, as suggested by Robinson, Gibbons, and Thatcher. A fundamental physiological difference thus appeared to exist between the two groups of organisms.

Shortly after our paper (1) appeared, the work of Egami and his co-workers (4, 15) was brought to our attention. These authors described an alkaline phosphomonoesterase from a moderate halophile, designated as *Pseudomonas* No. 101, which resembled the glycerol dehydrogenase of *P. salinaria* in being

[1]*Manuscript received July 3, 1956.*
Contribution from the Division of Applied Biology, National Research Laboratories, Ottawa, Canada. Issued as N.R.C. No. 4075.

most active at high salt concentrations. Later reports from Egami's laboratory described a salt-requiring glucose dehydrogenase from the same organism (14) and a salt-requiring cytochrome oxidase from a halotolerant micrococcus (13). Several other enzymes from the organisms studied by the Japanese workers do not require salt (3, 4, 14).

These facts suggested that the physiological difference between moderate and extreme halophiles may be less sharp than we had proposed. This paper describes the effects of salts on several enzymes from a moderate halophile, *Micrococcus halodenitrificans*, and several from *P. salinaria* in an attempt to assess the importance of salt-requiring enzymes in the adaptation of these organisms to life in brines.

Materials and Methods

Pseudomonas salinaria was grown as described previously (1). *Micrococcus halodenitrificans* was grown at room temperature on 0.5% proteose peptone and tryptone medium containing 8% sodium chloride. Cells were harvested in a Sharples supercentrifuge, washed with and finally suspended in sodium chloride solution of the same concentration as the growth medium, and broken by sonic vibration using a Raytheon 10 kilocycle sonic oscillator. The extracts were centrifuged in an M.S.E. high speed centrifuge at 10,000–11,000 r.p.m. and the supernatant stored at 4°–5° C. No attempt was made to purify any of the enzymes studied since partial purification of the glycerol dehydrogenase of *P. salinaria* (1) or of the halophilic phosphatase studied by Egami and his group (3) did not change the response of these enzymes to salt.

The activities of dehydrogenases not linked to DPN[2] or TPN[2] were measured manometrically in the Warburg apparatus by the ferricyanide technique of Quastel and Wheatley (6). The output of carbon dioxide in 30 min., corrected for the output in the absence of substrate, was used as a measure of enzyme activity. Substrate concentrations ranged from 0.003 to 0.01 *M*, the concentration chosen for each substrate being that found to give maximal activity. In the measurement of α-ketoglutaric dehydrogenase activity each vessel contained 0.5 mgm. TPP. The activity of the malic dehydrogenase of *P. salinaria* was also measured by the ferricyanide technique, in the presence of 1.0 mgm. DPN. All the enzymes of *M. halodenitrificans* were studied at 36° C., except the α-ketoglutaric dehydrogenase, which was studied at 29.5°. The enzymes of *P. salinaria* were also studied at 29.5° because the activity of the enzymes of this organism declined rather rapidly with time at the higher temperature. Experiments on the glycerol dehydrogenase of *P. salinaria* (unpublished) indicated that the salt concentration for maximal activity increased with increasing temperature. The effect of the difference in temperature will be to decrease rather than increase the difference between the enzymes of the two organisms.

[2]*The following abbreviations are used: DPN, diphosphopyridine nucleotide; TPN, triphosphopyridine nucleotide; TPP, thiamine pyrophosphate. DPN and TPN were obtained from the Sigma Chemical Company and TPP from Nutritional Biochemicals Corporation.*

The activities of malic dehydrogenase of *M. halodenitrificans*, and the isocitric dehydrogenases of both organisms were measured by following the increase in reduced DPN or TPN, as indicated by the increase in optical density at 340 mμ in the Beckman spectrophotometer. The increase in optical density in five minutes was taken as a measure of enzyme activity. Phosphate buffer (*M*/30, pH 7.3) was used with the isocitric dehydrogenase of *M. halodenitrificans*, and hydrazine buffer (0.3 *M*, pH 7.3) with the other enzymes.

Cytochrome oxidase activity was determined by measuring the oxygen uptake in the Warburg apparatus in the presence of *p*-phenylenediamine, 10 mgm./ml. The oxygen uptake in 30 min., corrected for the autoxidation of *p*-phenylenediamine, was taken as a measure of enzyme activity.

Results

Tricarboxylic Acid Cycle Dehydrogenases

The four dehydrogenases of the tricarboxylic acid cycle were present in extracts of *Micrococcus halodenitrificans*, but in *Pseudomonas salinaria* only three, succinic, malic, and isocitric dehydrogenases, could be demonstrated. Apart from their responses to salt, analogous enzymes from the two organisms were similar, and showed no unusual properties. Both malic dehydrogenases specifically required DPN, and both isocitric dehydrogenases TPN. The isocitric dehydrogenase of *M. halodenitrificans* required magnesium ion, but the corresponding enzyme from *P. salinaria* was fully active without the addition of magnesium. Possibly this enzyme contains sufficient magnesium for maximal activity in a firmly bound state. Neither succinic dehydrogenase could be shown to require any cofactor. The α-ketoglutaric dehydrogenase of *M. halodenitrificans* required TPP.

The effects of salts on corresponding enzymes from the two organisms are strikingly different (Fig. 1). Optimal salt concentrations for the enzyme of *M. halodenitrificans* ranged from less than 0.01 *M* (malic dehydrogenase) to about 0.6 *M* (α-ketoglutaric dehydrogenase), and there was little difference between the effects of sodium chloride and potassium chloride. The enzymes from *P. salinaria* on the other hand were most active in the presence of potassium chloride at concentrations of about 3 *M*, or higher. Sodium chloride was much less effective than potassium chloride in activating all the enzymes except malic dehydrogenase.

Other Dehydrogenases

M. halodenitrificans was found to contain a glutamic dehydrogenase which did not appear to require a coenzyme. It is most active at a salt concentration of about 0.3 *M* (Fig. 2). A corresponding enzyme could not be demonstrated in *P. salinaria*.

Both organisms were found to contain lactic dehydrogenases which do not require coenzymes. That from *P. salinaria* was similar to the other dehydrogenases of this organism (Fig. 2). The lactic dehydrogenase from *M. halodeni-*

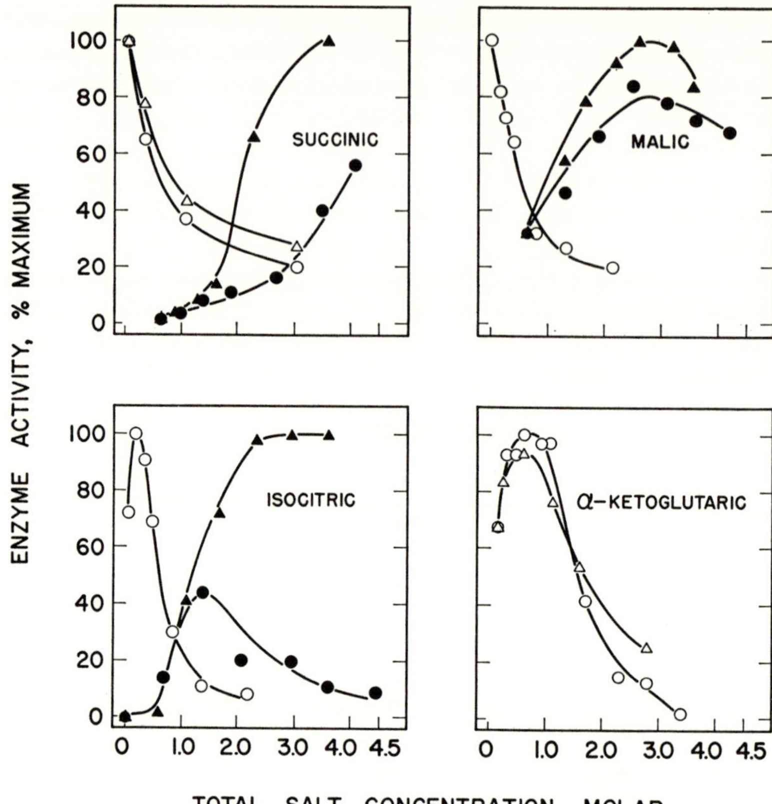

FIG. 1. Effects of sodium chloride and potassium chloride on the tricarboxylic acid cycle dehydrogenases of *Micrococcus halodenitrificans* and *Pseudomonas salinaria*. ○: *M. halodenitrificans*, NaCl; △: *M. halodenitrificans*, KCl; ●: *P. salinaria*, NaCl; ▲: *P. salinaria*, KCl.

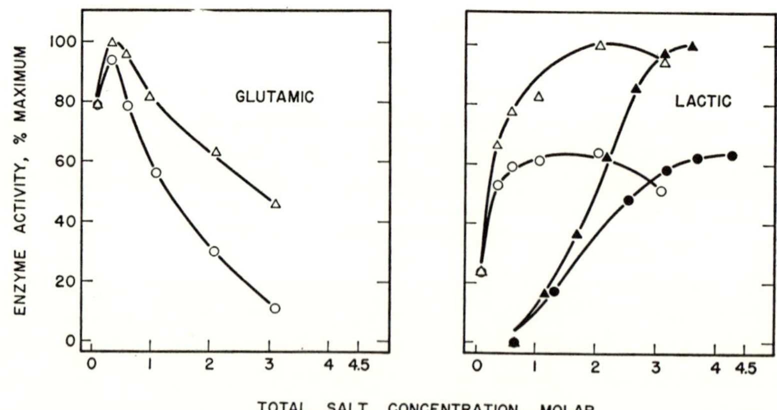

FIG. 2. Effects of sodium chloride and potassium chloride on glutamic and lactic dehydrogenase of *Micrococcus halodenitrificans*, and lactic dehydrogenase of *Pseudomonas salinaria*. ○: *M. halodenitrificans*, NaCl; △: *M. halodenitrificans*, KCl; ●: *P. salinaria*, NaCl; ▲: *P. salinaria*, KCl.

trificans (Fig. 2) was markedly different from the other dehydrogenases of the micrococcus, showing optimal activity at a potassium chloride concentration of about 2 *M* and little decline in activity even at 3 *M*. Sodium chloride was much less effective than potassium chloride in activating the enzymes of both organisms.

Cytochrome Oxidase

The cytochrome oxidases of both organisms were found to be halophilic, and differed from most other halophilic enzymes in being activated equally by either sodium or potassium chloride (Fig. 3). The responses of the two enzymes to salt were nearly the same, except that the enzyme from *M. halodenitrificans* showed relatively greater activity at low salt concentrations.

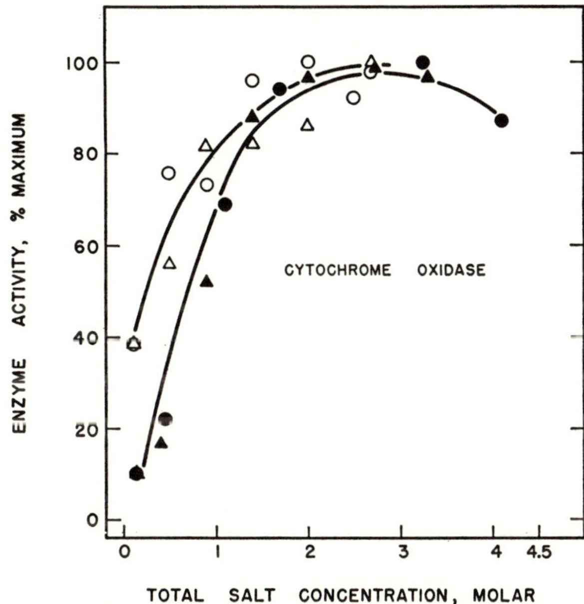

FIG. 3. Effects of sodium chloride and potassium chloride on cytochrome oxidase of *Micrococcus halodenitrificans* and *Pseudomonas salinaria*. ○: *M. halodenitrificans*, NaCl; △: *M. halodenitrificans*, KCl; ●: *P. salinaria*, NaCl; ▲: *P. salinaria*, KCl.

Discussion

Bacterial halophilism involves two phenomena, which are not necessarily related: tolerance of salt below a certain maximum and a requirement for salt above a certain minimum. There is evidence (10) that at low salt concentrations the permeability of the cell wall or cytoplasmic membrane of moderate halophiles is greatly increased allowing leakage of cytoplasmic constituents. Certain of the extreme halophiles, particularly rod-shaped forms, appear to dissolve completely at low salt concentrations. These observations provide at least a partial explanation of the salt requirement of halophiles. In extreme halophiles, salt is apparently also required for the activity of many, if not all, of the enzymes of the organism. Seven enzymes of *P. salinaria* have been described and all are halophilic (Table I).

The observation that the enzymes of *P. salinaria* are most active at high salt concentrations provides at the same time a partial explanation of the salt tolerance of the organism. Other macromolecular constituents of the organism may be similarly adapted to high salt concentrations. In particular the chromatin material may well differ from that of non-halophiles, which has been shown to be very sensitive to changes in the ionic environment (12). There seems to be no reason to modify our earlier view (1) that this organism possesses a high intracellular electrolyte concentration to which its vital processes are adapted, and that this is probably a general characteristic of extreme halophiles.

The salt tolerance of moderate halophiles cannot be adequately explained on this basis. Of 15 enzymes studied in three organisms of this type, only four are halophilic (Table I). It is perhaps significant that although the halophilic enzymes of *M. halodenitrificans* have about the same optimal salt concentrations as the corresponding enzymes of *P. salinaria*, their activity at suboptimal concentrations is relatively much greater. This is particularly true of the

TABLE I

THE SALT REQUIREMENTS OF ENZYMES OF HALOPHILIC BACTERIA

H: Optimal salt concentration $\geqq$ 1.5 *M*. N: Optimal salt concentration $\leqq$ 0.5 *M*
0: Not found. Figures in parentheses refer to references.

Enzyme	Organism: *Pseudomonas salinaria* (Extreme halophile)	*Micrococcus halodenitrificans* (Moderate halophile)	*Vibrio costicolus* (Moderate halophile)	*Pseudomonas* No. 101 (4) (Moderate halophile)
Dehydrogenases				
Isocitric	H	N	—	—
α-Ketoglutaric	0	N	—	—
Succinic	H	N	—	—
Malic	H	N	—	—
Lactic	H	H	—	—
Glucose	0	—	—	H (14)
Formic	0	0	—	N (14)
Glycerol	H (1)	0	N (1)	—
Glutamic	0	N	—	—
Nitrite reductase - lactic dehydrogenase	—	N (8)	—	—
Nitrate reductase - succinoxidase	—	—	N (7)	—
Cytochrome oxidase	H	H	—	—
Transaminase	H (9)	—	—	—
Protease	—	—	—	N (4)
Catalase	—	—	—	N (4)
Alkaline phosphomonoesterase	—	—	—	N (4)

lactic dehydrogenase (Fig. 2). At a salt concentration of 0.5 *M*, none of the halophilic or non-halophilic enzymes of *M. halodenitrificans* described in this paper operates at less than 55% of its maximum activity.

It also seems significant that one of the halophilic enzymes is cytochrome oxidase. It has often been postulated that the cytochrome system plays a role in ionic transfer in many organisms (e.g. (2)). If it is involved in ion transfer in *M. halodenitrificans*, the increase in its activity with increasing salt concentration may perhaps be regarded as a feedback mechanism tending to maintain a relatively constant internal environment. This also seems to be a reasonable explanation for the halophilic behavior of the cytochrome oxidase from a halotolerant organism (13), which would not be expected to contain a high concentration of salt.

On the basis of the foregoing discussion, the following seems a reasonable description of the behavior of *M. halodenitrificans*. Under optimal growth conditions, i.e. with 1 to 1.5 *M* salt in the medium, the cells contain about 0.5 *M* salt. At this concentration the various enzymes act at rates of from 60 to 100% of their maximum. The concentration of the various enzymes is probably such that under these conditions there is a maximal over-all rate of utilization of substrates. When salt enters the cell, driven by the steep concentration gradient across the cell wall, the activity of the cytochrome oxidase increases and the optimal concentration is restored. When the external salt concentration is too high, the cytochrome pump is unable to maintain the optimal internal concentration and a steady state is set up at a higher intracellular salt concentration where the activity of the non-halophilic dehydrogenases is inhibited with an over-all reduction in the rate of substrate utilization and growth. At very high salt concentrations the activity of the halophilic dehydrogenases and even of the cytochrome oxidase is reduced, so that eventually the maximum salt concentration at which growth may occur is reached when the cytochrome pump can no longer cope with the salt gradient.

The moderate halophiles then seem to be intermediate in the evolution of the extreme halophiles. For certain relatively halotolerant organisms saline environments provided an ecological niche in which potential competitors were unable to grow (11). The selection of mutants possessing a more halophilic cytochrome system and consequently a more efficient salt-transporting system gave rise to still more halotolerant forms. Further mutations which altered the cell walls and possibly other cellular components gave rise to moderate halophiles. However, in more concentrated brines, the expenditure of much energy was required to maintain a low internal concentration and eventually those forms were selected that could tolerate higher internal salt concentrations and thus conserve energy. Further mutation gave rise to the extreme halophiles in which all the enzymes are adapted to a high electrolyte concentration and the ion transporting mechanism is required only to concentrate potassium within the cell and to maintain very small differences in internal and external osmotic pressure.

References

1. Baxter, R. M. and Gibbons, N. E. The glycerol dehydrogenases of *Pseudomonas salinaria*, *Vibrio costicolus*, and *Escherichia coli* in relation to bacterial halophilism. Can. J. Biochem. Physiol. 32 : 206-217. 1954.
2. Conway, E. J. A Redox pump for the biological performance of osmotic work, and its relation to the kinetics of free ion diffusion across membranes. Intern. Rev. Cytol. 2 : 419-445. 1953.
3. Egami, F. Recherches biochimiques sur les bactéries halotolérantes et halophiles. Bull. soc. chim. biol. 37 : 207-217. 1955.
4. Egami, F., Yamada, T., and Shiio, I. Sur une méthode simple pour l'extraction des enzymes bactériens. Compt. rend. soc. biol. 147 : 1531. 1953.
5. Ingram, M. A theory relating the action of salts on bacterial respiration to their influence on the solubility of proteins. Proc. Roy. Soc. (London), B, 134 : 181-201. 1947.
6. Quastel, J. H. and Wheatley, A. H. M. Anaerobic oxidations. On ferricyanide as a reagent for the manometric investigation of dehydrogenase systems. Biochem. J. 32 : 936-943. 1938.
7. Robinson, J. A theory to explain microbial halophilism. Rev. can. biol. (Abstr.), 11 : 518. 1953.
8. Robinson, J., Gibbons, N. E., and Thatcher, F. S. A mechanism of halophilism in *Micrococcus halodenitrificans*. J. Bacteriol. 64 : 69-77. 1952.
9. Robinson, J. and Katznelson, H. Aspartate-glutamate transaminase in a red halophilic bacterium. Nature, 172 : 672. 1953.
10. Smithies, W. R. and Gibbons, N. E. The deoxyribose nucleic acid slime layer of some halophilic bacteria. Can. J. Microbiol. 1 : 614-621. 1955.
11. VanNiel, C. B. Natural selection in the microbial world. J. Gen. Microbiol. 13 : 201-217. 1955.
12. Whitfield, J. F. and Murray, R. G. E. The effects of the ionic environment on the chromatin structures of bacteria. Can. J. Microbiol. 2 : 245-260. 1956.
13. Yamada, T. and Asano, A. Cytochromeoxydase extraite d'une bactérie halotolérante. Compt. rend. soc. biol. 148 : 1310-1312. 1954.
14. Yamada, T. and Asano, A. Oxidation–reduction enzymes of a halophilic bacterium. J. Biochem. (Japan), 41 : 639-645. 1954.
15. Yamada, T., Shiio, I., and Egami, F. On the halophilic alkaline phosphomonoesterase. Proc. Japan Acad. 30 : 113-115. 1954.

15

Reprinted from *J. Gen. Microbiol.*, **20**(1), 27–31 (1959)

The Freezing Points of Bacterial Cells in Relation to Halophilism

By J. H. B. CHRISTIAN* and M. INGRAM

Low Temperature Station for Research in Biochemistry and Biophysics. University of Cambridge and Department of Scientific and Industrial Research

SUMMARY: The hypothesis that halophilic bacteria achieve a high degree of salt tolerance by exclusion of much of the external solutes of the growth medium has been tested. Organisms were grown in media containing from 1 to 4 M salts. For both halophilic and non-halophilic bacteria the freezing points of the cells were close to those of the media in which they were grown. It is concluded that halophilic bacteria do not maintain the intracellular aqueous phase at a water activity greater than in the surrounding growth medium.

One possible explanation for the phenomenon of bacterial halophilism is that the solutes in the aqueous phase within halophilic cells are much less concentrated than in the surrounding medium (Robinson, Gibbons & Thatcher, 1952). Such a system might be maintained, apart from mechanisms of accumulation and excretion, by a membrane impermeable to external solutes and a rigid cell boundary which resists plasmolysis. To examine this hypothesis, the freezing-point depressions of bacterial suspensions and of their growth media have been determined and compared. The bacteria included both halophilic and non-halophilic strains grown in media containing various concentrations of salts. No appreciable differences in the osmotic status of these two types of bacteria were observed.

METHODS

The four halophilic strains tested were *Micrococcus halodenitrificans*, *Vibrio costicolus*, *Halobacterium halobium* and *Sarcina littoralis* from the collection of Dr N. E. Gibbons, National Research Council, Ottowa. Other cultures used were *Micrococcus lysodeikticus* (NCTC 2665), *Staphylococcus aureus*, *Escherichia coli* and a salt-tolerant *Bacillus circulans* from the collection of the Low Temperature Research Station, Cambridge.

The red halophiles (*Halobacterium halobium* and *Sarcina littoralis*) were grown on the medium of Baxter & Gibbons (1954), but with 2 M-NaCl and 2 M-KCl. For all other bacteria the growth medium contained 0·5% (w/v) each of proteose peptone (Difco) and tryptone (Difco) with additions of NaCl, or of NaCl and KCl as required. The initial pH value was 7·0 in all media. These media were solidified by 2% (w/v) Bacto agar.

To decrease the amount of contaminating medium in suspensions of organisms, the bacteria were grown on solid media in Petri dishes. An inoculum

* Present address: Division of Food Preservation and Transport, Commonwealth Scientific and Industrial Research Organization, Homebush, N.S.W., Australia.

of 0·1 ml. was spread evenly over the surface. The plates were dried at the temperature of subsequent incubation until no free liquid remained on the surface. They were incubated inverted and without lids in sealed desiccators over salt solutions isotonic with the growth medium. The incubation temperature was 37° for red halophiles and 30° for other strains. The incubation period varied with the organism and with the concentration of salts.

The organisms so produced were carefully scraped from the medium by means of a thick wire and rapidly transferred to a tapered glass centrifuge tube, which was stoppered between transfers to limit evaporation. After weighing, the stoppered tube and its contents were autoclaved to break down the cell permeability barriers and any water lost was replaced from a fine capillary pipette. The wet weight of the samples ranged from 0·35 to 1·7 g.

Freezing points were determined with a copper-constantan thermocouple, and the temperature read directly from a Pye Scalamp galvanometer calibrated in degrees Centigrade. The tip of the thermocouple was wound into a tight spiral with the junction bent inwards, so that when inserted into the cell debris the junction was centrally situated away from the wall of the tube. The thermocouple leads were held in position by a cork stopper. In most experiments the freezing mixture was NaCl + ice, but for determinations with the red halophiles $CaCl_2$ + ice mixtures were necessary to give initial temperatures near −30°.

RESULTS

The initial freezing of the sample at a cooling rate of about 2°/min. involved substantial supercooling and the resulting freezing point was spuriously low. It was plainly essential to avoid such error. Accordingly, the sample was thawed and replaced in a freezing-mixture which gave a lower rate of freezing, so that subsequent supercooling was less and the freezing-point determination more accurate; this procedure was repeated, until the freezing point became constant. Values thus determined both during freezing and subsequent thawing never differed by more than 0·1°, indicating that supercooling was not occurring.

Because, as mentioned above, it was frequently necessary to restore water lost from the pellet of cells during the previous heat treatment, freezing points were recorded with the thermocouple in three different positions. When the three values differed by more than 0·1°, it was concluded that the solutes were imperfectly distributed in the aqueous phase, so the preparation was stirred with the thermocouple, and the observations repeated. The quoted values are the averages of the last three determinations.

Freezing points were similarly determined for all media before addition of agar and for all equilibrating salt solutions after growth. These two values agreed closely. Scott (1953) stated that the effect of the usual concentrations of agar on the water activity of a medium is very small.

The relationship between the freezing points of heated cell preparations and freezing points of the growth media are shown in Fig. 1, where the points refer to media with aggregate salt concentrations of 1, 2 or 4 M. For a given growth medium the values obtained with non-halophilic bacteria were generally slightly lower than those found for the moderate halophiles. However, within both groups the differences between species were too large for any significance to be attached to this observation. There were no differences between organisms grown on media containing additions of NaCl and KCl or NaCl alone.

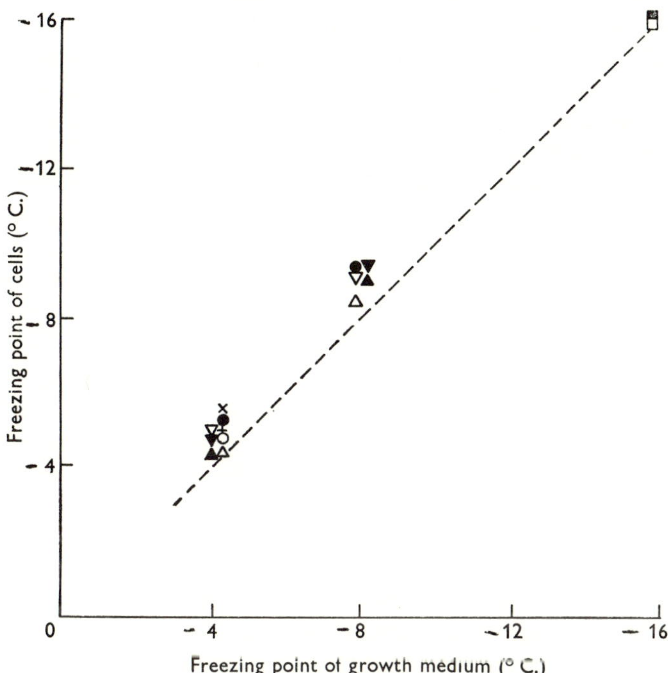

Fig. 1. Relationship between freezing points of heated organisms and growth medium. The broken line is isothermal. From media containing added NaCl only: ▲, *Micrococcus halodenitrificans*; ▼, *Vibrio costicolus*. From media containing equimolar additions of NaCl and KCl: ×, *Staphylococcus aureus*; △, *Micrococcus halodenitrificans*; +, *Escherichia coli*; ▽, *Vibrio costicolus*; ○, *Micrococcus lysodeikticus*; □, *Halobacterium halobium*; ●, *Bacillus circulans*; ■, *Sarcina littoralis*.

The amount of contaminating growth medium contributing to the freezing point of the samples is not known. However, since the freezing points of all organisms except *Halobacterium halobium* were lower than those of their growth medium, corrections for contamination would increase this deficit. Hence, the freezing points of the halophilic and non-halophilic organisms were equal to or lower than the freezing points of the media on which they were grown.

DISCUSSION

The use of cryoscopic data to determine the internal osmotic status of cells is open to criticism. A major objection is that in disrupting the cells considerable changes may occur in the distribution and association of intracellular substances. Thus constituents which are normally combined in an osmotically inactive form might be liberated and large complexes broken down to smaller, more osmotically active units. This criticism is largely answered by the data of Mitchell & Moyle (1956) and of Scott (1953). Mitchell & Moyle estimated the osmotic pressure within *Staphylococcus aureus* by two independent methods. By equilibrating washed organisms over sucrose solutions of known concentration they deduced an osmotic pressure of 20–25 atmospheres. From the weight of solutes extracted by trichloroacetic acid from organisms of known water content they calculated a value of 20–30 atmospheres. Scott obtained similar results by the cryoscopic method. From the freezing-point depression of unwashed, centrifuged and heated *S. aureus* an osmotic pressure of about 25 atmospheres was obtained after allowing for the contribution of contaminating growth medium.

The results in Fig. 1 show that in no case was the internal aqueous content of the halophilic or non-halophilic bacteria tested likely to be less concentrated than the growth medium. Hence, it is concluded that halophilic bacteria do not owe their salt tolerant properties to an ability to maintain a comparatively low internal solute concentration.

It is further interesting that no difference appeared in pattern of behaviour between the less extreme halophiles like *Micrococcus halodenitrificans* and the most extreme red types like *Sarcina littoralis* or *Halobacterium halobium*. Gibbons & Baxter (1953) concluded that the internal concentration of sodium chloride in *Micrococcus halodenitrificans* remained at 5 % in organisms grown in concentrations ranging from 8 to 22 %, whereas the concentration in *Sarcina littoralis* increased linearly from 10 to 19 % with increase of external concentration from 18 to 27 %. This encouraged Baxter & Gibbons (1954) to suggest that the group of organisms typified by *Micrococcus halodenitrificans* possesses a special mechanism, absent from the red group, for maintaining the intracellular concentration well below that of the environment. The apparent implication would be that, with organisms grown in about 4 M-NaCl, their freezing point should be about -5° for *M. halodenitrificans* but about -10°, roughly twice as great, for *Sarcina littoralis*. Our experiments, besides indicating freezing points lower than the above values, give no hint of any such difference in behaviour between the two groups of organisms. Because the conclusion of Gibbons & Baxter was based on estimations of chloride, however, the differences might well be due to the existence of additional solutes not accounted for by means of chloride estimations.

The nature and contribution of all the internal solutes is not known but Ingram (1938), and Yamada & Shiio (1953), indicated that there is substantial excess of Na^+ over Cl^-; and Christian (1956) found that halophilic organisms contained large amounts of potassium as well as sodium, the former probably predominating in the red extreme halophiles. Thus it is suggested that the salts of both these cations are largely responsible for the low internal water activities indicated by freezing-point determinations.

For a given strain grown in widely differing concentrations of salts there was a fairly constant difference between the freezing points of the cell contents and the growth medium. It has also been shown (Christian & Ingram, 1959) that the osmotic sensitivity of both halophilic and non-halophilic bacteria is proportional to the NaCl content of the medium during growth. From these data it is concluded that the osmotic status of bacterial cells is very largely determined by the water activity of the growth medium.

This work was carried out as part of the programme of the Food Investigation Organization of the Department of Scientific and Industrial Research during the tenure by one of us (J.H.B.S.) of an Overseas Studentship of the Commonwealth Scientific and Industrial Research Organization.

REFERENCES

BAXTER, R. M. & GIBBONS, N. E. (1954). The glycerol dehydrogenases of *Pseudomonas salinaria*, *Vibrio costicolus* and *Escherichia coli* in relation to bacterial halophilism. *Canad. J. Biochem. Physiol.* **32**, 206.

CHRISTIAN, J. H. B. (1956). The physiological basis of salt tolerance in halophilic bacteria. Dissertation, Cambridge.

CHRISTIAN, J. H. B. & INGRAM, M. (1959). Lysis of *Vibrio costicolus* by osmotic shock. *J. gen. Microbiol.* **20**, 32.

GIBBONS, N. E. & BAXTER, R. M. (1953). The relation between salt concentration and enzyme activity in halophilic bacteria. *Proc. VIth int. Congr. Microbiol.*, Rome, **1**, 210.

INGRAM, M. (1938). The effect of sodium chloride on a bacterial enzyme which destroys lactic acid. *Rep. Fd Invest. Bd Lond.*, p. 72.

MITCHELL, P. & MOYLE, J. (1956). Osmotic structure and function in bacteria. In *Bacterial Anatomy. Symp. Soc. gen. Microbiol.* **6**, 150.

ROBINSON, J., GIBBONS, N. E. & THATCHER, F. S. (1952). A mechanism of halophilism in Micrococcus halodenitrificans. *J. Bact.* **64**, 69.

SCOTT, W. J. (1953). Water relations of *Staphylococcus aureus* at 30° C. *Aust. J. biol. Sci.* **6**, 549.

YAMADA, T. & SHIIO, I. (1953). Effects of salt concentration on the respiration of a halotolerant bacterium. *J. Biochem., Tokyo*, **40**, 327.

(*Received* 24 *June* 1958)

16

Reprinted from *J. Bacteriol.*, **99**(1), 161–168 (1969)

Purification and Reversible Inactivation of the Isocitrate Dehydrogenase from an Obligate Halophile

JERRY S. HUBBARD AND ALAN B. MILLER

Bioscience Section, Jet Propulsion Laboratory, California Institute of Technology, Pasadena, California 91103

The nicotinamide adenine dinucleotide phosphate-specific isocitrate dehydrogenase of *Halobacterium cutirubrum* is rapidly inactivated at low NaCl levels. As much as 75% of the initial activity can be restored by dialyzing the inactive enzyme against 4 M NaCl. A mixture of 4 mM isocitrate and 10 mM $MnCl_2$ gives the same protection as 4 M NaCl but does not replace the NaCl requirement for reactivation. The reactivated and native enzymes have identical sedimentation rates on sucrose gradients, electrophoretic mobilities on polyacrylamide gels, and elution rates from Sephadex G-200. However, there are distinct differences between the active and inactive forms of the enzyme. Compared with the active enzyme, the inactive protein has a lower sedimentation rate, a lower electrophoretic mobility, and a faster elution rate from Sephadex. These differences indicate that inactivation causes a major conformational change in the protein. Presumably, the removal of NaCl permits the enzyme to expand into a less dense, inactive form. The isocitrate dehydrogenase was purified 69-fold by a procedure involving the following steps. When the enzyme is selectively protected with isocitrate and $MnCl_2$ at low ionic strength, most of the contaminating proteins are precipitated with $(NH_4)_2SO_4$ at 0.9 saturation. The enzyme in the supernatant fluid is then inactivated at low NaCl levels, precipitated with 0.5 saturated $(NH_4)_2SO_4$, and reactivated with 4 M NaCl. Minor impurities are removed by gel filtration on Sephadex G-200. The resulting preparation is more than 95% pure as judged by disc electrophoresis.

The enzymes of the extreme halophiles become rapidly inactivated if the salt concentration is reduced to low levels (3, 7, 8). Studies on the molecular basis of the salt dependency have been hampered by the fact that the halophilic enzymes are difficult to purify. This paper describes a method for the purification of the nicotinamide adenine dinucleotide phosphate (NADP)-specific isocitrate dehydrogenase (ICDH; EC 1.1.1.42) of *Halobacterium cutirubrum*. In addition, a comparison is made of certain physical properties of the active and inactive forms of the halophilic ICDH.

MATERIALS AND METHODS

Growth of cells. The strain of *H. cutirubrum* (No. 9) was obtained from N. E. Gibbons, National Research Council, Ottawa, Canada. Cultures were grown aerobically at 37 C for 24 hr in CSG medium (11) containing 4.2 M NaCl.

ICDH activity. The reduction of NADP was measured in a Calbiometer spectrophotometer by following the increase in absorbancy at 340 nm. The reaction was carried out in cuvettes of 1-cm light path at 30 C. Reaction mixtures contained 0.02 M tris(hydroxymethyl)aminomethane (Tris)-hydrochloride (*p*H 7.5), 1.33 mM $MnCl_2$, 0.167 mM NADP (sodium salt), 1.33 mM trisodium isocitrate, 0.5 to 0.65 M NaCl, and enzyme in a final volume of 3 ml. Determinations of absorbancy were made at 30-sec intervals after the addition of enzyme. The amount of enzyme was adjusted so that the initial rate of NADP reduction was linear with time and proportional to enzyme concentration. Specific activity is defined as micromoles of NADP reduced per minute per milligram of protein. Protein was determined by the method of Lowry et al. (9), with bovine serum albumin as the standard.

ICDH purification. All purification steps were performed at 4 C. In all cases, the Tris buffer refers to 0.02 M Tris-hydrochloride at *p*H 7.5. The harvest of cells from 4 liters of culture was suspended in 50 ml of Tris buffer containing 4.2 M NaCl. Cells were ruptured by treatment with a Sonifyer Cell Disruptor for 1 min, and the cell debris was removed by centrifugation at 18,000 × *g* for 30 min. The crude extract was dialyzed for 12 hr against 2 liters of Tris buffer containing 4.2 M NaCl and then was dialyzed for an additional 12 hr against 9 volumes of Tris buffer containing 4.4 mM isocitrate and 11 mM $MnCl_2$. Solid $(NH_4)_2SO_4$ was added to make the extract 0.8 saturated. After 15-min equilibration, the extract was centrifuged for 30 min at 18,000 × *g*. The supernatant fluid (S_1) was saved, and to the pellet (P_1) was added 25 ml of Tris buffer containing 0.42 M NaCl, 4 mM isocitrate, and 10 mM $MnCl_2$ which had been made 0.8 saturated with $(NH_4)_2SO_4$. The P_1 fraction was stirred intermittently for 1 hr and then centrifuged at 18,000 × *g* for 30 min. The supernatant fluid obtained (S_2) was combined with S_1. The combined fraction (S_1S_2) was made 0.9 saturated with the addition of solid $(NH_4)_2SO_4$, equilibrated for 15 min, and then centrifuged at 130,000 × *g* for 1 hr. The resulting supernatant fluid (S_3) was dialyzed for 12 hr against 2 liters of Tris buffer containing 4.2 M NaCl, transferred to a fresh dialysis tubing, concentrated over a bed of Carbowax 6000 (polyethylene glycol) to a final volume of 95 ml, and dialyzed for 12 hr against 2 liters of Tris buffer containing 4.2 M NaCl. This concentrated S_3 fraction was used as the source of partially purified enzyme. With 4.2 M NaCl present, the ICDH activity was stable for several months when stored at 4 C. A rapid loss of activity occurred when the extract was stored at −18 C.

For further purification, 60 ml of the concentrated S_3 was inactivated by dialysis for 2 hr against nine volumes of Tris buffer containing 2.2 mM β-mercaptoethanol. The inactivated enzyme was precipitated by slowly adding 1.25 volumes of Tris buffer containing 2 mM β-mercaptoethanol and 0.42 M NaCl which had been made 0.9 saturated with $(NH_4)_2SO_4$. After 15-min equilibration, the extract was centrifuged at 18,000 × *g* for 15 min. The pellet (P_4) was carefully drained and then dissolved in 10 ml of Tris buffer containing 2 mM β-mercaptoethanol and 0.42 M NaCl. Residual $(NH_4)_2SO_4$ was removed by dialyzing the P_4 twice for 1 hr against 200 ml of the same buffer. The ICDH in the P_4 was reactivated by the procedure described below. After reactivation, the volume of P_4 was 7.7 ml.

A 5-ml amount of reactivated P_4 was placed in a dialysis bag and concentrated over Carbowax to 1.8 ml. The concentrated P_4 was added to a Sephadex G-200 column (330 × 23 mm) which was equilibrated with Tris buffer containing 4.2 M NaCl. The ICDH was eluted from the column with Tris buffer containing 4.2 M NaCl, and the enzyme activity was measured on 0.1-ml samples of the eluant. The highest specific activity of ICDH was in the fractions comprising the trailing half of the activity peak, i.e., an elution volume of 78 to 88 ml. These fractions were combined in a dialysis bag and concentrated over Carbowax to 1.4 ml.

Inactivation and reactivation of ICDH. Unless otherwise indicated, the ICDH was inactivated by dialyzing the extract for 2 hr at 4 C against 100 to 200 volumes of 0.02 M Tris buffer (*p*H 7.5) containing 2 mM β-mercaptoethanol. Reactivation was performed by dialyzing the inactive ICDH against 100 volumes of 0.02 M Tris buffer containing 4.2 M NaCl (*p*H 7.5) for 1.5 hr at 31 C, and then dialyzing against a second change of the same buffer for 2 hr at 31 C.

Sucrose gradient centrifugation. Sucrose density gradients (5 to 20%) were prepared and sampled according to the procedure of Martin and Ames (10). For experiments with active ICDH, a buffer containing 0.02 M Tris (*p*H 7.5), 4 mM isocitrate, and 10 mM $MnCl_2$ was mixed with the sucrose solutions. Inactivated ICDH was analyzed on gradients prepared with sucrose solutions containing 0.02 M Tris (*p*H 7.5) and 2 mM β-mercaptoethanol. Samples were centrifuged for 8 hr at 39,000 rev/min in a Spinco model L centrifuge equipped with a swinging-bucket rotor (SW 39).

Disc gel electrophoresis. The disc electrophoresis of proteins was performed by the procedure of Ornstein and Davis according to the instructions provided by Canalco Co., Silver Spring, Md. The procedure was modified in that the 7% gel and the Tris-glycine developing buffers were supplemented with 4 mM isocitrate and 10 mM $MgCl_2$. Before electrophoresis, samples of active ICDH were dialyzed for 2 hr against 200 volumes of 0.02 M Tris buffer (*p*H 7.5) containing 4 mM isocitrate and 10 mM $MnCl_2$. Samples of inactivated ICDH were in 0.02 M Tris-2 mM β-mercaptoethanol buffer (pH 7.5). The electrophoretograms were stained with water-soluble Nigrosin (0.0125%) in methanol-water-acetic acid (4:5:1) for 3 hr, destained with the same solvent, and stored in 7% aqueous acetic acid (1).

RESULTS

Stabilization of ICDH. Figure 1 shows activity measurements after partically purified ICDH was incubated with various levels of NaCl. Complete stabilization was achieved when the buffer contained more than 3 M NaCl. At lower concentrations, the degree of protection was roughly proportional to the NaCl level. For example, with 2.2 M NaCl about 50% of the activity was lost in 20 hr. With 1.2 M NaCl essentially all the activity was lost in 3 hr.

Certain other salts can protect the ICDH (Table 1). The effectiveness appears to be related to both the anion and cation employed. At 1.5 M levels, the order of effectiveness of the cations is $Na^+ > K^+ > Rb^+ > Cs^+ > Li^+$ and NH_4^+. Of the anions tested, acetate and SO_4^{2-} were more effective than Cl^-, and Br^- was ineffective.

From experiments in which assays were made with low NaCl levels, it was apparent that some component(s) of the reaction mixture replaces the high salt requirement for stabilizing the ICDH activity. Table 2 shows that complete stability was not conferred by isocitrate, NADP, or the metal activator when tested individually. However, the ICDH activity was protected when both isocitrate and $MnCl_2$ were added. A synergistic effect was also noted with isocitrate and $MgCl_2$,

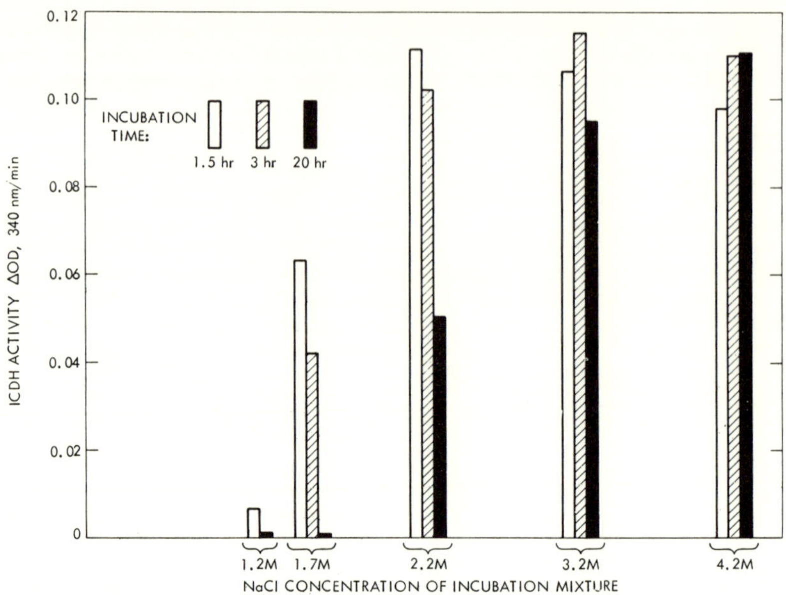

FIG. 1. *Stabilization of ICDH by various levels of NaCl. All incubations contained 0.02 M Tris (pH 7.5), 0.21 M NaCl, and enzyme (57 μg of protein) in a final volume of 0.4 ml. Incubations were performed at 0 C. Activity measurements were made on 0.1-ml samples of the incubation mixtures.*

TABLE 1. *Stabilization of ICDH by various salts*

Addition to incubation mixture[a]		Percentage of initial activity remaining after incubation for	
Salt	Concn (M)	1.5 hr	3 hr
None	—	0	0
NaCl	1.5	55	32
NaBr	1.5	0	0
Na_2SO_4	0.75	80	59
Na acetate	1.5	100	100
Na acetate	1.0	90	62
LiCl	1.5	0	0
LiCl	4.0	5	0
KCl	1.5	22	5
NH_4Cl	1.5	0	0
$(NH_4)_2SO_4$	0.75	25	8
RbCl	1.5	10	2
CsCl	1.5	4	2

[a] All incubation mixtures contained 0.02 M Tris (*p*H 7.5), 0.21 M NaCl, and enzyme (57 μg of protein) in a final volume of 0.4 ml. Incubations were at 0 C. Activity measurements were made at 30 C with the use of 0.1-ml samples of the incubation mixtures.

although a definite loss in activity occurred upon prolonged incubation. A slight protection was given by the mixture of NADP and isocitrate, but NADP did not act synergistically with isocitrate. The requirement for tricarboxylic acid appears specific for isocitrate. Other experiments showed that mixtures of $MnCl_2$ and sodium citrate or sodium *cis*-aconitate gave no more protection than $MnCl_2$ alone.

Reactivation of ICDH. The loss in activity caused by exposure to low ionic strength can be partially reversed by slowly increasing to high levels the concentration of NaCl. As much as 75% of the initial activity was recovered when the inactivated ICDH was dialyzed against 4 M NaCl (Table 3). Although a complete parametric study was not conducted, it appeared that the optimal conditions for reactivation involved dialysis at 31 C for 3 to 4 hr. Lower temperatures gave poorer recovery. Longer incubation at higher temperatures resulted in losses of activity which were presumably due to thermal denaturation. These conditions are quite similar to those reported for the reactivation of the halophilic malate dehydrogenase of *Halobacterium salinarium* (7). To insure good recovery, it was necessary to have β-mercaptoethanol present when the ICDH was inactivated. Presumably, this protects the sulfhydryl groups of the inactive protein from being oxidized and thereby causing permanent denaturation of the enzyme. Also, good recovery of activity was dependent upon the removal of the β-mercaptoethanol during the reactivation step. The mixture of isocitrate and $MnCl_2$ did

TABLE 2. *Stabilization of ICDH by substrates of the enzyme*

Addition to incubation mixture[a]	Percentage of initial activity remaining after incubation for	
	1.5 hr	20 hr
None	0	0
$MnCl_2$ (10 mM)	20	0
Isocitrate (4 mM)	0	0
$MnCl_2$ (10 mM) plus isocitrate (4 mM)	111	116
NADP (0.25 mM)	0	0
NADP (0.25 mM) plus $MnCl_2$ (10 mM)	64	6
NADP (0.25 mM) plus isocitrate (4 mM)	0	0
$MgCl_2$ (10 mM)	0	0
$MgCl_2$ (10 mM) plus isocitrate (4 mM)	105	77

[a] All incubation mixtures contained 0.02 M Tris (*p*H 7.5), 0.21 M NaCl, and enzyme (57 μg of protein) in a final volume of 0.4 ml. Incubations were at 0 C. Activity measurements were made at 30 C with the use of 0.1-ml samples of the incubation mixtures.

TABLE 3. *Reactivation of ICDH*

Reactivation buffer[a]	Percentage of initial activity recovered after dialysis at		
	4 C, 18 hr	31 C, 3.5 hr	31 C, 18 hr
None	0	0	0
NaCl (4 M)	22.9	74.9	53.4
Na acetate (4 M)	4.2	—	12.2
Isocitrate (4 mM) plus $MnCl_2$ (10 mM)	0	0	0

[a] Samples of inactivated ICDH were dialyzed against 160 volumes of 0.02 M Tris buffer (*p*H 7.5) containing the indicated additions.

not reactivate the ICDH (Table 3), even though this mixture effectively stabilizes the active enzyme (Table 2). Apparently, the latter effect is related to the ability of the substrates to hold the enzyme in its active conformation by virtue of their interaction at the active site. When the enzyme is in the inactive conformation, the substrate mixture cannot replace the high ionic strength needed to restore the active conformation. There is, however, some indication that $MnCl_2$ can interact with the inactive form of the enzyme. Other experiments showed that 10 mM $MnCl_2$ caused a 65% inhibition of the reactivation promoted by 4 M NaCl. Isocitrate did not affect the NaCl-promoted reactivation. Sodium acetate was less effective than NaCl in promoting reactivation (Table 3) even though the sodium acetate was effective in protecting the enzyme (Table 1).

The reversible inactivation could also be demonstrated by disc electrophoresis. In Fig. 2, gel F is a stained electrophoretogram of partially purified ICDH. The major protein bands were the ICDH and a nonhalophilic protein (NHP). Additional information on the NHP will be given in subsequent discussion. The identification of the other band as active ICDH was based on the detection of enzyme activity when the gels were sectioned and the enzyme was eluted with 4.2 M NaCl. To retain activity of the ICDH during electrophoresis, it was necessary to incorporate isocitrate and $MgCl_2$ in the gels and developing buffers. High levels of salt could not be used to stabilize the ICDH, because salts interfere with the polymerization of the polyacrylamide gels and with the development of electrophoretograms. When the ICDH was inactivated before electrophoresis, its electrophoretic mobility was altered. The inactive ICDH protein appeared concomitantly with the disappearance of the active ICDH (gel G). The lower mobility of the inactive protein suggests that the inactivation causes a major change in either the net charge or the shape of the protein molecule (12). The latter could be caused by aggregation of enzyme molecules to give a larger species or by a conformational change which increases the molecular radius of the protein. Gel H illustrates the partial reversal of the inactivation. After reactivation with 4.2 M NaCl, the band of inactive ICDH almost completely disappeared and the active ICDH band reappeared.

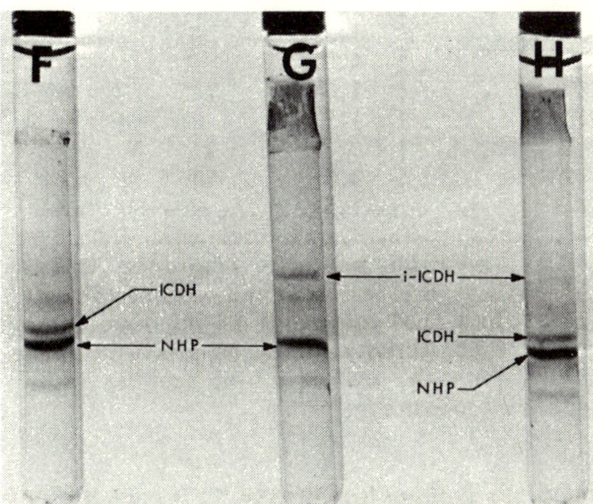

FIG. 2. *Disc electrophoresis of native (gel F), inactivated (gel G), and reactivated ICDH (gel H). NHP = nonhalophilic protein; i-ICDH = inactivated enzyme.*

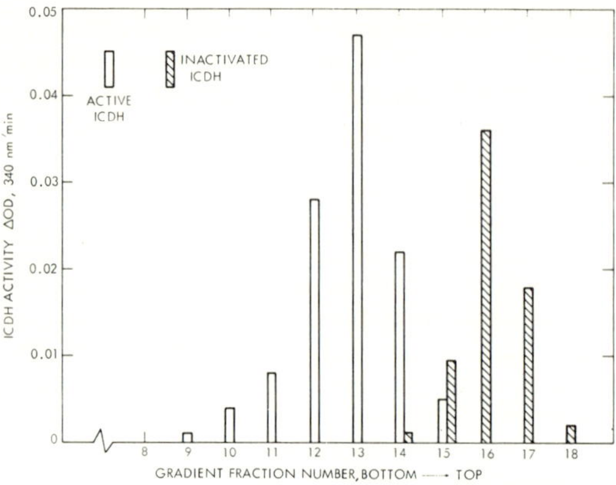

FIG. 3. *Sucrose gradient centrifugation of active and inactivated ICDH. Active and inactivated enzyme (0.1 ml, 0.22 mg of protein) were layered on 4.6-ml sucrose gradients. Samples from the gradient run with inactivated ICDH were reactivated before the activity was measured. In both experiments, 0.05 ml of the gradient fractions was assayed.*

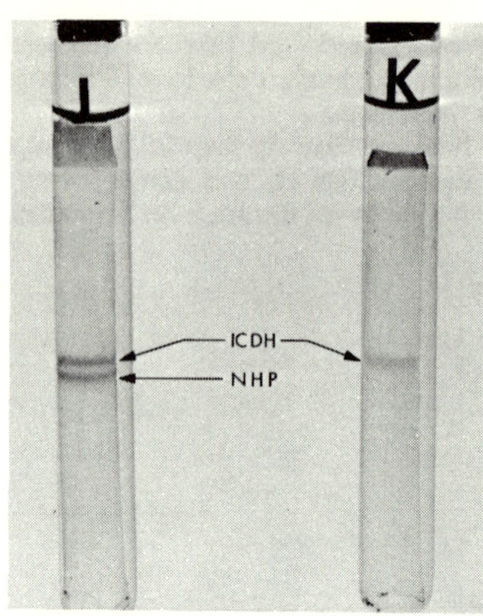

FIG. 4. *Disc electrophoresis of active peak fractions from sucrose gradients. Gel J is pooled fractions from a run with active ICDH. Gel K is ICDH which was inactivated, run on a sucrose gradient, reactivated, pooled, run on a second sucrose gradient, and then pooled for disc electrophoresis.*

Sedimentation of ICDH. Physical differences between the active and inactive forms of the enzyme could also be demonstrated by centrifuging the proteins on sucrose density gradients (Fig. 3). The inactivated ICDH was sedimented at a lower rate than the active enzyme. This suggests that the inactive protein is of lower molecular weight or is of lower density than active ICDH (15). In other experiments, the inactivated ICDH from sucrose gradients was reactivated and then centrifuged on a second sucrose gradient. The reactivated enzyme exhibited a sedimentation rate identical to that of the native enzyme. Figure 4 shows stained electrophoretograms of the ICDH recovered from sucrose gradients. When partially purified enzyme was sedimented in the active state (gel J), ICDH and NHP were detected in the same fractions. These two proteins must have similar sedimentation rates. When the inactivated ICDH was sedimented, reactivated, and run on a second sucrose gradient (gel K), the ICDH was the predominate protein. Apparently, the inactivated ICDH and the NHP had been separated because of the differences in sedimentation rates of the two proteins.

Gel filtration of ICDH. The two forms of the enzyme exhibit different rates of elution from Sephadex G-200 (Fig. 5). The more rapid elution of the inactive protein shows this species to be less restricted during its diffusion through the column of cross-linked gel particles. Apparently, the inactivated ICDH has a larger molecular radius than the active enzyme (15).

Purification of ICDH. The purification procedure is illustrated in Table 4 and Fig. 6. When the ICDH is selectively protected with isocitrate and $MnCl_2$ at low ionic strength, 0.9 saturated $(NH_4)_2SO_4$ precipitates large amounts of contaminating protein. Holmes and Halvorson (6) previously showed that halophilic proteins are easily precipitated with $(NH_4)_2SO_4$ once the proteins are denatured by the removal of NaCl. Thus, the components of the supernatant fluid should be either halophilic proteins which are protected by the isocitrate and $MnCl_2$ or nonhalophilic proteins which are not precipitated with 0.9 saturated $(NH_4)_2SO_4$. The effectiveness of this step is indicated by the 13-fold increase in specific activity with 68% recovery of activity.

After being inactivated, the ICDH can be precipitated with $(NH_4)_2SO_4$ at 0.5 saturation. The precipitated enzyme can then be partially reactivated by dialysis against 4.2 M NaCl (gel C). The NHP and other minor contaminants remained in the supernatant fluid. This poor recovery of activity at this step can be attributed to two causes. All of the inactivated ICDH was not precipitated and the efficiency of reactivation was low. The decrease in specific activity at step C (Table 4) is indicative of the incomplete reactivation. The relative intensities of the two forms of the enzyme (gel C) are not an accurate measure of the quantities of protein, since Nigrosin was found to stain the active ICDH more intensely than the inactive protein. Perhaps the $(NH_4)_2SO_4$ precipitation causes permanent inactivation of a large portion of the ICDH. Alternatively, small amounts of residual $(NH_4)_2SO_4$ may be inhibitory to the reactivation. Holmes and

Halvorson (6) reported that the reactivation of the halophilic malate dehydrogenase was inhibited by $(NH_4)_2SO_4$.

The final purification step involves separation of the inactive (gel D) and active forms (gel E) of the enzyme by gel filtration on Sephadex G-200.

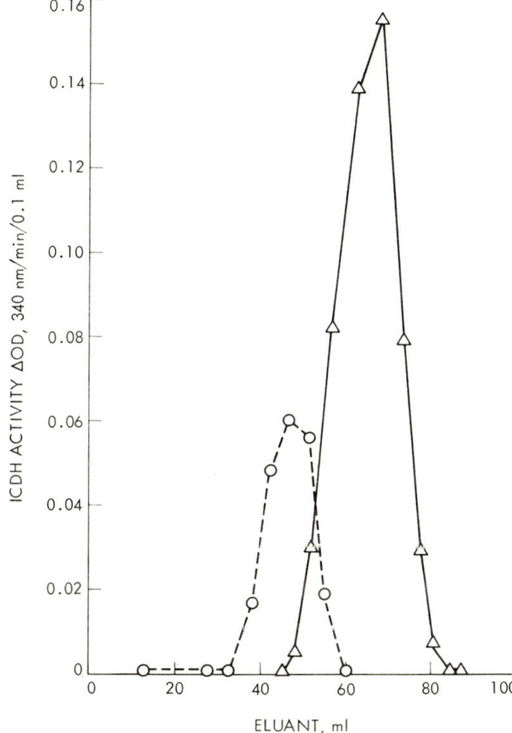

FIG. 5. *Gel filtration of active and inactivated ICDH on Sephadex G-200. The 23 × 225 mm column used for the active ICDH (△) was equilibrated and eluted with 0.02 M Tris (pH 7.5) containing 4 mM isocitrate and 10 mM $MnCl_2$. For the run with inactivated ICDH (○), the same column material was equilibrated and eluted with 0.02 M Tris (pH 7.5) containing 2 mM β-mercaptoethanol. The eluant fractions of the inactivated ICDH were reactivated before the enzyme activity was measured.*

Again, the inactivated ICDH was eluted from the column more rapidly than the active enzyme. The active ICDH (gel E) represents more than 95% of the material detected with the protein stain. The removal of the inactive ICDH and minor protein contaminants gave an eightfold increase in specific activity.

The entire procedure gave a 69-fold purification, and the resulting preparation had a specific activity of 13.8. This value is lower than the reported specific activities of 58 for purified ICDH from pig heart (14) and 21.5 for partially purified ICDH from *Brevibacterium flavum* (13).

DISCUSSION

On the basis of studies with the halophilic lactate dehydrogenase of *H. salinarium*, Baxter (2) proposed that lowering the salt concentration permitted the enzyme to expand to an inactive conformation. The salt was presumed to decrease the electrostatic repulsions between ionized groups within the enzyme molecule and thereby maintain the conformational structure which was catalytically active. Holmes and Halvorson (7) suggested a similar function for salt in the stabilization of the malate dehydrogenase of *H. salinarium*. Lowering the salt concentration caused a marked decrease in the sedimentation rate of malate dehydrogenase on sucrose gradients. The decreased sedimentation was interpreted as representing a decrease in the density of the enzyme. The alternate possibility that the inactivation caused a disaggregation of the malate dehydrogenase into inactive subunits was disregarded, since most of the enzyme activity could be regained by increasing the salt concentration. The latter assumption is unwarranted in view of findings that enzymes such as glutamine synthetase can be rapidly disaggregated and reaggregated with a near stoichiometric recovery of activity (16). The two interpretations, i.e., disaggregation or expansion, can be used to explain our demonstration that the inactivated ICDH has a slower sedimentation rate than the active

TABLE 4. *Purification of ICDH*

Fraction	Total activity	Total protein	Specific activity	Recovery corrected[a]
	units	*units*		%
(A) Dialyzed crude	627	3,120	0.2	100
(B) $(NH_4)_2SO_4$ supernatant fluid (0.9 saturated)	425	169	2.5	68
(C) $(NH_4)_2SO_4$ precipitate (0.5 saturated) of inactivated ICDH, reactivated	21.6	12.7	1.7	5.4
(E) Sephadex G-200 eluant (78–88 ml)	7.1	0.51	13.8	2.7

[a] Recovery values are corrected to the total volume of extract since only portions of fractions B and C were used in subsequent steps.

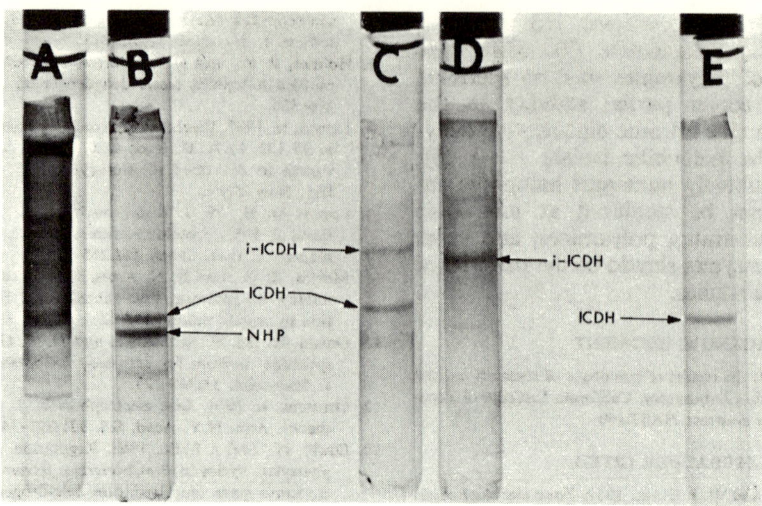

FIG. 6. *Disc electrophoresis of ICDH at different steps in the purification. Gels A, B, C, and E correspond to the purification steps in Table 4. Gel D shows the inactive form of ICDH (i-ICDH) eluted from the Sephadex G-200 column between 44 and 57 ml.*

enzyme. However, the additional data with gel filtration on Sephadex G-200 are more conclusive. The more rapid elution of the inactive enzyme indicates that the molecular radius of the protein has expanded either by aggregation of several molecules or by expansion of a single molecule. Thus, the interpretation that the removal of salt causes an expansion of the protein to a less dense state is consistent with the data on both sedimentation and gel filtration. Observation of the low electrophoretic mobility of the inactivated ICDH tends to support the molecular expansion theory, but the change in electrophoretic mobility could be explained in other ways.

It is probably an oversimplification to assume that salts stabilize the halophilic enzymes solely by the neutralization of repulsive, intramolecular charges. There is no apparent relationship between the protection afforded and the ionic radius or ionic mobility. Other factors must be involved. Moreover, different halophilic enzymes exhibit different specificities for salts. For example, the present study is the first report of NaCl being more effective than KCl in stabilizing a halophilic enzyme. Even more apparent are the differences in salt specificity in the activation of halophilic enzymes (4, 8). Other studies (Hubbard, *unpublished data*) have shown that the ICDH activity is stimulated by adding salts to the reaction mixture. For example, the activity with the optimal NaCl concentration of 0.5 M is 1.6 times that observed with 1 mM NaCl. An even greater degree of stimulation in activity is given with LiCl at its optimal concentration of 0.35 M. Again, the ICDH of *H. cutirubrum* differs from other halophilic enzymes in that KCl is a less effective activator than NaCl. Since LiCl was effective in activation and ineffective in stabilization, it would appear that salts occupy different roles in the two processes. A further indication that the functions are distinguishable comes from our observation and the observations of others (7) that different concentrations of a salt are needed for maximal activation and stabilization.

The use of isocitrate and $MnCl_2$ to selectively protect the ICDH was advantageous in our purification procedure. The number of purification steps that can be performed is quite limited when 4 M NaCl is present. However, with a buffer of low ionic strength the proteins can be fractionated by conventional methods. In the partial purification of halophilic malate dehydrogenase (6), the problem was circumvented by fractionating the enzyme in a salt-free state and then reactivating the enzyme. We were hesitant to use this technique for fear that the reactivated enzyme might be different from the native protein. Our attempts to purify the ICDH in its native state were unsuccessful, so one inactivation-reactivation step was used. However, the data from disc electrophoresis, sedimentation, and gel filtration experiments gave no indication that inactivation or reactivation altered the ICDH. Other studies (Hubbard, *unpublished data*) showed that native and once-reactivated enzyme had identical K_m values for isocitrate, $MnCl_2$, and NADP.

The ICDH of *H. cutirubrum* is not the only halophilic enzyme than can be stabilized with specific effectors as replacements for salt. The malate dehydrogenase of *H. salinarium* is partially stabilized by its coenzyme, reduced nicotinamide adenine dinucleotide (7). Millimolar concentrations of polyamines such as spermine and polylysine confer partial stability to the reduced nicotinamide adenine dinucleotide dehydrogenase of the halophilic isolate AR-1 (5). There are undoubtedly numerous halophilic enzymes which can be stabilized at low ionic strength with substrates, polyamines, and other effectors. Such enzymes should be the material of choice for future studies.

ACKNOWLEDGMENT

This paper presents the results of one phase of research carried out at the Jet Propulsion Laboratory, California Institute of Technology, under NASA contract NAS7-100.

LITERATURE CITED

1. Altschul, A. M., and W. J. Evans. 1967. Zone electrophoresis with polyacrylamide gel, p. 179–186. *In* C. H. W. Hirs (ed.), Methods in enzymology, vol. 11. Academic Press Inc., New York.
2. Baxter, R. M. 1959. An interpretation of the effects of salts on the lactic dehydrogenase of *Halobacterium salinarium*. Can. J. Microbiol. 5:47–57.
3. Baxter, R. M., and N. E. Gibbons. 1956. Effects of sodium and potassium chloride on certain enzymes of *Micrococcus halodenitrificans* and *Pseudomonas salinaria*. Can. J. Microbiol. 2:599–606.
4. Hochstein, L. I., and B. P. Dalton. 1968. Salt specificity of a reduced nicotinamide adenine dinucleotide oxidase prepared from a halophilic bacterium. J. Bacteriol. 95:37–42.
5. Hochstein, L. I., and B. P. Dalton. 1968. Factors affecting the cation requirement of a halophilic NADH dehydrogenase. Biochim. Biophys. Acta 167:638–640.
6. Holmes, P. K., and H. O. Halvorson. 1965. Purification of a salt-requiring enzyme from an obligately halophilic bacterium. J. Bacteriol. 90:312–315.
7. Holmes, P. K., and H. O. Halvorson. 1965. Properties of a purified halophilic malic dehydrogenase. J. Bacteriol. 90: 316–326.
8. Larsen, H. 1967. Biochemical aspects of extreme halophilism, p. 97–132. *In* A. H. Rose and J. F. Wilkinson (ed.), Advances in microbial physiology, vol. 1. Academic Press Inc., New York.
9. Lowry, O. H., N. J. Rosebrough, A. L. Farr, and R. J. Randall. 1951. Protein measurement with the Folin phenol reagent. J. Biol. Chem. 193:265–275.
10. Martin, R. G., and B. N. Ames. 1961. A method for determining the sedimentation behaviour of enzymes; application to protein mixtures. J. Biol. Chem. 236:1372–1379.
11. Onishi, H., M. E. McChance, and N. E. Gibbons. 1965. A synthetic medium for extremely halophilic bacteria. Can. J. Microbiol. 11:365–373.
12. Ornstein, L. 1964. Disc electrophoresis. I. Background and theory. Ann. N.Y. Acad. Sci. 121:321–349.
13. Ozaki, H., and I. Shiio. 1968. Regulation of the TCA and glyoxylate cycles in *Brevibacterium flavum*. I. Inhibition of isocitrate lyase and isocitrate dehydrogenase by organic acids related to the TCA and glyoxylate cycles. J. Biochem. 64:355–363.
14. Plaut, G. W. E. 1962. Isocitrate dehydrogenase (TPN-linked) from pig heart (revised procedure), p. 645–651. *In* S. P. Colowick and N. O. Kaplan (ed.), Methods in enzymology, vol. 5. Academic Press Inc., New York.
15. Siegel, L. M., and K. J. Monty. 1966. Determination of molecular weights and frictional ratios of proteins in impure systems by use of gel filtration and density gradient centrifugation. Application to crude preparations of sulfite and hydroxylamine reductases. Biochim. Biophys. Acta 112: 346–362.
16. Woolfolk, C. A., and E. R. Stadtman. 1967. Regulation of glutamine synthetase. IV. Reversible dissociation and inactivation of glutamine synthetase from *Escherichia coli* by the concerted action of EDTA and urea. Arch. Biochem. Biophys. 122:174–189.

17

Reprinted from *Biochim. Biophys. Acta,* **245**(1), 21–33 (1971)

STUDIES OF THE ELECTRON TRANSPORT CHAIN OF EXTREMELY HALOPHILIC BACTERIA

V. MODE OF ACTION OF SALTS ON CYTOCHROME OXIDASE

MICHAEL M. LIEBERMAN AND JANOS K. LANYI

Exobiology Division, Ames Research Center, National Aeronautics and Space Administration, Moffett Field, Calif. 94035 (U.S.A.)

(Received February 19th, 1971)

SUMMARY

1. Cytochrome oxidase from an extremely halophilic bacterium requires up to 5.0 M NaCl for maximal activity and stability. Differences in effectiveness among various salts in promoting enzyme activity were observed, but a minimal activity (about 30 %) could be supported by all salts tested including $MgCl_2$ and spermine at relatively low concentrations. The enzyme activity in the presence of low concentrations of $MgCl_2$ was shown to be more dependent on pH but less sensitive to hydrophobic bond-breaking agents, such as ethanol and especially *n*-propanol, than the enzyme activity in the presence of high concentrations of NaCl.

2. Spontaneous inactivation of the enzyme in the absence of salt demonstrated that the rate of inactivation of the $MgCl_2$-dependent activity has a greater temperature dependence than the rate of inactivation of the NaCl-dependent activity. The effect of temperature on the rate of inactivation of the NaCl-dependent activity was found to be greatest at neutral pH and became minimal at acid or alkaline pH. Inactivation of the enzyme in the presence of high concentrations of NaCl (by prolonged incubation) demonstrated that the enzyme was inactivated more rapidly at —10° than at 5° at pH 4.0, but not at pH 7.5.

3. These results suggest that hydrophobic forces predominate in promoting the major portion of the enzyme activity in the presence of high NaCl concentrations, while charge-shielding promotes partial enzyme activity in the presence of $MgCl_2$. Hydrophobic bonds also appear to be involved in the stability of the enzyme, particularly at acid and alkaline pH, where hydrogen bonding is expected to be less extensive.

INTRODUCTION

Previous studies from this laboratory have been concerned with the electron transport chain of an extreme halophile, *Halobacterium cutirubrum*. The cytochrome components of the chain have been identified[1] and the salt dependence of the NADH

Abbreviations: HEPES, *N*-2-hydroxyethylpiperazine-*N'*-2-ethanesulfonic acid; TMPD, *N,N,N',N'*-tetramethyl-*p*-phenylenediamine dihydrochloride.

oxidase system examined[2]. The mechanism of the effect of salt was studied in greater detail for the case of one of the components of the respiratory system, the NADH dehydrogenase (menadione reductase)[3,4], and it was found that the activity and stability of this enzyme reflects a requirement for extensive hydrophobic bonding between nonpolar side-chains in the protein[4]. It was proposed, therefore, that the reason for the high salt requirement is not merely the shielding of like charges, as suggested by others[5-7], but also the salting-out of solution of nonpolar side-chains, which are thus moved into the interior of the protein molecule and can form hydrophobic bonds.

Although both menadione reductase and the terminal component of the respiratory chain, cytochrome oxidase, are membrane-associated enzymes, the oxidase is much more tightly bound to the membrane than is the reductase[8]. Preliminary experiments in this laboratory revealed striking differences in the salt-dependence of binding to the membranes between menadione reductase and cytochrome oxidase. Thus, it is of special interest to compare the two enzymes with regard to the involvement of hydrophobic forces in their functional properties. Evidence is given in this report that hydrophobic bonding is primarily responsible for the activity, and to some extent the stability, of cytochrome oxidase.

MATERIALS AND METHODS

N,N,N',N'-Tetramethyl-*p*-phenylenediamine dihydrochloride (TMPD) and sodium azide were obtained from Eastman; Tris-HCl, spermine tetrahydrochloride and *N*-2-hydroxyethylpiperazine-*N'*-2-ethanesulfonic acid (HEPES) from Sigma; ascorbic acid (sodium salt) from Mann.

Preparation of enzyme

The growth and harvest of *H. cutirubrum* cells has been described before[1]. The crude, cell-free extract was prepared as described previously[1], except that in these studies the buffer used was 0.05 M HEPES, 3.4 M NaCl, pH 7.5. The crude extract was diluted 1:10 with the same buffer and centrifuged for 16 h at 150000 × *g* (av.) in the Spinco L2-65 ultracentrifuge. After centrifugation, three fractions were obtained: a reddish hard pellet; a clear, red, viscous liquid (soft pellet) at the bottom of the tube; and a colorless supernatant. The soft pellet was carefully withdrawn from the tubes, diluted about 25-fold with fresh HEPES buffer containing 3.4 M NaCl, and centrifuged as before. After the second centrifugation, both hard and soft pellets were collected and resuspended by homogenization with a Dounce-type tissue homogenizer (Kontes). This washed, particulate enzyme preparation was stored frozen at —78° and was used for all experiments except those involving release of the enzyme from the membrane. For the latter experiments the method of cell breakage and membrane preparation of STOECKENIUS AND ROWEN[9] was employed. For this preparation the HEPES buffer was used containing 3.4 M NaCl *plus* 2.0% (w/v) $MgCl_2$, 0.2% (w/v) KCl, and 0.02% (w/v) $CaCl_2 \cdot 2H_2O$.

Enzyme assay

Cytochrome oxidase was assayed by oxidation of TMPD and measurement of the accompanying increase in absorption at 610 nm. This method of assaying cytochrome oxidase was made possible by the fact that Wurster's Blue (extinction coefficient $1.2 \cdot 10^4$ $M^{-1} \cdot cm^{-1}$), the semiquinone free radical product of the one-electron oxidation of TMPD[10], is not oxidized further in the *H. cutirubrum* system, and is very stable in aqueous solution[10,11]. TMPD solutions (0.25 M), containing a small amount of ascorbic acid (about 0.05 % (w/v)) to prevent autoxidation, were prepared fresh before use and were kept on ice throughout the day. In each assay the autoxidation of 5 μl of the substrate solution in 3.0 ml of buffer was followed for 2–3 min in a Cary 14 recording spectrophotometer. This autoxidation rate was less than 2 % of the enzyme-dependent oxidation rate at high NaCl concentrations, 10–20 % of the enzyme rate at low salt concentrations or in $MgCl_2$, and up to 50 % of the enzyme rate at high concentrations of $NaClO_4$. The autoxidation rate was substracted from the rate obtained after addition of enzyme to obtain the enzyme-dependent rate. The net enzyme activity obtained was a linear function of enzyme concentration within the range of concentrations used. The final concentration of TMPD in the assay (0.417 mM) produced about 80–90 % of the maximal velocity obtainable and was from 4–10 times the K_m value, depending upon the conditions of assay. (The K_m value was increased in buffers with low salt concentrations or salts other than NaCl, as noted below.) The cytochrome oxidase activity by this assay method could be inhibited by 75 % with 2 mM sodium azide and by 90 % with 10 mM azide.

For calculation of specific enzymatic activities protein determinations were carried out by the biuret method[12] using a biuret reagent containing 1 % (w/v) deoxycholate. Specific activities are given in terms of nmoles of O_2 reduced per min per mg of protein. Protein concentrations in the assays were 0.5–0.7 mg/ml. The molar quantities of O_2 were calculated using an equivalence of 4 electrons (4 molecules of TMPD oxidized) for each O_2 reduced. The specific activities obtained by the new assay method calculated in this manner are about 2.5 times higher than the specific activities obtained by the conventional oxygen uptake assay method[13] under identical conditions. Experimental results, such as dependence of enzyme activity on salt concentration, however, were very similar with both assay systems.

Since CHEAH[14] has found multiple cytochrome oxidases in *H. cutirubrum* by spectrophotometric identification of their chromophores, it is probable that this procedure assays more than one oxidase.

Carbon monoxide difference spectra

Difference spectra were determined in a Phoenix Dual Beam Spectrophotometer. Suspensions of particulate preparations (containing 3–4 mg/ml protein) were reduced by either dithionite or TMPD (0.1 mM, containing a small amount of ascorbate) and were divided into two 10-mm pathlength cuvettes. One of the cuvettes was flushed with carbon monoxide in the dark for about one minute and the difference spectrum was recorded. When KCN was added (2 mM) the spectra were recorded after 10 min of incubation.

RESULTS

Dependence of cytochrome oxidase activity on salt

Cytochrome oxidase activity was determined in the presence of increasing concentrations of 3 different salts: NaCl, $NaNO_3$, and $NaClO_4$. The salt-dependence of enzyme activity in these experiments is shown in Fig. 1. It is apparent that maximal activity, about 33.4 nmoles O_2 reduced per min per mg protein, is obtained only with 5 M NaCl. $NaClO_4$ could support only about 30 % of the maximal activity, at 1–1.5 M, with no further stimulation or slight inhibition seen at higher concentrations. As shown in Fig. 1, $NaNO_3$ promotes an intermediate level of activity. The salt dependence curves show complex behavior at lower concentrations. Thus, there is a perceptible plateau in the $NaNO_3$ curve at 0.6 M to 1.0 M, but only a slight inflection in the NaCl curve at about 1 M, while the $NaClO_4$ curve begins to level off in this region. This plateau region will be discussed below.

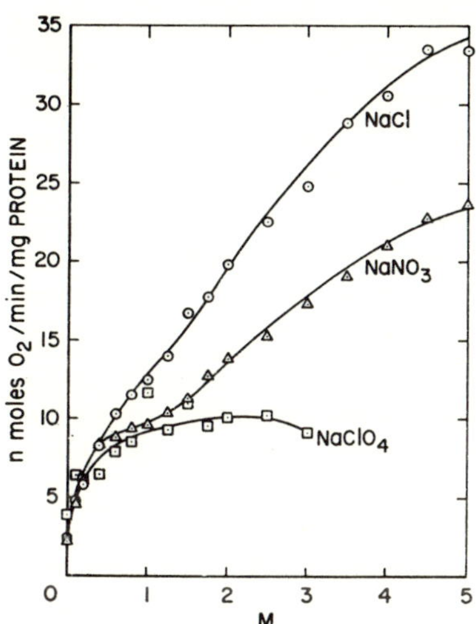

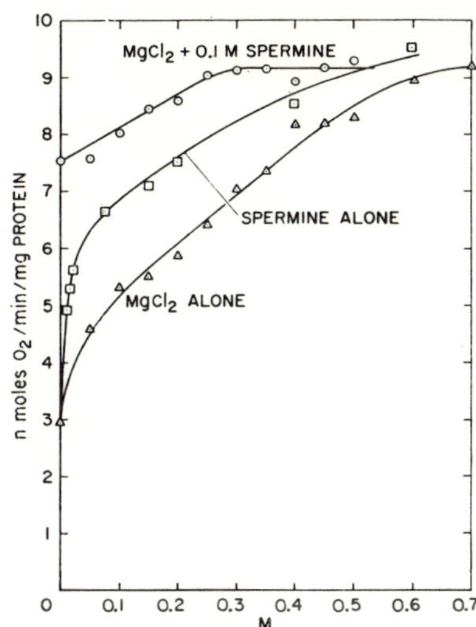

Fig. 1. Salt dependence and specificity of cytochrome oxidase activity. The buffer used was 0.05 M Tris–HCl, pH 7.5.

Fig. 2. Cytochrome oxidase activity in the presence of various concentrations of $MgCl_2$ alone and with 0.1 M spermine. The buffer used was 0.05 M Tris–HCl, pH 7.5.

It was found that $MgCl_2$ or spermine, a polyamine, could also promote enzyme activity (Fig. 2). The activity obtainable with either $MgCl_2$ (at 0.5–0.6 M) or spermine (at 0.1 M) is 20–30 % of the maximal activity obtainable with NaCl. Spermine causes a steep increase in activity up to 0.1 M; this activity (about 20 % of maximal) can be further increased to 30 % with spermine concentrations up to 1.0 M. In order to determine if $MgCl_2$ and spermine were in fact acting in the same manner on the enzyme, we examined their effects for additivity. If the polyvalent cations show nonspecific interactions with the enzyme, then the effects of $MgCl_2$ and spermine should be additive at low concentrations, but not at higher concentrations. In this experiment enzyme activity was measured (a) as a function of $MgCl_2$ concentration alone, and (b) as a function of $MgCl_2$ concentration in the presence of 0.1 M spermine. As shown in Fig. 2, the effects of $MgCl_2$ and spermine are additive up to 0.3 M $MgCl_2$, since the $MgCl_2$ plus 0.1 M spermine curve continues to increase up to 0.3 M $MgCl_2$. However, at this point the maximal level of activity obtainable with $MgCl_2$ alone is reached, and the $MgCl_2$ and spermine dependent activities are no longer found to be additive.

Comparison of NaCl- and $MgCl_2$-dependent activities

The pH dependence of the enzyme activity in the presence of NaCl was compared with that in the presence of $MgCl_2$. The results are given in Fig. 3 and show that the NaCl-dependent activity has a broad pH curve with a maximum at about pH 7.5. The $MgCl_2$-dependent activity has a much steeper pH curve with a maximum at pH 6.5. The ratio of activity in the presence of $MgCl_2$ to that in the presence of NaCl continuously decreases with increasing pH. At pH 5.5 the $MgCl_2$-dependent activity was 43 % of the NaCl-dependent activity, at pH 7.0 it is 32 %, while at pH 8.5 it is only 8 % of the NaCl-dependent activity.

Kinetic parameters for the enzyme were examined under various conditions and it was found that the K_m for TMPD was lowest when the enzyme was assayed in 5.0 M NaCl (about 50 μM TMPD) and was increased somewhat at low NaCl concentrations or when assayed in other salts including polyvalent cations. However, the increases were not great, with maximal differences less than two-fold. The inhibition constant K_i for sodium azide (60 μM in 5.0 M NaCl) showed differences in the same range, being about 2.5 times higher in 0.5 M $MgCl_2$ than in 5.0 M NaCl.

The sensitivity of the enzyme activity to inhibition by protein denaturants was investigated under various conditions. Ethanol and *n*-propanol were tested as inhibitors, in the presence of either 3.5 M NaCl or 0.5 M $MgCl_2$. In 16 % ethanol, the highest concentration used, 11.0 % of the $MgCl_2$-dependent activity but only 3.1 % of the NaCl-dependent activity remained. At 6 % *n*-propanol, the highest concentration used, 17.6 % of the $MgCl_2$-dependent activity but only 2.8 % of the NaCl-

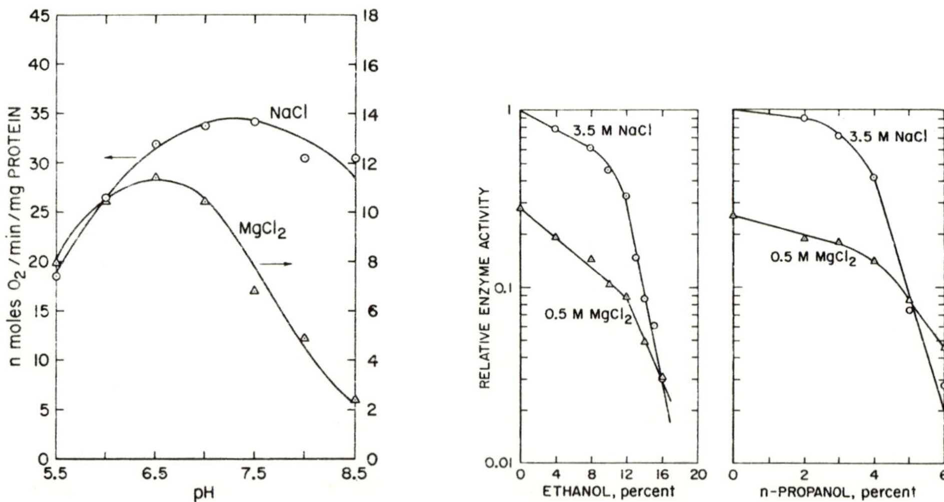

Fig. 3. pH dependence of enzyme activity in the presence of 5.0 M NaCl or $MgCl_2$ at the concentration which gave maximal activity at that particular pH. This concentration of $MgCl_2$ was 0.8–1.0 M for all pH values tested except pH 8.5 where 0.3 M $MgCl_2$ gave maximal activity. The buffers used were 0.05 M Tris–maleate from pH 5.5 to pH 7.0 and 0.05 M Tris–HCl from pH 7.5 to pH 8.5.

Fig. 4. Effect of ethanol and *n*-propanol on enzyme activity in the presence of 3.5 M NaCl or 0.5 M $MgCl_2$. Relative enzyme activities, based on the activity at 3.5 M NaCl in the absence of alcohol as 1.0, are plotted logarithmically against the concentration of the alcohol. The buffer used was 0.05 M Tris–HCl, pH 7.5.

dependent activity remained. Thus the activity at 3.5 M NaCl was found to be more sensitive to ethanol and especially to *n*-propanol than the activity at 0.5 M $MgCl_2$. In Fig. 4 the logarithm of relative enzyme activity in the presence of these agents is plotted against the concentration of the denaturant. The NaCl and $MgCl_2$ curves intersect at about 16% ethanol or 5% *n*-propanol; hence at these denaturant concentrations the salting-out effect of NaCl is apparently cancelled and high concentrations of NaCl no longer stimulate enzyme activity over the level of activity obtainable with low concentrations of $MgCl_2$.

Carbon monoxide difference spectra

Dithionite-reduced preparations exhibited a CO difference spectrum (Fig. 5a) similar in amplitude to that reported by CHEAH[14], for a given concentration of membranes, although the positions of the main peak and trough were somewhat different, 424 and 443 nm, respectively, for preparations in this study, as compared to the reported values of 419 and 442 nm.

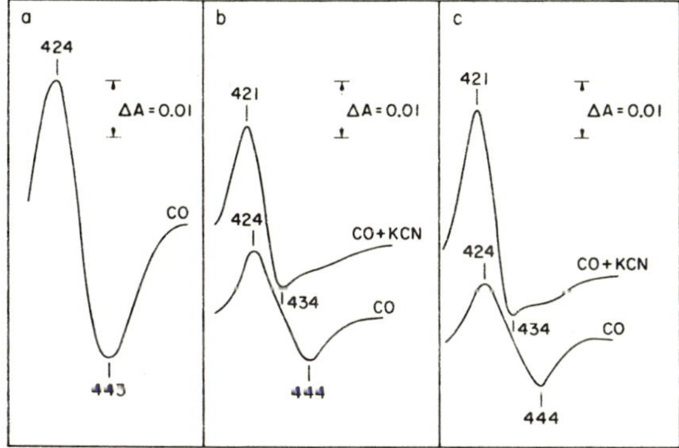

Fig. 5. Carbon monoxide difference spectra at room temperature. (a) dithionite + CO *minus* dithionite in 5.0 M NaCl, (b) TMPD + CO *minus* TMPD, after recording the difference spectra, KCN was added to both cuvettes and the CO + KCN difference spectra were recorded in 5.0 NaCl, (c) experimental conditions identical to (b) but performed in 0.2 M $MgCl_2$. The buffer used was 0.05 M Tris-HCl, pH 7.5.

The CO difference spectra of TMPD-reduced preparations were very similar when determined either in 5.0 M NaCl (Fig. 5b) or in 0.2 M $MgCl_2$ (Fig. 5c). Comparison with the dithionite-reduced CO-difference spectrum (Fig. 5a) indicated that (a) TMPD reduces only about a third of the CO-reactive pigments and (b) no significant differences in the positions of the peak and trough were observed in either solvent used, or between the TMPD-reduced and dithionite-reduced preparations.

Adding KCN to both cuvettes after the CO-difference spectra were recorded resulted in a shift of the peak from 424 to 421 nm and of the trough from 444 to 434 nm, leaving a shoulder at 444 nm. Such changes were reported in *Halobacterium halobium* by CHEAH[15] who interpreted them as revealing of the cytochrome o component of the oxidase in addition to the *a*-type cytochrome. No significant differences were observed between the CO–KCN spectra obtained from 5.0 M NaCl and 0.2 M $MgCl_2$.

Inactivation of the enzyme

Cytochrome oxidase undergoes a time-dependent irreversible inactivation in the absence of salt. The rate of inactivation is logarithmic until about 50 % inactivation has occurred, beyond which point inactivation becomes slower. The rates of inactivation given below are calculated from the initial (logarithmic) part of the time-dependent inactivation curves. Since cytochrome oxidase is more stable at lower salt concentrations than menadione reductase[4], the inactivation experiments with cytochrome oxidase were carried out over longer periods of time than was previously done for the other enzyme[4]. In order to study the inactivation of the NaCl-dependent activity separately from the inactivation of the $MgCl_2$-dependent activity, the enzyme was incubated in the absence of salt to inactivate it, then assayed at either high concentrations (2.5 M) of NaCl, or at low concentrations (0.5 M) of $MgCl_2$. The inactivation buffer was 0.05 M Tris–HCl, pH 7.5, and inactivation was stopped by adding an equal volume of the same buffer containing either 5.0 M NaCl or 1.0 M $MgCl_2$. The inactivation rates obtained in this experiment, carried out at different temperatures, are shown in Table I, where it is seen that at 0° the inactivation rates of both activities are similar, but at 30° the rate of inactivation of the $MgCl_2$-dependent activity is much greater than the rate of inactivation of the NaCl-dependent activity. Thus, the inactivation rate of the NaCl-dependent activity shows a lesser temperature dependence than the inactivation rate of the $MgCl_2$-dependent activity.

TABLE I

TEMPERATURE DEPENDENCE OF INACTIVATION OF NaCl AND $MgCl_2$-DEPENDENT ACTIVITIES

Inactivation was carried out without added NaCl at pH 7.5, for increasing periods of time at the given temperature, and terminated by addition of either NaCl or $MgCl_2$ (final concentrations 2.5 M and 0.5 M, respectively). Enzyme activities were determined at room temperature. Inactivation rates were calculated from logarithmic plots as described in the text.

Temperature (°)	*Rate of inactivation* (h^{-1})	
	Assayed in NaCl	*Assayed in* $MgCl_2$
0	0.68	0.74
23	2.8	4.2
30	5.7	31.2

The inactivation of the NaCl-dependent activity was investigated further as a function of pH at 5° and 23°. The enzyme was inactivated in the absence of salt at varying pH values at 5° or 23°, and then assayed in the presence of high NaCl concentrations at room temperature. At both 5° and 23° the rate of inactivation is less at neutral pH than at acid or alkaline pH values, but the effect of pH on inactivation is much greater at 5° than at 23°. In Fig. 6 the ratio of the rate of inactivation at 5° to the rate of inactivation at 23° is plotted as a function of pH. The curve is seen to have a minimum between pH 5.5 and 6.5 with a ratio of about 0.1. The ratio approaches 1.0 at both acid and alkaline pH values, leveling off at about 0.95 at pH 10.0 and reaching 0.7 at pH 4.0.

The enzyme can be inactivated even in the presence of high concentrations of salt when incubated over long periods of time. The effect of temperature on the inactivation of the enzyme was tested in the presence of 3.4 M NaCl at pH 7.5 and pH 4.0. The results, given in Fig. 7, show that in 3.4 M NaCl at pH 7.5 the enzyme is protected from inactivation by lowering the temperature. However, at pH 4.0, in 3.4 M NaCl, the enzyme is inactivated more rapidly at —10° than at 5°. Thus, at pH 4.0, but not at pH 7.5, the enzyme shows cold-sensitivity.

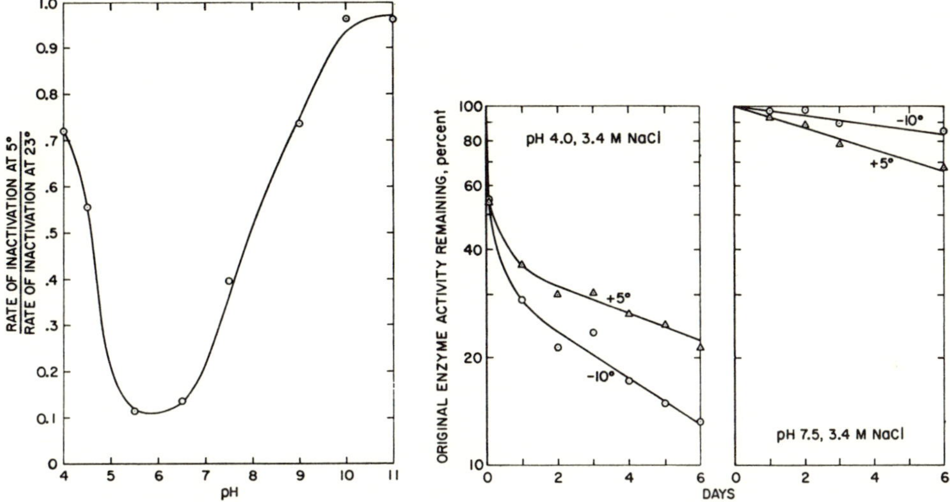

Fig. 6. Effect of pH on the temperature-dependence of inactivation of the enzyme in the absence of salt. The ratio of the initial rate of inactivation at 5° to the initial rate of inactivation at 23° is plotted against pH. The enzyme was inactivated by incubating in the following buffers (not containing added NaCl): 0.05 M acetate at pH 4.0 and 4.5, 0.05 M Tris–maleate at pH 5.5, 0.10 M NaH_2PO_4–K_2HPO_4 at pH 6.5 and 7.5, and 0.05 M glycine at pH 9.0, 10.0, and 11.0. The inactivation was terminated after appropriate incubation times by adding an equal volume of 0.10 M NaH_2PO_4–K_2HPO_4 containing 5.0 M NaCl and the pH was adjusted to neutrality with a pre-determined amount of either 0.5 M NaOH or 1.0 M HCl. Initial rates of inactivation were calculated as described in the text.

Fig. 7. Inactivation of the enzyme at high salt concentrations. The percent remaining activity is plotted logarithmically against time. The buffers used for inactivation were 0.05 M acetate containing 3.4 M NaCl, at pH 4.0, and 0.05 M HEPES containing 3.4 M NaCl, at pH 7.5. The samples were handled for assay as described in the legend to Fig. 6.

Release of the enzyme from the membrane vesicles

During the course of studies in this laboratory concerning the dissolution of membranes from *H. cutirubrum* at low salt concentrations, it became apparent that exposure to low salt concentrations could be a way of solubilizing the enzyme. Membrane vesicles prepared by the method of STOECKENIUS AND ROWEN[9] were exposed to various concentrations of NaCl, from 0.1 M to 0.8 M, for 60 min at 0°, and then centrifuged at 34000 × *g* (max.) for 15 min in the Sorvall centrifuge (which effectively sediments the membrane vesicles at high salt concentrations). An equal volume of 0.05 M Tris–HCl, pH 7.5, containing 5.0 M NaCl was added to the supernatants to stop inactivation and the same buffer was used for resuspension of the pellets. The enzyme activity of the pellet and supernatant fractions of each sample were assayed. The results are shown in Fig. 8, where the percent of the activity released from the membranes into the supernatant as well as the total activity recovered, are plotted as a function of NaCl concentration. It is evident that enzyme activity is readily released from the membranes at concentrations of NaCl below 0.4–0.5 M. When the enzyme released from the membranes was tested for inactivation in the absence of salt, the released activity was found to be inactivated at the same rate as the enzyme activity still associated with the membrane vesicles. Thus the inactivation of the enzyme at low salt concentrations is not associated with the observed detachment from the membrane vesicles, but appears to be an independent process.

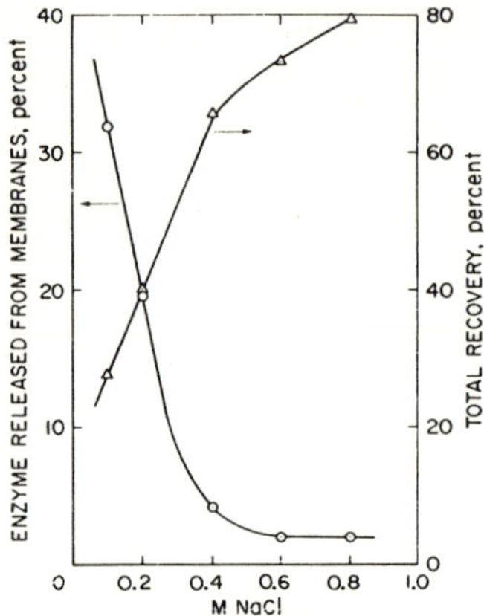

Fig. 8. Release of cytochrome oxidase from membrane vesicles. The percent enzyme activity released from the membranes and the percent total recovery after one hour incubation at 0° is plotted against concentration of salt. Experimental details are given in the text.

The sedimentation properties of released enzyme were studied because STOECKENIUS AND ROWEN[9] showed that the product of the low salt treatment of membranes still contained membrane fragments. A sample of the cytochrome oxidase preparation released from the membrane vesicles at 0.2 M NaCl that remained in the supernatant after centrifuging as described above, was adjusted to 2.6 M NaCl and centrifuged at 50000 $\times$ g (av.) for 2 h. All of the enzyme activity was found to sediment as a pellet, and thus the enzyme detached from membrane vesicles at low salt concentrations was either still associated with membrane fragments, or participates in membranous aggregates on addition of salt.

DISCUSSION

Cytochrome oxidase from *H. cutirubrum* has been shown to require high concentrations of sodium chloride for maximal activity[2,16]. Differences in effectiveness among various salts have been previously demonstrated for other membrane-bound enzymes, such as the complete NADH oxidase system of *Halobacterium salinarium*[17], and menadione reductase (NADH dehydrogenase) from *H. cutirubrum*[4]. In these systems, the enzymes are more selective among the anions than the cations. In the case of menadione reductase[4] the anion specificity was demonstrated to follow the lyotropic series Cl^-, $PO_4^{3-} > Br^- > NO_3^- > ClO_4^-$, SCN^-, the order of effectiveness in promoting enzyme activity being the same as the order of salting-out power. This observation, in addition to other evidence, led to the suggestion that the role of salt in activating and stabilizing menadione reductase is primarily due to the stabilization of hydrophobic interactions. In the case of cytochrome oxidase, reported here, the anion specificity is also consistent with the lyotropic series. The salt dependence curves (Fig. 1) show that at concentrations above 1 M the order of effectiveness of the salts tested was NaCl > $NaNO_3$ > $NaClO_4$. Below 1 M concentrations, however, little difference in effectiveness is found among these salts. The data suggest, therefore, that at lower salt concentrations the salts act in a nonspecific manner, perhaps by charge shielding, thereby supporting a relatively small amount of activity. At 1 M salt, charge shielding may be expected to be complete and at this concentration about 30 % of the maximal enzyme activity is obtained. The high concentrations of salt required for the major portion of the enzyme activity and the observed specificity suggest that the effect of salt on the enzyme consists mainly of promoting the formation of new hydrophobic bonds by the salting-out effect at high concentrations of salt. Thus the plateau in the $NaNO_3$ curve (Fig. 1) might reflect the region where charge shielding is complete, but salting-out is not yet effective. This plateau is not as pronounced with NaCl (Fig. 1), consistent with the fact that this salt is a better salting-out agent than $NaNO_3$. In this system $NaClO_4$, which generally salts in, may participate in charge shielding but not in salting-out. If this hypothesis concerning the action of salts is valid, agents such as $MgCl_2$, or spermine, which do not salt-out at low concentrations but effectively shield charges, should be able to support about the same level of activity as 0.5–1.0 M concentrations of NaCl, $NaNO_3$, or $NaClO_4$, or about 30 % of the maximal activity. This expectation is borne out by the data obtained (Fig. 2). The results suggest that, as in the case of menadione reductase[4], charge shielding by low concentrations of salt or by $MgCl_2$ causes only partial cytochrome oxidase activity, with the greater portion of enzyme activity dependent on hydrophobic bonds, stabilized by the presence of high concentrations of a salting-out type salt.

The complexity of halophilic cytochrome oxidases has been documented by spectrophotometric studies[14,15]. The existence of various cytochrome oxidase species were best revealed in a combination of CO and KCN difference spectra[15]. Since the possibility existed that the two plateaus in Fig. 1 were due to the participation of different portions of the cytochrome oxidase complex, the TMPD-reduced CO difference spectra, with and without KCN, were recorded in 0.2 M $MgCl_2$ and in 5 M NaCl. Since no detectable differences were seen in the spectra obtained in these solvents, it appears that TMPD reduces the same components of the oxidase complex even when the salt concentration is varied or when divalent salts are substituted. The differences in enzyme properties observed when the salt concentrations were changed must then have been due to the conformation of the oxidase or to the membrane environment surrounding the complex. At this time we are unable to distinguish between these effects. For brevity the two portions of the cytochrome oxidase activity, distinguished on the basis of the experiments discussed above, have been named NaCl-dependent and $MgCl_2$-dependent enzyme activity.

While the NaCl-dependent portion of the enzyme activity appears to involve hydrophobic bonding, all that can be said about the $MgCl_2$-dependent portion of activity at this stage is that negative charges must be shielded to obtain activity. However, we have sought to find differences in the behavior of the two parts of enzyme activity which might indicate the relative importance of hydrogen and hydrophobic bonds.

In contrast to hydrogen bonding[18], hydrophobic interactions should be little affected by the state of ionization of charged groups. EDELHOCH[19], in a study of the effects of ethanol and pH on the inactivation of pepsin, found that at lower ethanol concentrations the rate of inactivation was more dependent on pH than at higher ethanol concentrations. This was interpreted as indicative of the breaking of hydrogen bonds at lower ethanol concentrations and hydrophobic bonds as well at higher ethanol concentrations. Correspondingly, one might expect that in our system the $MgCl_2$-dependent enzyme activity would show more variation with pH than the NaCl-dependent activity. The results show, indeed, that the activity in the presence of $MgCl_2$ declines more rapidly with increasing pH above 6.5 than the activity in the presence of NaCl. The preferential loss of $MgCl_2$-dependent activity at higher pH values may reflect the loss of a proton from positively charged groups with a pK of 7.0–7.5, involved in the structure of the enzyme. In contrast, the NaCl-dependent activity does not show much pH dependence in this region.

Another approach in separating the two types of bonds presumed to be involved in cytochrome oxidase activity was to subject the enzyme preparation to agents which have less effect on hydrogen bonds than on hydrophobic interactions, such as ethanol and *n*-propanol[20–23]. The results showed that the contribution to enzyme activity by the presence of high NaCl concentrations was decreased relative to the $MgCl_2$-dependent activity in the presence of alcohols, indicating that the former portion of enzyme activity is more dependent on hydrophobic bonding. This suggestion is supported by the fact that the differences in sensitivity between the NaCl- and $MgCl_2$-dependent activities is greater for *n*-propanol than for ethanol, since the effect of *n*-propanol on hydrophobic bonds relative to its effect on hydrogen bonds is greater than the effect of ethanol. It is interesting to note that by manipulating the conditions of assay as discussed above, one can greatly reduce either the charge-shielding or the hydrophobic bonding-dependent portions of the enzyme activity. Thus, at pH 8.5 the $MgCl_2$-dependent activity is only 8 % of the NaCl-dependent activity, whereas at lower pH values $MgCl_2$ contributes a much larger proportion of the NaCl-dependent activity, up to 43 % at pH 5.5 (Fig. 3). On the other hand, by assaying the enzyme in the presence of 16 % ethanol or 5 % *n*-propanol, one can effectively inhibit any stimulation of the enzyme activity by high concentrations of NaCl over the level of activity obtainable with low concentrations of $MgCl_2$. Thus, the enzyme assayed at pH 8.5 appears to be virtually completely dependent on hydrophobic bonding for activity, whereas the enzyme assayed in the presence of 16 % ethanol or 5 % *n*-propanol behaves as if it were dependent mostly on charge shielding for activity.

Hydrophobic interactions are reportedly cold-sensitive[24–27]. In the previous paper of this series[4] it was demonstrated that menadione reductase shows decreasing temperature dependence of inactivation under conditions of increasing hydrophobic bonding, and becomes cold-sensitive in the presence of high NaCl concentrations. Cytochrome oxidase was examined for cold-lability also. This investigation was carried out in three parts: (a) separate inactivation of the NaCl-dependent and the $MgCl_2$-dependent activities at low salt concentrations, (b) pH dependence of the inactivation at low salt concentrations, and (c) inactivation of the enzyme at high concentrations of salt. In the first case the results demonstrated that the spontaneous inactivation of the $MgCl_2$-dependent activity at low salt concentrations has a much greater temperature dependence than the inactivation of the NaCl-dependent activity (Table I). As found with menadione reductase[4], the temperature coefficient of enzyme inactivation may be resolved into a hydrophobic bond component which has a negative sign and a hydrogen bond component with a positive sign. SHERAGA *et al.*[27] have shown that even though the stability of a protein may depend to a large extent on hydrophobic bonding, the temperature coefficient of inactivation may still be positive due to the influence of the positive enthalpy of breaking hydrogen bonds that are present. The lower temperature coefficient of the inactivation of the NaCl-dependent cytochrome oxidase activity, compared to the $MgCl_2$-dependent activity, may thus be interpreted as the consequence of the greater involvement of hydrophobic bonding. The pH dependence of the ratio of inactivation rates of the NaCl-dependent activity at 5° and 23° (Fig. 6) demonstrates that at acid or alkaline pH values, but not at neutral pH, change of temperature (in this range) has little effect on the inactivation of the enzyme. At extreme pH values the relative importance of hydrophobic bonds in the stability of the enzyme thus appears to be increased. The inactivation curves of the enzyme in the presence of high concentrations of NaCl (Fig. 7) show cold-lability at pH 4.0, but not at pH 7.5, indicating that at high salt concentrations and at acid pH, inactivation involves primarily breaking hydrophobic bonds.

The results of the studies on the inactivation of the enzyme suggest that at neutral pH hydrogen bonding is predominant in providing stability to the enzyme. At acid or alkaline pH and especially at high salt concentrations, however, the component of hydrophobic bonding is revealed by the temperature dependence of the inactivation process.

It has been assumed throughout this paper that the inactivation of cytochrome oxidase at low salt concentration is due to the unfolding of the enzyme, the effect thus being analogous to the generally observed instability of halophilic enzymes in the absence of salt[5–7]. In other systems, however, membrane lipids have been shown to oxidize on solubilization[28] and the resulting peroxides might be destructive to enzyme activity. In extreme halophiles, however, few or no unsaturated lipids were found[29], excluding lipid oxidation as a factor in the stability of cytochrome oxidase.

The experiments which show release of the enzyme from membrane vesicles (Fig. 8) are of interest in two respects. First, they demonstrate that unlike enzyme activity which is affected by salt concentrations of 0–5 M, the release of the enzyme takes place in a narrow concentration range of 0–0.5 M. Since the association of the enzyme with the membrane vesicles is complete at 0.5 M NaCl, it is suggested that salting-out has little effect on the attachment. Secondly, since the released enzyme activity was shown to be inactivated at the same rate as the membrane vesicle-associated enzyme activity, the inactivation observed at low salt concentrations in the first part of this paper cannot be ascribed to increased lability due to the detachment of the enzyme from the membrane vesicles.

It was not possible at this time, however, to study any effect of the immediate membrane environment on the low-salt stability of this halophilic enzyme since we had not been able to obtain it in a fully solubilized form.

ACKNOWLEDGMENT

M.M.L. is a National Academy of Sciences–National Research Council Research Associate.

REFERENCES

1 J. K. Lanyi, *Arch. Biochem. Biophys.*, 128 (1968) 716.
2 J. K. Lanyi, *J. Biol. Chem.*, 244 (1969) 2864.
3 J. K. Lanyi, *J. Biol. Chem.*, 244 (1969) 4168.
4 J. K. Lanyi and J. Stevenson, *J. Biol. Chem.*, 245 (1970) 4074.
5 R. M. Baxter, *Can. J. Microbiol.*, 5 (1959) 47.
6 H. Larsen, in I. C. Gunsalus and R. Y. Stanier, *The Bacteria*, Vol. 4, Academic Press, New York, 1962, p. 297.
7 H. Larsen, *Advan. Microbiol. Physiol.*, 1 (1967) 97.
8 R. G. Bartsch, *Ann. Rev. Microbiol.*, 22 (1968) 181.
9 W. Stoeckenius and R. Rowen, *J. Cell Biol.*, 34 (1967) 365.
10 L. Michaelis, M. P. Shubert and S. Granick, *J. Am. Chem. Soc.*, 61 (1939) 1981.
11 L. Michaelis and S. Granick, *J. Am. Chem. Soc.*, 65 (1943) 1747.
12 A. G. Gornall, C. J. Bardawill and M. M. David, *J. Biol. Chem.*, 177 (1949) 751.
13 T. Yonetani, *J. Biol. Chem.*, 235 (1960) 3138.
14 K. S. Cheah, *Biochim. Biophys. Acta*, 180 (1969) 320.
15 K. S. Cheah, *Biochim. Biophys. Acta*, 205 (1970) 148.
16 K. S. Cheah, *FEBS Letters*, 7 (1970) 301.
17 L. I. Hochstein and B. P. Dalton, *J. Bacteriol.*, 95 (1968) 37.
18 M. Joly, *A Physico-chemical Approach to the Denaturation of Proteins*, Academic Press, New York, 1965, p. 230–259.
19 H. Edelhoch, *J. Am. Chem. Soc.*, 80 (1958) 6648.
20 P. H. Von Hippel and K. Y. Wong, *Science*, 145 (1964) 577.
21 P. H. Von Hippel and K. Y. Wong, *J. Biol. Chem.*, 240 (1965) 3909.
22 P. H. Von Hippel and K. Y. Wong, *Biochemistry*, 1 (1962) 664.
23 S. Y. Gerlsma, *Eur. J. Biochem.*, 14 (1970) 150.
24 J. F. Brandts, *J. Am. Chem. Soc.*, 86 (1964) 4302.
25 E. A. Havir, H. Tamir, S. Rattner and R. C. Warner, *J. Biol. Chem.*, 240 (1965) 3079.
26 A. Rosenberg and R. Lumry, *Biochemistry*, 3 (1964) 1055.
27 H. A. Sheraga, G. Nemethy and I. Z. Steinberg, *J. Biol. Chem.*, 237 (1962) 2506.
28 J. Hatefi and W. G. Hanstein, *Proc. Natl. Acad. Sci. U.S.*, 62 (1969) 1129.
29 M. Kates, M. K. Wassef and D. J. Kushner, *Proc. 1967 Meeting Intern. Conf. on Biological Membranes*, North Holland Publishing Co., Amsterdam, 1968, p. 105.

Part IV
HYDROSTATIC PRESSURE

Editor's Comments on Papers 18 Through 24

Certes (Paper 18) collected numerous water and bottom-mud samples during the voyages of the *Travailleur* and the *Talisman* (1882–1883). Samples were collected from depths as great as 5100 m. Viable bacteria were found in 96 out of 100 samples, indicating that bacteria could survive at high hydrostatic pressure. No anaerobes were found. Certes believed he had surmounted the difficulties of sample collection from a predetermined depth and the attendant difficulties of aseptic technique while on board a ship under sail. The modern microbiologist might well consider what he would do in similar circumstances. This study appears to be the earliest report of microorganisms collected at great depths. In a second paper, Certes (1884) studied the effect on the viability of a number of microorganisms of applying pressures as high as 600 atm. The pressure of dry air at 0°C at sea level and 45° latitude is 1.0 atm. One atmosphere equals 14.696 lb $in.^{-2}$, 760 mm Hg, 1.0332 kg cm^{-2}, or 1.0133 bars.

The Galathea Deep Sea Expedition extended the upper pressure limits at which living organisms had been found to over 1000 atm when physiologically active bacteria were found in sediments of the Philippine Trench (ZoBell, 1952). Oppenheimer and ZoBell (Paper 19) determined the effect of hydrostatic pressure on 63 different species of marine bacteria, all of which had been taken from depths of less than 6000 m. ZoBell (1952) had not only demonstrated living organisms at depths of 10,000 ft., but also showed that some of these organisms, the obligate barophiles (the term "barophile" was coined by ZoBell and Johnson, 1949), depended on high pressure for survival. Oppenheimer and ZoBell found that only 7 of 63 species were able to reproduce at pressures of 600 atm. Fifteen of the 63 failed to reproduce at 200 atm. Those organisms which tolerated the higher pressures in general were usually found at greater depths in the sea. A number of morphological variations were induced in marine bacteria by incubation for 4 days at pressures ranging from 200 to 600 atm, including increases in cell size, formation of long filaments, and loss of motility.

Although Arrhenius (1889) developed a theory for the influence of temperature on chemical reaction rates that was soon applied to biological processes, no such theory existed for the effect of pressure on biological processes. Cattell (1936) reviewed the biological effects of pressure but failed to consider the interaction of temperature. Eyring's (1935) and Stearn and Eyring's (1941) theory of absolute reaction rates allowed both pressure and temperature to be considered. Johnson (Paper 20) discusses his and Eyring's work on the effects of pressure and temperature on biological processes and develops a rate equation for the process. The intensity of luminescence of *Photobacterium phosphoreum* and of cell-free extracts of *Achromobacter fischeri* was related to both temperature and pressure. Since luminescence requires an enzyme volume increase during activation, the combination of high pressure and low temperature tends to lower the luminescence. At temperatures that exceed the optimum, pressure shifts the equilibrium back in favor of the undenatured state, thereby increasing luminescence. Under a given set of conditions, increased pressure increased the optimum temperature.

The Galathea Report was published in 1959 and detailed the results of the 1950–1952 voyage. ZoBell and Morita (Paper 21) describe some of the results obtained on this voyage. Nine sediment samples were obtained under rigorous conditions at depths exceeding 10,000 m. These workers assayed the most probable numbers of several different physiological types. Both aerobic and anaerobic bacteria exceeded 10^5 viable cells/ml of sediment from the Philippine Trench. Only a few nitrifiers were found, but starch hydrolyzers, nitrate reducers, sulfate reducers, and ammonifiers usually exceed 100 cells/ml. The cultures were incubated at 1000 atm hydrostatic pressure. Apparently, these organisms not only grew at these pressures but also had significant physiological activity. Surface-dwelling nitrifiers are inhibited by pressures of 400 to 600 atm. An obligate halophilic sulfate-reducing bacterium was isolated from the Weber Deep (7250 m) and successfully transferred four times at 700 atm.

Paper 22 is the text of an Office of Naval Research Lecture at the Annual Meeting of the American Society for Microbiology. He discusses the effects of high hydrostatic pressure on DNA, bacteria, and bacteriophage. Pressure greater than 2700 atm protected transforming DNA from thermal inactivation. Both T_2 and T_4 phages of *Escherichia coli* B were inactivated by pressure. Heden proposes a model of cellular control mechanisms with hypothetical pressure targets. He discusses the results of his pressure experiments on the inactivation of phage and the effects of the multiplication of phage in terms of the model.

Numerous mechanisms have been proposed for the alteration of growth or physiological reaction rates of organisms by hydrostatic pressure. ZoBell and Kim (1972) list 13 possible mechanisms. Many of the mechanisms may act either singly or in various combinations, as is evident from the fact that organisms react so differently to pressure. Most terrestrial organisms are affected adversely by high hydrostatic pressure, but other organisms have been isolated for which it was necessary to coin the term "barophile." Furthermore, it is by no means certain that observed pressure effects on pure enzymes are necessarily the same as the effects on enzyme systems or intact cells.

Morphological changes in bacteria subjected to high hydrostatic pressure have been observed by many workers. ZoBell and Oppenheimer (1950) described the developments of elongated cells and bizarre forms of *Serratia marinorubra* when incubated at 400 to 500 atm. There was no evidence of multiplication or fission, and motility was lost. ZoBell and Cobet (1962, 1964) describe the effect of high pressure on cell division by *Escherichia coli.* At 200 to 475 atm, *E. coli* tended to form long filaments. An inverse relationship was found between the concentration of DNA and cell length. The authors interpreted these results as suggesting that failure of the DNA to be replicated at increased pressure may be responsible for the repression of cell division and consequent filament formation.

Ionization and associated effects on pH, solubility, and the like, may affect growth or physiological reaction rates. Owen and Brinkley (1941) determined the effect of pressure on ionic equilibria in pure and salt water. There were increases in the ionization constants of water and weak acids and in the solubility constants of several minerals as pressure was increased to 1000 bars. Owen and Brinkley predicted the change of K_w with pressure from the thermodynamic relation $RT(\sigma \ln K_w^* / \sigma P)_{T,m} = -\Delta V^{0*}$, where ΔV^0 is the algebraic difference between the partial molar volumes of the products and the reactants in their standard states. Their results were slightly in error because they used their only available value of 23.4 cm^3/mol at 25°C for ΔV^0. The actual value should have been 22.3 cm^3/mol. Hamann and Strauss (1955) measured the electrical conductance of five electrolytes at pressures up to 12,000 atm at 45°C. Hamann (1963) made the first direct measurements of the influence of pressure on K_w. His results agree with the prediction of Owen and Brinkley for the values of K_w when the correct value for ΔV^0 was used. Disteche and Disteche (1965) measured the effect of pressure on pH and dissociation constants with buffered and unbuffered electrode cells. Hamann and Linton (1969) studied the ionization of water at high shock pressures, and suggested that the ionization of water into H_3O^+ and OH^- ions becomes nearly complete at 200 kbar.

The effect of pressure on absolute reaction rates and the accompanying effects on biological processes was first suggested by Stearn and Eyring (1941). This study and many others to follow were the direct result of Eyring's (1935) theory of the activated complex in chemical reactions. Stearn and Eyring discuss the effect of pressure on the absolute reaction rate. Laidler (1951) and several earlier workers applied these concepts to biological systems. Laidler obtained theoretical expressions for the influence of pressure on low-temperature enzyme reactions, for both inhibited and noninhibited systems. The pressure effect depended upon whether the substrate concentration was high or low. Data for several uninhibited systems were analyzed and the suggestion was made that an unfolding and refolding of the enzyme occurred as the reaction proceeded. Johnson (Paper 20) reviews these and other papers.

Johnson, Eyring, and Polissar (1954), Suzuki and Kitamura (1963), and several others have demonstrated inactivation of enzymes by high pressures. Suzuki and Kitamura subjected α-amylase from *Bacillus subtilis* to pressures as great as 10,000 kg/cm^2. The rate of inactivation by pressure was increased proportionately from 6000 to 10,000 kg/cm^2. Mild pressures of 700 to 800 kg/cm^2 retarded the rate of acid (pH 4) inactivation of α-amylase. The pressure inactivation was accompanied by progressive denaturation of the enzyme and increasing susceptibility to digestion by pronase. Morita and ZoBell (1956) and Morita (1957) found that the succinic, formic, and malic dehydrogenases in resting cells of *E. coli* were sensitive to increased hydrostatic pressure (200 atm). Haight and Morita (1962) found activity of a cell-free preparation of an aspartase from *E. coli* to be stimulated at temperatures exceeding 50°C and inhibited at 45°C or lower. Hill and Morita (1964) examined the α-ketoglutaric acid and isocitric acid dehydrogenases in isolated mitochondria of *Allomyces macrogynus,* which is a shallow, freshwater mold. Increased hydrostatic pressure decreased mitochondrial dehydrogenase activity. Similar decreases in the activities of nitrate reduction (ZoBell and Budge, 1965) and phosphoenolpyruvate carboxylase (Izui, 1973) with increased hydrostatic pressures have been observed.

Not all enzyme systems are affected by hydrostatic pressure, however, and there are interactions between temperature and pressure. Morita and Haight (1962) found that malic dehydrogenase was active at 101°C at pressures exceeding 700 atm. The optimum pressure was 1300 atm. ZoBell and Hittle (1969) found active amylases in organisms recovered from depths exceeding 9500 m and continually cultured at 1000 atm. Starch hydrolysis by cell-free preparations, whole cultures, or washed cells grown at 1 atm were only moderately sensitive to increased hydrostatic pressure and retained 18 percent of their activity at 1000 atm. Thermal inactivation of the amylases was retarded by increased pressure. Albright and Morita (1972) found an interaction between hydrostatic pressure and incubation temperature of *Vibrio marinus.* In some cases activity was stimulated and in others retarded. Generally, hydrostatic pressure stimulated deamination of serine by washed cells at or above the temperature at which the cells were grown. At 4°C, for example, the optimal pressure was 130 atm. Penniston (1971) may have found the explanation for some of these reported inconsistencies. He reported that, when precautions were taken to saturate enzymes with their substrates and to prevent pressure denaturation, the activities of multimeric enzymes were inhibited and monomeric enzymes were stimulated by application of high pressure. Thus, deep-sea organisms must possess either enzymes that are mainly monomeric in nature or the protein multimers must be held together by stronger noncovalent interactions than those of similar enzymes of barophilic systems.

One of the best documented effects of high hydrostatic pressures on microorganisms is loss of ability to synthesize macromolecules such as DNA, RNA, and enzymes. Heden (Paper 22) speculated on the use of pressure as a tool in investigating microbial mechanisms. Pollard and Weller (1966) used this tool to investigate induced enzyme synthesis. An effect of hydrostatic pressure on the incorporation of thymine, proline, valine, and to a lesser extent of uracil was observed at 900 atm. Landau (1967) utilized the β-galactosidase system of *Escherichia coli* to study the effect of pressure on induction, transcription, and translation. Isopropyl thiogalactopyranoside was used as the inducer, and either proflavine or sodium borate was used to suppress RNA synthesis. Landau found that transcription was least affected by pressure and continued at 670 atm. Translation was totally inhibited at 670 atm and unaffected at 265 atm. Induction was totally inhibited above 255 atm. Albright and Morita (1968) found that a pressure of 1000 atm completely inhibited protein and DNA synthesis but allowed RNA synthesis to continue for a short time in *Vibrio marinus*. Pressures of 400 atm immediately lowered the rate of protein synthesis, whereas RNA synthesis was unaffected. Pressures of 400 to 500 atm probably lowered the rate of protein synthesis, which in turn may have lowered the rates of RNA and DNA synthesis. Yayanos and Pollard (1969) determined the kinetics of pressure effects on DNA, RNA, and protein synthesis in *E. coli* at pressures below 1000 atm.

Several workers have suggested that membrane permeability may be altered by hydrostatic pressure, resulting in leaking of cellular metabolites (Britten and McClure, 1962). Paul and Morita (Paper 23) found that the uptake of amino acids by a marine bacterium was inhibited by pressure. The data presented suggest that the uptake of amino acids is inhibited by high pressure, but that the respiratory enzymes necessary for subsequent metabolism of the amino acids are not significantly inhibited under the same conditions.

Cell-free systems have been used to determine the site of

inhibition of protein synthesis by hydrostatic pressure. Arnold and Albright (Paper 24) utilized cell-free ribosome suspensions prepared from *Escherichia coli* and found that pressure decreased the binding of aminoacyl tRNA to the ribosome, inhibited peptide bond formation, and probably reduced the stability of the mRNA ribosome complex. The results differed from those of Landau (1967) in that at 265 atm the binding of phenylalanine tRNA to ribosomes and dipeptide formation were inhibited in the cell-free preparation. Protein synthesis was completely inhibited in both studies at 600 atm pressure. Hildebrand and Pollard (1972) found that synthesis of polyphenylalanine was strongly inhibited above but stimulated below 100 atm pressure. Enzymatic activation of RNA was reduced by increased pressure in the range of 100 to 640 atm. They confirmed that the nonenzymatic attachment of phenylalanine tRNA to the polyuracil–ribosome complex and the stability of the phenylalanine tRNA, polyuracil–ribosome complex were decreased by high pressures (100 to 900 atm). Similar results were obtained by Schwarz and Landau (1972) and Harden and Albright (1974).

Marsland and Brown (1942) and Marsland (1942) suggested that pressure may alter the sol–gel equilibrium in protoplasm. These authors demonstrated the solation of myosin by pressure. The viscosity of water may also be altered by pressure. Horne and Johnson (1966) found that high hydrostatic pressures tend to destroy the structural regions in liquid water at extreme pressures. Viscosity is minimal at 1000 kg/cm^2.

The effect of high hydrostatic pressure on cell division and morphology probably was observed very early. ZoBell and Cobet (1962, 1964) described the formation of filamentous cells by *Escherichia coli* at pressures of 200 to 475 atm. An inverse relationship was found between the concentration of DNA and cell length. This suggested that the failure of DNA to replicate at increased pressure was responsible for cell division and consequent filament formation.

REFERENCES

Albright, L. J., and R. Y. Morita. 1968. Effect of hydrostatic pressure on synthesis of protein, ribonucleic acid, and deoxyribonucleic acid by the psychrophilic marine bacterium, *Vibrio marinus*. *Limnol. Oceanogr.* *13*(4): 637–643.

———, and R. Y. Morita. 1972. Effects of environmental parameters of low temperature and hydrostatic pressure on L-serine deamination by *Vibrio marinus*. *J. Oceanogr. Soc. Jap.* *28*(2): 63–70.

Arrhenius, S. 1889. Über die Reaktionsgeschwindigkeit bei der Inversion von Rohrzucker durch Säuren. *Z. Phys. Chem. 4*: 226.

Britten, R. J., and F. T. McClure. 1962. The amino acid pool in *Escherichia coli. Bacteriol. Rev. 26*(3): 292–335.

Cattell, McK. 1936. The physiological effects of pressure. *Biol. Rev. 11*: 441–476.

Certes, A. 1884. Note relative à l'action des hautes pressions sur la vitalité des micro-organismes d'eau douce et d'eau de mer. *Compt. Rend. Soc. Biol. Paris 36*: 220–222.

Disteche, A., and S. Disteche. 1965. The effect of pressure on pH and dissociation constants from measurements with buffered and unbuffered glass electrode cells. *J. Electrochem. Soc. 112*(3): 350–354.

Eyring, H. 1935. The activated complex in chemical reactions. *J. Chem. Phys. 3*(2): 107–115.

Haight, R. D., and R. Y. Morita. 1962. Interaction between the parameters of hydrostatic pressure and temperature on aspartase of *Escherichia coli. J. Bacteriol. 83*(1): 112–120.

Hamann, S. D. 1963. The ionization of water at high pressures. *J. Phys. Chem. 67*(10): 2233–2235.

———, and M. Linton. 1969. Electrical conductivities of aqueous solutions of KCl, KOH and HCl, and the ionization of water at high shock pressures. *Trans. Faraday Soc. 65*(8): 2186–2196.

———, and W. Strauss. 1955. The chemical effects of pressure. Part 3. Ionization constants at pressures up to 12000 atm. *Trans. Faraday Soc. 51*: 1684–1690.

Harden, M. J., and L. J. Albright. 1974. Hydrostatic pressure effects on several stages of protein synthesis in *Escherichia coli. Can. J. Microbiol. 20*(3): 359–365.

Hildebrand, C. E., and E. C. Pollard. 1972. Hydrostatic pressure effects on protein synthesis. *Biophys. J. 12*(10): 1235–1250.

Hill, E. P., and R. Y. Morita. 1964. Dehydrogenase activity under hydrostatic pressure by isolated mitochondria obtained from *Allomyces macrogynus. Limnol. Oceanogr. 9*(2): 243–248.

Horne, R. A., and D. S. Johnson. 1966. The viscosity of water under pressure. *J. Phys. Chem. 70*(7): 2182–2190.

Izui, K. 1973. Effects of high pressure on the stability and activity of allosteric phosphoenolpyruvate carboxylase from *Escherichia coli. J. Biochem. 73*(3): 505–513.

Johnson, F. H., H. Eyring, and M. J. Polissar. 1954. *The Kinetic Basis of Molecular Biology.* Wiley, New York.

Laidler, K. J. 1951. The influence of pressure on the rates of biological reactions. *Arch. Biochem. 30*(2): 226–236.

Landau, J. V. 1967. Induction, transcription and translation in *Escherichia coli.* A hydrostatic pressure study. *Biochim. Biophys. Acta 149*(2): 506–512.

Marsland, D. A. 1942. Protoplasmic streaming in relation to gel structure in the cytoplasm. In *The Structure of Protoplasm,* pp. 127–161. Iowa State College Press, Ames, Iowa.

———, and D. E. S. Brown. 1942. The effects of pressure on sol–gel equilibria, with special reference to myosin and other protoplasmic gels. *J. Cell. Comp. Physiol. 20*(3): 295–305.

Morita, R. Y. 1957. Effect of hydrostatic pressure on succinic, formic, and malic dehydrogenases in *Escherichia coli. J. Bacteriol. 74*(2): 251–255.

———, and R. D. Haight. 1962. Malic dehydrogenase activity at 101°C under hydrostatic pressure. *J. Bacteriol. 83*(6): 1341–1346.

———, and C. E. ZoBell. 1956. Effect of hydrostatic pressure on the succinic dehydrogenase system in *Escherichia coli. J. Bacteriol. 71*(6): 668–672.

Owen, B. B., and S. R. Brinkley. 1941. Calculation of the effect of pressure upon ionic equilibria in pure water and in salt solution. *Chem. Rev. 29*(3): 461–474.

Penniston, J. T. 1971. High hydrostatic pressure and enzymatic activity: inhibition of multimeric enzymes by dissociation. *Arch. Biochem. Biophys. 142*(1): 322–332.

Pollard, E. C., and P. K. Weller. 1966. The effect of hydrostatic pressure on the synthetic processes in bacteria. *Biochim. Biophys. Acta 112*(3): 573–580.

Schwarz, J. R., and J. V. Landau. 1972. Inhibition of cell-free protein synthesis by hydrostatic pressure. *J. Bacteriol. 112*(3): 1222–1227.

Stearn, A. E., and H. Eyring. 1941. Pressure and rate processes. *Chem. Rev. 29*(3): 509–523.

Suzuki, K., and K. Kitamura. 1963. Inactivation of enzyme under high pressure. Studies on the kinetics of inactivation of α-amylase of *Bacillus subtilis* under high pressure. *J. Biochem. (Tokyo) 54*(3): 214–219.

Yayanos, A. A., and E. C. Pollard. 1969. A study of the effects of hydrostatic pressure on macromolecular synthesis in *Escherichia coli. Biophys. J. 9*(12): 1464–1482.

ZoBell, C. E. 1952. Bacterial life at the bottom of the Philippine Trench. *Science 115*(2990): 507–508.

———, and K. M. Budge. 1965. Nitrate reduction by marine bacteria at increased hydrostatic pressures. *Limnol. Oceanogr. 10*(2): 207–214.

———, and A. B. Cobet. 1962. Growth, reproduction, and death rates of *Escherichia coli* at increased hydrostatic pressures. *J. Bacteriol. 84*(6): 1228–1236.

———, and A. B. Cobet. 1964. Filament formation by *Escherichia coli* at increased hydrostatic pressures. *J. Bacteriol. 87*(3): 710–719.

———, and F. H. Johnson. 1949. The influence of hydrostatic pressure on the growth and viability of terrestrial and marine bacteria. *J. Bacteriol. 57*(2): 179–189.

———, and J. Kim. 1972. Effects of deep-sea pressures on microbial enzyme systems. *Symp. Soc. Exptl. Biol. 26:* 125–146.

———, and L. L. Hittle. 1969. Deep-sea pressure effects on starch hydrolysis by marine bacteria. *J. Oceanogr. Soc. Jap. 25*(1): 36–47.

———, and C. H. Oppenheimer. 1950. Some effects of hydrostatic pressure on the multiplication and morphology of marine bacteria. *J. Bacteriol. 60*(6): 771–781.

18

ON THE CULTURE, PROTECTED FROM ATMOSPHERIC GERMS, OF WATER AND SEDIMENTS, COLLECTED BY THE EXPEDITIONS OF THE *TRAVAILLEUR* AND *TALISMAN*, 1882–1883

M. A. Certes

This article was translated expressly for this Benchmark volume by D. W. Thayer from "Sur la culture, à l'abri des germes atmosphériques, des eaux et des sédiments rapportés par les expéditions du Travailleur *et du* Talisman, *1882–1883,"* Compt. Rend., **98,** *690–694 (1884)*

According to all observers, neither plants nor animals associated with decomposition have been recovered from great depths by dredging. How can this be explained? Are there not microbes at the bottom of the sea like those which work daily under our eyes at the conversion of organic matter into inorganic matter?

The experiments that I have the honor to present to the Academy do not resolve the problem, let me hasten to say. Nevertheless, they have clarified a certain number of facts which it appeared useful to bring to your attention.

For nearly 2 years these experiments consisted essentially of the culture, protected from germs, of sediments collected in 1882 on the *Travailleur,* and of the water and sediments collected during October 1883, on the *Talisman.* Of the cultures in contact with the oxygen of the air, only four of more than 100 flasks inoculated with one drop of water or one particle of silt from great depths[1] failed to give any results. The anaerobic cultures, by contrast, have remained sterile to the present without exception. Anaerobic microbes are absent then at the bottom of the sea, but not aerobic microbes.

Seawater sterilized at between 120 and 128°C entered into the composition of most of the liquid nutrients. In addition, I employed sometimes a large amount, sometimes several drops, of veal or chicken broth. Of equal use were the liquids of Raulin and Cohn. Prior to inoculation the flasks are kept in an incubator for several days. In a word, regarding the standards for this note, one cannot neglect any of the precautions recommended by Mr. Pasteur for the prevention of the introduction of atmospheric or other germs. This condition, which is essential to all

[1] Sounding depths (m): *Travailleur:* 927, 1015, 1094, 2660, 3100, 4557, and 5100; *Talisman:* 500, 1918, 2638, 2685, 3175, and 3705.

experiments of this nature, is easily accomplished in a laboratory, but is very difficult to obtain aboard a ship at sea; therefore, because of a feeling of caution, which surely will be approved, I have not felt compelled to publish the results of the experiments completed in 1882 with sediments on the *Travailleur*. Thanks to Mr. Alph. Milne-Edwards, who successfully supervised the details of the delicate operations, all cause of error appears to have been removed on board the *Talisman*, even from that most difficult aspect concerning the tubes of water. The tubes were removed in advance and flamed at 200°C. By means of an ingenious device of Mr. Alph. Milne-Edwards, they open only at the desired depth under water and at a precise moment, when returning, the recording thermometer to which they were attached is caused to crack.

We know that, within the actual state of the science, it is more or less impossible to define clearly the species of the microorganisms, either from the point of view of their physiological role or especially from only their forms. I limited myself therefore to ascertaining in which of the following liquids (and the same with the nutritive liquids) many organisms grew in the cultures and in the same vessel. At the most rapid ordinary temperature for incubation, growth is very slow in certain media. The infusions of grass and the liquid of Raulin, for example, only became turbid at the end of 9 or 10 days. The molds appear last and only in milk, and very lightly in the broth and the liquid of Raulin, where they developed to the exclusion of all other organisms. In the milk, they appear only several days after the *Bacillus*, and probably only after the first culture has altered the composition.[2]

In the neutral liquids, the *Bacillus* that appear are usually motile, rather long, very wide, with large refractive spores. Very rarely, one encounters a large species of club-shaped vibrio; the organisms present the appearance of a simple or a double ring, and finally a few micrococci.

The cultures of water (500, 1918, 3975 m) offer the particular trait that the microbes are always the same; they are very much smaller and more motile than those from the mud. They form a pellicle on the surface, which occurs less frequently in the other cultures. These differences are to be noted, but considering the limited number of flasks inoculated with the water it would be presumptuous to conclude that the organisms of the water are always different from those of the mud. I did not find flagellated or ciliated infusoria in any of the cultures. It was otherwise in the Sargasso Sea water, which had been brought back by the Marquis of Folin and which I cultivated, protected from microorganisms, with the addition of a few drops of veal broth. In the flasks thus prepared, I have found, besides the usual rod-shaped bacteria and the characteristic diatoms, numerous amoeba that are remarkable for their small size, and flagellates, in small number, among which is one very unusual and probably new species.

[2] These molds have been sent to the *Talisman* exposition at the museum.

By successive culture, a few pure cultures were obtained, that is, cultures containing only a single species of organism. There was a large *Bacillus,* very abundant and fully sporulated. At my request, Professor Cornil inoculated guinea pigs. Even with a massive dose, these inoculations were never deleterious to the health of the animal, and the slight inflamation produced at the site of the inoculation always disappeared promptly without leaving a trace.

To summarize the presentation, it is legitimate to state that in the great depths of the ocean the water and the sediments contain microorganisms which, in spite of the enormous pressure that they must endure, have not lost the ability to multiply when placed in a favorable environment and temperature. Do the microorganisms form exclusively at the surface of the sea and settle slowly to the bottom? Is not one forced to consider the distinctive physiology of the cells that we already know? For the moment, we are not in a position to answer, but one can at least attempt to resolve this difficult question by new experiments. Thanks to the kindness of Mr. Cailletet, who placed at my disposition his clever apparatus, I am attempting at this moment a new series of experiments with cultures to which will be applied as closely as possible the conditions of pressure and temperature that were present at the great depths. These delicate experiments will require some time, and it is therefore that we have decided to place before the academy the results of our first experiments.[3]

The experiments of culture in air and in vacuum will be continued likewise with the numerous sediments that I have not yet had the time to use.[4]

[3] In a preliminary experiment, we have recovered living flagellated *Infusoria* with chlorophyll subjected for 7 hours to a pressure of 100 atm and for a few moments to 300 atm.

[4] These experiments have been completed in the laboratory of Pasteur, to whom I wish to express my thanks, as well as to his usual collaborators, Chamberland, Roux, and Loir.

19

Reprinted from *J. Marine Res. (Sears Found. Marine Res.),* **11**(1), 10–18 (1952)

THE GROWTH AND VIABILITY OF SIXTY-THREE SPECIES OF MARINE BACTERIA AS INFLUENCED BY HYDROSTATIC PRESSURE[1]

BY

CARL H. OPPENHEIMER AND CLAUDE E. ZOBELL

Scripps Institution of Oceanography
University of California
La Jolla, California

ABSTRACT

The hydrostatic pressure of sea water, which increases approximately 0.1 atmosphere per meter of depth, was found to affect the viability, reproduction, and morphology of 63 stock cultures of marine bacteria representing several genera. Many of the cultures were killed at 27° C by pressures ranging from 200 to 600 atm, although some few reproduced at 600 atm. Initial inoculum concentrations of the various bacteria appeared to influence their ability to reproduce or to tolerate high pressures. Pressures exceeding 400 atm inhibited the fission of certain bacteria without stopping their growth, thereby resulting in bizarre cells, some of which formed long filaments.

INTRODUCTION

The precursory observations of Certes (1884), Regnard (1891), ZoBell and Johnson (1949), and ZoBell and Oppenheimer (1950) indicate that hydrostatic pressures of the order of 200 to 600 atm have a pronounced effect on the vertical distribution and activities of bacteria in the sea. The significance of these observations is suggested by the fact that pressures exceeding 200 atm prevail in more than 90% of the area of the oceans. Approximately half of the oceans of the world are deeper than 4,000 m, at which depth the hydrostatic pressure is about 400 atm.

From the illustrative data in Table I it will be observed that the pressure of sea water increases with depth by approximately 0.1 atm/m, the pressure-depth gradient being influenced somewhat by latitude, temperature, and salinity. One atmosphere is equivalent to 14.696 lb/in^2, 760 mm Hg, 1.0332 kg/cm^2, or to 1.0133 bars.

Living bacteria have been found in virtually all samples of marine sediments examined for their presence regardless of water depth (ZoBell, 1946). Until recently the greatest depth at which bacteria

TABLE I. Selected Pressure-depth Gradients in Atmospheres Per Meter in the Sea for Different Depths, Salinities, Temperatures, and Latitudes (Adapted From Data Given By Bjerknes and Sandström, 1910)

Dynamic depth (m)	*Salinity* °/₀₀	*Temperature* °C	*Latitude 30°* atm/m	*Latitude 60°* atm/m
0	32	0	0.099,414	0.099,403
0	32	20	0.098,831	0.099,092
0	35	0	0.099,375	0.099,638
0	35	20	0.099,052	0.099,314
5,000	35	0	0.101,757	0.102,026
5,000	35	5	0.101,660	0.101,929
10,000	35	0	0.103,952	0.104,225

had been found was 5,800 m, but during the last year the Galathea Deep Sea Expedition extended the lower limits of the biosphere to 10,460 m. In several samples of sediments collected from depths exceeding 10,000 m in the Philippine Trench, ZoBell (1952) demonstrated the presence of bacteria that were physiologically active at pressures of the order of 1,000 atm. In fact, the well being of many bacteria collected from abyssal oceanic depths appears to depend upon high pressure. The purpose of this paper is to record the effect of pressure upon the activities of 63 different species of marine bacteria, all of which were taken from depths of less than 6,000 m.

EXPERIMENTAL METHODS

The bacteria were subjected to pressures up to 600 atm by means of apparatus described by ZoBell and Oppenheimer (1950). Essentially the apparatus consisted of thick-walled steel cylinders which have an inside diameter of 1 3/8 inches and which are long enough (11 inches) to accommodate two dozen small culture tubes. The latter were fitted with no. 000 neoprene stoppers which functioned as pistons to subject the aqueous contents of the tubes to virtually the same pressure as the hydraulic fluid filling the steel cylinder. Preliminary experiments proved that pressure up to 1,000 atm could be applied and released ten times successively by means of the connecting hydraulic pump without the contents of the closed tubes becoming contaminated.

The culture medium consisted of sea water enriched with 0.5% peptone, 0.1% yeast extract, and 0.01% ferric phosphate. Following autoclave sterilization the pH of the medium was 7.6. Sufficient volume of such sea water medium was inoculated with bacteria in their logarithmic phase of growth to fill the number of culture tubes required for any one experiment. This ensured having comparable numbers and kinds of bacteria at the beginning of each experiment. The inoculated medium was pipetted aseptically into sterile culture tubes, leaving approximately 0.5 ml of air at the top. The tubes were closed with sterile stoppers preparatory to placing them in pressure cylinders. After displacing the air from the cylinders by filling the space between the culture tubes with water, the cylinders were closed and immersed in a water bath at 27° C. Sufficient time (about 10 minutes) was allowed for the contents of the cylinders to attain this temperature. Then each cylinder in turn was connected to the hydraulic pump and the desired pressure applied. The valve was closed, the pump disconnected from the pressurized cylinders, and the latter were returned to the water bath to provide for the incubation of the bacterial cultures at 27° C.

Following four days' incubation the cylinders were reconnected to the hydraulic pump and gauge for checking the terminal pressure. After releasing the pressure, the culture tubes were removed from the cylinders for examination for evidence of growth as indicated by increased cloudiness or turbidity of the medium. Those tubes showing no evidence of bacterial growth under pressure were incubated an additional four days at room pressure in order to ascertain if the bacteria were still viable. The abundance and viability of bacteria in some tubes was estimated by standard plate count procedures. The contents of other tubes were examined by means of a phase microscope at a magnification of 970x.

RESULTS

In order to assess the effect of the abundance of bacteria in the inoculum upon the pressure tolerance of bacteria, four representative species were introduced into sea water medium to give initial populations ranging from 10 to 10^7 cells/ml. After four days' incubation at different pressures the bacterial population was determined by plate counts (Table II). When incubated at pressures of 1, 200, or 400 atm, the final bacterial population was not perceptibly affected by the initial number of cells present within the range of 10^2 to 10^6 cells/ml. Comparable results were not obtained when the initial bacterial population was less than 10^2 cells/ml. This suggests that an initial low bacterial population results in a more prolonged lag phase probably due to the presence of fewer pressure-tolerant cells in the smaller inocula. Therefore in subsequent experiments the medium was inoculated with enough cells to give initial bacterial populations of approximately 10^3 cells/ml.

As shown by the data in Table II, there was generally a decrease in

TABLE II. NUMBERS OF VIABLE CELLS PER ML OF (a) CULTURE NO. 643, (b) *Serratia marinorubra*, (c) *Sarcina pelagia*, AND (d) CULTURE NO. 38:172 AS INDICATED BY PLATE COUNTS AFTER SEA WATER BROTH CULTURES CONTAINING THE INDICATED INOCULA HAD BEEN INCUBATED AT DIFFERENT PRESSURES FOR FOUR DAYS AT 27° C

Cultural organism	*Inoculum cells/ml*	*Hydrostatic pressures in atmospheres* one	200	400	600
(a)	3×10^{0}	5×10^{7}	1×10^{7}	2×10^{8}	0
	3×10^{2}	2×10^{7}	5×10^{7}	9×10^{7}	8×10^{2}
	3×10^{4}	1×10^{7}	2×10^{7}	4×10^{7}	6×10^{4}
	3×10^{7}	3×10^{7}	3×10^{7}	7×10^{6}	4×10^{6}
(b)	2×10^{1}	5×10^{7}	9×10^{6}	6×10^{5}	7×10^{1}
	2×10^{3}	6×10^{7}	3×10^{7}	5×10^{6}	8×10^{4}
	2×10^{5}	9×10^{7}	5×10^{7}	7×10^{6}	8×10^{4}
	2×10^{7}	5×10^{7}	3×10^{7}	7×10^{7}	7×10^{6}
(c)	2×10^{-1}	1×10^{7}	3×10^{3}	1×10^{5}	0
	2×10^{1}	3×10^{7}	7×10^{7}	4×10^{7}	1×10^{3}
	2×10^{3}	1×10^{8}	5×10^{7}	3×10^{7}	3×10^{4}
	2×10^{5}	3×10^{7}	4×10^{7}	4×10^{7}	7×10^{6}
(d)	6×10^{0}	4×10^{7}	4×10^{7}	6×10^{3}	2×10^{1}
	6×10^{2}	9×10^{7}	4×10^{7}	5×10^{5}	2×10^{2}
	6×10^{4}	5×10^{7}	4×10^{7}	7×10^{7}	2×10^{3}
	6×10^{6}	1×10^{8}	5×10^{7}	5×10^{7}	6×10^{6}

the bacterial population of medium incubated at 600 atm. Employing turbidity as a criterion of growth or reproduction, it was found that only seven out of a total of 63 species tested reproduced at 600 atm. The pressure tolerance of the 63 species, representing 10 different genera, recorded in Table III, is summarized as follows:

Number of bacterial cultures which:	*Hydrostatic pressure at 27° C* 1 atm	200 atm	400 atm	600 atm
Reproduced	63	48	35	7
Showed no change	0	10	17	33
Were killed	0	5	11	23

In interpreting these results, it should be noted that the organisms had been maintained in the laboratory for several years since their isolation from the sea. Whether their pressure tolerance changed during this period is problematical, although significantly those tolerating the highest pressures in general are species commonly found at greater depths in the sea. Most of the organisms, described by ZoBell and Upham (1944), were originally isolated from shallow water

TABLE III. MULTIPLICATION OF MARINE BACTERIA AS INDICATED BY RELATIVE TURBIDITY IN SEA WATER BROTH AFTER EIGHT DAYS' INCUBATION AT DIFFERENT HYDROSTATIC PRESSURES AT 27° C

	Hydrostatic pressures in atmospheres			
*Name or number of culture**	*one*	*200*	*400*	*600*
Achromobacter stenohalis	++++	–	–	K
Achromobacter aquamarinus	++++	++++	+++	–
Achromobacter stationis	++++	++++	++++	–
Achromobacter thalassius	++++	K	K	K
Actinomyces halotrichis	++	–	–	–
Actinomyces marinolimosus	++++	++++	–	–
Bacillus imomarinus	++++	–	–	–
Bacillus cirroflagellosus	++++	++	–	–
Bacillus epiphytus	+++	++	++	K
Bacillus submarinus	++++	+++	+++	–
Bacillus thalassokoites	++++	+++	+	–
Bacillus filicolonicus	++++	–	K	K
Bacillus abysseus	++++	++++	+++	–
Bacillus borborokoites	++++	++++	++++	+++
Flavobacterium marinotypicum	++++	++++	++++	–
Flavobacterium marinovirosum	++++	++++	++	–
Flavobacterium neptunium	++	+	–	–
Flavobacterium okeanokoites	++++	++++	+	–
Micrococcus aquivivus	++++	++++	++++	++++
Micrococcus sedimenteus	++++	++++	++++	K
Micrococcus maripuniceus	++++	++++	–	–
Micrococcus infimus	+++	++++	–	–
Micrococcus sedentarius	++++	++++	++++	–
Micrococcus euryhalis	++++	K	K	K
Pseudomonas enalia	++++	+++	–	K
Pseudomonas neritica	++++	++++	+++	K
Pseudomonas azotogena	++++	K	K	K
Pseudomonas vadosa	++++	++++	++++	–
Pseudomonas oceanica	++++	+++	++++	–
Pseudomonas felthami	++++	++++	++	–
Pseudomonas aestumarina	++++	–	K	K
Pseudomonas membranula	+	–	K	K
Pseudomonas stereotropis	++++	+++	+	K
Pseudomonas coenobios	++++	++++	++++	–
Pseudomonas obscura	++++	++++	++	–
Pseudomonas pleomorpha	++++	++++	++	–
Pseudomonas marinopersica	++	K	K	K
Pseudomonas periphyta	++	–	K	K
Pseudomonas hypothermis	++++	K	K	K
Pseudomonas perfectomarinus	++++	++++	+++	+++
Pseudomonas xanthochrus	++++	–	–	–

TABLE III. (*Continued*)

*Name or number of culture**	*Hydrostatic pressures in atmospheres* one	200	400	600
Sarcina pelagia	++++	+++	+++	−
Serratia marinorubra	++++	++++	+++	−
Vibrio marinopraesens	++++	++++	+++	−
Vibrio algosus	++++	++++	++	+
Vibrio adaptatus	++++	+++	+++	K
Vibrio marinoflavus	+++	+++	+	K
Vibrio ponticus	++++	+++	++	K
Vibrio phytoplanktis	++++	++++	++++	+++
Vibrio haloplanktis	++++	++++	++++	+
Vibrio marinovulgaris	++++	+++	−	−
Vibrio marinofulvus	++++	++++	++	−
Vibrio marinagilis	++++	+	−	−
Vibrio hyphalus	++++	++++	K	K
Number 516	++++	++++	++	K
Number 549	++++	++++	+++	K
Number 595	++	−	−	−
Number 623	++	++	−	−
Number 632	++++	−	−	−
Number 633	++++	++++	−	−
Number 639	++++	++	K	K
Number 643	++++	++++	++++	++++
Number 689	++++	+++	−	K

* All except the numbered cultures have been described by ZoBell and Upham (1944).

++++ = good multiplication
+++ = fair multiplication
++ = less multiplication
+ = poor multiplication
− = no multiplication, but organisms not killed
K = all bacteria in culture killed

TABLE IV. Morphological Variations of Marine Bacteria Induced by Hydrostatic Pressures Ranging from 200 to 600 atm After Four Days' Incubation at 27° C

Organism	*Morphological variations*
Bacillus abysseus	Increase in cell size and spore formation
Bacillus borborokoites	General increase in cell size
Micrococcus aquivivus	General increase in cell size
Sarcina pelagia	Increase in size and marked pleomorphism
Serratia marinorubra	Formation of long filaments and loss of motility
Vibrio phytoplanktis	Longer pleomorphic rods, granular, loss of motility
Culture no. 643	Long filament formation

environments and none of them came from depths exceeding 6,000 m where pressures are greater than 600 atm.

Examination of the bacteria with a phase microscope shortly after the cultures were removed from the pressure cylinders showed that many had undergone morphological alterations (Table IV). The commonest change in organisms that failed to reproduce at elevated pressures was an increase in cell size, especially in length. Pressure to the limit of tolerance nearly always induced marked pleomorphism and rendered the cells nonmotile. Several seemed to grow in size or length without any evidence of cell fission or reproduction.

DISCUSSION

Although our knowledge of the effects of pressure on organisms is still woefully scant, data summarized by Cattell (1936) show that pressures such as those that occur in the sea may have multiple physiological effects. Slightly increased pressures may be stimulatory, whereas higher pressures cause the attenuation or death of organisms. According to literature reviewed by Bridgman (1946), pressure affects the rate and direction of many chemical reactions. By virtue of its effect on solubilities and dissociation constants, the pH and Eh values of complex solutions may be affected by pressure. These and other properties of solutions affected by pressure may combine to influence the physiology and ecology of marine organisms in many ways.

Very little is known about the ability of organisms to adjust themselves to higher pressures. In the marine environment, where pressures range from one to somewhat more than 1,000 atm, organisms may become acclimatized to the pressures of their habitat range. The results presented in this paper substantiate the observations of ZoBell and Johnson (1949) that the pressure tolerance of marine bacteria is related to their depth habitat. The intolerance of certain bacteria for pressures exceeding 200 atm, for example, suggests that such organisms are not active in the sea at depths exceeding 2,000 m. On the other hand, finding bacteria at depths of several thousand meters that grow at atmospheric pressure but not at pressures isobaric with the environment where they are found raises the question whether such bacteria were active in situ or were in a state of dormancy.

Such questions must be answered before the marine microbiologist can hope to appraise the importance of bacteria as biochemical or geochemical agents in the deep sea. The effect of temperature must be taken into account in seeking the answers to such questions, because there is evidence that the pressure tolerance of organisms is partly a function of the temperature. Although data to be detailed elsewhere show that the pressure tolerance of some bacteria is increased at elevated temperature, others seem to exhibit maximum pressure tolerance at the lowest temperature at which they are physiologically active.

While submarine pressures undoubtedly affect the distribution and physiological activities of bacteria, at no place in the sea will pressures prohibit the presence of bacteria, because mixed cultures have been shown to be alive and active at presssures higher than those characteristic of the most abyssal deeps.

SUMMARY

The optimal initial bacterial population in nutrient medium to give comparable results in pressure tolerance tests proved to be approximately 10^3 viable cells per ml.

Pressures ranging from 200 to 600 atm inhibited the normal growth of nearly all of the 63 different species of marine bacteria tested, although a few cultures of mixed microflora from deep sea sediments reproduced at pressures exceeding 1,000 atm. Out of 63 pure cultures tested, 15 failed to reproduce when subjected to a pressure of 200 atm at 27° C and 5 were killed in four days. Of the 28 that failed to reproduce at 400 atm 11 were killed, and of 56 that failed to reproduce at 600 atm 23 were killed by pressure.

The morphology of bacteria was affected by pressures ranging from 400 to 600 atm.

The distribution and physiological activities of bacteria in the sea are probably influenced by the hydrostatic pressures which increase per meter of depth by approximately 0.1 atm.

BIBLIOGRAPHY

Bjerknes, V. L. and J. W. Sandström

1910. Dynamic meteorology and hydrography. Part I. Statics. Publ. Carneg. Instn., No. 88: 3–146.

Bridgman, P. W.

1946. Recent work in the field of high pressures. Rev. mod. Phys., *18:* 1–93.

Cattell, M.

1936. The physiological effects of pressure. Biol. Rev., *11:* 441–476.

Certes, A.

1884. Sur la culture, a l'abri des germes atmosphériques, des eaux et des sédiments rapportés par les expéditions du "Travailleur" et du "Talisman." C. R. Acad. Sci., Paris, *98:* 690–693.

Regnard, M. P.

1891. Recherches expérimentales sur les conditions physiques de la vie les eaux. G. Masson, Paris. 500 pp.

ZoBell, C. E.
1946. Marine Microbiology. Chronica Botanica Co., Waltham, Mass. 240 pp.
1952. Bacterial life at the bottom of the Philippine Trench. Science (in press).
ZoBell, C. E. and F. H. Johnson
1949. The influence of hydrostatic pressure on the growth and viability of terrestrial and marine bacteria. J. Bact., *57:* 179–189.
ZoBell, C. E. and C. H. Oppenheimer
1950. Some effects of hydrostatic pressure on the multiplication and morphology of marine bacteria. J. Bact., *60:* 771–781.
ZoBell, C. E. and H. U. Upham
1944. A list of marine bacteria including descriptions of sixty new species. Bull. Scripps Inst. Oceanogr., *5:* 239–292.

20

Reprinted from *Symp. Soc. Gen. Microbiol.*, No. 7, 134–167 (1957)

THE ACTION OF PRESSURE AND TEMPERATURE*

FRANK H. JOHNSON

Biology Department, Princeton University, Princeton, N.J. and Marine Biological Laboratory, Woods Hole, Mass.

INTRODUCTION

Ecology deals with the mutual relationships between organisms and their environment, at all levels of complexity. The ultimate goal of ecology, as of other branches of biology, is to interpret all accessible phenomena in terms of the structure and behaviour of molecules. Since the states and reactivities of biological molecules are subject to the influence of the immediate chemical and physical environment, there is, so to speak, an 'ecology' of molecules which constitutes a basic aspect of the broader problems, and this aspect is emphasized in the discussions which follow.

From the present viewpoint, it is appropriate as well as convenient to consider the ecological significance of pressure and temperature under one heading. It would be appropriate to include also chemical factors such as pH, growth substances, salt concentration, etc., inasmuch as their effects are in principle or in fact subject to modification by pressure and temperature, and a full understanding of their action requires reference to all the significant parameters. Since they are dealt with at length elsewhere in the Symposium, discussion of these other factors in the present paper will be limited to a few data that seem especially relevant to the problems of temperature and pressure.

Interrelationships in the biological action of temperature and pressure have become apparent only within recent years. Even now, the importance of pressure as a variable of both fundamental theoretical interest and ecological importance remains to be fully appreciated, inasmuch as contemporary books and extended discussions pertaining to ecology, comparative physiology, oceanography, deep-sea life, etc., often include the influence of temperature on living organisms with scarcely more than a passing reference to pressure. The reasons for this situation are several, some obvious but others less so. In the interests of clarifying the present status as well as of envisioning likely avenues of future research, a brief review of the background and theory of biological pressure-temperature

* Aided in part by a contract (Nonr1353(00), Project NR165–233) between the Office of Naval Research and Princeton University.

relationships should be helpful. As a further point of clarity it is worth keeping in mind the kind of pressure, namely, purely hydrostatic, that is under discussion. Unlike pressure on gaseous systems, increased hydrostatic pressure on condensed systems, with no gas phase present, results in no change in concentration of dissolved substances beyond the extent of compressibility of the solvent. For water, throughout the range of pressures and temperatures that are of physiological interest, the compressibility is slight.

THE BACKGROUND OF BIOLOGICAL RESEARCH ON PRESSURE AND TEMPERATURE

For nearly half a century after Arrhenius (1889) published his theory concerning the influence of temperature on chemical reaction rates, and after Regnard (1891) coincidentally published a monograph concerning the influence of increased pressure on biological processes, research progressed independently in these seemingly unrelated fields of endeavour. The Arrhenius equation was soon applied to biological reactions (cf. Arrhenius, 1907, 1915; Bělehrádek, 1935), and because of the obvious importance of temperature, the availability of a quantitative theory of its influence on chemical rates, the relative ease of measuring its effects, and a virtually unlimited array of biological systems conveniently at hand, a vast literature on the subject accumulated. Pressure, on the other hand, enjoyed none of these inviting advantages. The only obvious natural environment where it might be significant was the sea, and there was no quantitative, physical chemical theory for interpreting its influence on reaction rates.

The discovery of living forms from depths of about 6000 m. in the sea inspired Regnard (1884*a*, *b*, *c*, *d*) and Certes (1884*a*, *b*, *c*) to undertake laboratory experiments in regard to the effects of increased hydrostatic pressure on a wide variety of biological processes which by 1891 included the activity and viability of aquatic organisms, enzyme action, fermentation, putrefaction, muscle contraction, nerve conduction, hatching of fish eggs, etc. Unfortunately, none of their experiments with increased pressure involved organisms from the deep sea; the various difficulties of obtaining and studying such organisms no doubt led these early investigators, as many subsequent ones, to concentrate their efforts on specimens from near the surface of the land or waters. Fortunately, the assembly of equipment employed early was unsuited for pressures exceeding those naturally occurring in the depths. As it turns out, this range of pressures, up to roughly 1000 atm., is of physiological interest, and although the early experiments did not take temperature into account, they demonstrated a number of phenomena of distinct interest. In general, they showed that under a few hundred atmospheres the rates of some processes were retarded, others accelerated; with rise in pressure, stimulation and then inhibition at still higher pressures sometimes occurred; and the effects were reversible on release of pressure provided the period of compression was not too long nor the pressure too high, in which event serious damage or death of organisms took place.

The general results described by Regnard and Certes were confirmed, extended and refined by later investigators, among whom was Fontaine (1930), who concluded a lengthy study of the effects of pressures up to 800 atm. on respiration, development, permeability, muscle contraction, photosynthesis, etc., of marine and other organisms, with the expressed hope that the importance of the results would lead to other researches, towards a solution of the mysterious and seductive problem of life in the great depths.

In the meantime, however, emphasis in research with pressure had begun to take a different turn. As methods became available, higher and higher pressures were studied, much beyond those to which any living organism, so far as anyone knows, is ever subjected in nature. Such pressures, of several thousand atmospheres and higher, are not *prima facie* of ecological or physiological interest. Yet, perhaps because of the added importance of possible practical applications, they have attracted the greater amount of attention and their effects, e.g. the denaturation of proteins at room temperature, the killing of bacteria, the inactivation of viruses, destruction of enzyme activity, modification of antigens, antibodies, nucleic acids, etc., are more widely known. These results are of much interest, but a more widespread familiarity with them than with those of lower pressures has had one unfortunate consequence, namely, a mistaken general impression that essentially the only results of increased hydrostatic pressure are of the kind just mentioned.

Although a detailed consideration of the very high pressures is beyond the scope of the present discussion* it is important to note that, with

* A summary review of the literature up to about 1950–1, along with discussion of the theory, is available in a recent book by Johnson, Eyring & Polissar (1954). Since 1950, most publications on the subject have come from investigators in France and in Russia. The former group includes Atanasiu, Basset, Barbu, Bordet, Chromé, Dubert, Joly, Macheboeuf, Rebeyrotte, Robert, Sclizewicz, Talwar and Vignais whose papers have appeared chiefly in the *Annales de l'Institut Pasteur*, the *Bulletin de la Société de Chimie Biologique* and the *Comptes Rendus hebdomadaires des Séances de l'Académie des Sciences*. The latter group includes Bresler, Finogenov, Frenkel, Glikina, Ivanov, Kasatochkin, Koniov, Selezneva and Tongur, whose papers have appeared chiefly in *Biokhimiya*, *Doklady Akademii Nauk*, and *Izvestiia Akademii Nauk, S.S.S.R., Seriia Fizicheskaia.*

respect to biological materials and processes, there are two different ranges of increased hydrostatic pressures, roughly those below and those above about 1000 atm. respectively, whose net effects may be entirely different and take place via different mechanisms. Under certain conditions, for example, protein denaturation may be retarded or reversed by the lower pressures, as discussed presently. Between 1000 and 5000 atm. the two types of effects probably overlap, the extent depending on the materials involved and the conditions, including temperature, of the experiment. Only the lower pressures will be dealt with in the remainder of this paper.

Had the Arrhenius (1889) equation included an explicit basis for interpreting the influence of pressure on chemical reaction rates, the picture might have been different when Bělehrádek (1935) published a detailed, critical monograph on the biological effects of temperature with no reference to pressure, and Cattell (1936) published a review of the biological effects of pressure with all but no reference to temperature. It was not until 1935, however, that advances in quantum physics and other fields led to a fully rational theory—the theory of absolute reaction rates—for understanding the influence of pressure as well as temperature on rate processes in general (Eyring, 1935; Polanyi & Evans, 1935; Glasstone, Laidler & Eyring, 1941). At the same time, Brown (1934, 1935) was inquiring into the pressure-temperature relationships of muscle contraction and found not only that the effects of increased pressure could be modified by temperature, which had been observed a short while earlier (Cattell & Edwards, 1930, 1932), but that these effects were correlated with the specific temperature-activity relationship of the muscle involved, i.e. similar effects were noted within different temperature ranges which more or less corresponded to those prevailing in the normal habitats of the respective organisms from which the muscles were obtained. Research on luminous bacteria (Johnson, Brown & Marsland, 1942*a*; Brown, Johnson & Marsland, 1942; Eyring & Magee, 1942) revealed similar phenomena in the process of bioluminescence and provided data suitable for application of the modern rate theory. A new relationship was also encountered, namely, between hydrostatic pressure and the action of narcotics (Johnson, Brown & Marsland, 1942*b*).

Pressure, temperature, biological specificity and chemical environment of living cells were thus brought together as fundamentally interrelated factors, along with a quantitative physical chemical theory. In the ensuing years, pressure-temperature relationships have been investigated in regard to some representative biological processes, including specific enzyme activity *in vivo* and *in vitro*, protoplasmic sol-gel equilibria, denaturation of proteins and bacteriophage, bacterial reproduction and disinfection, cell division of marine eggs, cardiac rhythmicity and others (cf. Johnson & Lewin, 1946*a, b, c, d*; ZoBell & Johnson, 1949; ZoBell & Oppenhimer, 1950; Johnson, Eyring & Polissar, 1954; Marsland, 1950, 1956). On the basis of suggestive evidence, and in anticipation of the possibility that microbial organisms from the deep sea or from deep oil-well brines might be found to exhibit specific adaptations for life under increased pressure, the word 'barophilic' was coined to describe such organisms (Johnson, 1948; ZoBell & Johnson, 1949). In due course, modifications such as 'stenobarophilic' and 'eurybarophilic' may become useful.

THE PHYSICAL CHEMICAL THEORY OF PRESSURE AND TEMPERATURE IN BIOLOGICAL REACTIONS

According to the theory of absolute reaction rates, any chemical rate process proceeds through the formation of an unstable intermediate complex of the reactants, referred to as the activated complex or the transition state, having a lifetime of the order of 10^{-13} sec. Once formed, the activated complex decomposes at a universal rate which is the same for all reactions and is given by the expression kT/h, where k is the Boltzmann constant, i.e. the gas constant R over Avogadro's number N, T is the absolute temperature and h is Planck's constant. The probability that the activated complex will decompose to products of the reaction rather than to a reconstitution of the reactants is given by the transmission coefficient κ, which for most reactions is equal to unity or nearly so. The specific reaction rate k' is determined by the fraction of the population of molecules in the activated state at any moment multiplied by the universal rate of decomposition and by the probability of successful reaction, and this fraction is a function of a quasi-equilibrium, whose constant is designated $K^{\ddagger}$, between the normal and activated states. Although the equilibrium of activation is conceptually different from the thermodynamic equilibrium of a reversible reaction, as well as from that of the Arrhenius theory, it behaves in the manner of an ordinary equilibrium constant and for all practical purposes can be treated as such. Thus, we have $K^{\ddagger}=e^{-\Delta F^{\ddagger}/RT}$, analogous to $K=e^{-\Delta F/RT}$ of thermodynamics, $\Delta F^{\ddagger}$ and ΔF being the free energy of activation and of reaction respectively. The specific reaction rate is therefore given by the following relation:

$$k'=\kappa\frac{kT}{h}K^{\ddagger}=\kappa\frac{kT}{h}e^{-\Delta F^{\ddagger}/RT}. \qquad (1)$$

Moreover, since $\Delta F^{\ddagger}=\Delta H^{\ddagger}-T\Delta S^{\ddagger}$ and $\Delta H^{\ddagger}=\Delta E^{\ddagger}+p\Delta V^{\ddagger}$, we have

$$k'=\kappa\frac{kT}{h}\mathrm{e}^{-\Delta H^{\ddagger}/RT}\,\mathrm{e}^{\Delta S^{\ddagger}/R}=\kappa\frac{kT}{h}\mathrm{e}^{-\Delta E^{\ddagger}/RT}\,\mathrm{e}^{-p\Delta V^{\ddagger}/RT}\,\mathrm{e}^{\Delta S^{\ddagger}/R}. \qquad (2)$$

In (2), $\Delta H^{\ddagger}$, $\Delta S^{\ddagger}$, $\Delta E^{\ddagger}$ and $\Delta V^{\ddagger}$ are the heat, entropy, energy and volume changes, respectively, of activation, and p is pressure.

Equation (2) provides the quantitative basis for the influence of temperature and pressure on reaction rates. The numerical value of $\Delta H^{\ddagger}$ differs only slightly, by the factor RT, from that of μ in the Arrhenius equation (3):

$$v=Ae^{-\mu/RT}, \qquad (3)$$

in which v is the velocity of reaction and A is an empirical constant which was replaced in the extended Arrhenius theory by PZ, P representing a probability or steric factor and Z the number of collisions. In equation (2) the fact that various chemical reactions characterized by very different activation energies may nevertheless take place at similar rates at a given temperature is accounted for by differences in the entropy of activation. Thus at normal pressure the constant A of the Arrhenius equation is approximately equal to $\kappa\frac{kT}{h}\mathrm{e}^{\Delta S^{\ddagger}/R}$. For reactions to proceed at measurable rates at ordinary temperatures, the value of $\Delta F^{\ddagger}$ must fall roughly between 20 and 30 kcal.

Biological reactions are catalysed or ultimately controlled by large molecules, viz. enzymes, proteins and nucleic acids. While an enzyme lowers the effective activation energy, it introduces additional factors through which the overall rate may be modified by pressure and temperature. The chief points are as follows.

First, although ordinary chemical reactions are usually accompanied by only small volume changes, of the order of a few c.c./mole at most, and their rates are therefore affected only slightly by pressures up to 1000 atm., reactions involving large molecules may be accompanied by large volume changes, amounting to as much as 100 c.c./mole or more, and their rates are then markedly influenced by such pressures. The large volume changes could result either from the summation of electrostriction associated with the ionization of a number of groups, or from more or less drastic changes in molecular configuration, as in a partial unfolding of protein superstructure that possibly accompanies the combination of certain enzymes with their substrates and also the process of protein denaturation. Moreover, the initial action of proteases apparently involves a loosening of the native configuration of the substrate in a manner akin to denaturation (Linderstrøm-Lang, 1950). The initial change may involve a volume increase (cf. ribonuclease action; Chantrenne, Linderstrøm-Lang & Vandendriessche, 1947). The subsequent cleavage of peptid bonds is accompanied by a volume decrease of reaction (Linderstrøm-Lang & Jacobsen, 1941).

Secondly, any enzyme-catalysed process involves at least three reactions,

$$E+S \underset{k_2}{\overset{k_1}{\rightleftharpoons}} ES \overset{k_3}{\rightarrow} E+P, \tag{4}$$

in which the overall velocity v may be expressed, in accordance with the Michaelis-Menten (1913) theory, as follows:

$$v=\frac{k_3k_1(S)\,(E_0)}{k_2+k_3+k_1(S)}. \tag{5}$$

Here (S) is substrate concentration, (E_0) is the total amount of enzyme $(E)+(ES)$ and P the products of the reaction. From (5) it is evident that the net effect of pressure may undergo fancy variations, depending not only on the values of $\Delta H^{\ddagger}$, $\Delta S^{\ddagger}$ and $\Delta V^{\ddagger}$ pertaining to the three rate constants, but also, as Laidler (1951) has pointed out, depending on the concentration of substrate. Similarly, complications in temperature relationships may occur.

Thirdly, the amount of E_0 in equation (5) is subject to the influence of pressure, temperature and inhibitors or other chemical agents which combine with the enzyme in a manner affecting its catalytic activity. Any enzyme is subject to thermal inactivation, the temperature required for an appreciable effect depending both on the specific enzyme and its chemical environment. In a number of instances it has been shown that the thermal inactivation is reversible on cooling, and may be properly represented as a reversible protein denaturation with equilibrium constant K_1 between the catalytically active E_n and inactive E_d states respectively. The reversible

$$E_n \overset{K_1}{\rightleftharpoons} E_d \tag{6}$$

denaturation reaction is usually associated with, and may be obscured by, a rate process of irreversible destruction. Both reactions are likely to have high heats and entropies, of reaction and activation respectively, much higher than those of the enzyme reaction itself.

On very short exposures to relatively high temperatures, the equilibrium change, whereby the amount of active enzyme is reversibly reduced, is primarily responsible for the decrease in observed rate of enzyme action at temperatures exceeding the 'optimum'. At these higher temperatures, when the change from native to denatured states is accompanied by a large volume increase, the rate of the observed reaction may be increased several fold by raising the hydrostatic pressure from atmospheric to a few hundred atmospheres, as in bacterial luminescence (Johnson *et al.* 1942*a*, *b*; Strehler & Johnson, 1954), or to a somewhat less extent in some other systems such as invertase (Eyring, Johnson & Gensler, 1946), salivary amylase (Schneyer, 1952) and trypsin (Fraser & Johnson, 1950). Likewise, where the irreversible destruction is characterized by a large volume increase of activation, the rate is reduced by increased pressure, as has been shown to occur in certain proteins, e.g. serum globulin (Johnson & Campbell, 1945, 1946), and tobacco mosaic virus (Johnson, Baylor & Fraser, 1948). With bacteriophage, however, opposite effects of pressure on the rate of thermal inactivation have been noted, that of coli phage T5 being decreased but that of T7 being increased (Foster, Johnson & Miller, 1949). Moreover, in the presence of chemical agents, such as urea, that promote denaturation at room temperature, the effects of pressure are more complicated; the net effect may be nil, or an increase or a decrease in rate or amount of denaturation (Schlegel & Johnson, 1949; Wright & Schomaker, 1948; Simpson & Kauzmann, 1953).

At temperatures considerably below that of maximum activity of an enzyme, the value of K_1 is generally insignificant and equation (5) is applicable. Under conditions wherein the substrate concentration is essentially constant and the rate is limited by the amount of active enzyme, the overall velocity may be expressed as follows:

$$v = bk'(S)(E), \tag{7}$$

in which b is a proportionality constant and k' the rate constant as given be equation (2). As long as neither (S) nor (E) vary, and appreciable complications do not arise from equation (5), the rate changes with temperature and pressure in proportion to the change in k'. When, with rise in temperature, K_1 of the reversible denaturation becomes significant, equation (7) must be extended as follows:

$$v = \frac{bk'(S)(E_0)}{1+K_1} = \frac{b\kappa \frac{kT}{h} e^{-\Delta F^{\ddagger}/RT}(S)(E_0)}{1+e^{-\Delta F_1/RT}}. \tag{8}$$

This equation is sufficient to describe with some accuracy the quantitative relation between temperature and rates of certain processes limited by enzyme activity, from temperatures well below to those well above the optimum. Similarly, a reversal in the observed effects of pressure at relatively low and high temperatures, respectively, is understandable on the basis of a difference in the volume change of activation pertaining to k' and the volume change of reaction pertaining to K_1, the latter being insignificant at the low temperatures.

Certain inhibitors, X, such as sulphanilamide acting on bacterial luminescence, combine reversibly with the limiting enzyme through an equilibrium, whose constant may be represented as K_2, independently of the denaturation equilibrium with constant K_1. Others, U, such as urethane or alcohol, establish one or more equilibrium combinations with the enzyme, and often catalyse a rate process of irreversible destruction in addition. For simplicity, the equilibrium constants involving U may be designated as K_3. Letting r equal the average number of molecules of X, and s the average number of molecules of U that combine per enzyme molecule, equation (8) becomes

$$V = \frac{bk'(S)(E_0)}{1+K_1+K_2(X)^r+K_1K_2(X)^r+K_1K_3(U)^s}, \quad (9)$$

provided there is no interaction between X and U (cf. Johnson, Eyring & Kearns, 1943). Inhibitory actions illustrated by X and U have been referred to as type I and type II, respectively (Johnson, Eyring & Williams, 1942).* Typically, the former leads to a pronounced increase in observed activation energy and a slight increase in the temperature of maximum rate, whereas the latter leads to a pronounced decrease in observed activation energy and a marked lowering of the temperature of maximum rate, the extent of these effects in all cases depending, of course, on concentration of the inhibitor.

Formulations useful for determining the numerical values of the constants in equation (9) have been derived and applied to data obtained with intact cells and isolated systems. Although undoubtedly over-simplified when applied to such processes as respiration, growth, luminescence, etc., of living cells, in many instances complicated biological processes vary under the influence of temperature or pressure in remarkable conformance to the simple theory which postulates a few limiting reactions that vary in importance with the conditions involved. While the actual number of constants in equations (8) or (9) may seem large, chemical theory requires that they be no fewer for the reactions postulated, and it is unlikely that the constants are so numerous *per se* as to permit the quantitative description of biological rates which arise from unrelated mechanisms. Moreover, amid the multitude of reactions that govern the distribution and activities of micro-organisms in natural

* A detailed discussion with the pertinent literature references is given by Johnson *et al.* (1954).

environments, it is reasonable to expect that, under given conditions, the operation of a relatively few reactions in the manner formulated above may constitute the major limiting factors.

The capacity and mechanisms of adaptation to special conditions is another problem, but one to which the basic physical-chemical theory applies no less. Furthermore, experimental evidence indicates that the rate of microbial mutation—one of the important aspects of adaptation—is subject to modification by pressure as well as temperature, e.g. the influence of pressure on mutagenic action of nitrogen mustards on *Neurospora* (McElroy & de la Haba, 1949), and the influence of temperature on spontaneous mutation in *Phytomonas stewartii* (Lincoln, 1947) and on the mutagenic action of Fe^{2+} on *Escherichia coli* (Catlin, 1953).

THE PRESSURE-TEMPERATURE RANGE OF NATURAL HABITATS

It is a fairly safe assumption that micro-organisms occur in nature wherever there is any other form of life, as well as in some environments where no other form of life can long survive. The extremes of pressure and temperature in places from which bacteria have been isolated range from arctic areas in which the temperature is far below freezing most of the time, never much higher, and the pressure is normally of no significance (McLean, 1918; Darling & Siple, 1941; McBee & McBee, 1956), to deep oil-well brines where the temperature is perpetually high and pressures reach several hundred atmospheres (Bastin, 1926; Bastin & Greer, 1930; Ginter, 1930; Müller & Schwartz, 1949; Ekzertsev, 1951; ZoBell, 1952*a*). Hot springs at 89–90° under somewhat less than 1 atm. (Setchell, 1903; Copeland, 1936; Egorova, 1938) and cold abysses of the ocean at less than 3° under several hundred to more than 1000 atm. of pressure, are again opposite extremes where living bacteria have been found (Certes, 1884*a*; Fischer, 1894; ZoBell, 1952*b*).

The limits of these extremes as natural habitats cannot be precisely stated, inasmuch as time as well as pressure and temperature is a factor. Moreover, it is not always possible to distinguish with certainty between organisms that represent survivors of accidental dispersal and those of a native population. The occurrence of thermophiles in the Arctic (Egorova, 1938; McBee & McBee, 1956) and in deep ocean-bottom cores (Bartholomew & Rittenberg, 1949) as well as their widespread distribution in soils where the temperature is generally below the minimum for their cultivation in the laboratory, is a case in point (Gaughran, 1947; Clegg & Jacobs, 1953; Allen, 1953). Temperature relationships of growth, however, may be modified by nutritional factors or vice versa (Robbins & Kavanagh, 1944; Barnett & Lilly, 1948; Borek & Waelsch, 1951; Ware, 1951; Maas & Davis, 1952; Kasai, 1953; Campbell, 1954; Campbell & Williams, 1953*a*, *b*), and possibly the growth of thermophiles in nature may take place very slowly at temperatures lower than in artificial media (Koch & Hoffman, 1912; Black & Tanner, 1928; Hansen, 1933). Conceivably, therefore, organisms with the same physiological potentialities can develop at a low temperature under certain conditions in nature, such as the presence of unknown chemical influences and perhaps absence of unknown chemical inhibitors, but only at a relatively high temperature with the usual methods of laboratory cultivation.

Although the vast majority of living microbes probably exist under the more or less familiar conditions of moderate temperature and no significant pressure, the largest ecological unit in the world, comprising more than half the total area of the earth, lies between 3000 and 6000 m. deep in the oceans, at pressures between 300 and 600 atm., where there is no light except for an unknown amount of bioluminescence (cf. Clark & Wertheim, 1956), and where the temperature is always less than 4° (Bruun, 1951, 1956). Below this is the 'hadal' zone (Bruun, 1956), making up about 1 % of the total area of the earth, under pressures of 600–1100 atm. The actual number and varieties of micro-organisms, as well as their rates of growth and metabolic activities in oceanic depths, constitute challenging problems about which little is yet known. While essentially the same cold temperatures may exist at the bottoms of some fresh-water lakes, the deepest one known, Lake Baikal in Siberia, has a maximum pressure of less than 200 atm. at a depth of 1741 m. (Brooks, 1950).

Laboratory studies indicate an enormous variation in the extent to which different species of micro-organisms are more or less permanently restricted to conditions comparable to the environment in which they are found, as witnessed, on the one hand, by familiar examples of the 'fastidiousness' of certain haemophilic pathogens, and on the other hand by the extraordinary capacity of *Desulfovibrio desulfuricans* (now *Sporovibrio desulfuricans*) reversibly to acquire the ability to grow with the characteristics of a halophile at ordinary temperatures and of a thermophile in ordinary media, respectively (Kluyver & Baars, 1932; Starkey, 1938). Kluyver & Baars discussed evidence for the laboratory existence of 'physiological artefacts', i.e. organisms isolated by selective factors or special enrichments but unable to grow in nature with the same characteristics. In a similar sense, morphological artefacts no doubt occur also, inasmuch as this organism in its mesophilic state consists morphologically of small, asporogenous vibrios or spirals but in its thermophilic state consists of larger, granulated vibrios that produce spores. Temperature and pressure are as much instruments of the enrichment method as are specific nutrients or other chemical factors. Thus, the proportion of described species that in one respect or another constitute true artefacts is a moot question.

Even though the example of *Desulfovibrio* is unusual in recorded experience, it serves to emphasize the potentialities of microbial adaptation to different temperature ranges, in this instance over a relatively short period of time. Much slower mechanisms of adaptation were evidently involved in another classic example, that of the flagellates which Dallinger (1887), over a period of 7 years, succeeded in acclimatizing to a temperature of 70° instead of their usual best temperature of *c.* 15°. The adapted organisms would quickly die when transferred to a nutritive medium at 15°. Given sufficient time for the various mechanisms of adaptation to operate at slow as well as rapid rates in nature, there is nothing incredible about indigenous populations in the environmental extremes referred to above. Indeed, evidence exists (ZoBell, 1956, personal communication) that bacteria recovered from oil-well brines at a depth of *c.* 3000 m. can grow in the laboratory at 104° under 1000 atm. of pressure.

TEMPERATURE RELATIONSHIPS AT NORMAL PRESSURE

In general, the range of temperatures favourable to growth and metabolism of a micro-organism are at least roughly correlated with its habitat. Growth and certain other processes, however, e.g. adaptive enzyme formation (Knox, 1950) are apt to have a somewhat narrower range than that for enzyme activity. With specific enzymes the difference is sometimes quite marked, e.g. the formic dehydrogenase of *Escherichia coli*, which has a maximum rate at about 80°, some 35° higher than the maximum temperature for growth (Johnson & Lewin, 1946*a*). Conversely, various enzymes may function normally at temperatures much below those at which growth is measurable, e.g. at 13° in thermophiles (Gaughran, 1949). Bacterial luminescence continues perceptibly at −11 to −20° (Harvey, 1913, 1940; Morrison, 1924–5) which is considerably lower than the several below-freezing minimum temperatures that have been reported for growth of micro-organisms (Rubentschik, 1925; Haines, 1930, 1931; Butkevitsch, 1932; Berry & Magoon, 1934; ZoBell & Feltham, 1935). In a given species of luminous bacteria the temperature of maximum light intensity is lower than that of maximum respiratory activity (Root, 1932, 1934–5), and in some instances is lower than the maximum temperature for growth, at which maximum non-luminous cells develop (Beijerinck, 1916; Root, 1934–5).

Among different species of luminous bacteria, under similar extra-cellular conditions, the maximum intensity of luminescence in non-proliferating cells differs by 10 to 15°, as illustrated by the species represented in Fig. 1 (Johnson *et al.* 1942*a*). In this figure the

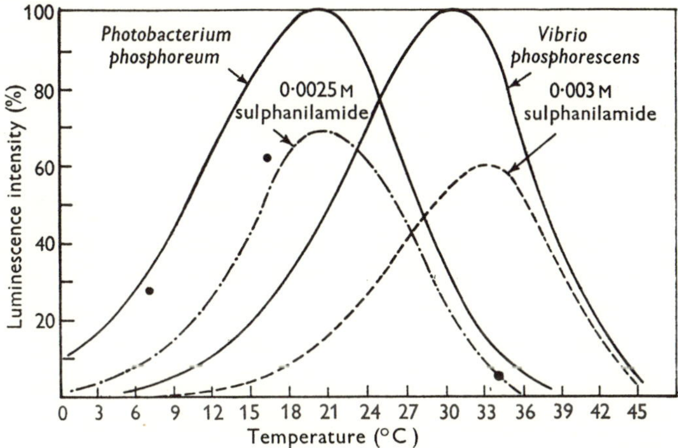

Fig. 1. Temperature-activity curves for the intensity of luminescence in a psychrophilic, marine species (*Photobacterium phosphoreum*), and a mesophilic, 'fresh-water' species (*Vibrio phosphorescens*) of luminous bacteria, in phosphate buffered salt solution believed to be isotonic for the respective organisms. Broken lines indicate the luminescence intensity of aliquot suspensions of cells containing moderately inhibitory concentrations of sulphanil-amide for the respective species. The three large solid points indicate the intensity during rapid cooling of a suspension of *P. phosphoreum* cells after a brief exposure to 34° (Johnson *et al.* 1942*a*).

maximum intensity for each species has arbitrarily been expressed as 100 %, although the average brightness per cell differed. The curves can be described fairly well by equation (8)* on the assumption that the rate of a limiting enzyme reaction increases with temperature in accordance with the Arrhenius equation (3) or with equation (2), and the overall brightness goes through a maximum ('optimum')

* In application to data on processes in living cells, the numerical value of $\Delta S^{\ddagger}$ cannot be readily obtained, but on the assumption that it does not vary much with temperature, it can be lumped together with κ, k, h, and the proportionality constant b, into an unknown constant c, yielding a useful form of the equation, as follows:

$$v = \frac{cTe^{-\Delta H^{\ddagger}/RT}}{1 + e^{-\Delta H_1/RT}\, e^{\Delta S_1/R}} = \frac{cTe^{-\Delta E^{\ddagger}/RT}\, e^{-p\Delta V^{\ddagger}/RT}}{1 + e^{-\Delta E_1/RT}\, e^{-p\Delta V_1/RT}\, e^{\Delta S_1/R}}.$$

because of the more rapid increase, with rise in temperature, in the amount of reversibly denatured enzyme. Differences in values of heats and entropies pertaining to k' and K_1, respectively, are found for the two curves with different temperature optima.

Fig. 1 illustrates also the extent to which luminescence is inhibited by a given concentration of sulphanilamide. The fact that the concentrations employed are so similar for these two species is probably a coincidence, inasmuch as the luminescence of other species varies in sensitivity to this drug at the optimal temperatures. The influence of temperature on the amount of inhibition, however, is similar for each of the several species that have been studied (Johnson, Eyring & Williams, 1942), i.e. the heats and entropies pertaining to K_2 (equation (9)) do not vary greatly. In general, the temperature relative to the biological temperature-activity curve of the system in the species involved is significant, rather than the actual temperature itself, in understanding the influence of enzyme inhibitors and activators. In the present instance, an essentially similar relationship in the two species is shifted a few degrees on the temperature scale. The effects of pressure, also, vary with respect to the specific temperature-activity curve, as discussed presently.

Rates of growth or cell division among several bacterial species and one mould, with pronounced differences in their respective optimal temperatures (each taken arbitrarily as 100 %), are plotted as the logarithm against $1/T$ in Fig. 2.* The actual maximum rates, in terms of generation time, are subject to considerable variation in each species, depending upon the nature of the culture medium and conditions of the experiment. For example, the minimum generation time of *Escherichia coli* according to Barber's (1908) data is 17 min., while according to Johnson & Lewin's (1946*c*) data it is 40 min. Barber's experiments were all started with a single cell in a broth medium, whereas Johnson & Lewin's were started with a few thousand cells per ml. in a chemically defined, semi-synthetic medium. In both cases the rate of reproduction was uniform over the period of observation, and the number of cells was too small throughout to cause any appreciable change in the medium. These are important considerations for experiments on growth rates. Information based on the growth of cultures over long periods of time and wide differences in numbers of cells is useful in other respects, for it generally involves profound changes in the chemical environment of the cells as well as complicated changes in numerous rates of individual reactions.

* The points in Figs. 2, 4, 5 and 6 for the most part represent averaged values, in a number of instances estimated from smooth curves of published figures.

The short periods of observation in Johnson & Lewin's experiments made it possible to demonstrate a quantitative reversal, on cooling, of thermal bacteriostasis at 45°. Application of equation (8) was thus justified, and the calculated curve (Fig. 2), throughout the range of temperature from about 18 almost to 45°, conformed to the data within the limits of experimental error.

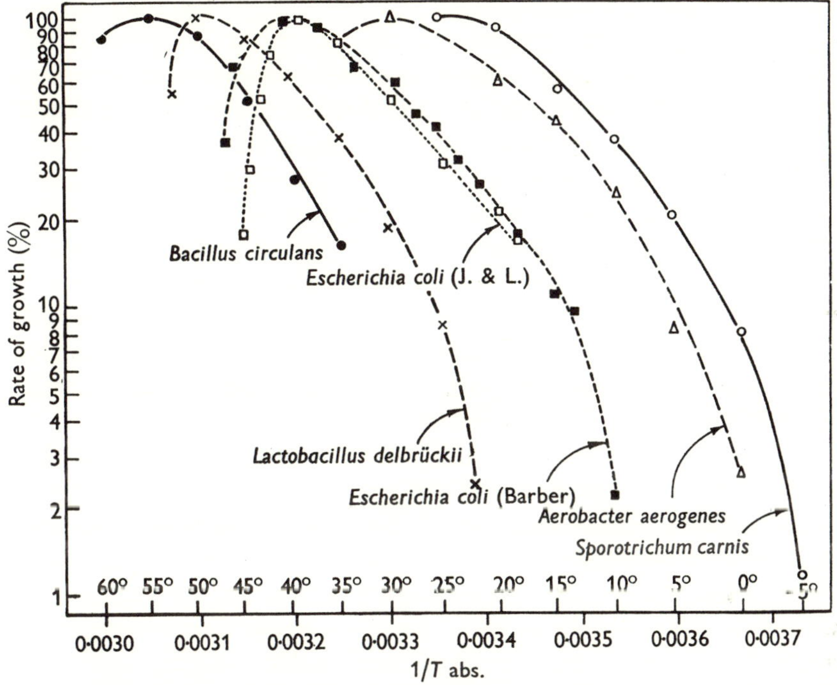

Fig. 2. Influence of temperature on rates of growth, relative to the observed maximum as 100 in each, of different species of bacteria and a mould. The data were obtained from the following sources: Allen (1953), *Bacillus circulans*; Slator (1916), *Lactobacillus delbrückii*; Barber (1908), *Escherichia coli* (solid squares); Johnson & Lewin (1946*c*), *E. coli* (hollow squares); Greene & Jezeski (1954), a strain of *Aerobacter aerogenes*; Haines (1930), *Sporotrichum carnis*.

The positions of the optima on the temperature scale (Fig. 2) and the slopes of the curves on either side of these optima are subject to modification as in luminescence (Johnson, Eyring, Steblay, Chaplen, Huber & Gherardi, 1945) by pH and other chemical factors (Johnson & Lewin, 1947) as well as by pressure (Johnson & Lewin, 1946*d*; see below).

The general similarity of the curves in Fig. 2 is not surprising. They show that the activation energy for growth in favourable media, throughout the range of below-optimal temperatures permitting between about

10 and 90 % of the rate at the optimum, is not very different among various species and is within the range commonly encountered for enzyme action.

An impressive feature of Fig. 2 is the fairly sudden increase in slope of the line for each species at the relatively low temperatures, i.e. below those temperatures at which the rate is approximately 10 % of the maximum, regardless of the actual temperature at which the change takes place. The steep slopes indicate activation energies much higher than are usually associated with enzyme activity. With *Lactobacillus delbrückii*, for example, the slope of a straight line between the points at 25 and 22° gives a μ value of some 77,000 calories. Extrapolation of this line to lower temperatures would indicate generation times of roughly 6 hr. at 15°, 67 days at 10°, and 26 years at 5°, at which point, if not above, the cell might die of old age, for there is no doubt that optimum temperatures for longevity exist in general.

The reasons for the sharp drop in growth rate at temperatures low relative to the optimum are largely conjectural and they probably stem from more than one cause. A contributory factor possibly consists in differences, with temperature of growth, in the degree of saturation of fatty constituents synthesized (Bělehrádek, 1935; Gaughran, 1947). In various organisms the lipid constituents are more saturated and have higher melting-points, after growth at higher than at lower temperatures. It is reasonable to believe that growth and cell division would be more difficult with intracellular lipids in a solid than in a more fluid state. For this hypothesis to suffice in the present instance, however, it would be necessary to postulate a similar degree of unsaturation in the lipids of *Sporotrichum carnis* growing at 0° and in *Lactobacillus delbrückii* growing at 25°. An alternative hypothesis seems worth suggesting, namely, that inability to develop at the low temperatures results in part from solation of systems whose gelation is required in the process of cell division and perhaps growth. This hypothesis stems from the critical pressure-temperature relationship of cytoplasmic gel strength which is directly correlated with the ability of egg cells to divide (Marsland, 1950, 1956), as discussed in the next section. Analogous relationships possibly exist with respect to sol-gel changes in the degree of polymerization of nucleic acids. Consistent with this notion is the observation that bacterial nuclei round up, reversibly, under the influence of increased concentration of salts in the medium surrounding the cells, as well as at low temperatures or in the presence of metabolic inhibitors (Johnson & Gray, 1949; Whitfield & Murray, 1956). It will be interesting to find out if bacterial nuclei round up under increased pressures, analogous to the reversible rounding and fusion of *Tradescantia* chromosomes, at certain phases of meiosis, under 200–400 atm. (Pease, 1946). The evidence indicates that in bacteria the nuclear substances behave like an anionic gel whose state of aggregation or dispersal varies, without necessarily damaging the viability of the cell, with the internal ionic environment that is regulated by metabolic devices controlling the influx and efflux of electrolytes, as in cells of higher plants and animals (Whitfield & Murray, 1956; Cowie & Roberts, 1955; Shanes, 1955). It will be interesting to learn, perhaps with ultra-micro-electrodes no larger than Wamoscher's (1930) needles, about the electrical polarization across the bacterial cell membrane.

An effort to account, at the molecular level, for differences among biological species in rates of growth at low temperatures thus invokes a wide array of interrelationships: metabolism, properties and reactions of specific molecules, chemical environment, temperature and pressure. The same is true for growth and viability at the upper limits of temperature.

Diminution in growth rates at temperatures above an optimum are evidently attributable primarily to the thermal instability of proteins, enzymes and nucleic acids, in part reversible but in part not reversible, and characterized as a rule by high heats of reaction or of activation. Because of the unusually high temperatures involved, the problem in 'extreme' thermophiles is of especial interest, although there is a virtually imperceptible gradation between the optimum temperatures of these organisms and those of mesophiles (Morrison & Tanner, 1924; Smith, Gordon & Clark, 1952; Allen, 1950, 1953). At high temperatures it is to be expected that essentially all reaction rates will be relatively fast, even though the net rates of enzyme reactions may become slow again because of reversible and irreversible decreases in the amount of active catalysts. Allen (1950, 1953) has stressed the dynamic nature of thermophily, i.e. rapid synthesis and repair along with rapid destruction. The same concept applies, at lower temperatures and lower absolute rates, to mesophiles and psychrophiles also, inasmuch as the rate of growth at maximal temperatures in all cases is limited by differences in the effects of temperature upon constructive and destructive reactions, and the point at which the latter overtake the former depends upon the activity and thermal stability of the systems. While the stability is in part dynamic, the stability of non-metabolizing systems is subject also to profound modification by slight changes in the chemical environment. For example, coli phage T5, with no metabolism of its own, is rapidly inactivated at temperatures only above 70° in broth, but is rapidly inactivated at 30° in 0·1 M-NaCl. At 50° the rate of inactivation is a million times faster in 0·1 M-NaCl than in broth or in 2 M-NaCl or in 0·003 M-Ca^{2+} or Mg^{2+} (Adams, 1949; Lark & Adams, 1953). In view of the potent effect of Ca^{2+} in this instance, it appears likely that the greater concentration of Ca^{2+} in bacterial spores than in vegetative cells is a fundamentally important factor in the thermal resistance of spores (Curran, Brunstetter & Myers, 1943).

Here again, complex, though in principle understandable, interrelationships between biologically specific molecular structures and the chemical environment are evident in the influence of temperature. The significance of genetically controlled molecular structure is apparent in certain 'temperature mutants' of *Neurospora*, of which a considerable number are now known (Catcheside, 1951; Horowitz & Leupold, 1951). Also in bacteria, a mutant of *Escherichia coli* requiring addition of pantothenate for growth above 30° produces at lower temperatures an altered, excessively heat-labile pantothenate-synthesizing enzyme. In extracts, its activity and thermal stability are unaffected by the presence of extracts containing the more heat-resistant, corresponding enzyme of the wild type (Maas & Davis, 1952). Similarly, the thermal stability of tyrosinase in *Neurospora* is under genic control (Horowitz & Fling, 1953). Very likely the same is true of the greater heat stability of certain enzymes from thermophilic bacteria (Militzer, Sonderegger, Tuttle & Georgi, 1949, 1950; Militzer, Tuttle & Georgi, 1951; Militzer & Tuttle, 1952; Marsh & Militzer, 1956*a*, *b*), although association of the enzyme with particulate matter is apparently a factor in some of these instances, as it is in the greater thermal stability of alanine racemase from spores than from vegetative cells of *Bacillus subtilis* (Stewart & Halvorson, 1954).

The thermal stability of nucleic acids, because of their fundamental role in synthesis (cf. Gale, 1953, 1955), is of no less importance to growth and viability than that of proteins and enzymes. In a favourable ionic environment, desoxyribonucleic acid (DNA) from animal and microbial sources is stable at temperatures up to between 80 and 90°, above which it denatures at rapidly increasing rates as the temperature is raised further (Goldstein & Stern, 1950; Zamenhof, Alexander & Leidy, 1953; Zamenhof, Griboff & Marullo, 1954; Doty & Rice, 1955). The stability is markedly influenced, however, by salt concentration, pH and the buffer system, as well as other factors (Zamenhof *et al.* 1953, 1954; Reichmann, Bunce & Doty, 1953). Few data are yet available regarding a possible correlation between the thermal stability of nucleic acids and the temperature range of growth or of the habitat of different organisms. The fact that DNA from starfish testes denatures irreversibly in very dilute NaCl at 35 to 40° (Thomas, 1954) is suggestive evidence of such a correlation. Moreover, yeast DNA is more stable to acid, down to pH 3·7 (at the temperature of the experiment), than calf thymus DNA, which is stable down to only pH 5·8, possibly indicating a correlation with intracellular pH (Zamenhof & Chargaff, 1950).

PRESSURE-TEMPERATURE-INHIBITOR RELATIONSHIPS

Interrelations between pressure, temperature and the action of certain chemical agents on a typical biological process directly under enzyme control can be conveniently illustrated with bacterial luminescence, for which the available data are most extensive. Moreover, it has recently become possible (Strehler, 1953) to study this process in cell-free extracts. Luminescence of extracted enzyme preparations requires flavine mononucleotide (FMN) and a long-chain aliphatic aldehyde; addition of reduced diphosphopyridine nucleotide (DPNH) then results in a sustained, steady-state luminescence which, in all important respects, is like that of the living cells (Strehler, 1955; Hastings & McElroy, 1955).

Fig. 3 shows, for comparison, the variation in intensity of steady-state luminescence in cells of *Photobacterium phosphoreum* (Brown *et al.* 1942) and in extracts of *Achromobacter fischeri* (Strehler & Johnson, 1954) at different pressures and temperatures and in the absence of known inhibitors. The former species has a 'normal' optimum temperature for luminescence a few degrees lower than that of the latter species, and it will be noted that the pressure-temperature relationships are essentially similar in the cells and extracts, except that a given effect occurs at a lower temperature in the species having the lower optimum. All the effects illustrated in this figure are reversible, provided the exposures to high temperatures or pressures are only brief.

At low temperatures, increasing pressure causes an exponential decrease in luminescence intensity, whereas at the high temperatures pressure causes an increase in observed intensity, relative to the control arbitrarily taken as 100% at normal pressure and all temperatures. The actual intensity of this control varied with temperature, of course, in the manner usually found for enzyme reactions, as illustrated in Fig. 4, which includes the variation of intensity with temperature under an increased pressure of 7000 lb./sq.in. Fig. 4 also shows the temperature-activity curves for the intensity at normal and 7000 lb./sq.in. pressure after adding 0·5 M alcohol to the cell suspension.*

* The reason for the slight 'stimulation' by alcohol at 0° is obscure. It occurs also in the extracted system (Strehler & Johnson, 1954).

The relationships shown in Figs. 3 and 4 have been interpreted as follows. At low temperatures the major effect of pressure is on the limiting enzyme reaction, which in accordance with equation (8) proceeds with a volume increase of activation $\Delta V^{\ddagger}$ of 51·5 c.c./mole at 0° (Eyring & Magee, 1942). With rise in temperature near to and exceeding the optimum, the amount of reversibly denatured enzyme becomes rapidly more important as a limiting factor in the overall rate of light

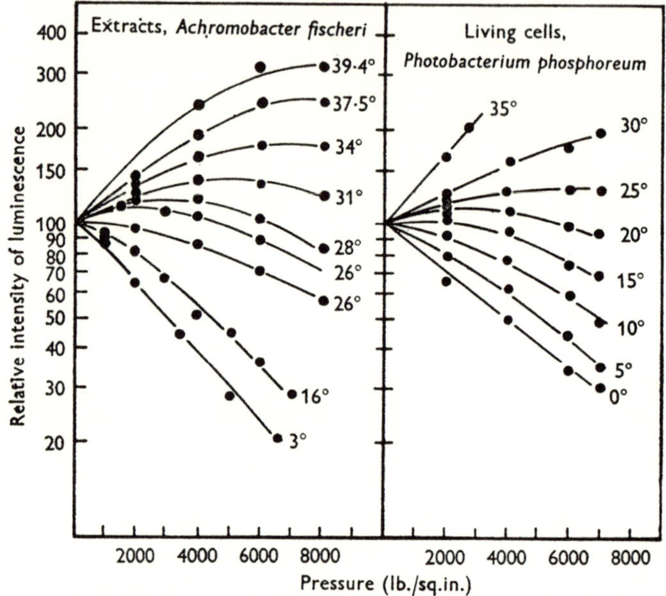

Fig. 3. Pressure-temperature relationships for the steady-state intensity of luminescence in cell suspensions of *Photobacterium phosphoreum* and in cell-free extracts of *Achromobacter fischeri*, respectively. The intensity at normal pressure is arbitrarily taken as 100 at each temperature (Strehler & Johnson, 1954; data for cells of *P. phosphoreum* from Brown *et al.* 1942).

emission. Under pressure, the equilibrium shifts in favour of the undenatured state, thus increasing the amount of active enzyme and therefore increasing the intensity of luminescence. The volume increase of reaction ΔV_1 according to equation (8) amounts to 64·6 c.c./mole at 35°. Equation (8) is sufficient to describe with considerable accuracy the intensity of luminescence from 0 to 36° at pressures between normal and 7000 lb./sq.in., provided a temperature dependence of $\Delta V^{\ddagger}$ and ΔV_1 is taken into account (Eyring & Magee, 1942). While this temperature dependence of $\Delta V^{\ddagger}$ and ΔV_1 could be real, it probably indicates that the theoretical equation is over-simplified.

At temperatures above the optimum at atmospheric pressure, irreversible destruction of the luminescent system takes place at rates which increase with temperature in the manner typical of irreversible protein denaturation. These rates are retarded by increased pressure to an extent indicating a volume increase of activation amounting to

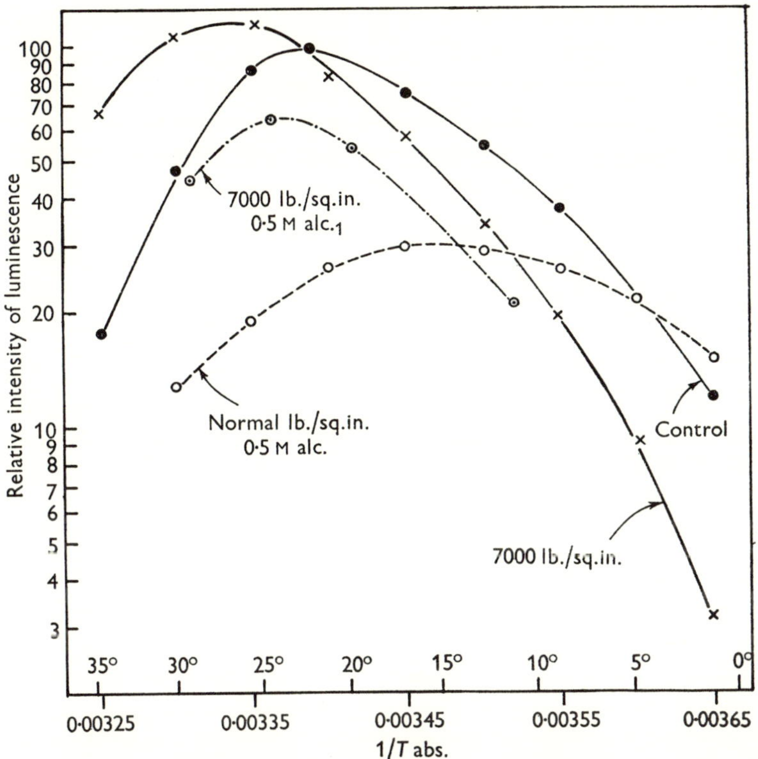

Fig. 4. Temperature-activity curves for the intensity of luminescence, at normal pressure and 7000 lb./sq.in. respectively, in cells of *Photobacterium phosphoreum* suspended in buffered salt solution, with and without the addition of 0·5 M ethyl alcohol. The control curve, for normal pressure without the addition of alcohol, represents the average of several curves with many points obtained in repeated experiments. Points on the other curves represent the same percentage difference, with respect to the averaged control, as with respect to the control in the experiment involved. The data are from figures in Johnson *et al.* (1954).

71 c.c./mole in the formation of the activated complex of the molecules whose destruction results in the loss of luminescence (Johnson *et al.* 1945).

Relationships similar to those of Fig. 3 were found by Brown (cf. Johnson *et al.* 1954) for muscle tension. In both these examples, and to some extent in regard to specific enzyme reactions *in vitro*, increasing the pressure has been found to have the same kind of effect as lowering the temperature, at temperatures both above and below the optimum. A note of caution is in order, however, against generalizing that pressure and cold have the same effect. Their influence on bacterial growth rates, for example, is more complicated (*vide infra*), and as yet the data on biological pressure-temperature relationships are too few to justify broad generalizations.

The presence of alcohol or other narcotics of similar nature markedly alters the pressure-temperature relationships of luminescence. At normal pressure the inhibition tends to increase rapidly with rise in temperature. The effect of pressure is to reduce the amount of inhibition. This effect is most conspicuous at the higher temperatures represented in Fig. 4, but the statement holds also for the lower temperatures where the amount of inhibition under pressure appears to be greater than that at normal pressure (between 10 and 15° in Fig. 4). At the lower temperatures the total effect includes a large contribution of the action of pressure in retarding the limiting enzyme reaction rate. The total reduction in intensity under pressure in the presence of alcohol at low temperatures is less than the sum of the alcohol inhibition at normal pressure plus the inhibition due to pressure in absence of alcohol. Variations in the amount of pressure and concentration of alcohol at different temperatures lead to complex variations in intensity which, however, are generally understandable on the basis of equation (9). The equation is over-simplified in not taking into account more than a single equilibrium between the enzyme and alcohol, whereas the evidence indicates that several equilibria are established, as discussed in detail elsewhere (Johnson *et al.* 1954).

Qualitatively, the action of alcohol may be pictured as promoting a reversible protein denaturation of the limiting enzyme. At higher temperatures, or with higher concentrations of alcohol at lower temperatures, alcohol catalyses irreversible thermal denaturations of certain enzymes and proteins, through reactions that proceed with large volume increases of activation as shown by their retardation under pressure (Johnson *et al.* 1954).

The action of urethane on luminescence is much like that of alcohol, and it resembles also the action of unfavourable alkalinity (Fig. 5). Sulphanilamide or moderate acidity, however, act differently on this system; the inhibitor apparently dissociates with rise in temperature, K_2 of equation (9) decreasing independently of K_1's increasing, with the results that the amount of inhibition decreases with rise in temperature up to and beyond the normal optimum, the observed activation energy for luminescence becomes greater, and the temperature of maximum intensity becomes slightly higher.

Figs. 4 and 5 illustrate how greatly the observed activation energy and temperature of maximum observed activity of a process directly under enzyme control may vary with pressure and the chemical environment. Such variations would be expected to be significant, or in some instances perhaps the determining factors of metabolic activities and growth in natural environments.

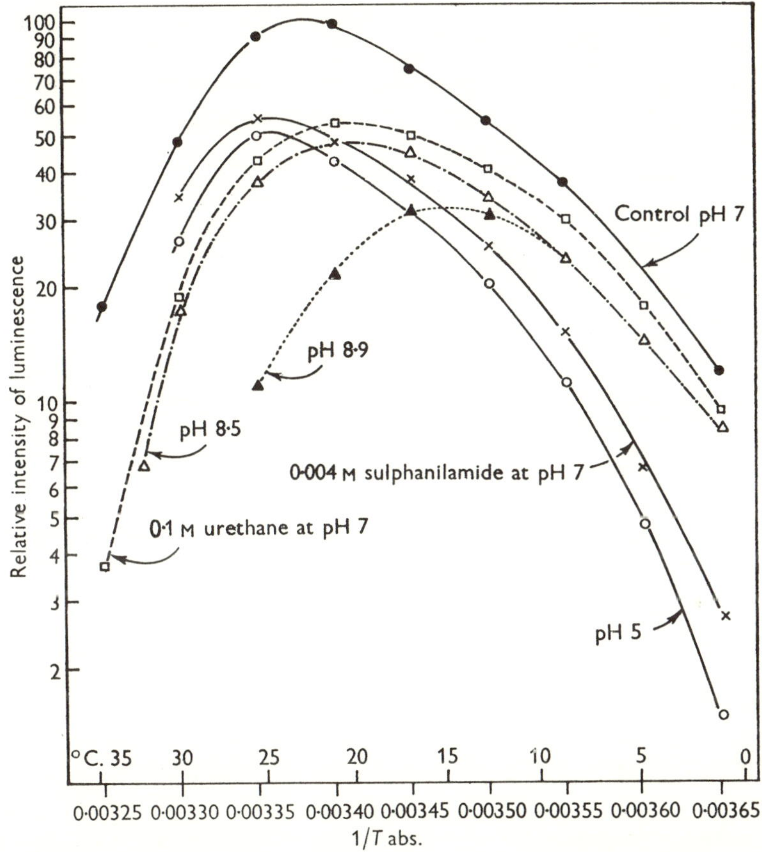

Fig. 5. Influence of inhibitory concentrations of sulphanilamide, urethane, hydrogen ions, and hydroxyl ions on the temperature-activity curve for luminescence intensity of cells of *Photobacterium phosphoreum* suspended in phosphate buffered NaCl solution. The data are from Johnson *et al.* (1954) treated as in Fig. 4.

Growth and reproduction are manifestly subject to control by factors beyond the activity of enzymes. In consequence, the pressure-temperature relationships of growth are more complicated than those of a simpler process such as luminescence. For example, with *Escherichia coli* in the early logarithmic growth phase, a pressure of 1000 lb./sq.in., during short periods of time,* depresses the rate of reproduction at temperatures below, and accelerates the rate at temperatures above the optimum, as in luminescence, whereas a pressure of 5000 lb./sq.in. depresses the rate at all temperatures (Fig. 6). At temperatures leading to a measurable rate of thermal disinfection either in growing cultures

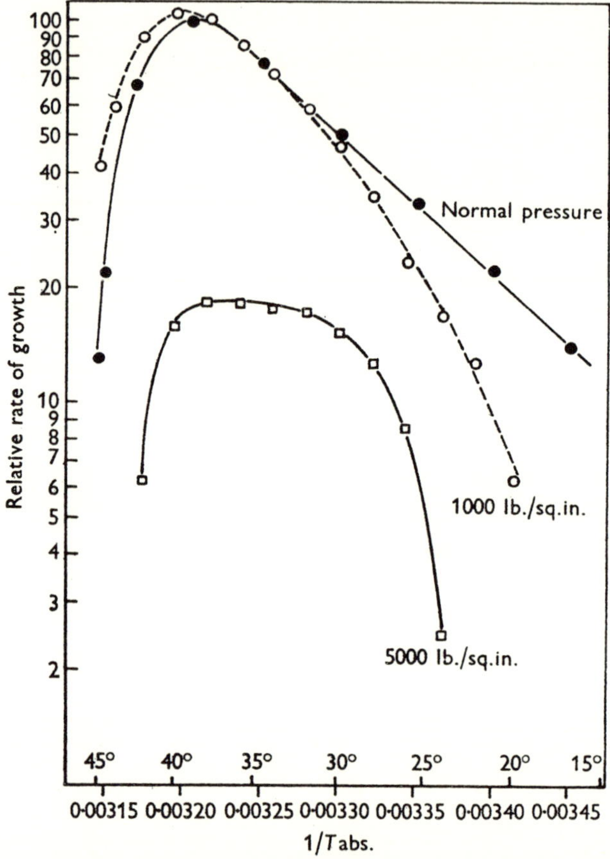

Fig. 6. Influence of pressure on the temperature activity curve for the rate of reproduction of *Escherichia coli* in an inorganic salts, glucose, asparagine medium. The data are replotted, in the manner of Figs. 4 and 5, from Johnson & Lewin (1942*c*, *d*).

* The influence of pressure on uniform growth rates of aerobic bacteria, or of facultative anaerobes whose growth rate is influenced by oxygen tension, can be satisfactorily determined only with small populations of cells in the early logarithmic phase. In order to include an initial supply of oxygen adequate to insure aerobic growth of a culture maintained continuously under pressure, from the time of inoculation with a relatively small inoculum to the development of visible turbidity, an air or oxygen bubble included in the culture tube at the start would very probably have to be larger in volume than that of the medium. Moreover, the abnormally high initial oxygen tension might be expected of itself to inhibit growth of some organisms, depending on their sensitivity to higher than normal tensions (Bean, 1945) and on the temperature (Thaysen, 1934).

Johnson & Lewin, 1946*d*) or in a non-proliferating state (Johnson & Lewin, 1946*b*), the rate is retarded under pressure (Fig. 7). Similarly, the rate of spore disinfection at temperatures above 90° is retarded by pressure (Johnson & ZoBell, 1949*a*). Small concentrations of urethane accelerate disinfection at high temperatures, and pressure retards the rate in the presence as well as absence of urethane (Johnson & ZoBell, 1949*b*). Although the kinetic data indicate that more than a single limiting reaction is involved in the effects of pressure on disinfection,

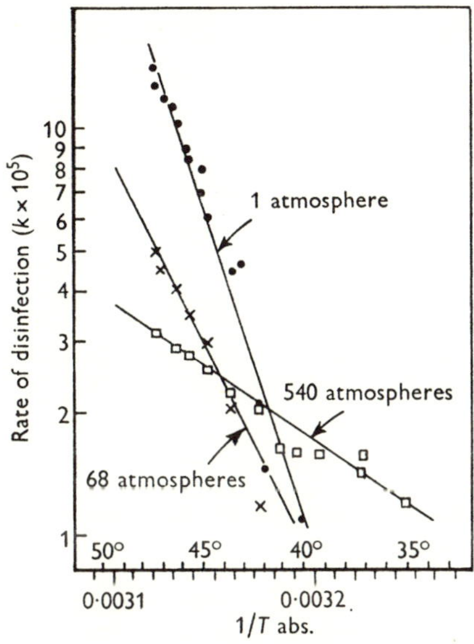

Fig. 7. Rate of thermal disinfection of non-proliferating cells of *Escherichia coli* at normal and increased pressure (Johnson & Eyring, 1948; original data of Johnson & Lewin, 1946*b*). Each point represents the velocity constant in reciprocal seconds for a first order rate of decrease in number of viable cells per ml.

this action is probably related to the retardation of thermal denaturation of proteins under pressure. It is an interesting possibility that similar mechanisms are responsible, in part, for the ability of bacteria (sulphate reducers) to survive and perhaps grow in deep oil-well brines mentioned above.

While the nature of the reactions accounting for the greater sensitivity of growth than of enzyme activity to pressure remains largely a matter of speculation, two possibilities suggest themselves, namely, synthetic reactions and sol-gel changes.

In regard to the effects of pressure on biosynthesis, practically no data are yet available pertaining to isolated systems.* Suggestive evidence resides in the fact that phage synthesis in *Escherichia coli* appears to be stopped, in part reversibly, by 600 atm. at 38° (Foster & Johnson, 1951). At this temperature the net rate of enzyme reactions in the infected cell would not be expected to be seriously reduced under this pressure, by analogy with the pressure-temperature data available with respect to luminescence and some other systems. In addition, pressure might be expected to interfere with biosynthesis on general principles. For template mechanisms must be involved in the synthesis of biologically specific molecules, and both complexly folded proteins (Pauling & Corey, 1951 *a*, *b*, *c*, *d*, *e*, *f*; Linderstrøm-Lang, 1952) and plectonemically coiled nucleic acids (Watson & Crick, 1953; Dekker & Schachman, 1954) must exist in at least partially unfolded or uncoiled states in the process. Such changes in states very likely involve volume changes, and pressure would then tend to stabilize one state whereas both are required. Temperature, of course, would likewise have an effect in accordance with the difference in heat and entropy of the two states, and the temperature relationship would again be subject to modification by the chemical environment.

In regard to the role of sol-gel changes, a fundamental pressure-temperature relationship has been found with respect to gel strength and ability of egg cells to divide (Marsland, 1950, 1956; Marsland & Landau, 1954). The cytoplasmic gels undergo solation at low temperatures or high pressures. As the temperature is raised, the amount of pressure required for a given extent of solation increases. A minimum gel strength is required for cell division, and both vary with temperature and pressure in the manner illustrated in Fig. 8. Division of the cell can take place only at those combinations of temperature and pressure that are represented below the respective lines; above these lines cytoplasmic division is blocked or reversed. Variations in the quantitative relationships among eggs of different animals are evident in differences in the slopes of the lines and in the position of these lines with respect to the temperature scale. Such differences are no doubt under genetic control and have

* Interpretation of the synthesis of proteins under pressures of 5000–6000 atm. at 37–38° reported by Bresler and his co-workers (Bresler, Glikina, Konikov, Selezneva & Finogenov, 1949, and later papers) is not clear. The same results were not obtained in similar experiments by other investigators (Talwar & Macheboeuf, 1954). Relevant to the problem is the observation that an even lower pressure of *c.* 700 atm. causes purified serum globulin, denatured at 65°, to go partly back into aqueous solution slowly at room temperature (Johnson & Campbell, 1946; cf. also Tongur, 1952). Moreover, an optimum pressure at 7500 atm. occurs for the rate of denaturation of tobacco mosaic virus (TMV), as judged by insolubilization (Lauffer & Dow, 1941). Denatured, insoluble TMV was formed only slowly at either 5000 or 10,000 atm., although loss of infectivity took place throughout this range of pressures. Evidently, denaturation and solubility of the product are both influenced by pressure.

some adaptive significance, but they can be modified to a greater or lesser extent by the chemical environment. The addition of small concentrations of adenosine triphosphate, for example, raises the pressure needed to block division of *Arbacia* or *Chaetopterus* eggs by about 500 lb./sq.in. at each temperature (Landau, Marsland & Zimmerman, 1955).

Studies similar to the above are lacking with respect to bacteria but they are not unfeasible and they should be rewarding, especially because the range of temperature and pressure in microbial physiology so far

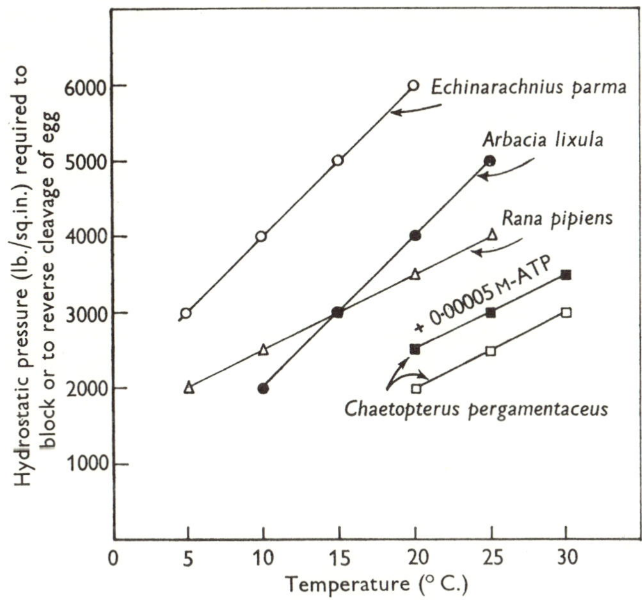

Fig. 8. Pressure-temperature relationship for blocking or reversing the cleavage of egg cells of different animals. The data are from Marsland & Landau (1954).

exceeds that of any other organisms. Data of this kind pertaining to deep-sea bacteria and to extreme thermophiles, for example, would be particularly interesting.

A main objective in the foregoing discussions has been to indicate the usefulness of studying not only the various individual factors of ecological significance, at both the level of molecules and organisms, but also the various parameters of these factors, in an effort to understand their interrelationships. While the ultimate goal of ecology—interpretation of all interrelationships in terms of the structure and behaviour of molecules—is still only a faint smudge on a distant horizon, the progress of modern physics, chemistry and biology contains an encouraging hint of eventual success.

REFERENCES

ADAMS, M. H. (1949). The stability of bacterial viruses in solutions of salt. *J. gen. Physiol.* **32**, 579.

ALLEN, MARY B. (1950). The dynamic nature of thermophily. *J. gen. Physiol.* **33**, 205.

ALLEN, MARY B. (1953). The thermophilic aerobic sporeforming bacteria. *Bact. Rev.* **17**, 125.

ARRHENIUS, S. (1889). Über die Reaktionsgeschwindigkeit bei der Inversion von Rohrzucker durch Säuren. *Z. phys. Chem.* **4**, 226.

ARRHENIUS, S. (1907). *Immunochemistry.* New York: Macmillan.

ARRHENIUS, S. (1915). *Quantitative Laws in Biological Chemistry.* London: Bell.

BARBER, M. A. (1908). The rate of multiplication of *Bacillus coli* at different temperatures. *J. infect. Dis.* **5**, 379.

BARNETT, H. L. & LILLY, V. G. (1948). The interrelated effects of vitamins, temperature and pH upon vegetation growth of Sclerotinia camelliae. *Amer. J. Bot.* **35**, 297.

BARTHOLOMEW, J. W. & RITTENBERG, S. C. (1949). Thermophilic bacteria from deep ocean bottom cores. *J. Bact.* **57**, 658.

BASTIN, E. S. (1926). The presence of sulfate reducing bacteria in oil field waters. *Science*, **63**, 21.

BASTIN, E. S. & GREER, F. E. (1930). Additional data on sulfate-reducing bacteria in soils and waters of Illinois oil fields. *Bull. Amer. Ass. Petrol. Geol.* **14**, 153.

BEAN, J. W. (1945). Effects of oxygen at increased pressure. *Physiol. Rev.* **25**, 1.

BEIJERINCK, M. W. (1916). Die Leuchtbakterien der Nordsee im August und September. *Folia microbiol.* **4**, 15.

BĚLEHRÁDEK, J. (1935). *Temperature and Living Matter.* Berlin: Borntraeger.

BERRY, J. A. & MAGOON, C. A. (1934). Growth of microorganisms at and below 0°. *Phytopathology*, **24**, 780.

BLACK, L. A. & TANNER, F. W. (1928). A study of thermophilic bacteria from the intestinal tract. *Zbl. Bakt.* (2 Abt.), **75**, 360.

BOREK, E. & WAELSCH, H. (1951). The effect of temperature on the nutritional requirement of microorganisms. *J. biol. Chem.* **190**, 191.

BRESLER, S. E., GLIKINA, M. V., KONIKOV, N. A., SELEZNEVA, N. A. & FINOGENOV, P. A. (1949). (Synthesis of proteins and peptides under pressure; cf. *Chem. Abstr.* **43**, 7988*b*.) *Inzvest. Akad. Nauk. S.S.S.R.*, Serr. Fiz., **13**, 392.

BROOKS, J. L. (1950). Speciation in ancient lakes. *Quart. Rev. Biol.* **25**, 30, 131.

BROWN, D. E. (1934). The pressure-tension-temperature relation in cardiac muscle. *Amer. J. Physiol.* **109**, 16.

BROWN, D. E. (1935). Cellular reactions to high hydrostatic pressure. *Annual Report of Tortugas Lab., Carnegie Inst. Wash.* p. 76.

BROWN, D. E., JOHNSON, F. H. & MARSLAND, D. A. (1942). The pressure-temperature relations of bacterial luminescence. *J. cell. comp. Physiol.* **20**, 151.

BRUUN, A. F. (1951). The Philippine Trench and its bottom fauna. *Nature, Lond.*, **168**, 692.

BRUUN, A. F. (1956). The abyssal fauna: Its ecology, distribution and origin. *Nature, Lond.*, **177**, 1105.

BUTKEVITSCH, V. S. (1932). Methodik der bakteriologischen Meeresuntersuchungen und einige Angaben über die Verteilung der Bakterien im Wasser und in den Boden des Barents Meeres. *Trans. Oceanogr. Inst. Moscow*, **2** (no. 2), 5. (Russian, with German summary.)

CAMPBELL, L. L. (1954). The growth of an 'obligate' thermophile at 36° C. *J. Bact.* **68**, 505.

CAMPBELL, L. L. & WILLIAMS, O. B. (1953*a*). The effect of temperature on the nutritional requirements of facultative and obligate thermophilic bacteria. *J. Bact.* **65**, 141.

CAMPBELL, L. L. & WILLIAMS, O. B. (1953*b*). Observations on the biotin requirement of thermophilic bacteria. *J. Bact.* **65**, 146.

CATCHESIDE, D. G. (1951). *The Genetics of Microorganisms.* London: Pitman.

CATLIN, B. W. (1953). Response of *Escherichia coli* to ferrous ions. I. Influence of temperature on the mutagenic action of Fe^{2+} for a streptomycin-dependent strain. *J. Bact.* **65**, 413.

CATTELL, McK. (1936). The physiological effects of pressure. *Biol. Rev.* **11**, 441.

CATTELL, McK. & EDWARDS, D. J. (1930). The influence of hydrostatic pressure on the contraction of cardiac muscle in relation to temperature. *Amer. J. Physiol.* **93**, 97.

CATTELL, McK. & EDWARDS, D. J. (1932). Conditions modifying the influence of hydrostatic pressure on striated muscle, with special reference to the role of viscosity changes. *J. cell. comp. Physiol.* **1**, 11.

CERTES, A. (1884*a*). Sur la culture, à l'abri des germes atmosphériques, des eaux et des sédiments rapportés par les expéditions du 'Travailleur' et du 'Talisman'; 1882–1883. *C.R. Acad. Sci., Paris*, **98**, 690.

CERTES, A. (1884*b*). Note relative à l'action des hautes pressions sur la vitalité des microorganismes d'eau douce et d'eau de mer. *C.R. Soc. Biol., Paris*, **36**, 220.

CERTES, A. (1884*c*). De l'action des hautes pressions sur les phénomènes de la putréfaction et sur la vitalité des microorganismes d'eau douce et d'eau de mer. *C.R. Acad. Sci., Paris*, **99**, 385.

CHANTRENNE, H., LINDERSTRØM-LANG, K. & VANDENDRIESSCHE, L. (1947). Volume change accompanying the splitting of ribonucleic acid by ribonuclease. *Nature, Lond.*, **159**, 877.

CLARK, G. L. & WERTHEIM, G. K. (1956). Measurements of illumination at great depths and at night in the Atlantic Ocean by means of a new bathyphotometer. *Deep-Sea Res.* **3**, 189.

CLEGG, L. F. L. & JACOBS, S. E. (1953). Environmental and other aspects of adaptation in thermophiles. *Adaptation in Micro-organisms. Third. Symp. Soc. gen. Microbiol.* p. 306.

COPELAND, J. J. (1936). Yellowstone thermal myxophyceae. *Ann. N.Y. Acad. Sci.* **36**, 1.

COWIE, D. B. & ROBERTS, R. B. (1955). Permeability of micro-organisms to inorganic ions, amino acids, and peptides. In *Electrolytes in Biological Systems* (A. M. Shanes, ed.), pp. 1–34. Washington: Amer. Physiol. Soc.

CURRAN, H. R., BRUNSTETTER, B. C. & MYERS, A. T. (1943). Spectrochemical analysis of vegetative cells and spores of bacteria. *J. Bact.* **45**, 485.

DALLINGER, W. H. (1887). The President's Address. *J.R. micr. Soc.* **7**, 185.

DARLING, C. A. & SIPLE, P. A. (1941). Bacteria of Antarctica. *J. Bact.* **42**, 83.

DEKKER, C. A. & SCHACHMAN, H. (1954). The macromolecular structure of desoxyribonucleic acid: an interrupted two-strand model. *Proc. nat. Acad. Sci., Wash.*, **40**, 894.

DOTY, P. & RICE, S. A. (1955). Denaturation of desoxypentose nucleic acid. *Biochim. biophys. Acta*, **16**, 446.

EGOROVA, A. A. (1938). Thermophile bacteria in Arctic. *C.R. Acad. Sci. U.R.S.S.* **19**, 649.

EKZERTSEV, V. A. (1951). (Microscopic examination of bacterial flora in oil-bearing facies of the Secondary Baku; cf. *Chem. Abstr.* **46**, 9830*i*.) *Microbiologya, Moscow*, **20**, 324.

EYRING, H. (1935). The activated complex in chemical reactions. *J. chem. Phys.* **3**, 107.

EYRING, H., JOHNSON, F. H. & GENSLER, R. L. (1946). Pressure and reactivity of proteins, with special reference to invertase. *J. phys. Chem.* **50**, 453.

EYRING, H. & MAGEE, J. L. (1942). Application of the theory of absolute reaction rates to bacterial luminescence. *J. cell. comp. Physiol.* **20**, 169.

FISCHER, B. (1894). Die Bakterien des Meeres nach den Untersuchungen der Plankton-Expedition. *Erbebn. Plankton-Exped. Humb. Stiftung*. Kiel and Leipsig: Lipsius and Tischer.

FONTAINE, M. (1930). Recherches expérimentales sur les réactions des êtres vivants aux fortes pressions. *Ann. Inst. Océanogr. Monaco*, N.S., **8**, 1.

FOSTER, RUTH A. C. & JOHNSON, F. H. (1951). Influence of urethane and of hydrostatic pressure on the growth of bacteriophages T2, T5, T6 and T7. *J. gen. Physiol.* **34**, 529.

FOSTER, R. A. C., JOHNSON, F. H. & MILLER, V. K. (1949). The influence of hydrostatic pressure and urethane on the thermal inactivation of bacteriophage. *J. gen. Physiol.* **33**, 1.

FRASER, D. & JOHNSON, F. H. (1951). The pressure-temperature relationship in the rate of casein digestion by trypsin. *J. biol. Chem.* **190**, 417.

GALE, E. F. (1953). Assimilation of amino acids by Gram-positive bacteria and some actions of antibiotics thereon. *Advance. Protein Chem.* **8**, 287.

GALE, E. F. (1955). From amino acids to proteins. In *Symposium on Amino Acid Metabolism* (W. D. McElroy and B. Glass, eds.). Baltimore: Johns Hopkins University Press.

GAUGHRAN, E. R. L. (1947). The thermophilic microorganisms. *Bact. Rev.* **11**, 189.

GAUGHRAN, E. R. L. (1949). Temperature activation of certain respiratory enzymes of stenothermophilic bacteria. *J. gen. Physiol.* **32**, 313.

GINTER, R. L. (1930). Causative agents of sulphate reduction in oil-well waters. *Bull. Amer. Ass. Petrol. Geol.* **14**, 139.

GLASSTONE, S., LAIDLER, K. J. & EYRING, H. (1941). *The Theory of Rate Processes.* New York: McGraw-Hill.

GOLDSTEIN, G. & STERN, K. (1950). Experiments on the sonic, thermal and enzymic depolymerization of desoxyribosenucleic acid. *J. Polym. Sci.* **5**, 687.

GREENE, V. W. & JEZESKI, J. J. (1954). Influence of temperature on the development of several psychrophilic bacteria of dairy origin. *Appl. Microbiol.* **2**, 110.

HAINES, R. B. (1930). The influence of temperature on the rate of growth of *Sporotrichum carnis* from $-10°$ to $+30°$ C. *J. exp. Biol.* **8**, 379.

HAINES, R. B. (1931). The influence of temperature on the rate of growth of saprophytic actinomyces. *J. exp. Biol.* **9**, 45.

HANSEN, P. A. (1933). The growth of thermophilic bacteria. *Arch. Mikrobiol.* **4**, 23.

HARVEY, E. N. (1913). The temperature limits of phosphorescence of luminous bacteria. *Biochem. Bull.* **2**, 456.

HARVEY, E. N. (1940). *Living Light.* Princeton: Princeton University Press.

HASTINGS, J. W. & MCELROY, W. D. (1955). Purification and properties of bacterial luciferase. In *The Luminescence of Biological Systems* (F. H. Johnson, ed.). Washington: Amer. Ass. Adv. Sci.

HOROWITZ, N. H. & FLING, M. (1953). Genetic determination of tyrosinase thermostability in *Neurospora. Genetics*, **38**, 360.

HOROWITZ, N. H. & LEUPOLD, U. (1951). Some recent studies bearing on the one gene-one enzyme hypothesis. *Cold Spr. Harb. Symp. quant. Biol.* **16**, 65.

JOHNSON, F. H. (1948). Bioluminescence: a reaction rate tool. *Sci. Mon., Wash.*, **67**, 225.

JOHNSON, F. H., BAYLOR, M. B. & FRASER, D. (1948). The thermal denaturation of tobacco mosaic virus in relation to hydrostatic pressure. *Arch. Biochem.* **19**, 237.

JOHNSON, F. H., BROWN, D. E. & MARSLAND, D. A. (1942*a*). A basic mechanism in the biological effects of temperature, pressure and narcotics. *Science*, **95**, 200.
JOHNSON, F. H., BROWN, D. E. & MARSLAND, D. A. (1942*b*). Pressure reversal of the action of certain narcotics. *J. cell. comp. Physiol.* **20**, 269.
JOHNSON, F. H. & CAMPBELL, D. H. (1945). The retardation of protein denaturation by hydrostatic pressure. *J. cell. comp. Physiol.* **26**, 43.
JOHNSON, F. H. & CAMPBELL, D. H. (1946). Pressure and protein denaturation. *J. biol. Chem.* **163**, 689.
JOHNSON, F. H. & EYRING, H. (1948). The fundamental action of pressure, temperature, and drugs on enzymes, as revealed by bacterial luminescence. *Ann. N.Y. Acad. Sci.* **49**, 376.
JOHNSON, F. H., EYRING, H. & KEARNS, W. (1943). A quantitative theory of synergism and antagonism among diverse inhibitors, with special reference to sulfanilamide and urethane. *Arch. Biochem.* **3**, 1.
JOHNSON, F. H., EYRING, H. & POLISSAR, M. J. (1954). *The Kinetic Basis of Molecular Biology*. London: Chapman and Hall.
JOHNSON, F. H., EYRING, H., STEBLAY, R., CHAPLIN, H., HUBER, C. & GHERARDI, G. (1945). The nature and control of reactions in bioluminescence. With special reference to the mechanism of reversible and irreversible inhibitions by hydrogen and hydroxyl ions, temperature, pressure, alcohol, urethane, and sulfanilamide in bacteria. *J. gen. Physiol.* **28**, 463.
JOHNSON, F. H., EYRING, H. & WILLIAMS, R. W. (1942). The nature of enzyme inhibitions in bacterial luminescence: sulfanilamide, urethane, temperature and pressure. *J. cell. comp. Physiol.* **20**, 247.
JOHNSON, F. H. & GRAY, D. H. (1949). Nuclei and large bodies of luminous bacteria in relation to salt concentration, osmotic pressure, temperature and urethane. *J. Bact.* **58**, 675.
JOHNSON, F. H. & LEWIN, I. (1946*a*). The action of quinine on dehydrogenases of *E. coli*. *J. cell. comp. Physiol.* **28**, 1.
JOHNSON, F. H. & LEWIN, I. (1946*b*). The disinfection of *E. coli* in relation to temperature, hydrostatic pressure and quinine. *J. cell. comp. Physiol.* **28**, 23.
JOHNSON, F. H. & LEWIN, I. (1946*c*). The growth rate of *E. coli* in relation to temperature, quinine and coenzyme. *J. cell. comp. Physiol.* **28**, 47.
JOHNSON, F. H. & LEWIN, I. (1946*d*). The influence of pressure, temperature and quinine on the rates of growth and disinfection of *E. coli* in the logarithmic growth phase. *J. cell. comp. Physiol.* **28**, 77.
JOHNSON, F. H. & LEWIN, I. (1947). The rates of growth and disinfection of *Escherichia coli* in relation to pH, quinine and temperature. *Leeuwenhoek J. Microbiol. Seerl.* **12**, 177.
JOHNSON, F. H. & ZOBELL, C. E. (1949*a*). The retardation of thermal disinfection of *Bacillus subtilis* spores by hydrostatic pressure. *J. Bact.* **57**, 353.
JOHNSON, F. H. & ZOBELL, C. E. (1949*b*). The acceleration of spore disinfection by urethan and its retardation by hydrostatic pressure. *J. Bact.* **57**, 359.
KASAI, G. J. (1953). Growth response of microorganisms to vitamins at different temperatures. *J. infect. Dis.* **92**, 58.
KLUYVER, A. J. & BAARS, J. K. (1932). On some physiological artefacts. *Proc. K. Akad. Wet. Amst.* **35**, 370.
KNOX, R. (1950). Tetrathionase: the differential effect of temperature on growth and adaptation. *J. gen. Microbiol.* **4**, 388.
KOCH, A. & HOFFMAN, C. (1912). Über die Verschiedenheit der Temperaturansprüche thermophiler Bakterien im Boden und in künstlichen Nährsubstraten. *Zbl. Bakt.* (2. Abt.), **31**, 433.

LAIDLER, K. J. (1951). The influence of pressure on the rates of biological reactions. *Arch. Biochem.* **30**, 226.

LANDAU, J. V., MARSLAND, D. A. & ZIMMERMAN, A. M. (1955). The energetics of cell division: effects of adenosine triphosphate and related substances on the furrowing capacity of marine eggs (*Arbacia* and *Chaetopterus*). *J. cell. comp. Physiol.* **45**, 309.

LARK, K. G. & ADAMS, M. H. (1953). The stability of phages as a function of the ionic environment. *Cold Spr. Harb. Symp. quant. Biol.* **18**, 171.

LAUFFER, M. A. & DOW, R. B. (1941). The denaturation of tobacco mosaic virus at high pressures. *J. biol. Chem.* **140**, 509.

LINCOLN, R. E. (1947). Mutation and adaptation of *Phytomonas stewartii*. *J. Bact.* **54**, 745.

LINDERSTRØM-LANG, K. (1950). Structure and enzymatic breakdown of proteins. *Cold Spr. Harb. Symp. quant. Biol.* **14**, 117.

LINDERSTRØM-LANG, K. (1952). Proteins and enzymes. *Stanf. Univ. Publ., Univ. Ser., Med. Sci.* **6**, 1.

LINDERSTRØM-LANG & JACOBSEN, C. F. (1941). The contraction accompanying enzymatic breakdown of proteins. *C.R. Lab. Carlsberg*, **24**, 1.

MAAS, W. K. & DAVIS, B. D. (1952). Production of an altered pantothenate-synthesizing enzyme by a temperature-sensitive mutant of *Escherichia coli*. *Proc. nat. Acad. Sci., Wash.*, **38**, 785.

MCBEE, R. H. & MCBEE, V. H. (1956). The incidence of thermophilic bacteria in Arctic soils and waters. *J. Bact.* **71**, 182.

MCELROY, W. D. & DE LA HABA, G. (1949). The effect of pressure on induction of mutations by nitrogen mustard. *Science*, **110**, 640.

MCLEAN, A. L. (1918). Bacteria of ice and snow in Antarctica. *Nature, Lond.*, **102**, 35.

MARSH, C. & MILITZER, W. (1956*a*). Thermal enzymes. VII. Further data on an adenosinetriphosphatase. *Arch. Biochem. Biophys.* **60**, 433.

MARSH, C. & MILITZER, W. (1956*b*). Thermal enzymes. VIII. Properties of a heat-stable inorganic phosphatase. *Arch. Biochem. Biophys.* **60**, 439.

MARSLAND, D. A. (1950). The mechanisms of cell division; temperature-pressure experiments on the cleaving eggs of Arbacia punctulata. *J. cell. comp. Physiol.* **36**, 205.

MARSLAND, D. (1956). Protoplasmic contractility in relation to gel structure: temperature-pressure experiments on cytokinesis and amoeboid movement. *Internat. Rev. Cytol.* **5**, 199.

MARSLAND, D. A. & LANDAU, J. V. (1954). The mechanisms of cytokinesis: temperature-pressure studies on the cortical gel system in various marine eggs. *J. exp. Zool.* **125**, 507.

MICHAELIS, L. & MENTEN, M. L. (1913). Die Kinetik der Invertinwirkung. *Biochem. Z.* **49**, 333.

MILITZER, W., SONDEREGGER, T. B., TUTTLE, L. C. & GEORGI, C. E. (1949). Thermal enzymes. *Arch. Biochem.* **24**, 75.

MILITZER, W., SONDEREGGER, T. B., TUTTLE, L. C. & GEORGI, C. E. (1950). Thermal enzymes. II. Cytochromes. *Arch. Biochem.* **26**, 299.

MILITZER, W. & TUTTLE, L. C. (1952). Thermal enzymes. IV. Partial separation of an adenosinetriphosphatase from an apyrase fraction. *Arch. Biochem. Biophys.* **39**, 379.

MILITZER, W., TUTTLE, L. C. & GEORGI, C. E. (1951). Thermal enzymes. III. Apyrase from a thermophilic bacterium. *Arch. Biochem. Biophys.* **31**, 416.

MORRISON, E. L. & TANNER, F. W. (1924). Studies on thermophilic bacteria. *Bot. Gaz.* **77**, 171.

MORRISON, T. F. (1924–5). Studies on luminous bacteria. II. The influence of temperature on the intensity of the light of luminous bacteria. *J. gen. Physiol.* **7**, 741.

MÜLLER, A. & SCHWARTZ, W. (1949). Untersuchungen zur Erdölbakteriologie. I. *Arch. Mikrobiol.* **14**, 291.

PAULING, L. & COREY, R. B. (1951*a*). The structure of synthetic polypeptides. *Proc. nat. Acad. Sci., Wash.*, **37**, 241.

PAULING, L. & COREY, R. B. (1951*b*). The pleated sheet, a new layer configuration of polypeptide chains. *Proc. nat. Acad. Sci., Wash.*, **37**, 251.

PAULING, L. & COREY, R. B. (1951*c*). The structure of feather rachis keratin. *Proc. nat. Acad. Sci., Wash.*, **37**, 256.

PAULING, L. & COREY, R. B. (1951*d*). The structure of hair, muscle, and related proteins. *Proc. nat. Acad. Sci., Wash.*, **37**, 261.

PAULING, L. & COREY, R. B. (1951*e*). The structure of fibrous proteins of the collagen-gelatin group. *Proc. nat. Acad. Sci., Wash.*, **37**, 272.

PAULING, L. & COREY, R. B. (1951*f*). The polypeptide-chain configuration in hemoglobin and other globular proteins. *Proc. nat. Acad. Sci., Wash.*, **37**, 282.

PEASE, D. C. (1946). Hydrostatic pressure effects upon the spindle figure and chromosome movement. II. Experiments on the meiotic divisions of *Tradescantia* pollen mother cells. *Biol. Bull., Woods Hole*, **91**, 145.

POLANYI, M. & EVANS, M. G. (1935). Some applications of the transition state method to the calculation of reaction velocities, especially in solution. *Trans. Faraday Soc.* **31**, 875.

REGNARD, P. (1884*a*). Note sur les conditions de la vie dans les profondeurs de la mer. *C.R. Soc. Biol., Paris*, **36**, 164.

REGNARD, P. (1884*b*). Note relative à l'action des hautes pressions sur quelques phénomènes vitaux (mouvement des cils vibratiles, fermentation). *C.R. Soc. Biol., Paris*, **36**, 187.

REGNARD, P. (1884*c*). Sur la cause de la rigidité des muscles soumis aux très hautes pressions. *C.R. Soc. Biol., Paris*, **36**, 310.

REGNARD, P. (1884*d*). Effet des hautes pressions sur les animaux marins. *C.R. Soc. Biol., Paris*, **36**, 394.

REGNARD, P. (1891). *Recherches expérimentales sur les conditions physiques de la vie dans les eaux.* Paris: Librairie Acad. Med.

REICHMANN, M. E., BUNCE, B. H. & DOTY, P. (1953). The changes induced in sodium desoxyribonucleate by dilute acid. *J. Polym. Sci.* **10**, 109.

ROBBINS, W. J. & KAVANAGH, F. (1944). Temperature, thiamine and growth of *Phycomyces*. *Bull. Torrey bot. Cl.* **71**, 1.

ROOT, C. W. (1932). The relation between respiration and light intensity of luminous bacteria, with special reference to temperature. I. Temperature and light intensity. *J. cell. comp. Physiol.* **1**, 195.

ROOT, C. W. (1934–5). The relation between respiration and light intensity of luminous bacteria with special reference to temperature. II. Temperature and oxygen consumption. *J. cell. comp. Physiol.* **5**, 219.

RUBENTSCHIK, L. (1925). Ueber die Lebenstätigkeit der Urobakterien bei einer Temperatur unter 0° C. *Zbl. Bakt.* (2. Abt.), **64**, 166.

SCHLEGEL, F. MCK. & JOHNSON, F. H. (1949). The influence of temperature and hydrostatic pressure on the denaturation of methemoglobin by urethanes and salicylate. *J. biol. Chem.* **178**, 251.

SCHNEYER, L. H. (1952). Effects of hydrostatic pressure and selected sulfhydryl inhibitors on salivary amylase. *Arch. Biochem. Biophys.* **41**, 345.

SETCHELL, W. A. (1903). The upper temperature limits of life. *Science*, **17**, 934.

SHANES, A. B. (ed.) (1955). *Electrolytes in Biological Systems.* Washington: Amer. Physiol. Soc.

SIMPSON, R. B. & KAUZMANN, W. J. (1953). Kinetics of protein denaturation. I. Behavior of the optical rotation of ovalbumin in urea solutions. *J. Amer. chem. Soc.* **75**, 1539.

SLATOR, A. (1916). The rate of growth of bacteria. *J. chem. Soc.* **109**, 2.

SMITH, N. R., GORDON, R. E. & CLARK, F. E. (1952). *Aerobic Sporeforming Bacteria.* U.S. Dep. Agric., Agric. Monograph, no. 16.

STARKEY, R. L. (1938). A study of spore formation and other morphological characteristics of *Vibrio desulfuricans. Arch. Mikrobiol.* **9**, 268.

STEWART, B. T. & HALVORSON, H. O. (1954). Studies on the spores of aerobic bacteria. II. The properties of an extracted heat-stable enzyme. *Arch. Biochem. Biophys.* **49**, 168.

STREHLER, B. L. (1953). Luminescence in cell-free extracts of luminous bacteria and its activation by DPN. *J. Amer. chem. Soc.* **75**, 1264.

STREHLER, B. L. (1955). Factors and biochemistry of bacterial luminescence. In *The Luminescence of Biological Systems* (F. H. Johnson, ed.). Washington: Amer. Ass. Adv. Sci.

STREHLER, B. L. & JOHNSON, F. H. (1954). The temperature-pressure-inhibitor relations of bacterial luminescence in vitro. *Proc. nat. Acad. Sci., Wash.*, **40**, 606.

TALWAR, G. P. & MACHEBOEUF, M. (1954). A propos des travaux de Bresler sur la synthèse des protéines sous hautes pressions. *Ann. Inst. Pasteur*, **86**, 169.

THAYSEN, A. C. (1934). Preliminary note on the action of gases under pressure on the growth of microorganisms. I. Action of oxygen under pressure at various temperatures. *Biochem. J.* **28**, 1330.

THOMAS, R. (1954). Recherches sur la dénaturation des acides desoxyribonucléïques. *Biochem. Biophys. Acta.* **14**, 231.

TONGUR, V. S. (1952). (Regeneration of egg albumins under pressure; cf. *Chem. Abstr.* **47**, 643*h*.) *Biochemistry, Leningrad*, **17**, 495.

WÁMOSCHER, L. (1930). Versuche über die Struktur der Bakterienzelle. *Z. Hyg. InfecktKr.* **111**, 422.

WARE, G. C. (1951). Nutritional requirements of *Bacterium coli* at 44°. *J. gen. Microbiol.* **5**, 880.

WATSON, J. D. & CRICK, F. H. C. (1953). A structure for desoxyribose nucleic acid. *Nature, Lond.*, **171**, 737.

WHITFIELD, J. F. & MURRAY, R. G. E. (1956). The effects of the ionic environment on the chromatin structures of bacteria. *Canad. J. Microbiol.* **2**, 245.

WRIGHT, G. G. & SCHOMAKER, V. (1948). Studies on the denaturation of antibody. IV. The influence of pH and certain other factors on the rate of inactivation of staphylococcus antitoxin in urea solutions. *J. biol. Chem.* **175**, 169.

ZAMENHOF, S., ALEXANDER, H. E. & LEIDY, G. (1953). Studies on the chemistry of the transforming principle. I. Resistance to physical and chemical agents. *J. exp. Med.* **98**, 373.

ZAMENHOF, S. & CHARGAFF, E. (1950). Studies on the diversity and the native state of desoxypentose nucleic acids. *J. biol. Chem.* **186**, 207.

ZAMENHOF, S., GRIBOFF, G. & MARULLO, N. (1954). Studies on the heat resistance of desoxyribonucleic acids to physical and chemical factors. *Biochim. biophys. Acta*, **13**, 459.

ZOBELL, C. E. (1952*a*). Part played by bacteria in petroleum formation. *Sediment. Petrol.* **22**, 42.

ZOBELL, C. E. (1925*b*). Bacterial life at the bottom of the Philippine trench. *Science*, **115**, 507.

ZOBELL, C. E. & FELTHAM, CATHERINE B. (1935). The occurrence and activity of urea-splitting bacteria in the sea. *Science*, **81**, 234.

ZOBELL, C. E. & JOHNSON, F. H. (1949). The influence of hydrostatic pressure on the growth and viability of terrestrial and marine bacteria. *J. Bact.* **57**, 179

ZOBELL, C. E. & OPPENHEIMER, C. H. (1950). Some effects of hydrostatic pressure on the multiplication and morphology of marine bacteria. *J. Bact.* **60**, 771.

21

Reprinted from *Galathea Report,* Vol. I, Dansk Videnskabs Forlog, A/S, Copenhagen, 1959, pp. 139–154

DEEP-SEA BACTERIA

By CLAUDE E. ZOBELL *and* RICHARD Y. MORITA

Professor of Microbiology, University of California, La Jolla, California, U.S.A.

Research Assistant, University of California. Presently Assistant Professor of Biology, University of Houston, Houston, Texas, U.S.A.

SUMMARY[1]

The presence of living bacteria in some of the deepest parts of the ocean was demonstrated on the Galathea Expedition. Sediment samples taken from the bottom of the Philippine Trench at depths exceeding 10 000 meters were found to contain from 10^4 to 10^6 living bacteria per ml. Large bacterial populations were also detected in sediment samples from the floor of the Kermadec-Tonga Trench (6790 to 9820 meters), the Sunda Deep (7020 meters) in the Java Trough, and the Weber Deep (7250 meters) in the Banda Sea. Several hundred microbial analyses were made on nine sediment samples taken from depths exceeding 10 000 meters.

The occurrence of bacteria in deep-sea sediment samples was demonstrated by direct microscopic examinations shortly after the samples were hauled aboard the Galathea. Growth or reproduction of the bacteria in nutrient sea water media proved that the bacteria were alive. Their indigenity was indicated by three lines of evidence: (1) Every possible precaution was exercized to prevent the contamination of the deep-sea samples, (2) More bacteria were found in the sediment samples than in the overlying water, and (3) The deep-sea bacteria were unique in their ability to grow preferentially or exclusively at *in situ* hydrostatic pressures (700 to 1 000 atmospheres) and low temperatures (3 to 5°C).

Apparatus is described which was designed to maintain sediment samples and bacterial cultures at any desired pressure in the ship's refrigerator. This apparatus made it possible to send by Air Express deep-sea sediment samples held at 1 000 atm to our laboratory at Scripps Institution for further observation.

The deep-sea bacteria appear to be even more heat-sensitive than pressure-sensitive. Many were killed by holding them for ten minutes at 30°C. Less than 20 per cent of the deep-sea bacteria survived for ten minutes at 40°C.

The unique pressure, temperature, and nutrient requirements of the deep-sea bacteria presents an enigma in bacterial taxonomy, because the bacteria cannot be characterized by conventional or standard method cultural procedures. Morphologically the deep-sea bacteria resemble common soil and water forms, but physiologically and culturally they show differences which suggest that most deep-sea bacteria are new and undescribed species or genera.

Though predominantly aerobic, some of the deep-sea bacteria develop anaerobically. Among the physiological types that were present and active at *in situ* pressures and temperatures were starch hydrolyzers, nitrate reducers, sulfate reducers, ammonifiers, and nitrifiers. The types of bacteria present as well as the organic and sulfate content of profile series of mud cores from the Kermadec-Konga Trench suggests that the bacteria have been active *in situ* in altering organic compounds and in reducing sulfate.

Besides affecting the non-conservative chemical composition of sea water and sediment, the deep-sea bacteria are believed to contribute to the nutrition of benthic fauna. The "standing crop" of organic carbon in the cells of living bacteria is estimated at from 0.2 to 2.0 mg per liter of sediment at the mud-water interface. There may be from 10 to 100 "crops" or bacterial generations per year. The bacteria are believed to obtain their carbon and energy requirements from organic detritus settling to the sea floor and to a much larger extent from dissolved or colloidal organic compounds carried by the movements of water masses.

---

1. Contribution from the Scripps Institution of Oceanography, University of California, La Jolla, California, U.S.A., New Series No. 926. These investigations were supported in part by grants from the Office of Naval Research, U.S.A. Department of Navy, and the Rockefeller Foundation.

INTRODUCTION

Observations on the Galathea Expedition in 1951 demonstrated for the first time the occurrence of living bacteria in some of the deepest parts of the ocean. Prior to this time, 5942 meters was the greatest depth at which bacteria had been found. Most microbiologists questioned whether bacteria could exist in oceanic abysses, although an abundant microflora has been shown to occur in near-shore sediments (ZoBell, 1946).

On the Talisman Expedition, Certes (1884a) found a few bacteria in bottom deposits taken from a depth of 5100 meters. The indigenous nature of these bacteria was indicated by their ability to grow in nutrient medium at hydrostatic pressures that were approximately isobaric with the depth from which they were taken (Certes, 1884b). While on the Humboldt Plankton Expedition to the West Indies, Fischer (1894) found a few bacterial colonies on nutrient agar plates inoculated with deep-sea sediment samples, one of which came from a depth of 5280 meters, but there was nothing to indicate whether these bacteria were indigenous or adventitious species.

In 1948, we obtained from a 5800-meter deep off Bermuda a mud sample in which numerous barophilic bacteria were found. The term "barophilic" was coined by ZoBell and Johnson (1949) to characterize microbes which grow preferentially or exclusively at high hydrostatic pressures. During the 1950 Mid-Pacific Expedition to the Marshall Islands, large bacterial populations were demonstrated in pelagic sediments, including 14 samples taken from depths exceeding 4000 meters; one of these samples containing bacteria came from a depth of 5942 meters (Morita and ZoBell, 1955).

Cultural methods, supplemented by direct microscopic observations, were employed to investigate the numbers and kinds of bacteria in marine materials collected by the Galathea. Immediately after being hauled aboard the ship, deep-sea sediment and water samples were transferred aseptically to steel vessels for compression to hydrostatic pressures approximately isobaric with the depth from which taken, and placed in a refrigerator at 3-5°C. This operation ordinarily required from 5 to 15 minutes. Portions of each sample were removed for detailed examination as time and facilities permitted.

Temperature Tolerance of Deep-Sea Bacteria.

Haste in handling the deep-sea samples proved to be of utmost importance owing to the temperature sensitivity of the bacteria. About 25 per cent of the marine bacteria collected off the coast of California by ZoBell and Conn (1940) were killed in 10 minutes when warmed to 30°C and only 20 per cent survived for 10 minutes at 40°C. Deep-sea bacteria have proved to be even more sensitive to heat than those from shoal waters.

The temperature of most of the deep-sea samples ranged from 5° to 10°C when hauled aboard the Galathea, having warmed up from about 3°C while being brought to the surface. The air temperature in the tropics, like that of the surface water, ranged from 28° to 30°C. Room temperature in the Galathea laboratory was usually a few degrees higher, and the temperature in the stateroom used as a microbiological laboratory often exceeded 40°C. Consequently we were working near or beyond the threshold of temperature tolerance of many marine microbes.

In spite of haste in handling materials and precautionary measures designed to keep them cool, many of the more sensitive organisms were probably killed by the heat. Lacking a cold room in which to make the necessary transfers involving minute amounts of inocula (often only 0.01 ml), part of the material got warmed up to near laboratory temperature during the measuring, transfer, dilution, and inoculation of the bacteria therein. Fortunately, a good many of the deep-sea bacteria did survive the extreme change in climate to which they were subjected during sampling and analytical operations, but we do not know how many may have perished. Data summarized in Table I illustrate the loss of viability of bacteria in deep-sea samples intentionally subjected to different temperatures for 10 minutes prior to plating on nutrient agar.

Only a small percentage (2.3 to 4.7) of the bacteria survived heating to 50°C for 10 minutes and less than 20 per cent survived for 10 minutes at 40°C. Most of those that survived proved to be spore formers. From the data in Table I, it might appear that most of the deep-sea bacteria tolerated a temperature of 30°C for 10 minutes, but this may not be true because, even at "room" temperature, the samples were subjected for several seconds to temperatures between 30° and 40°C or higher. Perhaps all of

Table I. Relative numbers of bacteria in sediment samples from Philippine Trench which developed on nutrient agar after small portion of sample was held for 10 minutes at stated temperature. The results are expressed as percentages of the number of colonies which developed from portions of the same samples subjected to "room" temperature during rapid routine manipulations.

Station Number	Water depth	Rapid room	Held 10 minutes at		
			30° C	40° C	50° C
	meters	per cent	per cent	per cent	per cent
413	10 060	100	91.6	18.0	4.7
418	10 190	100	82.5	7.4	2.3
419	10 210	100	93.7	11.8	3.0
420	10 160	100	78.4	6.5	2.8

the more heat-sensitive microbes were lost. This should stress the importance of providing properly air-conditioned rooms for microbiological manipulations on subsequent deep-sea expeditions. Maintaining the organisms at *in situ* temperatures and pressures during their transposition from the deep-sea floor to the surface may also be a useful innovation.

Effect of Hydrostatic Pressure on Marine Microbes.

Although bacteria appear to require for growth or reproduction a hydrostatic pressure which is near that of their native environment, it seems likely that few if any are killed by the decompression resulting from their transposition from the deep-sea floor to the surface. Such decompression may actually be advantageous to the organisms, because the resultant adiabatic cooling (about 1.6°C from a depth of 10 000 meters to the surface) helps counteract warming during transposition through warmer surface water. Of course, microbes may be killed mechanically by the sudden release of pressure in a system supersaturated with gas, but in spite of the pressure, there is no more gas dissolved in deep-sea water than in surface water.

Prolonged storage of deep-sea sediment samples at refrigeration temperatures has resulted in the gradual death of the bacteria at atmospheric pressure. Very few living bacteria could be detected in sediment samples from the Philippine Trench (depth 10 000 meters) following six weeks' storage at 3° to 5°C at atmospheric pressure, although large numbers of barophilic bacteria have persisted for 30 months in similar samples maintained in the refrigerator at 1000 atm. Whether this die-off is directly attributable to reduced pressure or is an indirect effect on the metabolism and nutrition of the bacteria under the experimental conditions can be determined only by further investigations.

Working with freshly collected deep-sea material (less than an hour after it was hauled aboard the Galathea), we found many more bacteria developing in nutrient media incubated at *in situ* pressures than at atmospheric pressure. The barophilic property of deep-sea bacteria is illustrated by representative results with samples taken from the Philippine Trench (Table II).

Table II. Most probable numbers (MPN) of bacteria per gram wet weight demonstrated in sediment samples from the Philippine Trench by the minimum dilution method in nutrient medium incubated in refrigerator at different pressures.

Station Number	Location of station		Water depth	Incubation pressure	
	Latitude	Longitude		1 atm.	1000 atm.
	North	East	meters	MPN	MPN
413	10° 20′	126° 36′	10 060	2 300	760 000
418	10° 13′	126° 43′	10 190	930	3 500 000
419	10° 19′	126° 39′	10 210	680	210 000
420	10° 24′	126° 40′	10 160	8 400	920 000
421	10° 29′	126° 05′	1 000	540 000	0
422	10° 49′	126° 01′	1 960	2 300 000	0
424	10° 28′	126° 39′	10 120	5 900	2 800 000

The most probable numbers of bacteria were estimated by "Standard Methods for the Examination of Water and Sewage" (American Public Health Assoc., 1936), as amplified by HALVORSEN and ZIEGLER (1933) and by PRESCOTT *et al.* (1947). Methods of inoculating the nutrient medium and of applying pressure are outlined below.

From the data in Table II, it will be observed that from 10 to 100 times as many bacteria from a depth of about 10 000 meters grew in nutrient medium incubated at 1000 atm as the number which grew at 1 atm. We now know that the effect of hydrostatic pressure on bacteria is partly a function of the composition of the medium, suggesting that with proper nutrients there may be as many bacteria from the deep sea that would grow at 1 atm as at 1000 atm, but we have not yet learned what these proper nutrients might be.

Although data like those summarized in Table II indicate the existence of strict barophiles (high pressure dependent bacteria), as this progress report is written we must admit that the pressure picture appears to be somewhat more complicated. Be this as it may, finding more bacteria in deep-sea sediments growing at *in situ* pressures (1000 atm) than at 1 atm is regarded as conclusive evidence that most, if not all, of the bacteria were indigenous to the deep and that they are physiologically active at high pressures. This conclusion is supported by finding in nearby shallow sediments from the sides of the Philippine Trench large numbers of bacteria which grew at 1 atm and none which grew at 1000 atm. This latter observation agrees with the findings of ZoBell and Johnson (1949) and ZoBell and Oppenheimer (1950), who reported that surface-dwelling microbes fail to grow at deep-sea pressures and many are slowly killed in nutrient media when held at pressures ranging from 400 to 600 atm. According to Oppenheimer and ZoBell (1952), pressures ranging from 200 to 600 atm inhibited the normal growth of nearly all of 63 different species of marine bacteria, mostly from shoal water habitats. Five of these were killed in four days by compression to 200 atm, 11 were killed at 400 atm, and 23 were killed in four days at 600 atm.

Nutrient Media and Enumeration Procedures.

The nutrient medium employed routinely for demonstrating the most probable number of bacteria by the minimum dilution method had the following composition:

Peptone (Difco)	5.0 gm
Ferric phosphate	0.1 gm
Potassium nitrate	1.0 gm
Soluble starch	2.0 gm
Yeast extract (Difco)	1.0 gm
Sea Water	1000.0 ml

Following autoclave sterilization its pH was 7.7. It was dispensed in 9.0-ml quantities in 15-ml screw-cap bottles.

Into the first bottle of sterile medium in each series was introduced aseptically 1.0 gm of the sample. After vigorously shaking to distribute the sediment (or water) sample evenly throughout the medium, a 1.0-ml aliquot was transferred, with a semi-automatic pipette, to the second bottle of sterile medium. This gave a 1:10 dilution or provided a 0.1-gm inoculum. After similarly shaking this second bottle, 1.0 ml was removed to inoculate the third bottle of sterile medium to give a 1:100 dilution or a 0.01-gm inoculum. The dilution procedure was repeated seriatim by powers of 10 until the original sample was diluted to 1:1 000 000.

Assuming a random distribution of the bacteria in the sediment sample and throughout the dilution series, it follows that bacterial growth in the 1:10 000 and all lower dilutions but no growth in the 1:100 000 dilution, for example, would indicate the presence of at least 10 000 viable bacteria per gram of sample but not 100 000 per gram. The most probable number (MPN) of bacteria in certain samples was determined with greater accuracy by preparing quintuplets of each series. For example, finding growth in all 5 tubes of medium inoculated with the 1:10 series, 3 positive in the 1:100 series, and none of the 5 positive in the 1:1000 series would indicate a bacterial population (MPN) of 79 per gram.

For incubation at atmospheric pressure, the diluted material in the nutrient medium was left in the screw-cap bottles. These bottles provided about 5 ml of air space for aerobes. To estimate the abundance of anaerobic bacteria in the sediment samples, the inoculated bottles were filled to capacity with O_2-free medium and the caps screwed tightly in place to exclude atmospheric oxygen. Following incubation at the desired temperature and pressure, the medium was examined for evidence of bacterial growth as manifested by increased turbidity of the medium or by finding large numbers of bacteria by direct microscopic examinations.

Ammonia liberation from peptone by ammonifiers:

$$R\text{-}NH_2 + HOH \rightarrow R\text{-}OH + NH_3$$

was detected by nesslerizing 0.2 ml of medium in a spot plate.

The presence of nitrate-reducing bacteria was indicated by the appearance of nitrite or by the disappearance of nitrate in the inoculated nutrient medium. Iodine was added to aliquots placed in a spot plate to test for the terminal presence of starch; its absence indicated the activity of starch-hydrolyzing bacteria. Sulfate-reducing bacteria were demonstrated in medium M-10-E:

Calcium lactate	3.5 gm
Magnesium sulfate, hydrated	0.2 gm
Potassium phosphate, dibasic	0.2 gm
Sodium sulfite	0.1 gm
Ferrous ammonium sulfate	0.1 gm
Ascorbic acid	0.1 gm
Peptone (Difco)	1.0 gm
Yeast extract (Difco)	1.0 gm
Bacto-agar (Difco)	3.0 gm
Sea water	1000.0 ml

It was adjusted to pH 7.5 following autoclave sterilization. The medium was dispensed in 15-ml screw-cap bottles, which were filled to capacity to displace all air, since the sulfate reducers are strict anaerobes. Following inoculation by serial dilutions and incubation, the presence of sulfate reducers was indicated by the blackening of the medium, resulting from the formation of ferrous sulfide:

$$SO_4^{--} + 10H \rightarrow H_2S + 4H_2O$$
$$H_2S + FeSO_4 \rightarrow FeS + H_2SO_4$$

Sea water enriched with 0.1 per cent ammonium phosphate (dibasic) and buffered with 0.2 per cent calcium carbonate was used to detect the presence of nitrifying bacteria. Following inoculation and incubation at the desired temperature and pressure, the enriched sea water was examined for the appearance of nitrite:

$$NH_4^+ + 2O_2 \rightarrow NO_2^- + 2H_2O$$

Nitrifiers which oxidize ammonium to nitrite were demonstrated in deep-sea sediment samples only when the inoculated medium was incubated at high pressures.

Inoculated media or samples to be held at high pressures were transferred to small glass vials, size 10 × 50 mm. After filling to capacity (5 ml) each vial was closed with a tapered No. 000 Neoprene stopper. When subjected to pressure in a steel vessel filled with hydraulic fluid, the stopper functioned as a piston, compressing the contents of the vial to approximately the same pressure as was built up in the steel vessel. Numerous tests have established that the compressed Neoprene piston stoppers provide an effective seal against contamination of any kind, the higher the pressure the better the seal.

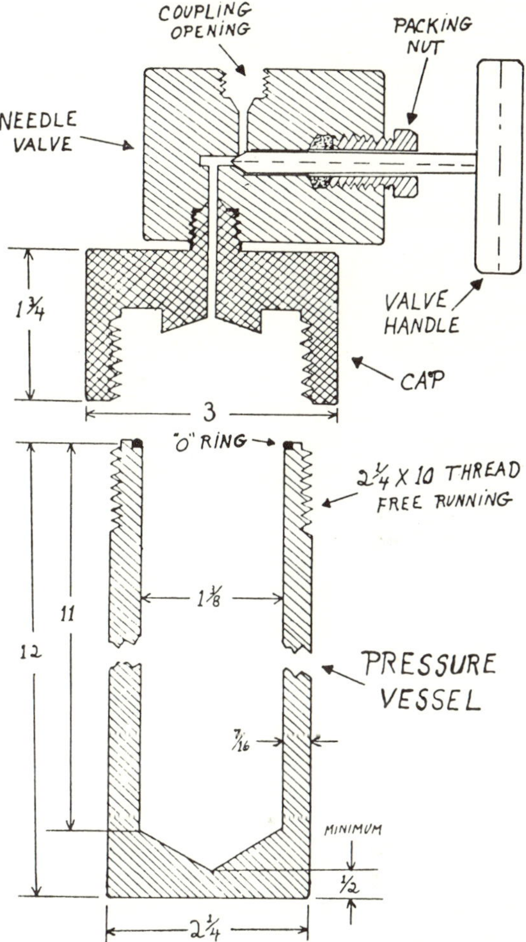

Fig. 1. Cross-section of stainless steel pressure vessel with its connecting cap and needle valve used for maintaining deep-sea sediment samples and bacterial cultures at high hydrostatic pressures. Dimensions are given in inches.

High Pressure Apparatus.

The pressure cylinders (Figure 1) were constructed of 18-8 type 303 stainless steel. Such a vessel having an inside diameter of 1-3/8 inches and an inside length of 11 inches holds 25 10 × 50-mm stoppered culture tubes and weighs only 6 kg. The vessels were fabricated by drilling a radially central hole, with tapered bottom, in 12-inch lengths of 2-1/4-inch steel bar stock. The caps were prepared by machining 3-inch lengths of 3-inch steel bar stock as illustrated. Machining also provided for a male-threaded top extension for the permanent attachment of a needle valve and a perforated cone-shaped ceiling for exhausting air from the vessel. When fitted with an "O" ring and the vessel is filled level full with hydraulic fluid, the latter is displaced through the cap and attached needle valve when the cap is screwed on the cylinder.

In practice the pressure vessel, equilibrated to the exact temperature at which it is to be incubated, is loaded with piston-stoppered culture tubes and filled level full with water at the same temperature. Hastily the cap is screwed on, an operation that can be performed by hand in a few seconds without the use of a wrench, and connected to the hydraulic pump (Figure 2). A few strokes with the handle of the pump produce a pressure up to 1000 atm in 20 to 30 seconds. After closing the needle valve, the pressurized vessel is disengaged from the pump for storage in the refrigerator or other thermostat. Con-

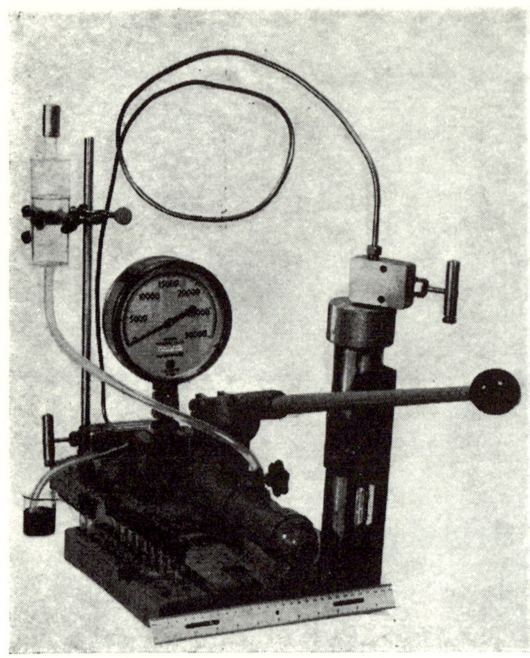

Fig. 2. High pressure vessel connected by semi-flexible steel tubing to hydraulic pump and Bourdon gauge. A rack of small culture tubes with piston stoppers is shown in foreground. ($^{1}/_{10}$ natural size).

stant pressure is maintained almost indefinitely provided the temperature is kept constant. The pressure changes between 6 and 7 atm per 1 °C change in temperature. Adiabatic heating amounts to about 2 °C during compression to 1000 atm and it requires about 30 minutes for temperature equilibrium to be established by heat conduction in the thermostat, the half-life of the adiabatic temperature increase being between 5 and 10 minutes.

The portable pump pictured in Figure 2 was prepared from a hydraulic truck jack having a capacity of 30 tons or about 4000 atm. The pump is mounted on a block of wood along with the pressure gauge, high-pressure valves, hydraulic fluid reservoir, and semi-flexible steel tubing for connecting the pump to pressure vessel. The clamp on the corner of the assembly is for holding the pressure vessel in an upright position during pressurization. Immediately after applying pressure the needle valve on the pressure vessel is closed and the vessel is returned to the thermostat after disconnecting the top coupling.

The pumps were obtained from the Blackhawk Mfg. Co., Milwaukee, Wisconsin. Needles valves, steel tubing, couplings, and pressure gauges were supplied by the American Instrument Co., Silver Springs, Maryland. "O" rings were provided by The Parker Appliance Co., Cleveland, Ohio. The pressure vessels were prepared by the Special Developments Division, Scripps Institution of Oceanography, La Jolla, under the supervision of James M. Snodgrass. The hydraulic fluid employed in the pump was a 1:1 mixture of water and glycerol.

The pressure had to be released from the vessels before the latter could be opened for the examination of cultures or stored samples. For depressurization the vessel was reconnected to the pump and gauge assembly. Then pressure was applied by operating the pump until the pressure registered on the Bourdon gauge corresponded with the pressure initially applied to the vessel. If the pressure indicator did not move upon opening the needle valve, it showed that the pressure in the vessel was the same as that initially applied. Should the terminal pressure in the vessel be either more or less than that initially applied, the gauge would register an increase or decrease as the case might be. This procedure was always used to check the terminal pressures in the vessels. Interestingly, in thousands of cases the terminal pressure was always found to be the same as that initially applied, provided the temperature had been kept constant.

Air Express Shipment of Pressurized Samples.

Most of the microbiological analyses were made on the Galathea. However, because of their great significance, several sub-samples of sediment from the Philippine Trench were shipped by air express from Cebu to California, where they were examined in our laboratories at the Scripps Institution of Oceanography. This unique operation was made possible only through the cooperation of personnel from several organizations, including the officers and men on the Galathea, the East Asiatic Company, Royal Danish Consulate Service, Philippine Airlines, and the University of California.

Indispensable services were rendered by Mr. Kjeld Danoe, who had our high pressure (1000 atm) vessels, containing sediment samples recently taken from the bottom of the Philippine Trench, insulated and packed in dry ice at Cebu for sending by air express to Manila. There this unique package was received two hours later by Gustav Halberg and K. Raaschou Nielsen, who made arrangements for its immediate transfer to a refrigerator on a Philippine Airlines plane. Thirty hours later the P.A.L. hostess placed the refrigerated pressurized vessel in the hands of a colleague, who met the plane in San Francisco, thanks to Royal Danish Navy radiograms that had forwarded instructions from the Galathea. Neither the temperature nor the pressure of the vessels changed perceptibly from the time they left the Galathea until they were in the laboratory at La Jolla. Examination by direct microscopic and cultural procedures confirmed the presence of numerous living bacteria in sediment samples from depths exceeding 10 000 meters.

Direct Microscopic Observations.

As soon as time permitted on the Galathea, small sub-samples of deep-sea sediment were examined with a phase contrast microscope at a magnification of 970 diameters. By means of a sterile platinum loop designed to deliver approximately 0.01 ml, portions were spread evenly over an area of 1.0 cm^2 on a clean glass slide and protected with a cover glass. Each bacterium observed in the area (0.0002 cm^2) covered by one field of the microscope represents 500 bacteria per 0.01 ml or 50 000 per ml of the original sample. With samples containing fewer than 50 000 per ml, it was necessary to scan several fields in order to find any bacteria. It so happened, however, that a good many of the sediment samples had bacterial populations exceeding 10^5 per ml. Thus, in spite of the obscuring effects of sediment particles with which the bacteria were associated, several bacteria were observed in some fields.

The first significant microscopic observation was made on a sample of silty clay dredged from a depth of 8870 meters in the Philippine Trench on 13 July 1951 at Station No. 412. Twelve sub-samples were prepared for microscopic examination; 50 or more fields of each being scanned. In nearly every field were found definite discrete bacterial cells, including ten different morphological types. Most common were non-sporulating rods with rounded ends, size about 0.5 microns in width by from 2 to 3 microns in length. Fairly abundant were capsulated ovoid rods having an average width of 0.8 microns and a length of 2 microns. Slender rods, 0.6 microns in width and up to 6 microns in length were found attached to sediment particles. Also recognized were several small vibrio, two different types of sporulating bacilli, two kinds of spirilla, and a few diplococci. Motility was not observed in wet preparations. Ostensibly the deep-sea bacteria lose their powers of locomotion due to the change in climate (lower pressure and higher temperature), because it was established by subsequent observations that the vibrio and some of the rods are flagellated.

These microscopic observations on the sediment sample from Galathea Station No. 412 were considered as presumptive evidence for the occurrence of bacteria from a depth of 8870 meters. However, conclusive evidence that they were indigenous species in this, and, eventually, in sediment samples from even greater depths had to await cultural tests, the first of which were completed 24 July 1951. Also observed in this sediment sample were a few empty diatom tests and some fish scales with attached bacteria. There was no evidence of cellulose fibers or of any kind of particulate organic matter besides bacterial cells. The absence of plankton organisms helped to rule out the possibility of the bacteria's having been picked up inadvertently while the dredge was passing through the photosynthetic zone, where the bacterial population was of the order of 10^3/ml.

The microscope also revealed the presence of bacteria in the topmost strata of a clay core collected 15 July 1951 at Station No. 413 from a depth of 10 060 meters. The core was 75 cm long. Detecting no bacteria in the lower strata of the core either by cultural or by direct microscopic examination is highly significant; negative results proved that sediment samples could be collected and analyzed on the Galathea without contamination. The topmost 0.1 cm layer of sediment was estimated to contain at least 3×10^7 bacteria per ml. Cultural methods showed the presence of 7.6×10^5 bacteria which grew at 1000 atm and only 2.3×10^3 which grew at 1 atm.

Predominating in the clay sample from Station No. 413 were rod-shaped bacteria with only a few helicoidal-shaped (vibrio and spirilla) cells, and still fewer spherical-shaped (cocci) bacteria. A good many of the bacteria occured in pairs or short chains, suggesting that the bacteria were reproducing. No protozoans, or other form of microscopic life other than bacteria, were observed.

It was extremely difficult to distinguish bacteria from sediment particles in clay samples dredged from depths of 10 190 and 10 210 meters at Stations No. 418 and 419 respectively. The direct microscopic examination of material dissected from the interior of balls of clay revealed the presence of fewer than 10^5 bacteria per ml, but exterior material was primarily from the mud-water interface. None of the bacteria observed in wet preparations exhibited any motility. Most of the bacteria were small, 1 to 2 microns long by 0.5 to 0.8 microns in diameter.

That the majority of them came from the deep-sea floor was indicated by cultural tests; the MPN at 1000 atm being 10^5 to 10^6 per ml as compared with MPN of only 10^2 per ml at 1 atm (see Table II).

The microscopic examination of greyish-green clay from a depth of 10 120 meters at Station No. 424 was more rewarding. Besides showing the presence of numerous bacilli (order of 10^7 per ml) there were many large (1.6×8 microns) spirilla unlike any bacteria previously described. None of the latter appeared among the many bacteria that grew in nutrient media incubated at 1000 atm (MPN 2.8×10^6 per ml). The microscope also showed many bodies believed to be fungous spores. None of these germinated in nutrient medium incubated at 1000 atm, but in similar medium incubated at 1 atm bacteria were overgrown by fungi. This observation suggests the occurrence of viable fungous spores on the deep-sea floor. Failure to find fungous mycelial material in any of the deep-sea samples indicates that fungi are not growing on the deep-sea floor. Four different kinds of diatom tests were observed with no evidence of protoplasm.

A good many of the minimum dilution cultures prepared for MPN determinations were examined microscopically to confirm bacterial growth. Most of the deep-sea bacteria were morphologically similar to those from soil and shallow water environments with which we are more familiar, showing even less variety among the forms found. Short rods predominated. About 5 per cent were vibrio. Cocci were encountered only occasionally. Differential staining procedures proved that the majority of the deep-sea bacteria are Gram-negative. Capsulated forms were common. Endospores were observed in about 20 per cent of the bacilli.

Physiological Characteristics of Bacteria From Philippine Trench.

The deep-sea bacteria exhibited considerable physiological versatility in differential media at 1000 atm (Table III). Apparently there were almost as many anaerobes as aerobes and a good many reduced nitrate. Some seemed to be able grow in the absence of free oxygen by utilizing nitrate as the hydrogen acceptor. The sulfate reducers were the only obligate anaerobes demonstrated.

The deep-sea nitrifiers differ in their activity at high pressure from surface-dwelling forms as do the nitrate reducers. Nitrate reduction by several shoal water species, which have been tested by ZoBell and Budge (1954), is retarded by pressures of 400 to 600 atm and at 1000 atm their nitratase system is slowly inactivated.

Table III. Minimum numbers of different physiological types of bacteria detected per ml of sediment from the Philippine Trench.

Station number	418	419	420	424
Water depth in meters	10 190	10 210	10 160	10 120
Total aerobic bacteria	10^6	10^5	10^5	10^6
Total anaerobic bacteria	10^5	10^5	10^5	10^5
Starch hydrolyzers	10^3	10^2	10^2	10^3
Nitrate reducers	10^5	10^4	10^5	10^5
Sulfate reducers	10^2	10	10^2	0
Ammonifiers	10^5	10^4	10^5	10^5
Nitrifiers	10	0	0	10

Bacteria in Indian Ocean Deeps.

Two sediment samples, suitable for bacteriological analyses, were obtained from depths exceeding 7000 meters in the Indian Ocean. Large numbers of viable bacteria were found in both samples, but surprisingly not nearly as many as in sediment samples taken from the Philippine Trench at depths exceeding 10 000 meters. Moreover, almost as many bacteria grew in nutrient medium at 1 atm as at 700 atm, and only a few from the Sunda and Weber Deeps grew at 1000 atm (Table IV).

Finding in these deeps starch hydrolyzers, nitrate reducers, sulfate reducers, ammonifiers, and nitrifiers which were active in differential media at 700 atm indicates that they are indigenous species which are probably active *in situ*. The activity of the deep-sea bacteria *in situ* is probably limited by the nutrient content of sea water and sediment and not by the high hydrostatic pressure or low temperature.

Bacteria in Kermadec-Tonga Trench.

Barophilic bacteria were found in 4 different sediment samples from the Kermadec-Tonga Trench (Table V).

Unlike the microflora found predominating in the Philippine Trench, nearly as many and in one sediment sample from the Kermadec-Tonga Trench (Station No. 686) more bacteria developed in nutrient media incubated at 1 atm than at high *(in situ)* pressures when isolates were examined. The inability of these bacteria recovered from the deep-sea floor to reproduce at *in situ* pressures suggests that

Table IV. Most probable numbers (MPN) of bacteria per gram wet weight demonstrated in sediment samples from Indian Ocean by minimum dilution method in nutrient media incubated in refrigerator at different pressures.

Station number	463 (Sunda Deep)	492 (Weber Deep)
Station location	10°16′ S × 109°51′ E	5°31′ S × 131°01′ E
Water depth	7020 meters	7250 meters
MPN at 1 atm	690 000	810 000
MPN at 700 atm	1 050 000	2 300 000
MPN at 1000 atm	4 800	17 000
Starch hydrolyzers at 700 atm	1 000	1 000
Nitrate reducers at 700 atm	10 000	10 000
Sulfate reducers at 700 atm	100	10
Ammonifiers at 700 atm	100 000	10 000
Nitrifiers at 700 atm	0	10

Table V. Minimum numbers of bacteria per gram of sediment from the Kermadec-Tonga Trench as determined by minimum dilution method in differential media incubated in refrigerator at different pressures.

Station number	649	650	658	686
Latitude	35°15′ S	32°20′ S	35°51′ S	28°30′ S
Longitude	178°40′ W	176°54′ W	178°31′ W	176°53′ W
Water depth (meters)	8300	6620	6720	9820
Total aerobes at 1 atm	10^5	10^6	10^4	10^6
Total aerobes at near *in situ* pressure*	10^6	10^6	10^6	10^5
Total anaerobes at 1 atm	10^4	10^6	10^5	
Starch hydrolyzers at 1 atm	0	10		
Nitrate reducers at 1 atm	0	0		
Sulfate reducers at 1 atm	0	0		

* Since the pressure-depth gradient in the sea approximates 0.1 atm/M, nutrient media inoculated with these four sediment samples were incubated at about 850, 680, 790. and 1000 atm respectively.

Table VI. Minimum numbers of bacteria per gram wet weight of sediment from different core depth detected in nutrient media incubated in refrigerator at 1 atm.

Station number	677		686	
Location of station	28°30′ S × 175°53′ W		20°53′ S × 173°31′ W	
Water depth	9190 meters		9820 meters	
Core depth	Aerobes	Anaerobes	Aerobes	Anaerobes
0-5 cm	1 000 000	1 000	1 000 000	1 000 000
5-10 cm	1 000 000	100 000	1 000 000	1 000 000
10-15 cm	100 000	10 000	1 000 000	100 000
15-20 cm	1 000 000	100 000	1 000 000	100 000
20-25 cm	100 000	10 000	1 000 000	10 000
25-30 cm	10 000	1 000	1 000 000	1 000
45-50 cm			10 000	10 000
85-90 cm			10 000	1 000

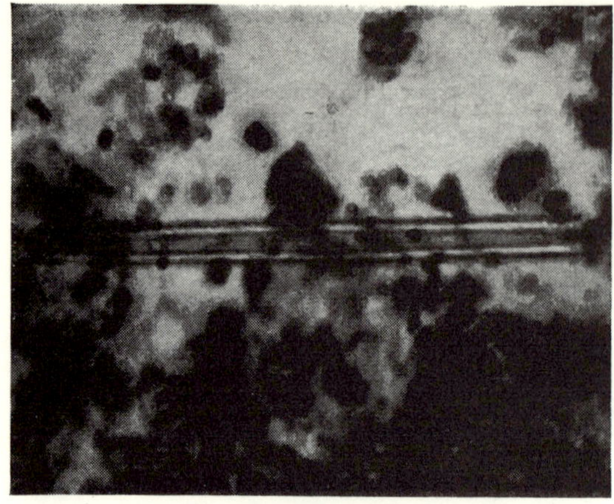

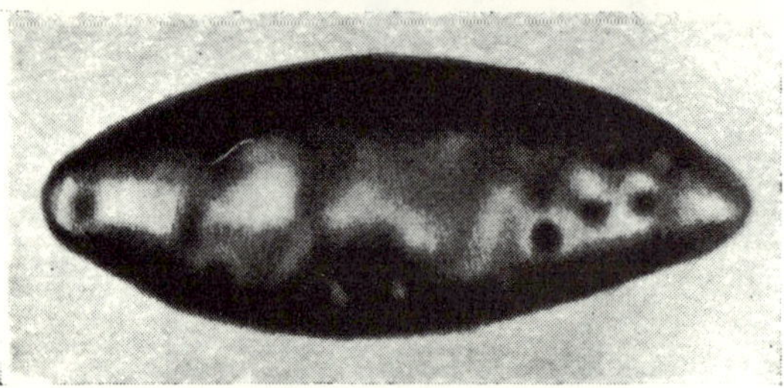

Fig. 3. Photomicrographs of two apparently intact diatoms found in deep-sea sediment following two years' storage in refrigerator at a pressure of 1000 atm.

they are not native to the deep-sea floor. The implication is that they have sunk to the sea floor where they may remain dormant for long periods of time. Such forms that tolerate high pressure without being able to reproduce are termed *baroduric*.

What appeared to be baroduric foraminifera and diatoms were detected in some of the Kermadec-Tonga Trench sediment samples examined by phase contrast microscope. This was confirmed by E. J. F. Wood (1956) working in our laboratory at La Jolla two years after the samples were taken. Since their collection the sediment samples had been stored in piston-stoppered tubes under conditions which were approximately isothermic and isobaric with the environment from which originally taken. Every one in the Scripps Institution laboratory who examined these preparations agreed that certain *Globigerina* contained protoplasm as did the *Coscinodiscus* and *Navicula* specimens. Extensive efforts to cultivate these foraminifera and diatoms have been unsuccessful. They are believed to be transient or adventitious forms which, like some of the baroduric bacteria, have sunk to the deep-sea floor. Regardless of their origin or significance, it is noteworthy that they have been preserved in refrigerated pressure vessels for many months. Two of the diatoms are shown in Figure 3.

The prolonged survival of baroduric bacteria buried at considerable depth was indicated by finding them in adundance to the bottom of cores collected from the Kermadec-Tonga Trench (Table VI). The vertical distribution of aerobes as well as anaerobes appeared to be sporadic, the general trend being decreasing abundance with core depth. Finding numerous aerobes buried at depths where there is probably no free oxygen constitutes further evidence of their dormancy. Some activity of bacteria is indicated by changes with core depth in the nonconservative chemical content of the sediment.

Chemical Content of Deep-Sea Cores.

Portions of two cores from the Kermadec-Tonga Trench were preserved in sealed bottles and brought back to the Scripps Institution laboratory for chemical analysis. Water content was determined by subtracting the constant dry weight (at 105°C) from the initial wet weight. The dried sediment was used for determining organic carbon, carbonate, Kjeldahl nitrogen, ammonia nitrogen, and sulfate.

Organic carbon was estimated by oxidizing at boiling point with a mixture of concentrated sulfuric acid, potassium iodate, and C.P. oxygen after first liberating carbonate carbon by treatment with phosphoric acid and flushing with a stream of pure nitrogen. The carbon dioxide resulting from both treatments was collected (separately) in standard barium hydroxide solution which was back-titrated with N/100 hydrochloric acid. The method of Bien (1952) for determining carbonate and organic carbon was used. This method employs a closed system which eliminates erroneous results from atmospheric CO_2.

Ammonia was driven off weighed sediment samples by sodium hydroxide treatment. Micro-Kjeldahl procedures were employed for determining organic nitrogen. Weighed samples were washed several times with distilled water accompanied by mechanical agitation to leach out sulfate, which was then determined gravimetrically as its barium salt. Results of the chemical analyses are summarized in Table VII.

The low carbonate content was expected, since carbonate in sediments usually decreases with depth in the ocean (Kuenen, 1950). This may be due to the increased solubility of calcium carbonate at the lower pH of sea water associated with the high hydrostatic pressure (Buch and Gripenberg, 1932).

Table VII. Certain non-conservative chemical constituents found in different strata of two mud cores from the Kermadec-Tonga Trench. Values are recored as milligrams per gram of dried (at 105°C) sediment.

Core from Galathea Station No. 677; water depth 9190 meters.

Core depth	Water content	Carbonate content	Organic carbon	Ammonia nitrogen	Kjeldahl nitrogen	Sulfate content
cm	per cent	mg/gm	mg/gm	mg/gm	mg/gm	mg/gm
0-5	46.2	0.7	0.5	0.8	0.5	1.9
5-10	43.8	0.0	0.2	0.2	0.3	1.5
10-15	40.1	0.2	0.1	0.1	0.3	1.2
15-20	40.0	0.6	0.2	0.1	0.3	1.6
20-25	41.8	0.2	0.2	0.1	0.4	1.2
25-28	38.4	0.4	0.1	0.1	0.3	0.9

Core from Galathea Station No. 686; water depth 9820 meters.

Core depth	Water content	Carbonate content	Organic carbon	Ammonia nitrogen	Kjeldahl nitrogen	Sulfate content
cm	per cent	mg/gm	mg/gm	mg/gm	mg/gm	mg/gm
0-10	53.7	0.2	0.3	0.1	0.2	2.0
15-20	51.7	0.5	0.3	–	0.2	1.6
25-30	50.7	0.5	0.2	0.2	0.2	1.3
35-40	52.5	0.3	0.3	0.1	0.3	1.1
45-50	53.0	–	0.1	0.0	0.1	–
55-60	50.8	0.2	0.2	0.1	0.3	0.2
65-70	53.4	0.4	0.2	0.0	0.4	0.4
75-80	52.0	0.2	0.2	–	0.2	0.2
85-90	47.7	–	0.2	0.1	0.2	

The organic carbon as well as the organic nitrogen content of the deep-sea sediments was very low; in the deeper core strata the C/N ratio was 1:1. Total contents and the C/N ratios are considerably lower than in near-shore sediments, conditions which have been noted by WISEMAN and BENNETT (1940) and by ARRHENIUS (1952). Arrhenius postulates that the low C/N ratios in deep-sea clay deposits may be caused by some property, e.g., low calcium carbonate content, of the clay favoring conservation of nitrogen during the low rate of organic matter deposition, or it might be a reflection of differences in the way in which organic matter is decomposed in the deep sea. Most likely the relatively abundant ammonia-nitrogen content of the deep-sea deposits has resulted from the decomposition of proteinaceous materials. Bacteria (ammonifiers) which liberate ammonium from proteinaceous material were detected in virtually all deep-sea sediments tested for their presence.

There seems to be a general trend of decreasing organic content with core depth, which suggests its slow oxidation. In the absence of any obvious supply of free oxygen below the mud-water interface, the organic matter must be attacked by anaerobes, if by any kind of bacteria. The rate of attack appears to be very slow. Sulfate-reducing bacteria could account for the slow oxidation of organic matter in strictly anaerobic environments. Sulfate-reducing bacteria have been found in deep-sea sediments and, interestingly, there is a definite decrease in the sulfate content of the deep-sea cores.

Barophilic Sulfate-Reducing Bacteria.

Because of their importance as geochemical agents, special effort was devoted to the detection of sulfate-reducing bacteria in deep-sea sediment samples collected by the Galathea. Besides reducing sulfate to sulfide and oxidizing a great variety of organic compounds, certain sulfate reducers fix nitrogen (SISLER and ZOBELL, 1951), and utilize molecular hydrogen (ZOBELL, 1947; SISLER and ZOBELL, 1950). Sulfate-reducing bacteria, which commonly occur in near-shore sediments (ZOBELL and RITTENBERG, 1948), were also found by MORITA (1954) in several ocean basins. Populations exceeding 100 per gram of wet sediment were demonstrated in collections from a depth of 5032 meters at Mid-Pacific Station No. 36 (MORITA and ZoBELL, 1955).

As shown in Table III above, barophilic sulfate-reducing bacteria were demonstrated in Philippine Trench sediment samples from depths exceeding 10 000 meters. Such bacteria which grew at 700 atm were also demonstrated in bottom deposits from the Sunda Deep and the Weber Deep (see Table IV). Sulfate reducers from the Weber Deep have been maintained in pressure vessels at 3-5°C and 700 atm for more than 4 years and are still under observation at the Scripps Institution. Upon decompressing and opening the pressure vessel for the examination of some of the original sediment sample, a strong odor of hydrogen sulfide emanated, indicating activity of the organisms during storage. Small quantities of the mud were used to inoculate medium M-10-E in piston-stoppered tubes for enrichment in the refrigerator at 700 atm. After 60 days' incubation there was no evidence of sulfate reduction, but after 10 months sulfide had been produced in all tubes of medium M-10-E inoculated with this stored mud from the Weber Deep (7250 meters). The uninoculated controls were negative. There was no evidence of sulfate reduction in similar medium after 17 months' incubation at 1 atm in medium inoculated either with some of the original mud sample or with enrichment cultures that developed at 700 atm.

The obligate barophilic sulfate-reducing bacteria from the Weber Deep have been transplanted four successive times and kept growing at 3-5°C and 700 atm. With the cooperation of Dr. JAMES W. BARTHOLOMEW, electron micrographs were made. They show a small ovoid-shaped microbe about 0.3 microns in width by 0.5 to 0.8 microns in length, which differs in both shape and size from other known sulfate reducers commonly classified as *Desulfovibrio* species. Though highly pleomorphic, known *Desulfovibrio* species are usually comma-shaped organisms 0.5 to 1.0 by 3 to 5 microns in size. The barophilic sulfate reducer from the Weber Deep is believed to be a new genus, but further studies must precede its naming. Like most barophilic bacteria, it grows very slowly (ZoBELL and MORITA, 1957).

An Enigma in Bacterial Taxonomy.

In their unique ability to grow at high hydrostatic pressures, a good many of the bacteria isolated from the deep sea differ from known or described species whose habitat is soil, sewage, near-shore sediments, and other surface or shallow-water environments. However, pressure tolerance is probably not a sufficiently distinctive characteristic to delineate species, particularly since the deep-sea bacteria may have become acclimatized to high pressure while sinking from the surface, or their pressure tolerance may be associated with the composition of the medium.

Ordinarily, the taxonomic position of bacteria is based upon their physiological, cultural, and morphological characteristics as determined at 1 atm. Since some of these characteristics are altered and others are indeterminate at high pressure, we are confronted by an enigma in bacterial taxonomy. At least superficially a good many of the deep-sea bacteria appear to be new species or even new genera, but this can be established only by much more study.

The only deep-sea microbe to which we have applied a generic name is a small (0.4 micron) Gram-negative sphere, usually occurring singly but occasionally in pairs. It has exhibited no evidence of motility and does not form endospores. It liquefies gelatin and liberates ammonia from peptone in sea water medium in the refrigerator at 1000 atm. It oxidizes glucose with the formation of acid; sucrose, lactose, and starch are not attacked. In the refrigerator (3-5°C) it survives at 1 atm, but is killed in a few minutes when warmed to 25°C. It was isolated from two different mud samples from the Philippine Trench. Tentatively it has been named *Bathycoccus galathea*. The Greek suffix *bathy-* denotes great depth and *coccus* indicates that the cells are spherical.

Oceanographic Significance of Deep-Sea Bacteria.

What is the ecological and oceanographic significance of the bacterial populations found in oceanic deeps? Are they primarily passive inhabitants on the deep-sea floor or is their rate of metabolism and reproduction rapid enough for them to affect substantially chemical or biological conditions? Our observations prove only the presence of viable bacteria in some of the deepest parts of the ocean and show that these deep-sea bacteria are physiologically active in the laboratory under experimental conditions designed to duplicate deep-sea conditions, including high hydrostatic pressure and low temperature. Prerequisite to assessing the oceanographic significance of these deep-sea bacterial populations is much more information on the circulation of water masses, the amounts of organic matter reaching the deep-sea floor, the non-conservative chemical composition of the water and bottom deposits, abundance and growth rates of other organisms inhabiting the deeps, and more information on the characteristics of the bacteria.

An estimate of an average bacterial population of 10^6 per ml in the topmost cm of deep-sea sediment is believed to be conservatively low. Approximately this many were demonstrated in nutrient medium by the minimum dilution method, which does not detect all living microbes, since no one medium can provide for the nutrient requirements and essential environmental conditions for the growth of all bacteria. Moreover, groups of bacteria occurring in agglomerates or attached to solid particles register only as single individuals in the minimum dilution method. Extensive experience by bacteriologists who have been analyzing soil, sewage, milk, and similar materials for several decades shows that conventional cultural methods usually detect only about 10 per cent of the viable bacterial population. Actually bacterial populations of the order of 10^7 per ml of deep-sea sediment were indicated by the direct microscopic method, but the microscope fails to distinguish living from dead bacteria.

Taking 10^6 per ml as the order of magnitude of the bacterial population and considering these bacteria to contain an average of 2×10^{-13} gm of organic carbon per cell, it follows that 1.0 ml of such sediment would contain 2×10^{-7} gm of bacterial carbon, or 0.2 mg per liter. In a layer 1 cm thick there would be 2 mg of bacterial carbon per square meter.

Such a standing crop of bacteria may contribute substantially to the nutrition of benthic animals. Protozoans, worms, sponges, filter feeders, and mud eaters are among the animals known to ingest and digest bacteria (ZoBELL and FELTHAM, 1938). Certain animals can survive and thrive on an exclusive diet of bacteria (ZoBELL and LANDON, 1937), and some can reduce the bacterial population of sea water to 10^2 to 10^3/ml. Bacteria consist largely of easily digestible proteins and lipids. Many are rich in vitamins and other accessory growth factors which may be beneficial to benthic fauna far removed from products of photosynthetic activity.

KRISS (1954) has stressed the importance of bacteria as food for marine animals and in the organic cycles in the sea. According to KRISS and RUKINA (1952), the microbial biomass in marine sediments and overlying water column amounts to from 0.2 to 1.4 gm per square meter.

The extent to which bacteria contribute to the nutrition of deep-sea animals cannot be estimated from the standing crop or population of either, but must be based upon reproduction or growth rates. The population is only an expression of a dynamic balance between the rate of reproduction and the rate of death. Laboratory observations indicate that deep-sea bacteria reproduce (by transverse fission) once every 2 to 20 hours in nutrient medium incubated at 1000 atm and 3°C; somewhat slower in deep-sea sediment or sea water to which no nutrients have been added. They may be dormant *in situ*, as indeed many appear to be in the lower strata of cores from the Kermadec-Tonga Trench (Table VI), but at the mud-water interface they are probably reproducing. The low content of organic matter in deep-sea sediments and immediately overlying water must be attributable to bacterial activity, and there is pretty good evidence for the activity of sulfate-reducing bacteria. Should the rate of reproduction *in situ* be comparable to that found in the laboratory under conditions designed to simulate the *in situ* environment, we may think with confidence of an *in situ* reproduction rate which could provide a hundred or more bacterial "crops" per year, but other dynamic factors must be evaluated before concluding that this is a reasonable estimate.

The maximum amount of bacterial cell substance producible per unit of time is limited by the amount of available energy source. This is believed to consist largely of dissolved or colloidal organic matter reaching the bacteria. Chemosynthetic autotrophs may fix some carbon on the deep-sea floor, but such bacteria appear to play a very minor role. Of course, bacteria are highly efficient in recycling the organic waste products of animals, but there must be an influx of energy to keep the processes going. This energy, originating as fixed carbon in the photosynthetic zone, must reach the deep-sea floor primarily by settling or by the movements of water masses.

Ordinarily an average of one-third of the organic carbon assimilated by heterotrophic bacteria in dilute media is converted into bacterial cell substance, the other two-thirds being oxidized to CO_2. With similar efficiency a bacterial population of 10^6/ml representing a standing crop of 0.2 mg bacterial carbon per liter would require 0.6 mg organic carbon per liter and about the same amount of O_2. The complete oxidation of organic matter requires somewhat more than its own weight in O_2; 1.0 mg of various carbohydrates requiring 1.07 to 1.18 mg O_2, common fatty acids 1.06 to 2.97 mg O_2, triglycerides 2.40 to 2.92 mg O_2, amino acids and proteins 0.64 to 2.58 mg O_2/mg.

Replacing a bacterial population of 10^6/ml 100

times per year would require 60 mg of organic carbon per liter and somewhat more than this amount of O_2. This is much more than occurs in sea water at any one time, but throughout the year organic matter on the sea floor may be replenished by the movements of water masses and to a lesser extent by sedimentation; O_2 by diffusion and water movements. Apparently bacterial populations of the order of 10^6/ml occur only in the immediate vicinity of the mud-water interface, falling off rapidly with distance off the bottom. Only a few samples of sea water from depths exceeding 7000 meters and a meter or more off the bottom have been examined for the presence of bacteria, but all of these, like hundreds of water samples collected from depths exceeding 1000 meters, have bacterial populations of less than 10/ml. If this is representative of conditions throughout deep water masses, organic matter and O_2 from a water column of considerable thickness may be supplying the active biotic zone at the mud-water interface.

Besides oxidizing most kinds of organic matter with the consumption of O_2 and contributing to the nutrition of benthic fauna, bacteria may affect the non-conservative chemical composition of the deep-sea floor. The decomposition of organic matter probably results in the formation of complex marine humus and the liberation of phosphate, sulfate or sulfide, and nitrogenous compounds. Ammonifiers liberate ammonia from certain nitrogenous substances and nitrifiers may oxidize the ammonia to nitrite or nitrate. In the absence of free oxygen, sulfate reducers may produce hydrogen sulfide. If the latter diffuses into oxygenated waters, it may be oxidized with the formation of sulfate or sulfur. Though not detected on the Galathea Expedition, other bottom-dwelling bacteria may fix nitrogen or oxidize hydrogen or methane, which often result from the fermentation of organic compounds.

Bacteria in Water.

Although attention was concentrated on the microbial content of deep-sea sediments, several samples of sea water were examined for the presence of bacteria. Aseptic samples were collected in sterile evacuated bottles hermetically sealed with glass capillary tubes which could be sheared off at any desired depth to provide for the aspiration of water (ZoBell, 1941). Near-surface samples were collected in glass bottles tossed forward, with a hand line, off the bow of the Galathea in order to avoid contamination from the ship. Water samples from depths of 7000 to 8000 meters were collected in 300 ml collapsible rubber bulbs attached in properly spaced series to the hydrographic wire (Figure 4). At greater depths the elasticity of the rubber was adversely affected by the high pressure as has been described by ZoBell (1954).

From 12 July to 15 August 1951 a total of 44 surface water samples collected over the Philippine Trench were examined for bacteria by the minimum dilution method. In every one of these samples were detected from several hundred to a few thousand bacteria per ml, which grew in nutrient medium incubated at air temperature (ca. 30°C). There was no growth in similarly inoculated medium incubated at 1000 atm, and very little growth at 3-5°C, unlike the bacteria found in Philippine Trench sediments.

In organic-rich waters from Philippine Island bays and harbors visited by the Galathea were found

Table VIII. Minimum numbers of bacteria per ml of surface sea water collected on Galathea at different locations. The bacteria were demonstrated in nutrient medium incubated at air temperature and atmospheric pressure.

Total samples	General location of stations from which samples were collected	Date of collection	Bacteria per ml
2	Manila Bay	9-10 July	10^4 to 10^5
11	Over Philippine Trench	12-15 July	10^2 to 10^3
3	Tubajon Bay off Dinagat Island	17-19 July	10^3 to 10^4
12	Over Philippine Trench	21-24 July	10^2 to 10^3
2	Cebu Harbor	25 July	10^4 to 10^6
9	Over Philippine Trench	26-29 July	10^2 to 10^3
1	Kanlanuk Bay off Bucas Grande	30 July	10^4
2	Candos Bay off Lapinigan Island	31 July	10^4
8	Over Philippine Trench	2-8 August	10^2 to 10^3
6	Cebu Harbor	10-12 August	10^4 to 10^5
4	Over Philippine Trench	14-15 August	10^3

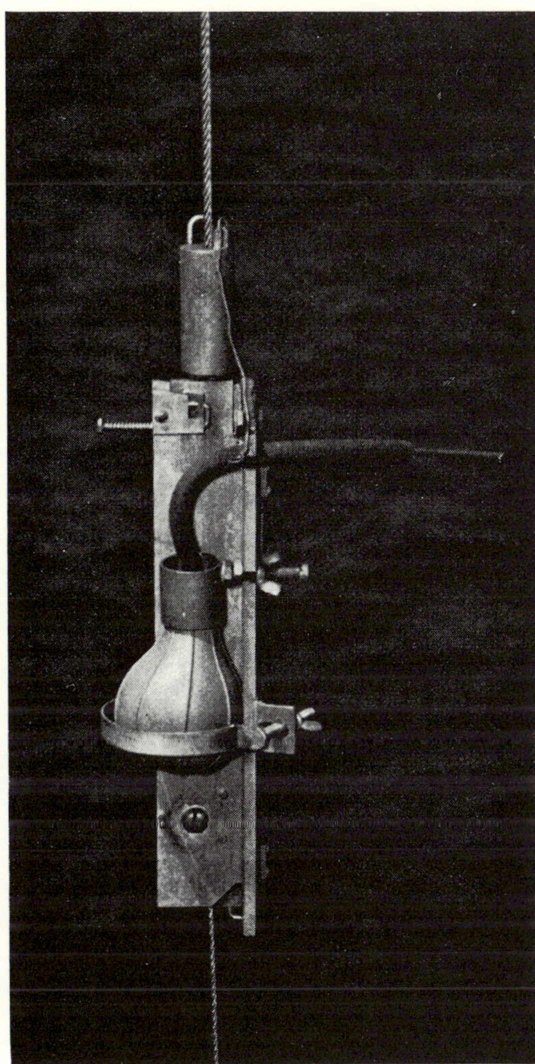

Fig. 4. Bacteriological sampler attached to hydrographic wire; at left the sterile rubber bulb is shown collapsed ready to be lowered into the water. Upon reaching the desired depth a messenger is dropped which shears off the capillary glass tubing thus permitting the thick-walled rubber tubing to flip out to the position shown at the right so water is aspirated from an area considerable distance from the apparatus in order to minimize self-contamination. Simultaneously the messenger suspended from the apparatus (left) is released, permitting it to drop for activating similar or other hydrographic apparatus at greater depth.

from 10 to 1000 times as many bacteria as in near-surface water collected over the Philippine Trench (Table VIII).

Throughout the photosynthetic zone over the Philippine Trench to a depth of 100 to 200 meters, the bacterial population ranged from 10^2 to 10^3 per ml. Not enough samples were analyzed to determine whether the vertical abundance of bacteria corresponded with vertical distribution of phytoplankton, as has been found to be the case in certain areas (ZoBell, 1946). At depths exceeding 1000 meters fewer than 10 bacteria per ml were found. In northern Pacific Ocean deeps Kriss (1952) found few bacteria in water samples, but in bottom sediments he found from 10^6 to 10^8 bacteria per gram.

Acknowledgments.

The authors appreciate the financial support of the University of California Research Committee (Grant No. 1250), the U.S.A. Office of Naval Research (contract N6onr-27518), and The Rockefeller Foundation. It is also a pleasure to acknowledge the invaluable assistance of Dr. ANTON F. BBUUN together with all the officers, crew members, and scientists on the Galathea, who contributed in many ways to the success of the bacteriological investigations. Special thanks are due to Cmdr. TAGE FEDDERSEN, Lt. L. FERDINAND, Capt. SVEN GREVE, Cmdr. SIGURD BARFOED, TORBEN WOLFF, POUL JACOBSEN, ALF KIELERICH, and P. ANDREASEN for technical help. A special debt of gratitude is due the Royal Danish Navy and the people of Denmark who subsidized the Expedition.

REFERENCES

AMERICAN PUBLIC HEALTH ASSOCIATION, 1946: Standard Methods for the Examination of Water and Sewage. Amer. Pub. Health Assoc., New York City, 9th Ed., 286 pp.

ARRHENIUS, G., 1952: Sediment cores from the East Pacific. Part I. Properties of the sediment and their distribution. Rept. Swedish Deep-Sea Expedition, 5: 1-90.

BIEN, G. S., 1952: Marine Chemical Analytical Methods. Scripps Institution Oceanography, Progress Report 52-58: 1-9.

BUCH, K., and GRIPENBERG, S., 1932: Über den Einfluss des Wasserdruckes auf pH und das Kohlensäuregleichgewicht in grösseren Meerestiefen. Jour. du Conseil, 7: 233-245.

CERTES, A., 1884a.: Sur la culture, à l'abri des germes atmosphériques, des eaux et des sédiments rapportés par les expéditions du Travailleur et du Talisman; 1882-1883. Compt. rend. Acad. Sci., 98: 690-693.

– 1884b.: De l'action des hautes pressions sur les phénomènes de la putréfaction et sur la vitalité des micro-organismes d'eau douce et d'eau de mer. Compt. rend. Acad. Sci., 99: 385-388.

FISCHER, B., 1894: Die Bakterien des Meeres nach den Untersuchungen der Plankton-Expedition unter gleichzeitiger Berücksichtigung einiger älterer und neuerer Untersuchungen. Ergebnisse der Plankton-Expedition der Humboldt-Stiftung, 4: 1-83.

HALVORSEN, H. O., and ZIEGLER, N. R., 1933: Quantitative Bacteriology. Burgess Pub. Co., Minneapolis, Minn., 64 pp.

KRISS, A. E., 1952: Microbial life in the ocean depths. "Advances in Modern Biology." Moscow, pp. 194-218 (Russian article).

– 1954: The basic tasks of sea and ocean microbiology. Vestnik Akad. Nauk SSSR, 8: 22-34 (Russian article).

KRISS, A. E., and RUKINA, E. A., 1952: Microorganisms in the bottom deposits of ocean regions. Isvestiya Akad. Nauk SSSR, 6: 67-79 (Russian article).

KUENEN, P. H., 1950: Marine Geology. John Wiley & Sons, New York City, 568 pp.

MORITA, R. Y., 1954: Occurrence and significance of bacteria in marine sediments. Doctorate Dissertation, University of California, Los Angeles, 80 pp.

MORITA, R. Y., and ZOBELL, C. E., 1955: Occurrence of bacteria in pelagic sediments collected during the mid-Pacific Expedition. Deep-Sea Research, 3: 66-73.

OPPENHEIMER, C. H., and ZOBELL, C. E., 1952: The growth and viability of sixty-three species of marine bacteria as influenced by hydrostatic pressure. Jour. Mar. Res., 11: 10-18.

PRESCOTT, S. C., WINSLOW, C. A., and MCGRADY, M. H., 1946: Water Bacteriology. John Wiley & Sons, New York City, 368 pp.

SISLER, F. D., and ZOBELL, C. E., 1951: Nitrogen fixation by sulfate-reducing bacteria indicated by nitrogen/argon ratios. Science, 113: 511-512.

WISEMAN, J. D., and BENNETT, H., 1940: The distribution of organic carbon and nitrogen in sediment from the Arabian Sea. Sci. Rept. John Murray Expedition, 3: 160-200.

WOOD, E. J. F., 1956: Diatoms in the ocean deeps. Pacific Science, 10: 377-381.

ZOBELL, C. E., 1941: Apparatus for collecting water samples from different depths for bacteriological analysis. Jour. Mar. Res., 4: 173-188.

– 1946: Marine Microbiology. Chronica Botanica, Waltham, Mass., 240 pp.

– 1947: Microbial transformation of molecular hydrogen in marine sediments. Bull. Amer. Assoc. Petrol. Geol., 31: 1709-1751.

– 1952: Bacterial life at the bottom of the Philippine Trench. Science, 115: 507-508.

– 1952: Occurrence and importance of bacteria in Indonesian waters. Jour. Scient. Res. Indonesia, 4: 2-6.

– 1952: Dredging life from the bottom of the sea. Research Reviews, Office of Naval Research, October, pp. 14-20.

– 1953: Bakterier under tusind atmosfaerers tryk. Vor Viden, Copenhagen, No. 96: 563-568.

– 1954: Some effects of high hydrostatic pressure on apparatus observed on the Danish Galathea Deep-Sea Expedition. Deep-Sea Research, 2: 24-32.

ZOBELL, C. E., and BUDGE, K. M., 1954: Effect of high hydrostatic pressure on activity of marine nitrate-reducing bacteria. O. N. R. Progress Report, No. 7: 3-13.

ZOBELL, C. E., and CONN, J. E., 1940: Studies on the thermal sensitivity of marine bacteria. Jour. Bact., 40, 223-238.

ZOBELL, C. E., and FELTHAM, C. B., 1938: Bacteria as food for certain marine invertebrates. Jour. Mar. Res., 1: 312-327.

ZOBELL, C. E., and JOHNSON, F. H., 1949: The influence of hydrostatic pressure on the growth and viability of terrestrial and marine bacteria. Jour. Bact., 57: 179-189.

ZOBELL, C. E., and LANDON, W. A., 1937: Bacterial nutrition of the California mussel. Proc. Soc. Exper. Biol. & Med., 36: 607-609.

ZOBELL, C. E., and MORITA, R. Y., 1953: Dybhavets bakterieliv. Galatheas Jordomsejling 1950, under redaktion af A. F. Bruun, Sv. Greve, H. Mielche, R. Spärck, Copenhagen: 199-206. (Translated: Bacteria in the Deep Sea. The Galathea Deep Sea Expedition 1950-1952, described by Members of the Expedition, edited by Anton F. Bruun, Sv. Greve, Hakon Mielche and Ragnar Spärck, London 1956: 202-210).

– 1957: Barophilic bacteria in some deep sea sediments. Jour. Bact., 73: 563-568.

ZOBELL, C. E., and OPPENHEIMER, C. H., 1950: Some effects of hydrostatic pressure on the multiplication and morphology of marine bacteria. Jour. Bact., 60: 771-781.

ZOBELL, C. E., and RITTENBERG, S. C., 1948: Sulfate reducing bacteria in marine sediments. Jour. Mar. Res., 7: 602-617.

22

Reprinted from *Bacteriol. Rev.*, **28**(1), 14–29 (1964)

EFFECTS OF HYDROSTATIC PRESSURE ON MICROBIAL SYSTEMS[1]

CARL-GÖRAN HEDÉN

Department of Bacteriology, Karolinska Institutet, Stockholm, Sweden

Introduction

Given the opportunity to choose the topic for this paper, I was torn between a wish to remain within my main area of interest, which in recent years has been the biotechnology of fermentation, and a desire to make a tangential excursion out of this field. I decided to take the second alternative, because it gives emphasis to an important theme: that biotechnology should function as a link between basic and applied research. This thesis is so far from being generally accepted that it is well worth special attention.

Usually I define biotechnology as the study of the *establishment and maintenance of artificial environments in which living cells or labile biological substances are produced, preserved, selectively destroyed, or modified as desired.* This definition clearly indicates that the subject is focused on application, but obviously it would become sterile if it had no room for basic studies. These may then involve "toolmaking," and this paper is concerned with the establishment of one, highly artificial environment: that attained under high hydrostatic pressure. Such techniques may provide useful tools both in the mapping of microbial control mechanisms and in the manipulation of those mechanisms with practical aims.

Much of the fundamental work on pressure effects on microbial systems has been performed in the United States by Johnson, Haight, Morita, ZoBell, and many others (for references, see 53), but here I will not consider enzyme kinetics, luminescence, or barophilic bacteria. Rather, I will draw from our own limited experience to illustrate the points I want to make.

Action of Pressure on Frozen Cells

My personal interest in subjecting microorganisms to high pressure was initiated by a practical problem of biotechnology: how to disrupt the bacteria in a frozen paste at very low temperatures. This was of importance for the removal of labile toxins from *Bordetella pertussis* (10), so the problem was studied in some detail, the initial point being some experiments aimed at improving existing methods, among them the classical freezing and thawing procedure. This, of course, has the drawback of involving many exposures to relatively high temperatures (9).

Considering the disruption caused by cycling across the phase-boundary between liquid and ice I (Fig. 1), it seemed reasonable to assume that passages across the boundary between ice I and ice III would have a similar effect, the crystal structure of ice III being almost 20%

[1] The Office of Naval Research Lecture by the same title, presented at the Annual Meeting of the American Society for Microbiology, at Cleveland, Ohio, in May 1963, formed the basis of this contribution. I should like to express my sincere thanks to the Office of Naval Research and to the Program Committee of the Society for making the lecture possible. This paper is not intended to be a comprehensive review, but rather aims at focusing attention on some interesting phenomena which we have recently studied in Stockholm under U.S. Public Health Service grant C 5799 from the National Cancer Institute. In preparing the paper, I have drawn freely from the data emerging from the lines of study followed by the different members of a team consisting of the following individuals besides myself: E. Hammarsten and T. Lindahl (biochemists), L. Rutberg (microbial geneticist), A. Rupprecht (physical chemist), and I. Toplin (chemical engineer).

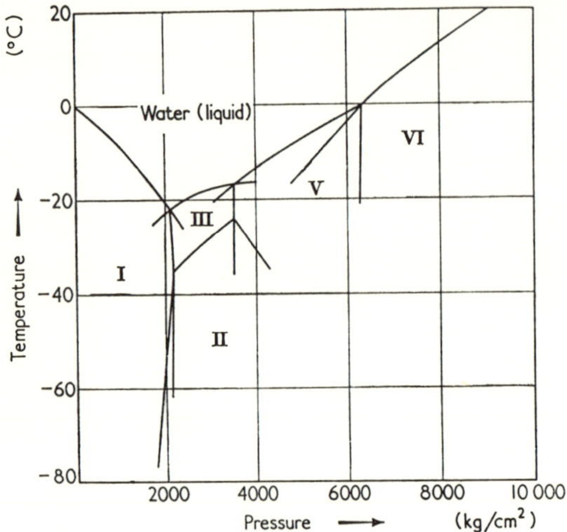

FIG. 1. *Different forms of ice.*

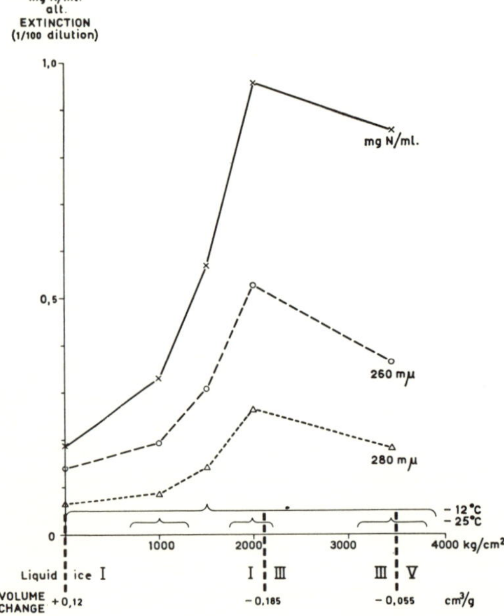

FIG. 2. *Disintegration effect of ice-phase transitions. The measurements were made on the supernatant from frozen and pressure-exposed cells of Escherichia coli B (10-ml suspension of 10^{11} cells per ml of buffer) centrifuged at 25,000 × g for 30 min, after extraction for 2 hr at 0 C with 0.1 M phosphate-citrate buffer. The brackets indicate the regions for ten pressure cycles.*

more compact than that of normal (I) ice (12, 13, 102). As indicated in. Fig. 1, ice III cannot, however, exist at normal pressures, so a special chamber was built for testing the hypothesis.

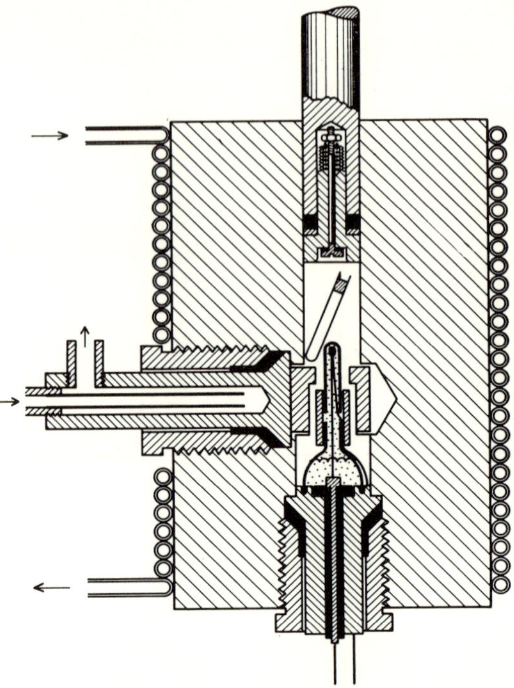

FIG. 3. *Cross section of cylindrical pressure chamber with slanting specimen tube in axial water-filled hole. The tube rests on the copper portion of a "cooling finger" inserted from the right-hand side. A small thermistor probe, housed in a glass capillary filled with paraffin oil, used as an electrical insulator, is located near the specimen. Nylon is used for all gaskets except for the auxiliary Neoprene rubber gasket below the nylon ring around the plunger. Nylon is also used for electrical insulation wherever needed.*

With this apparatus, it was found that when the pressure was pulsated up and down in the region of 2,000 atm at −25 C, disruption occurred as expected (Fig. 2), yielding cells which looked quite empty under an electron microscope in spite of the fact that the envelope was not too badly damaged (25). When Edebo added the shearing forces also used by Hughes (49), and forced the frozen material back and forth through a narrow cylindrical channel (24), the practical device needed for our work became available. It permitted subzero disintegration even down to a degree where cell-wall antigens become accessible to immunodiffusion analysis, and it has been used with success for the low-temperature preparation of cell walls from *B. pertussis* and staphylococci (10, 108, 109).

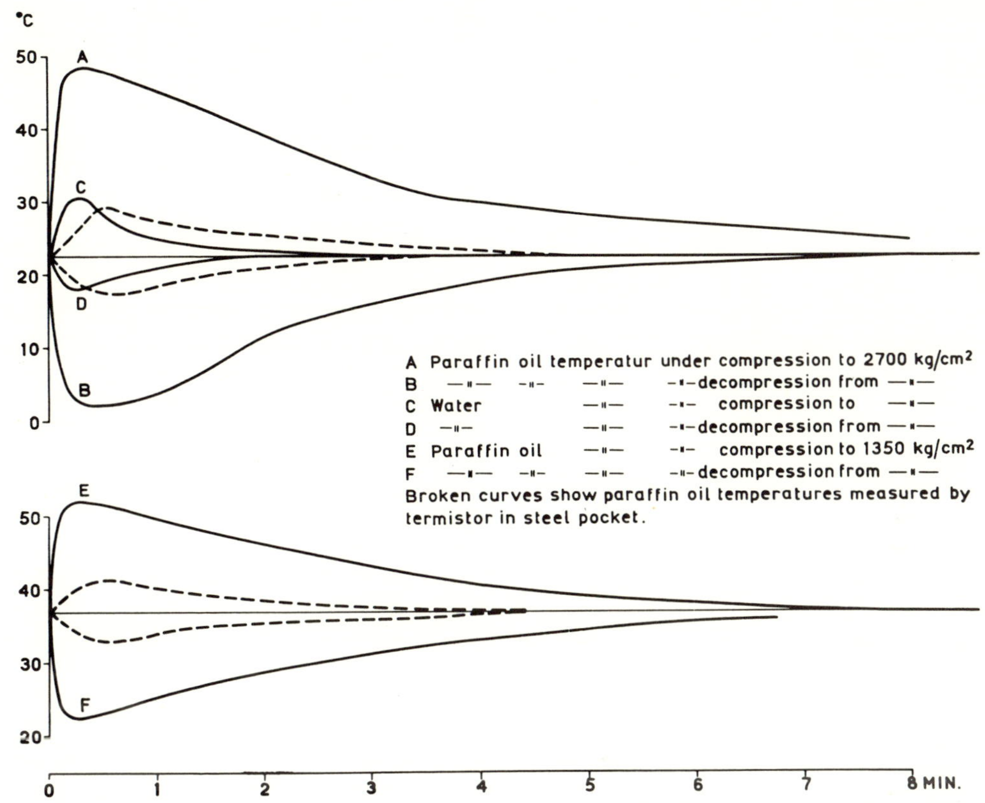

FIG. 4. *Temperature pulses associated with compression-decompression of paraffin oil and water. The measurements were made both with a free thermistor and with the same probe "delayed" by housing in a 1-mm thick pocket of tempered steel (Uddeholm, Sweden; quality Arne).*

GENERAL EFFECTS OF PRESSURE AT PHYSIOLOGICAL TEMPERATURES

Nucleic acid metabolism has been a department interest for many years (18, 42, 43, 70, 85), and with the high-pressure equipment available it was tempting to investigate such phases of the cellular control mechanisms, where large volume changes could be expected to occur. The basis is, of course, that pressure will accelerate processes leading to a volume decrease, whereas any synthesis involving a volume increase will be inhibited. The particular interest of pressure data in regard to biological processes derives from the fact that, whereas reactions involving only small molecules are likely to have small volume changes, of the order of a few milliliters per mole, the reactions involving large molecules, such as protein, may exhibit very large volume changes, of the order of 100 ml per mole (54).

Affecting volume changes is, however, not the only way in which pressure may act on the control systems of the cell. These systems might also be affected by increased solation of the cytoplasm (72), which could cause acceleration of particle interactions, or by denaturation of critical enzymes, or changes in pH (53, 76). Obviously, we could anticipate difficult problems of interpretation, but we also ran into some difficult technical problems.

The pressure equipment originally used for disrupting frozen cells proved inadequate, because it became necessary to measure pressure and temperature directly in the chamber. Figure 3 shows a cross section of one of the devices now used, and Fig. 4 illustrates the value of being able to measure the temperature pulses associated with compression and decompression (78). It is shown as a warning to those who might tend to forget this effect. We did so ourselves in the early work on phage and bacteria (91, 92), where the expos-

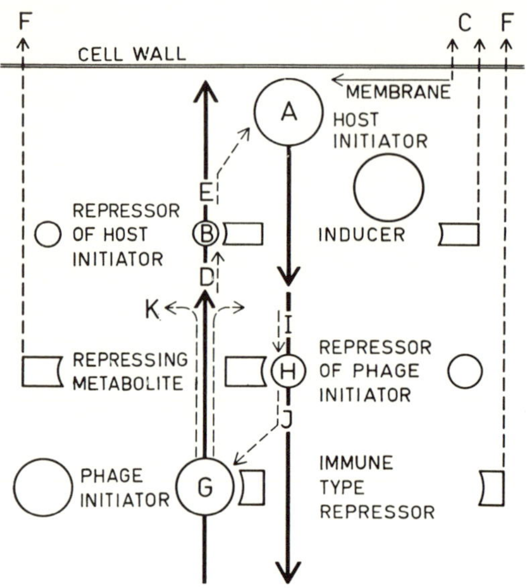

FIG. 5. *Model of cellular control mechanisms with hypothetical pressure "targets" indicated by letters: A, B, G, H (= desensitization); C, F (= leakage) and D, E, I, J, K (= direct effects on DNA). Considering the evidence for linear processes both in replication (86) and protein synthesis (57) and also the theoretical (106) and experimental data (56, 77), which indicate opposed orientation of the phosphodiesterbonds in the two coils making up the double helix, the two strands are shown as heavy vertical arrows. Initiators and repressors (51) are shown as spheres not only free in the cytoplasm but also attached to the chromosome. When located at specific sites (replicator and operator) on the chromosome, they could be influenced by small molecule effectors (inducers or repressing metabolites). Those have been drawn between the strands to illustrate an intimate relation to the transcription process and a function of the nontranscribed strand (41) as part of an anchoring or "switch" mechanism, which could be either reinforced or weakened. Synthesis of repressors and initiators is shown as fine arrows which indicate a control by specific loci (D, E, I, J). Replication and protein synthesis in general are supposed to follow similar linear patterns, except for the fact that the synthesis of labile messenger RNA would require both strands (15, 27, 80, 100, 107). DNA synthesis, on the other hand, would proceed according to the "semiconservative" scheme, i.e., the chains would separate without breakage and each would serve as a template for the formation of a complementary chain (22, 33, 62, 74). The initiator can, for instance, be visualized as a specific depolymerase (60). It, or an electrical event which it triggers, would expose two primer strands (23, 71, 93), and they would both (101) attract complementary bases and participate in a polymerization event, which might proceed over a four-stranded structure (16, 17) held together by unreplicating chromosome segments by the DNA attached proteins, which may take over the interbase hydrogen bonds still keeping the helix in a "quasi-one-strand-helix" grip (98), or by some other mechanism. The finishing of one replication would not necessarily require protein synthesis (40, 68), but a new cycle would require at least so much RNA or protein synthesis, or both, that the initiator level could be maintained. An element of competition between protein and DNA synthesis (83, 96) would seem natural, at least in the small portion of the chromosome, perhaps 10% (59), which is supposed to be single stranded at any one time during the DNA synthesis proceeding during most of the cell cycle (1, 67, 94, 110). The diagram, of course, does not exclude that the balance between replication and synthesis of messenger RNA may depend on the rate of deoxyribonucleotide formation from ribonucleotides (20, 84), the synthesis of thymidylic acid (36), or the phosphorylation of the deoxyribonucleotides (61). The phage genome has been drawn in line with the host genome, but this does, of course, not exclude the more probable types of attachment.*

ures were done in paraffin oil. However, for instance in the studies on transforming deoxyribonucleic acid (DNA), which will be described below, water was used for transmitting the pressure, and the samples were carefully precooled to compensate for the unavoidable small temperature increase (45).

SPECULATIVE MODEL FOR DISCUSSING SOME POSSIBLE PRESSURE TARGETS

As mentioned above, high hydrostatic pressure may act on many different levels in the cell, so it might be useful to consider some possible targets in terms of some concepts of biochemical genetics. It must be realized, however, that this necessarily involves a lot of speculation; therefore, the ideas condensed in Fig. 5 should only be regarded as a hypothesis formulated to link up the experiments reported later.

I use the terminology of Jacob and Monod (51), who suggest that two types of receivers for signals from specific regulator genes (regulators) are located on the chromosome, one triggering replication (the replicator) and the other protein synthesis (the operator). In Fig. 5, both a host and a phage genome are included. The heavy arrows are DNA strands, and the fine ones indicate signals to the receivers. The latter can be considered as specific points on the strands. They are covered by spheres corresponding to cytoplasmic switch molecules, which activate replication (the initiator) or block the operator (the repressor).

In the cell, both the initiators and the repressors obviously occur attached to specific codes on the chromosome, as well as in a diffusible form. Let us now assume that the DNA-attached molecules are of special significance, and that they themselves—or the electrical events which they trigger (46)—travel along one particular nucleotide chain in a direction determined by the polarity of that strand.

The prerequisite necessary for an electrical event to run as indicated seems to be that the strands are unlinked by hydrogen bonds. Such a situation, unlikely as it might seem, was actually considered by Spitkovsky (98) as an explanation for certain spectrophotometrical data and the great mechanical elasticity of nucleoprotein. He suggested that protein may take over the interbase bonds and hold the DNA in a "quasi-one-strand-helix" configuration. Such

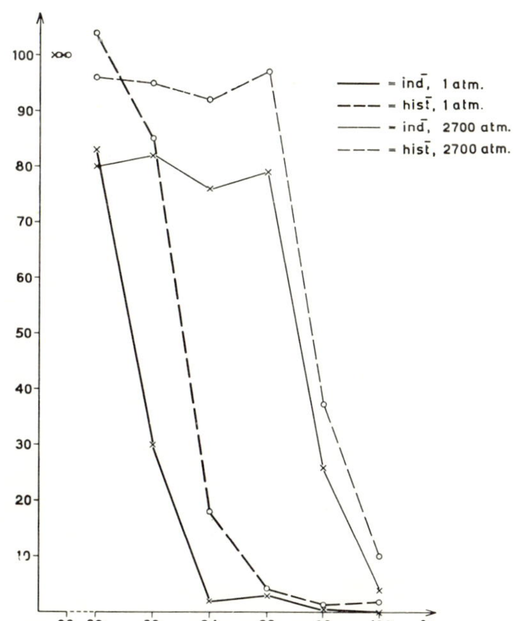

FIG. 6. *Residual transforming activity of Bacillus subtilis DNA in 0.15 M NaCl and 0.015 M sodium citrate (pH 7.9) heated, at a concentration of about 2 μg/ml, to different temperatures for 30 min (see 45).*

a structure might actually have been very useful for the development of the molecule, because it would tend to preserve the genetic code against "proton tunneling" (63), making tautomeric bases in both strands at the same time. If proteins, or ribonucleic acid (RNA) for that matter, took over the hydrogen bonds, except during the brief periods of replication, this would greatly stabilize the genetic message.

Let us suppose that the molecules or charges traveling along a DNA strand are uninterrupted by interpositioned switches or blocking mechanisms (inducers and repressors?). They would then eventually reach a "linker" or a special code, perhaps one complementary to that characteristic for their point of origin, and would there find the conditions for switching over, either to the other strand of the original double helix, or to a newly synthesized strand. Löwdin (63) pointed out that, in the case of transcription, the possibility for a "return journey" along the complementary strand suggests a simple explanation for the base ratios found in RNA and for the unwinding of messenger RNA. Another interesting consequence of the strand-switching model is that an accumulation of molecules or charges might occur at both ends of the double helix, but on different strands. This could be significant for the formation of ring structures (32, 52, 99) and the type of tertiary organization which requires alternating polarities of the replicons making up some chromosomes (35, 58, 81, 103).

When considering high pressure, which certainly may affect orbital overlaps and π-electron pathways, it becomes tempting to think in terms of electrical events, but in Fig. 5 the targets have been illustrated on the physiological level. They fall into three categories: (i) desensitization of trigger molecules attached to DNA (A, B, G, H), (ii) leakage (C, F), and (iii) direct effects on DNA (D, E, I, J, K).

We may take them in the reverse order and start by asking: what effect does pressure have on isolated, transforming DNA?

EFFECTS ON DNA

To determine the effect of pressure on transforming DNA (45), we studied two unlinked markers (indole and histidine) in the DNA from *Bacillus subtilis* (2). This DNA melts between 86 and 92 C, and above that temperature the biological activity rapidly disappears. A pressure

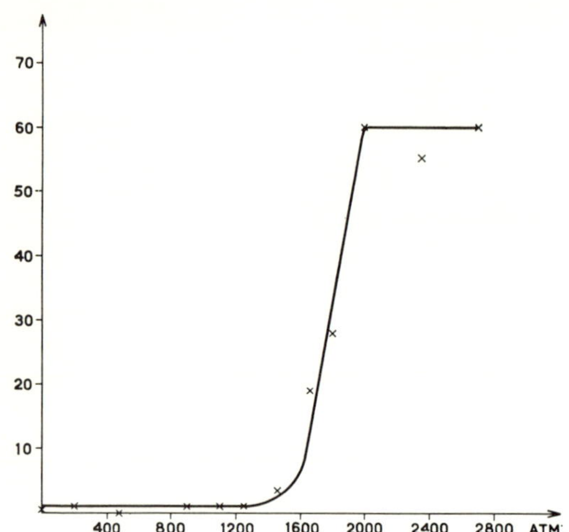

FIG. 7. *Residual transforming activity of Bacillus subtilis DNA in 2 M NaCl and 0.001 M sodium citrate after heating to 100 C for 30 min at various pressures (see 45).*

of 2,700 atm, however, increases the melting temperature and protects the biological activity against thermal inactivation (Fig. 6). Similar effects were observed earlier with enzymes (6, 39, 76). Even pressures lower than 2,700 atm have an effect on DNA, provided that a certain threshold is passed. Figure 7 shows that at 100 C this is close to 2,000 atm.

If the loss of transforming activity with time is studied, the protective function of pressure becomes conspicuous (Fig. 8); the slow first-order decrease at 100 C and 2,700 atm probably illustrates the hydrolytic splitting off of purine bases, a phenomenon known to occur at high temperatures (38).

If we assume that the double helix is preserved under pressure, it is also reasonable to assume that in such a structure the bases are partly protected from contact with the solvent. In the single strand, they would not be similarly protected, which would explain the higher rate of breakdown.

Pressure obviously opposes single-stranding, but will it also affect DNA more directly, for instance, by causing permanent changes in the bases at more physiological temperatures than around 100 C? Considering the induction effects, which will be described later, we rather expected that the pressure levels used in our experiments

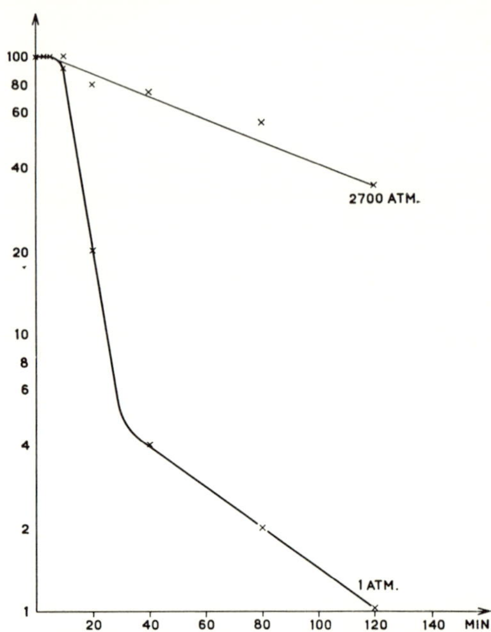

FIG. 8. *Residual transforming activity of Bacillus subtilis DNA in 2 M NaCl and 0.001 M sodium citrate after heating to 100 C for various times (see 45).*

would be mutagenic; however, when Holme and Edebo (47) tested this by the streptomycin technique described by Bertani (8), they could detect no effects at 2,000 or 3,000 atm applied for 30 sec or at pressures ranging from 150 to 700 atm with longer exposure times. Actually, an antimutational effect of pressure had earlier been reported by McElroy (66), who regarded it as an indication of a volume increase in the chromosome caused by certain mutagens.

EXPERIMENTS ON BACTERIA AND EXTRACELLULAR PHAGE

The brief pressure period employed in the experiments just described does not have much effect on viability. On the other hand, longer exposures kill many bacteria. The literature concerning this phenomenon goes back almost a century and will not be reviewed here, but a few experiments on *Escherichia coli* will be mentioned, because this is the host for the viruses which will be discussed later.

In the case of *E. coli*, Johnson and Lewin (54, 55) showed that the temperature inactivation

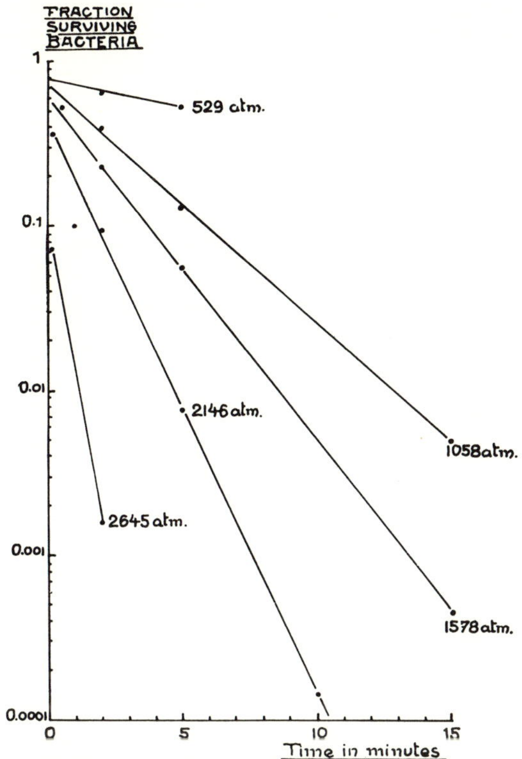

FIG. 9. *Inactivation rates of Escherichia coli B at different pressures (see 88)*

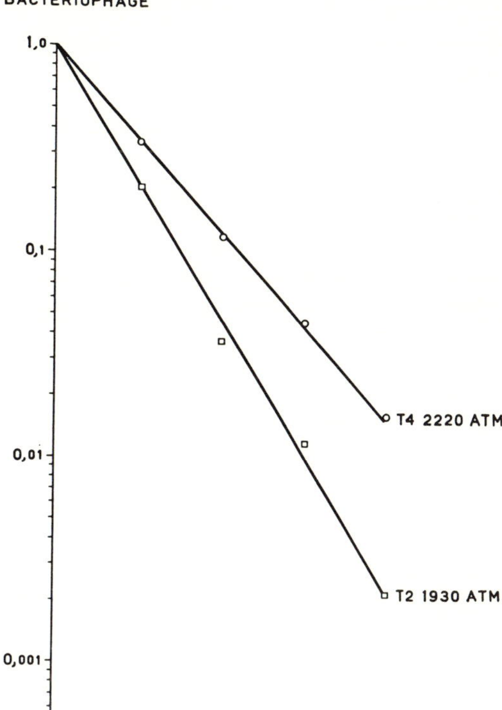

FIG. 10. *Inactivation rates of phages T_2 and T_4 (see 89).*

is opposed by pressure; later, Foster and Johnson (34) found that it also reduces the ability of infected cells to support phage growth as measured by burst sizes and latent periods.

In our group, Rutberg (88, 92) studied the effects on *E. coli* B in some detail, and found an exponential increase in the killing effect with time, the rate being related to the pressure level (Fig. 9).

We found similar curves for the phages T_2 and T_4 (Fig. 10), the former being the most sensitive (89, 92). This is actually the same relation as found with ultraviolet exposure, which might perhaps be regarded as an argument for a DNA target at very high pressure levels and long exposure times.

It has been shown that a hybrid from a cross $T_2 \times T_4$—having the ultraviolet sensitivity of T_4 and backcrossed nine times with T_2, selecting for the ultraviolet sensitivity of T_4—has the pressure sensitivity of T_4 (89). Marker rescue can also be demonstrated with T-even pressure-inactivated phage. These experiments strongly support the idea that the target for very high pressures is the same as that for ultraviolet irradiation.

If we want to think of the DNA as the "target" in the low-pressure range also, it becomes necessary to think of the sensitive element as being exposed in its labile, single-stranded form.

Let us then consider two types of replication, that associated with cell division and that required for phage multiplication.

With regard to cell division of normal terrestrial organisms, this was shown by ZoBell and his co-workers to be blocked at very moderate pressures, around 50 to 200 atm, and to be resumed when the pressure is released (111, 112, 113). All these authors worked with unaerated cultures, but we also obtained the effect in actively aerated cultures (44). In this case, one must of course consider the toxic effects of the diradical oxygen, but under purely hydrostatic pressure such an effect does not seem very likely. However, such pressures were found by Britten and McClure (14) to cause a conspicuous leakage of the amino acid pool, and Berger (7) found an accumulation in the medium of building blocks for the cell walls of *E. coli* grown under 50 to 400 atm.

In the model referred to earlier (Fig. 5), leakage is indicated by the letter C. It will be noted that the chromosome is illustrated as being intimately related to the cell wall, a situation suggested by Jacob and Monod as an explanation for the physiological synchronization of replication and cell division. Obviously, a leakage of material required for finishing the cell membrane might prevent the initiator release and, consequently, also the replication required for cell division. However, the cell metabolism would continue, and swollen forms and "snakes" would appear, just as was actually found by several investigators (7, 113).

Other alternatives for the blocking of chromosome replication would be a more or less direct DNA effect influencing the initiator-repressor synthesis (D, E), or an interference with the physiological single-stranding. The low pressures required, and the fact that phage and bacteria are not killed by such pressures, however, seem to make the former alternative unlikely, and the fact that other types of replication, such as phage multiplication, are unaffected at the pressure levels in question argues against the second possibility. This would leave us with essentially one alternative apart from leakage: desensitization, i.e., partial denaturation of the cytoplasmic switch molecules of Jacob and Monod (51).

These critical control molecules (initiators and repressors) are supposed to be subjected to delicate allosteric influences by inducers or repressing metabolites (75). The latter modify the signals which reach the chromosome triggers. The allosteric influence is pictured as a desensitization similar to the loss of sensitivity toward the inhibitor, without the loss of activity toward substrate which is known to exist in enzymes subjected to partial inactivation or denaturation (19, 37, 73). Such phenomena are well-known effects of the application of hydraulic pressure (53, 69).

EFFECTS ON THE MULTIPLICATION OF VIRULENT AND TEMPERATE PHAGE

Much information is available concerning a very active and highly specialized nucleic acid

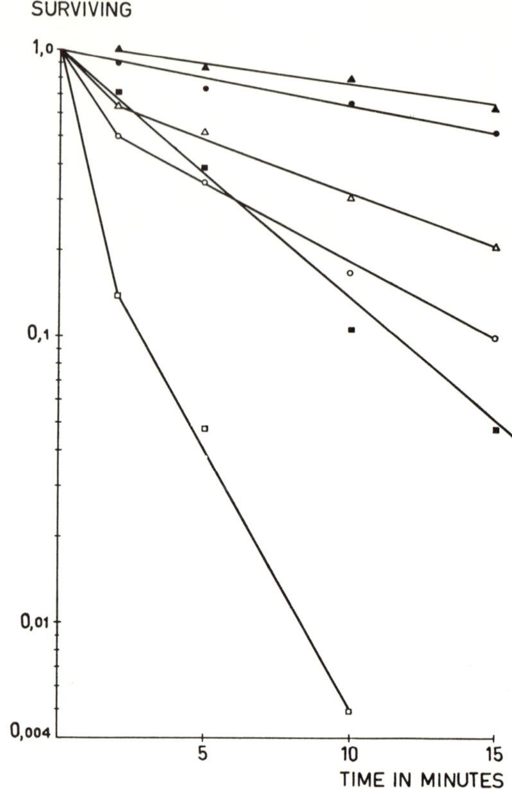

FIG. 11. *Inactivation rates of Escherichia coli B at different pressures measured by capacity to form colonies (△ at 740 atm, ○ at 1,040 atm, and □ at 1,480 atm) and to support T_2 multiplication (▲ at 740 atm, ○ at 1,040 atm, and ■ at 1,480 atm). (See 88.)*

protein synthesis, namely phage multiplication. What can such a system tell us about the intracellular target of high pressure? How does phage multiplication, for instance, proceed in bacteria that are first pressure-treated and then infected?

Figure 11 shows that the capacity to support T_2 multiplication falls off with increasing pressure at a rate which is not too different from that showing survival of uninfected cells (88, 92). This is actually one of the many facts which indicate that pressure and ultraviolet irradiation have different targets, ultraviolet treatment being much more potent in killing the cells than in destroying the above capacity (3).

A second problem concerns the effect on phage multiplication, if the pressure is applied *after* infection. In this case, Rutberg was able to demonstrate a sensitivity to 2-min exposures of

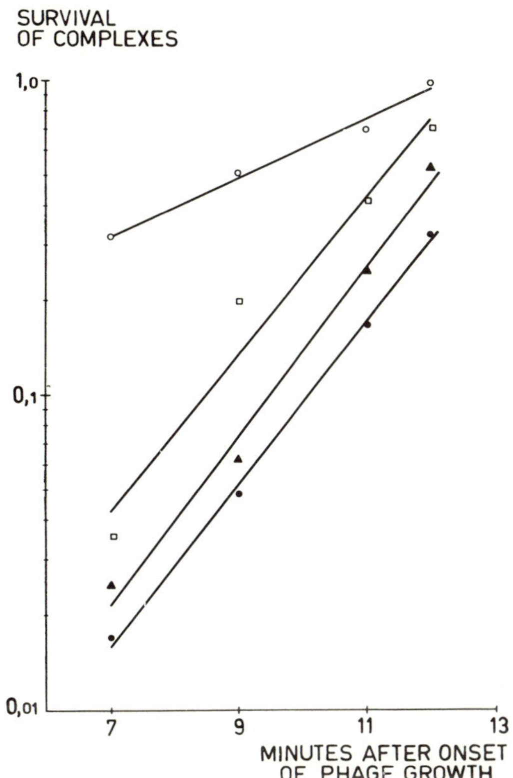

FIG. 12. *Effect of 1,260 atm, applied for different lengths of time, on intracellular multiplication of T_2 phage. Symbols:* ○ = *1 min,* □ = *2 min,* ▲ = *4 min,* ● = *6 min. (See 88.)*

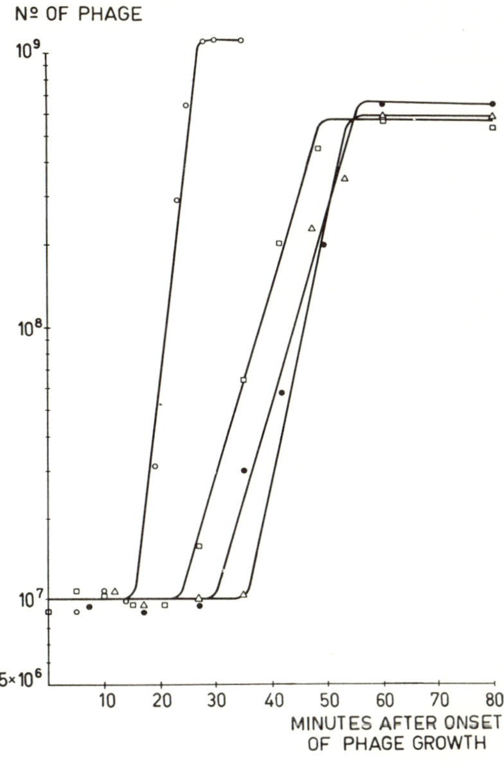

FIG. 13. *Effect of 2-min exposure to 740 atm on the latency period and burst size measured on Escherichia coli B infected with T_2 phage. Pressure-treated at time 0,* □; *at 7 min,* ●; *and at 12 min,* △ *(control,* ○*). (See 88.)*

the intracellular phage, which was much greater than that exhibited by mature particles (88). As shown in Fig. 12, the sensitivity is particularly high during the first 7 min. The one-hit type curves for the different pressure levels then tend to converge. After 10 to 12 min, mature particles could be detected by lysis from without, and bursts then started to appear, confusing the phage counting. It seems likely that the extrapolated curves give the time required for the particles to become resistant to the pressure level used. The fact that the sensitivity of the developing complexes shows a pattern different from that observed for ultraviolet treatment (5, 64) is an additional argument for the assumption that ultraviolet treatment and low pressure have different targets.

Figure 13 shows that 2-min pressure exposures (740 atm) during the period of intracellular phage development reduce the burst sizes somewhat (88). It also demonstrates that the latency period is largely constant after the pressure exposure, irrespective of whether this takes place immediately after infection or 7 or 12 min later. This fact, together with the fact that the late exposures do not yield bigger bursts than the early ones, makes it reasonable to assume that pressure stops phage multiplication, which has to begin from the start when the pressure is released. In terms of the model described earlier, this can hardly be explained as simple leakage of building blocks of low molecular weight, and more profound effects, such as destruction of initiator, a reversal of single-stranding, or both, seem more likely.

The existence of such profound changes is also indicated by another pressure phenomenon, phage induction, which was observed in connection with some early T_2 burst-size studies (91). In this case, we noticed an irregular appearance of an unknown phage after treatment of *E. coli* B (26). The system was, however, very difficult

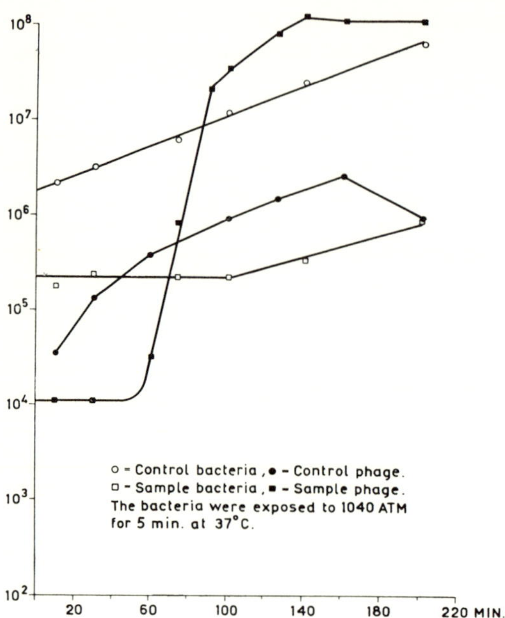

FIG. 14. *Induction of phage lambda in Escherichia coli K-12 by means of pressure. (See 90.)*

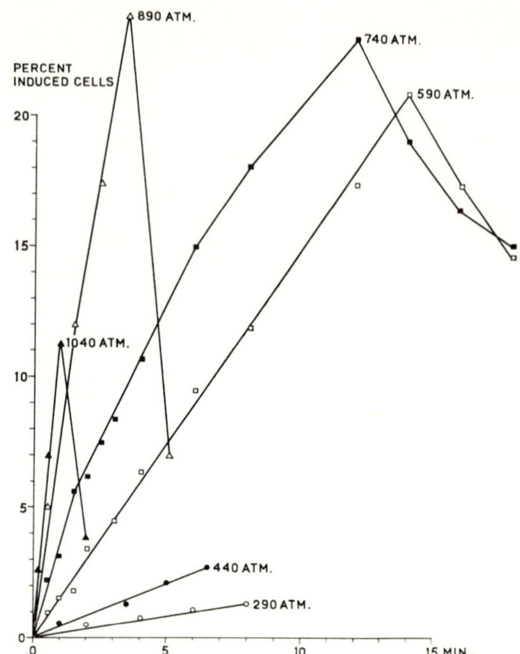

FIG. 15. *Induction of phage lambda by different pressure levels. (See 73.)*

to work with, so induction will be illustrated here with *E. coli* K-12, where Rutberg had studied the lambda phage (90). This was a fortunate choice, because the system has proved to be unusually apt for this type of induction.

As shown in Fig. 14, *E. coli* K-12 lyses after the same latency period as observed in the case of ultraviolet induction, i.e., about 60 min. By applying the spraying technique of Six (97), which depends on blocking the noninduced cells by spraying the plates with streptomycin (using, of course, a streptomycin-resistant strain for plating), Rutberg then determined the actual number of induced cells. The response followed a pattern quite different from that seen with ultraviolet treatment (Fig. 15). It is proportional to the time of exposure at any one pressure level. These levels can be quite low, and, by a special technique, it was actually possible to show induction of 2-min exposures down to a pressure of 150 atm. This should be a memento in ultracentrifugation, where such pressures are easily attained (Fig. 16).

In the high-pressure range, optimal induction obviously requires short exposure times; otherwise one may get into the sharp-decline range. On the one hand, too short an exposure might return the released phage to an environment of immune-type repression, and, on the other, an extended pressure treatment might interfere with the escape mechanisms. In both cases, there would be a reduced induction.

The model used in this paper as a basis for discussing pressure effects (Fig. 5) indicates that leakage of a repressing metabolite or an immune-type repressor could be an explanation (F), but it seems unlikely that leakage would increase and decrease with pressure in a way which could explain the results just described. An inactivation of repressing metabolites or immune-type repressor might, however, be an alternative explanation. Earlier, a series of arguments against direct DNA effects of relatively low pressures were indicated, and in this particular case one might add that cell suspensions which were aerated in buffer for 2 hr lost their aptitude for ultraviolet induction but could still be induced with pressure.

Pressure actually seems to belong to a different category than the mutagens and carcinogens normally used for induction purposes (50, 65). Unfortunately, at the present time, we can only guess at the mechanism involved. However, considerable volume changes take place. Estimates of about 220 ml per mole can actually be

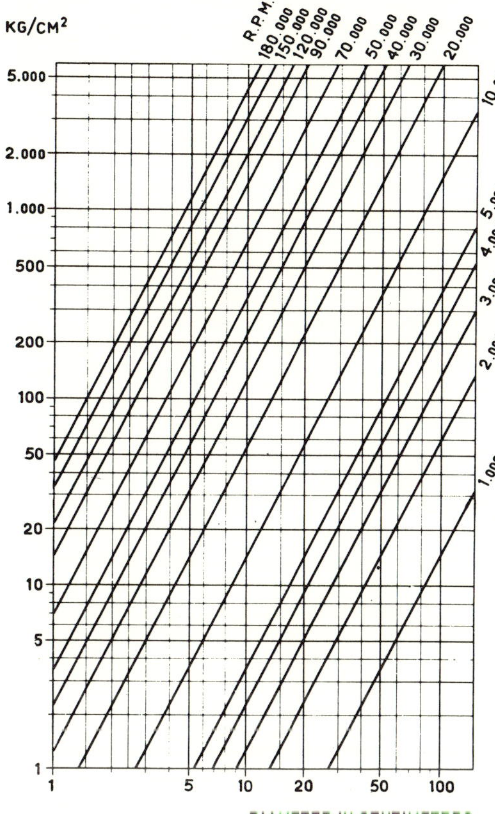

FIG. 16. *Nomogram illustrating the centrifugally produced liquid pressures in water.*

made on the basis of induction data obtained at 30 and 37 C. Such volume changes might perhaps be expressions of a desensitization of the type mentioned earlier or of steric alterations in the replication process (K).

The latter possibility would include alterations in newly formed single strands of DNA or RNA. Such strands are probably rather vulnerable before being stabilized by double-helix formation, and they might tend to fold to gain a smaller volume. It is interesting to note that such a folding or looping process might give us structures similar to transfer RNA, and that segments with secondary structure in messenger RNA have been assumed to be of importance for their biological function (79).

Concluding Remarks

As mentioned earlier, the DNA-attached molecules might be important for ring formation and, consequently, also for the formation of the type of secondary structure just mentioned. Stedman and his group proposed that histones act as gene inhibitors (21; *see also* 4), and later such proteins were actually shown to repress DNA-dependent RNA synthesis (48). However, much remains to be learned about the DNA-attached proteins before one can start to consider the pressure sensitivity of the DNA-protein complex. A start in this direction is being made in our laboratory by Hammersten and his co-workers, who developed counterflow electrophoresis in the presence of ethylene glycol as a means of preparing "functional" nucleoprotein without a denaturation, which might irreversibly attach extraneous proteins to the molecule complex (28, 29, 30, 31). Nucleoprotein isolated in this manner contains less than 0.1% protein, but this has a strong nuclease activity. The finding of Macheboeuf and his group (104, 105), that both polymerization and depolymerization of RNA varies with the pressure and with the size of the macromolecules exposed, adds interest to studies aimed at determining the effect of pressure on the type of nucleoprotein mentioned.

DNA in the cell exists in the form of nucleoprotein and, as pointed out before, we must keep this high degree of organization in mind when discussing pressure effects. The electron-transport pathways in nucleoproteins may, of course, be pressure-sensitive, but information must be obtained about the electrical properties of the simple deproteinized DNA molecule before one can consider the nucleoproteins from this point of view. In this area, our own efforts have so far been limited to the development of a technique for bulk preparation of oriented DNA with retained transforming activity (87).

Hoffman and Ladik (46) regard DNA as a semiconductor, which has completely filled valence bands and completely empty conduction bands; i.e., it behaves as an insulator. These authors think that electron donors, thermal excitation, or irradiation may turn the molecule into a conductor by moving electrons into the conduction bands. Under the influence of intracellular electrostatic fields, these electrons might then migrate to the ends of the double helix and initiate single-stranding by a process of electrostatic repulsion. This hypothesis, which in fact can easily be fitted into the model presented earlier, would indicate that pressure, which is known to increase the electrical conductivity of organic semiconducting polymers (82), might have a very direct effect on the molecule. It is of interest that electrical fields recently were shown to stimulate cell division (11).

Apart from the approaches just mentioned, there are certainly many other roads which may lead to a better understanding of the mechanisms of pressure effects. At present, the interpretations are, of necessity, as loosely knit as indicated by this paper. Actually, the only general conclusion which can be drawn at the present time is that the effects of very low pressures are compatible with the hypothesis of leakage, that higher pressures add involvements of cellular control mechanisms, and that the highest pressure levels may cause irreversible DNA changes. However, the effects on the cellular control mechanisms are obscure, and when I have focused on replication and DNA transcription into RNA it is little more than a guess, which must be proved or disproved by future experiments. In any event, such experiments will certainly further emphasize one fact, clearly shown by the investigations of Johnson et al., namely, that pressure may be a very useful tool in molecular biology. It might some day also be of value in the applied areas, where it might be put to use in blocking cell division, without necessarily stopping the metabolism, or in releasing defective viruses, for example.

However, not only the microbiologist, but also other scientists, might find it rewarding to remember to include pressure among their parameters. Schramm and his co-workers (95) synthesized polynucleotides at atmospheric pressures in a nonenzymatic environment, and it is, of course, well known that pressure accelerates many polymerizations. Temperature denaturation and protection against this effect by pressure may also have been two important links in a development process of such macromolecules as nucleic acid in the abysmal depths of the oceans on our primitive earth.

ACKNOWLEDGMENTS

I am indebted to my colleagues at the Bacteriology Department, Karolinska Institutet, for their help and criticism. I also want to thank G. Bertani and T. A. Hoffman for the guidance and advice they have given me.

LITERATURE CITED

1. ABBO, F. E., AND A. B. PARDEE. 1960. Synthesis of macromolecules in synchronously dividing bacteria. Biochim. Biophys. Acta **39**:478–485.
2. ANAGNOSTOPOULOS, C., AND J. SPIZIZEN. 1961. Requirements for transformation in *Bacillus subtilis*. J. Bacteriol. **81**:741–746.
3. ANDERSON, T. F. 1948. The growth of T2 virus on ultraviolet-killed host cells. J. Bacteriol. **56**:403–410.
4. BARTALOS, M. 1963. A possible chemical term for the operator gene. Nature **198**:109.
5. BENZER, S. 1952. Resistance to ultraviolet light as an index to the reproduction of bacteriophage. J. Bacteriol. **63**:59–72.
6. BERGER, L. R. 1958. Some effects of pressure on a phenylglycosidase. Biochim. Biophys. Acta **30**:522–528.
7. BERGER, L. R. 1959. The effect of hydrostatic pressure on cell wall formation. Bacteriol. Proc., p. 129.
8. BERTANI, G. 1951. A method for detection of mutations, using streptomycin dependence in Escherichia coli. Genetics **36**:598–611.
9. BILLAUDELLE, H., AND P. O. BRUNÉR. 1959. An automatic apparatus for extraction of cell suspensions by repeated freezing and thawing. J. Biochem. Microbiol. Technol. Eng. **1**:229–238.
10. BILLAUDELLE, H., L. EDEBO, E. HAMMARSTEN, C.-G. HEDÉN, B. MALMGREN, AND H. PALMSTIERNA. 1960. Studies on the chemical and immunological structure of Bordetella pertussis. Acta Pathol. Microbiol. Scand. **50**:208–224.
11. BOZOKY, L., G. KISZELY, T. A. HOFFMAN, AND J. LADIK. 1963. Effects of electrostatic fields on cell mitosis. Nature **199**:1306.
12. BRIDGMAN, P. W. 1912. Water in the liquid and five solid forms under pressure. Proc. Am. Acad. Arts Sci. **47**:439–559.
13. BRIDGMAN, P. W. 1937. The phase diagram of water to 45000 kg/cm^2. J. Chem. Phys. **5**:964–966.
14. BRITTEN, R. J., AND F. T. MCCLURE. 1962. The amino acid pool in *Escherichia coli*. Bacteriol. Rev. **26**:292–335.
15. BURMA, D., H. KROGER, S. OCHOA, R. WARNER, AND J. WEILL. 1961. Further studies on deoxyribonucleic acid-dependent enzymatic synthesis of ribonucleic acid. Proc. Natl. Acad. Sci. U.S. **47**:749–752.
16. CAVALIERI, L. F., AND B. H. ROSENBERG. 1961. The replication of DNA. I. Two molecular classes of DNA. II. The number of polynucleotide strands in the conserved unit of DNA. Biophys. J. **1**:317–336.
17. CAVALIERI, L. F., AND B. H. ROSENBERG. 1961. The replication of DNA. III. Changes in the number of strands in E. Coli DNA

during its replication cycle. Biophys. J. **1**:337–351.
18. CAVANNA, R., E. B. CHAIN, C.-G. HEDÉN, B. MALMGREN, N. A. ELIASSON, E. HAMMARSTEN, AND S. ÅQVIST. 1955. Interrelation of protein and polynucleotide synthesis in Escherichia coli. Rend. Ist. Super. Sanita (English Ed.) **17**:168–186.
19. CHANGEUX, J. P. 1961. The feedback control mechanism of biosynthetic L-threonine deaminase by L-isoleucine. Cold Spring Harbor Symp. Quant. Biol. **26**:313–318.
20. COHEN, S. S., H. D. BARNER, AND J. LICHTENSTEIN. 1961. The conversion of a phage-induced ribonucleic acid to deoxyribonucleotides in vitro. J. Biol. Chem. **236**: 1448–1457.
21. CRUFT, H. J., C. M. MAURITZEN, AND E. STEDMAN. 1954. Abnormal properties of histones from malignant cells. Nature **174**:580–585.
22. DELBRÜCK, M., AND G. S. STENT. 1957. On the mechanism of DNA replication, p. 699–736. *In* W. D. McElroy and B. Glass [ed.], The chemical basis of heredity. Johns Hopkins Press, Baltimore.
23. DOUNCE, A. L., N. K. SARKAR, AND E. R. KAY. 1961. The possible role of DNA-ase I in DNA replication. J. Cellular Comp. Physiol. **57**:47–54.
24. EDEBO, L. 1960. A new press for the disruption of microorganisms and other cells. J. Biochem. Microbiol. Technol. Eng. **2**:453–479.
25. EDEBO, L., AND C.-G. HEDÉN. 1960. Disruption of frozen bacteria as a consequence of changes in the crystal structure of ice. J. Biochem. Microbiol. Technol. Eng. **2**: 113–120.
26. EDEBO, L., AND L. RUTBERG. 1960. Electron-microscopical observations on a pressure-induced phage. J. Ultrastruct. Res. **4**: 88–91.
27. EISENSTADT, J. M., T. KAMEYAMA, AND G. D. NOVELLI. 1962. A requirement for gene-specific deoxyribonucleic acid for the cell-free synthesis of β-galactosidase. Proc. Natl. Acad. Sci. U.S. **48**:652–659.
28. ELIASSON, R., E. HAMMARSTEN, AND T. LINDAHL. 1962. Separation of polynucleotides and proteins from cell extracts at low temperatures by counter flow electrophoresis. Arch. Biochem. Biophys. Suppl. 1, p. 139–146.
29. ELIASSON, R., E. HAMMARSTEN, AND T. LINDAHL. 1962. Use of ethylene glycol in counter flow electrophoresis and in cold concentration of DNA and protein solutions. Biotechnol. Bioeng. **4**:53–56.
30. ELIASSON, R., E. HAMMARSTEN, T. LINDAHL, I. BJÖRK, AND T. C. LAURENT. 1963. The stability of deoxyribonucleic acid in glycol solution. Biochim. Biophys. Acta **63**:234–239.
31. ELIASSON, R., E. HAMMARSTEN, T. LINDAHL, AND H. PALMSTIERNA. 1961. Preliminary notes on the separation of nucleic acids and proteins by counter flow electrophoresis. Acta Chem. Scand. **15**:570–574.
32. FIERS, W., AND R. L. SINSHEIMER. 1962. The structure of the DNA of bacteriophage ϕ X174. J. Mol. Biol. **5**:408–434.
33. FORRO, F., AND S. A. WERTHEIMER. 1960. The organization and replication of deoxyribonucleic acid in thymine-deficient strains of Escherichia coli. Biochim. Biophys. Acta **40**:9–21.
34. FOSTER, R. A. C., AND F. H. JOHNSON. 1950. Influence of urethane and of hydrostatic pressure on growth of bacteriophages T2, T5, T6 and T7. J. Gen. Physiol. **34**:529–550.
35. FREESE, E. 1958. The arrangement of DNA in the chromosome. Cold Spring Harbor Symp. Quant. Biol. **23**:13–18.
36. FRIEDKIN, M. 1959. Speculations on the significance of thymidylic acid synthesis in the regulation of cellular proliferation, p. 97–103. *In* F. Stohlman [ed.], The kinetics of cellular proliferation. Grune and Stratton, New York.
37. GERHARDT, J. C., AND A. B. PARDEE. 1962. The enzymology of control by feedback inhibition. J. Biol. Chem. **237**:891–896.
38. GINOZA, W., AND B. ZIMM. 1961. Mechanisms of inactivation of deoxyribonucleic acids by heat. Proc. Natl. Acad. Sci. U.S. **47**: 639–652.
39. HAIGHT, R. D., AND R. Y. MORITA. 1962. Interaction between the parameters of hydrostatic pressure and temperature on aspartase of *Escherichia coli*. J. Bacteriol. **83**:112–120.
40. HANAWALT, P. C., O. MAALØE, D. I. CUMMINGS, AND M. SCHAECHTER. 1961. The normal DNA replication cycle. II. J. Mol. Biol. **31**:156–165.
41. HAYASHI, M., S. SPIEGELMAN, R. E. FRANKLIN, AND S. E. LURIA. 1963. Separation of the RNA message transcribed in response to a specific inducer. Proc. Natl. Acad. Sci. U.S. **49**:729–736.
42. HEDÉN, C.-G. 1951. Studies of the infection of E. coli B with the bacteriophage T2.

Acta Pathol. Microbiol. Scand. Suppl. **88**:1–126.

43. HEDÉN, C.-G., T. HOLME, AND B. MALMGREN. 1955. Application of the continuous cultivation technique to problems connected with growth rate and nucleic acid synthesis. Acta Pathol. Microbiol. Scand. **37**:50–57.
44. HÉDEN, C.-G., AND A. S. MALMBORG. 1961. Aeration under pressure and the question of free radicals. Sci. Rept. Ist. Super. Sanita **1**:213–221.
45. HÉDEN, C.-G., T. LINDAHL, AND I. TOPLIN. 1964. The stability of DNA solutions under high pressure. Acta Chem. Scand. *in press.*
46. HOFFMAN, T. A., AND J. LADIK. 1964. Quantum mechanical properties of DNA. *In* J. J. Duchesne [ed.], Advances in chemical physics, vol. 6, Structure and properties of biomolecules and biological systems. Interscience Publishers, Inc., New York.
47. HOLME, T., AND L. EDEBO. 1960. Undersökningar över mutagen effekt av vissa fysikaliska agentia. Rept. R-318/1960, Swedish Med. Res. Council.
48. HUANG, R. C., AND J. BONNÉR. 1962. Histone, a suppressor of chromosomal RNA synthesis. Proc. Natl. Acad. Sci. U.S. **48**:1216–1222.
49. HUGHES, D. E. 1951. A press for disrupting bacteria and other microorganisms. Brit. J. Exptl. Pathol. **32**:97–109.
50. JACOB, F. 1954. Les bactéries lysogène et la notion de provirus, p. 1–176. Monographies de l'Institut Pasteur, Masson et Cie, Paris.
51. JACOB, F., AND J. MONOD. 1963. Elements of regulatory circuits in bacteria, p. 1–24. UNESCO Symposium on biological organization. Biological organization at the cellular and supercellular level.
52. JACOB, F., AND E. WOLLMAN. 1957. Analyse des groupes de liaison génétique de différentes souches donatrices d'Escherichia coli K 12. Compt. Rend. **245**:1840–1843.
53. JOHNSON, F. H., et al. 1954. Hydrostatic pressure and molecular volume changes, p. 286–368. *In* F. H. Johnson, H. Eyring, and M. J. Polissar [ed.], The kinetic basis of molecular biology. John Wiley & Sons, Inc., New York.
54. JOHNSON, F. H., AND I. LEWIN. 1946. The disinfection of E. coli in relation to temperature, hydrostatic pressure and quinine. J. Cellular Comp. Physiol. **28**:23–45.
55. JOHNSON, F. H., AND I. LEWIN. 1946. The influence of pressure, temperature, and quinine on the rates of growth and disinfection of E. coli in the logarithmic growth phase. J. Cellular Comp. Physiol. **28**: 77–97.
56. JOSSE, J., A. D. KAISER, AND A. KORNBERG. 1961. Enzymatic synthesis of deoxyribonucleic acid. VIII. Frequencies of nearest neighbour base sequences in deoxyribonucleic acid. J. Biol. Chem. **236**:864–875.
57. KANO, S., AND S. SPIEGELMAN. 1962. Evidence for a nonrandom reading of the genome. Proc. Natl. Acad. Sci. U.S. **48**:1942–1949.
58. KELLENBERGER, E. 1960. The physical state of the bacterial nucleus. Symp. Soc. Gen. Microbiol. **10**:39–66.
59. LARK, K. G., L. F. CAVALIERI, AND B. H. ROSENBERG. 1963. Cellular control of DNA biosynthesis. Molecular biology, part I, ref. p. 184. Academic Press, Inc., New York.
60. LASKOWSKI, M., SR. 1961. Deoxyribonucleases, p. 123–147. *In* P. D. Boyer, H. Lardy, and K. Myrbäck [ed.], The enzymes, vol. 5. Academic Press, Inc., New York.
61. LEHMAN, I. R., M. J. BESSMAN, E. S. SIMMS, AND A. KORNBERG. 1958. Enzymatic synthesis of deoxyribonucleic acid. J. Biol. Chem. **233**:163–170.
62. LEVINTHAL, C. 1956. The mechanism of DNA replication and genetic recombination in phage. Proc. Natl. Acad. Sci. U.S. **42**: 394–404.
63. LÖWDIN, P.-O. 1962. Quantum genetics and the aperiodic solid. Some aspects on the biological problems of heredity, mutations, ageing and tumors in view of the quantum theory of the DNA molecule. Preprint No. 85, Quantum Chemistry Group, Uppsala, Sweden.
64. LURIA, S. E., AND R. LATARJET. 1947. Ultraviolet irradiation of bacteriophage during intracellular growth. J. Bacteriol. **53**:149–163.
65. LWOFF, A. 1953. Lysogeny. Bacteriol. Rev. **17**:269–337.
66. MCELROY, W. D. 1952. Evidence for the occurrence of intermediates during mutation. Science **115**:623–626.
67. MCFALL, E., AND G. STENT. 1959. Continuous synthesis of deoxyribonucleic acid in *Escherichia coli*. Biochim. Biophys. Acta **34**:580.
68. MAALØE, O., AND P. C. HANAWALT. 1961. Thymine deficiency and the normal DNA replication cycle. I. J. Mol. Biol. **3**:144–155.
69. MACHEBOEUF, M., J. BASSET, E. BARBU, M.

LE SAGET, AND G. NUNEZ. 1950. Influence de la pression sur l'hydrolyse des protéines par les acides ou les alcalis. Compt. Rend. Soc. Biol. **144**:962–964.

70. MALMGREN, B., AND C.-G. HEDÉN. 1947. Studies on the nucleotide metabolism of bacteria. Acta Pathol. Microbiol. Scand. **24**:417–504.
71. MANTSAVINOS, R., AND E. S. CANELLAKIS. 1959. Studies on the biosynthesis of DNA by cell-free extracts of mouse leukemic cells. Cancer Res. **19**:1239–1243.
72. MARSLAND, D. A., AND D. E. BROWN. 1942. The effects of pressure on sol-gel equilibria, with special reference to myosin and other protoplasmic gels. J. Cellular Comp. Physiol. **20**:295–305.
73. MARTIN, R. G. 1963. The first enzyme in histidine biosynthesis: the nature of feedback inhibition by histidine. J. Biol. Chem. **238**:257–268.
74. MESELSON, M., AND F. W. STENT. 1958. The replication of DNA in Escherichia coli. Proc. Natl. Acad. Sci. U.S. **44**:671–682.
75. MONOD, J., AND F. JACOB. 1961. General conclusions. Teleonomic mechanisms in cellular metabolism, growth, and differentiation. Cold Spring Harbor Symp. Quant. Biol. **26**:389–401.
76. MORITA, R. Y., AND R. D. HAIGHT. 1962. Malic dehydrogenase activity at 101 C under hydrostatic pressure. J. Bacteriol. **83**:1341–1346.
77. NAGATA, T. 1962. Polarity and synchrony in the replication of DNA molecules of bacteria. Biochem. Biophys. Res. Commun. **8**:348–351.
78. NATHAN, M. I., AND W. PAUL. 1962. Effect of pressure on the energy levels of impurities in semiconductors. II. Gold in silicon. Phys. Rev. **128**:38–42.
79. NIERENBERG, M. W. 1963. The genetic code. Sci. Am. **208**:80–94.
80. OCHOA, S., D. BURMA, H. KROGER, AND J. WEILL. 1961. Deoxyribonucleic acid-dependent incorporation of nucleotides from nucleoside triphosphates into ribonucleic acid. Proc. Natl. Acad. Sci. U.S. **47**:670–679.
81. PAPPAS, G. D., AND P. W. BRANDT. 1960. Helical structure in the nuclei of free-living amebas. Intern. Kongr. Elektronenmikroscopie, 2, Berlin, 1958, p. 244–246.
82. POHL, H. A., A. REMBAUM, AND A. HENRY. 1962. Effects of high pressure on some organic semiconducting polymers. J. Am. Chem. Soc. **84**:2699.
83. PRESCOTT, D. M., AND R. F. KIMBALL. 1961. Relation between RNA, DNA, and protein synthesis in the replication nucleus of Euplotes. Proc. Natl. Acad. Sci. U.S. **47**:686–693.
84. REICHARD, P., Z. N. CANELLAKIS, AND E. S. CANELLAKIS. 1961. Studies on a possible regulatory mechanism for the biosynthesis of deoxyribonucleic acid. J. Biol. Chem. **236**:2514–2519.
85. REICHARD, P., AND L. RUTBERG. 1960. Formation of deoxycytidine 5'-phosphate from cytidine 5'-phosphate with enzymes from Escherichia coli. Biochim. Biophys. Acta **37**:554–555.
86. RUDNER, R. 1960. Mutation as an error in base pairing. Biochem. Biophys. Res. Commun. **3**:275–280.
87. RUPPRECHT, A. 1963. Preparation of oriented DNA in large amounts. Biochem. Biophys. Res. Commun. **12**:163.
88. RUTBERG, L. 1964. On the effects of high hydrostatic pressure on bacteria and bacteriophage. I. Action on the reproduction of bacteria and their ability to support growth of bacteriophage T2. Acta Pathol. Microbiol. Scand. *In press.*
89. RUTBERG, L. 1964. On the effects of high hydrostatic pressure on bacteria and bacteriophage. II. Inactivation of bacteriophage. Acta Pathol. Microbiol. Scand. *In press.*
90. RUTBERG, L. 1964. On the effects of high hydrostatic pressure on bacteria and bacteriophage. III. Induction with high hydrostatic pressure of E. coli K lysogenic for phage lambda. Acta Pathol. Microbiol. Scand. *In press.*
91. RUTBERG, L., AND C.-G. HEDÉN. 1960. The activation of prophage in E. coli B by high pressure. Biochem. Biophys. Res. Commun. **2**:114–116.
92. RUTBERG, L., AND C.-G. HEDÉN. 1963. Effecten av Höga Tryck på Backterier och Bakterievirus. Rept. 1962, Swedish Cancer Society, p. 290–293.
93. SARKAR, N. K., AND A. L. DOUNCE. 1961. A spectroscopic study of the reaction of formaldehyde with deoxyribonucleic and ribonucleic acids. Biochim. Biophys. Acta **49**:160–169.
94. SCHAECHTER, M., M. W. BENTZON, AND O. MAALØE. 1959. Synthesis of deoxyribonucleic acid during the division cycle of bacteria. Nature **183**:1207–1208.
95. SCHRAMM, G., H. GRÖTSCH, AND W. POLLMAN. 1962. Nicht-enzymatische Synthese von Polysacchariden, Nucleosiden und Nu-

cleinsäuren und die Entstehung selbstvermehrungsfähiger Systeme. Angew. Chem. **74**:53–92.

96. SISKEN, J. E. 1959. The synthesis of nucleic acids and proteins in the nuclei of tradescantia root tips. Exptl. Cell Res. **16**: 602–614.
97. SIX, E. 1959. The rate of spontaneous lysis of lysogenic bacteria. Virology **7**:328–346.
98. SPITKOVSKY, D. M. 1963. To the problem of quasisinglecoil condition of DNA in DNP (to the mechanism of dNA reduplication). Biofizika **8**:140–141.
99. STREISINGER, G., R. S. EDGAR, AND G. HARRAR. 1961. The organization of DNA in bacteriophage and bacteria by C. A. Thomas, Jr., ref. p. 132. *In* J. H. Taylor [ed.], Molecular biology, part I. Academic Press, Inc., New York.
100. STRELZOFF, E., AND F. J. RYAN. 1962. The necessary involvement of both complementary strands of DNA in the specification of messenger RNA. Biochem. Biophys. Res. Commun. **7**:471–476.
101. SWARTZ, M. N., T. A. TRAUTNER, AND A. KORNBERG. 1962. Enzymatic synthesis of deoxyribonucleic acid. XI. Further studies on nearest neighbour base sequences in deoxyribonucleic acids. J. Biol. Chem. **237**:1961–1967.
102. TAMANN, N. 1903. Kristallisieren und Schmelzen, p. 315–344. Bart, Leipzig.
103. TAYLOR, J. H. 1962. Chromosome reproduction. Intern. Rev. Cytol. **13**:39–73.
104. VIGNAIS, P., E. BARBU, J. BASSET, AND M. MACHEBOEUF. 1951. Modifications de l'état des acides ribonucléiques sous hautes pressions en présence ou en absence de nucléodépolymérase. Compt. Rend. **232**:2364–2366.
105. VIGNAIS, P., AND M. MACHEBOEUF. 1954. Influence des hautes pressions hydrostatiques sur l'action enzymatique de la ribonucléase. Ann. Inst. Pasteur **86**:180–188.
106. WATSON, J. D., AND F. H. C. CRICK. 1953. Molecular structure of nucleic acids. A structure for deoxyribose nucleic acid. Nature **171**:737–738.
107. WOOD, W. B., AND P. BERG. 1962. The effect of enzymatically synthesized ribonucleic acid on amino acid incorporation by a soluble protein-ribosome system from Escherichia coli. Proc. Natl. Acad. Sci. U.S. **48**:94–104.
108. YOSHIDA, A., AND C.-G. HEDÉN. 1962. Some serological properties of staphylococcal cells and cell walls. J. Immunol. **88**:389–393.
109. YOSHIDA, A., C.-G. HEDÉN, B. CEDERGREN, AND L. EDEBO. 1961. A method for the preparation of undigested bacterial cell walls. J. Biochem. Microbiol. Technol. Eng. **3**:151–159.
110. YOUNG, I. E., AND P. C. FITZ-JAMES. 1959. Pattern of synthesis of deoxyribonucleic acid in Bacillus cereus growing synchronously out of spores. Nature **183**:371–373.
111. ZOBELL, C. E., AND F. H. JOHNSON. 1949. The influence of hydrostatic pressure on the growth and viability of terrestrial and marine bacteria. J. Bacteriol. **57**:179–189.
112. ZOBELL, C. E., AND R. Y. MORITA. 1957. Barophilic bacteria in some deep sea sediments. J. Bacteriol. **73**:563–568.
113. ZOBELL, C. E., AND C. H. OPPENHEIMER. 1950. Some effects of hydrostatic pressure on the multiplication and morphology of marine bacteria. J. Bacteriol. **60**:771–781.

23

Reprinted from *J. Bacteriol.*, **108**(2), 835–843 (1971)

Effects of Hydrostatic Pressure and Temperature on the Uptake and Respiration of Amino Acids by a Facultatively Psychrophilic Marine Bacterium[1]

KALA L. PAUL AND RICHARD Y. MORITA

Departments of Microbiology and Oceanography, Oregon State University, Corvallis, Oregon 97331

Studies of pressure and temperature effects on glutamic acid transport and utilization indicated that hydrostatic pressure and low temperature inhibit glutamate transport more than glutamate respiration. The effects of pressure on transport were reduced at temperatures near the optimum. Similar results were obtained for glycine, phenylalanine, and proline. Pressure effects on the transport systems of all four amino acids were reversible to some degree. Both proline and glutamic acid were able to protect their transport proteins against pressure damage. The data presented indicate that the uptake of amino acids by cells under pressure is inhibited, which is the cause of their inability to grow under pressure.

It is well recognized that marine bacteria live under varying hydrostatic pressure, but many of the bacteria isolated from the surface waters of shallow sediments can be adversely affected by pressure. Oppenheimer and ZoBell (16) subjected 63 species of marine bacteria to various hydrostatic pressures and found that they all grew better at 1 atm than at 200 atm or higher. When these 63 species were subjected to 400 atm for 8 days at 27 C in nutrient medium, 28 failed to grow, and 11 were killed. At 600 atm, under the same conditions, 23 of the 56 species that failed to grow were killed. There are many reasons presented as to why hydrostatic pressure adversely affects bacteria, including the interruption of energy production due to pressure (3), molecular volume changes that affect enzyme reaction rates (2, 5, 6, 9, 12, 13, 15), and loss of the ability to synthesize macromolecules (1, 10, 11, 17). Molecular volume decreases with decreasing temperature and with increasing pressure. Both environmental stresses occur as terrestrial forms descend through the water column. The data presented here, we believe, may indicate the primary basis for the adverse affect of pressure on many bacteria.

[1] This paper was taken in part from a dissertation by the senior author submitted in partial fulfillment of the requirements for the Master of Science degree at Oregon State University, Corvallis. Published as technical paper no. 3143, Oregon Agricultural Experiment Station. A portion of this paper was presented at the 71st Annual Meeting of the American Society for Microbiology, Minneapolis, Minn., 2 7 May 1971.

MATERIALS AND METHODS

A facultatively psychrophilic marine bacterium, designated MP-38, isolated from station NH-4 (Department of Oceanography, Oregon State University, R/V Acona, Cruise 6406) by R. D. Haight, was used for experiments on pressure and temperature effects on transport. This organism is a *Vibrio* species, and its taxonomic characteristics have been determined by R. R. Colwell (*personal communication*). MP-38 exhibits a wide temperature tolerance range; it is able to grow below 0 C and above 30 C. The optimum growth temperature [temperature at which the greatest optical density at 600 nm (OD_{600}) is reached in 8.5 hr] is about 22 C. Stock cultures of this organism were kept on agar slants at 5 C and transferred bimonthly.

Cells were grown in a dilute nutrient saline medium (NSM) containing (in g/liter): Rila Marine Salts (Rila Products Co., Teaneck, N.J.), 35; succinic acid, 0.2; polypeptone (BBL), 0.5; yeast extract (Difco), 0.3. The nutrients were kept in a sterile concentrated solution, *p*H 7.4, and diluted 1:10 with artificial seawater (ASW) before being dispensed. ASW was adjusted to *p*H 7.4 with NaOH. The dispensed medium was autoclaved at 120 C for 10 to 20 min, depending on volume, and cooled to 15 C before inoculation. ASW, in all cases, refers to a solution of Rila salts of 35 g/liter, *p*H 7.4. When a solid medium was needed, NSM was solidified with 1.5% agar (Difco).

Cells were grown for 12 hr in NSM using a 1% (v/v) inoculum from a 12-hr culture. The cells were harvested by centrifugation for 5 min at $10,400 \times g$ in a Sorvall RC2-B refrigerated centrifuge, washed twice with membrane filtered (MF) ASW, and suspended to an OD_{600} of 0.25, corresponding to approximately 6.0×10^8 to 8.0×10^8 cells/ml. Viable count versus OD was established by plate count versus OD_{600} on a spectrophotometer (Bausch & Lomb Spectronic 20). The suspension was diluted 1:10^3 with MF ASW, and ^{14}C-amino acids were added to a concentration of 200 μg/liter and 0.02 μCi/ml. The radioactive amino acids, L-glycine-*UL*-^{14}C (Amersham/Searle), L-phenylalanine-*UL*-^{14}C, L-methionine-*UL*-^{14}C, L-proline-*UL*-^{14}C, and L-glutamic acid-*UL*-^{14}C (International Chemical and Nuclear Corp.), were made up to a concentration of 2 μg/ml (0.2 μCi/ml) with cold DL-amino acid (Nutritional Biochemicals Corp.) and stored frozen at −10 C. The specific activity of each solution was 100 μCi/mg. After addition of the substrate, 10-ml portions of the iced cell suspension were drawn into plastic syringes, tightly capped, and kept on ice until pressurized. The syringes were placed in pressure cylinders (19) and pumped up to the desired pressure. The blank was treated as a 500-atm sample, only up to temperature equilibration time, after which the cylinder was depressurized. The syringe contents were then emptied into 50-ml serum bottles containing 0.6 ml of 0.1 N H_2SO_4 to fix the cells and release CO_2. The bottles were quickly stoppered with rubber serum caps equipped with a plastic bucket containing a piece of fluted Whatman no. 1 filter paper (25 by 57 mm). At least 10 min after fixing, 0.2 ml of phenylethylamine was introduced with a syringe onto the filter paper through the serum cap for $^{14}CO_2$ adsorption. All serum bottles were left sealed at room temperature at least 1 hr before uncapping. (4). The fluted filter papers were placed in scintillation vials and counted in 5 ml of scintillation fluor containing 5 g of 2,5-diphenyloxazole (PPO) per liter and 0.3 g of triphenyldioxazole (POPOP) in toluene. Twelve hours of incubation of the filter papers in fluor at 15 C was necessary to permit clearing of the fluor and stabilization of counts. Cells were collected by filtration through membrane filters (0.45 μm, 25 mm diameter), air dried at 60 C for 2 hr, and counted in 2 ml of fluor in a liquid scintillation counter (Nuclear Chicago Mark I). The counting efficiency was determined from channels ratio by using toluene-*UL*-^{14}C (International Chemical and Nuclear Corp.) quenched with chloroform.

The viability of MP-38 under pressure was checked by plate count to make certain that no substantial decrease in viable cell number occurred during the course of the experiments. Experiments on the effect of substrate concentration or exposure time were set up in the same manner described above, with varied parameters. Experiments were performed to determine whether preincubation with substrate could protect the transport system from pressure inhibition. Samples were prepared as above, incubated with ^{14}C-glutamate or ^{14}C-proline for 3 hr at 5 C, 1 atm, and pressurized to 500 atm. Cells were depressurized and fixed at 3 and 7 hr after pressurization. To determine whether pressure damage was reversible or irreversible, samples in syringes were incubated at 5 C under 500-atm pressure for 5 hr. The pressure was then released, and the samples were allowed to incubate further at 1 atm for 6 hr. Sampling was done on release of pressure and during atmospheric incubation.

In these experiments, uptake of ^{14}C material into the cell was measured in disintegrations per minute. The rate of uptake is measured in disintegrations per minute of ^{14}C material taken up per hour. Relative uptake refers to uptake under pressure, taken as a percentage of the uptake of an identical sample at atmospheric pressure.

RESULTS AND DISCUSSION

Figure 1 presents the uptake of a variety of amino acids by MP-38, indicating both the specificity of transport proteins and their differing substrate affinities. These data include a CO_2 correction for uptake. MP-38 shows a marked preference for amino acids over the other organic substances presented.

Most pressure studies with MP-38 were done with glutamic acid because of the key role this amino acid plays in cellular metabolism. Pressure-viability studies over the temperature range of 5 to 25 C showed that MP-38 remained viable for at least 5 hr at 5 C up to 400 atm. At 20 C, the organism grew within 3 hr at pressures below 200 atm but started to expire at 400 atm after 2 hr. At 25 C, MP-38 grew at atmospheric pressure but died at 200 and 400 atm after 3 hr. The following experiments were run for a period of time at each temperature such that there was no change of cell numbers during the time of exposure.

Glutamate uptake by MP-38 was linear with concentration up to 200 μg/liter, the highest concentration used (Fig. 2). Uptake was also linear with time for the time interval and over the temperature and pressure range used in these experiments (Fig. 3 and 4). This information, taken with the viability studies, indicates that there were no substantial numerical alterations in the bacterial population during the course of the experiment and that decreases in uptake resulted from the effects of temperature and pressure. The linearity of uptake with time also indicates that the event or events which accounted for decreased uptake with application of pressure happened at the outset of the timed period. There was no progressive pressure effect indicated for the time period observed; the uptake rate is constant for each time interval.

Figure 5 presents a comprehensive picture of glutamic acid transport over the temperature range 5 to 25 C and the pressure range of 1 to 500 atm. It can be seen readily that increased pressure decreased the total amount of glutamate taken up by the cells in a given amount of time. The continued production of CO_2 at a constant or increasing proportion of the total indicates that the cells respired under pressure. It also indicates that glutamate respiration and incorporation into cellular material are affected differently by pressure. The proportion of CO_2 increased as pressure increased at all temperatures up to 400 atm.

At temperatures below 25 C, the percentage of

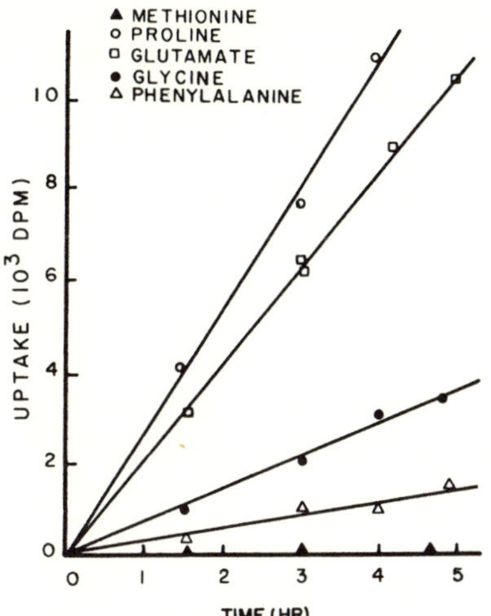

FIG. 1. *Time course of total uptake of* ^{14}*C-methionine,* ^{14}*C-glycine,* ^{14}*C-phenylalanine,* ^{14}*C-glutamate, and* ^{14}*C-proline by marine bacterium MP-38 at 5 C, 1 atm. All amino acid concentrations were 200 μg/liter with an activity of 0.02 μCi/ml in 10 ml of artificial seawater; the cell concentration was 6.0 × 10^5/ml.*

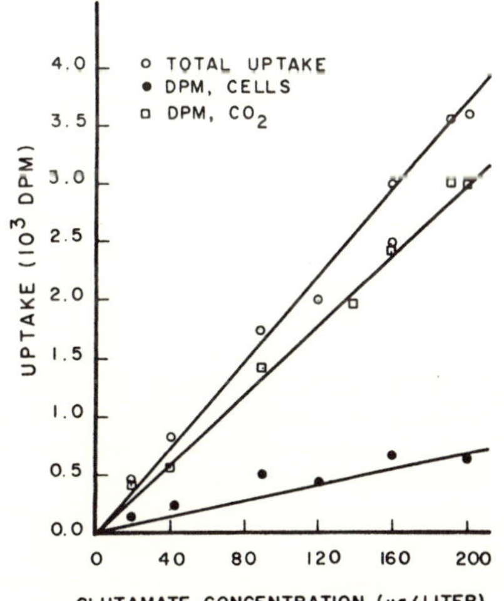

FIG. 2. *Uptake of* ^{14}*C-glutamic acid by marine bacterium MP-38 versus concentration at 15 C, 1 atm. Cell concentration was approximately 6.0 × 10^5/ml in artificial seawater.*

CO_2 dropped or remained constant between 400 and 500 atm; but at 25 C, the proportion of CO_2 increased at 500 atm. The overall increase in CO_2 production may be due to the failure of the cell to maintain a balance in the proportion of substrate which goes into any of the metabolic or synthetic pathways. The factors which alter the distribution of radioactivity may reflect the different effects of temperature and pressure on the enzymes of those pathways. Any set of conditions which favors the activity of the enzymes of the respiratory pathway or inhibits those of the synthetic pathways, or both, would result in an increased proportion of CO_2. If pressure depressed incorporation more than respiration, the availability of substrate for the respiratory enzymes would be effectively increased; that is, the competition for glutamate would be decreased. It is possible that the effects of pressure on enzyme structure, solvent structure, substrate ionization, etc., might result in increased enzyme-substrate

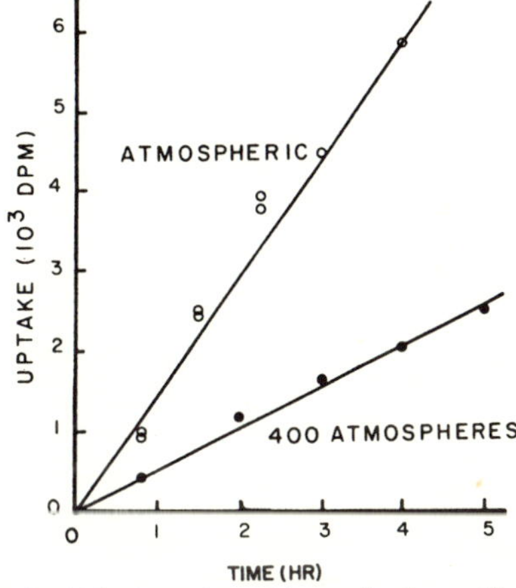

FIG. 3. ^{14}*C-glutamic acid uptake in artificial seawater by marine bacterium MP-38 at 5 C, 1 atm and 400 atm. Cell concentration was approximately 6.0 × 10^5/ml; glutamic acid concentration was 200 μg/liter with an activity of 0.02 μCi/ml in 10 ml.*

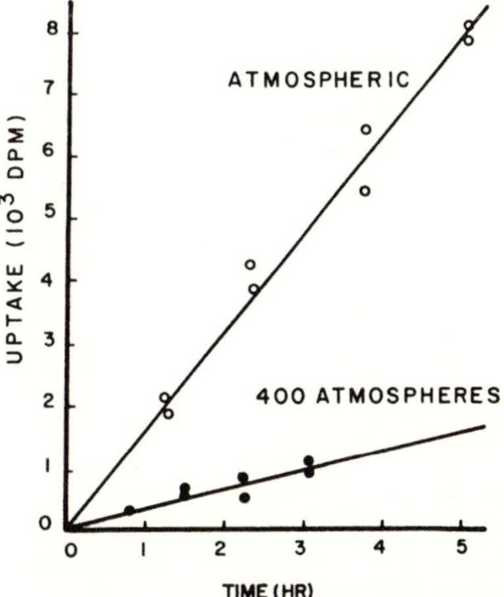

FIG. 4. ^{14}*C-glutamic acid uptake in artificial seawater by marine bacterium MP-38 at 20 C, 1 atm and 400 atm. Cell concentration was approximately 6.0 × 10^5/ml; glutamic acid concentration was 200 μg/liter with an activity of 0.02 μCi/ml in 10 ml.*

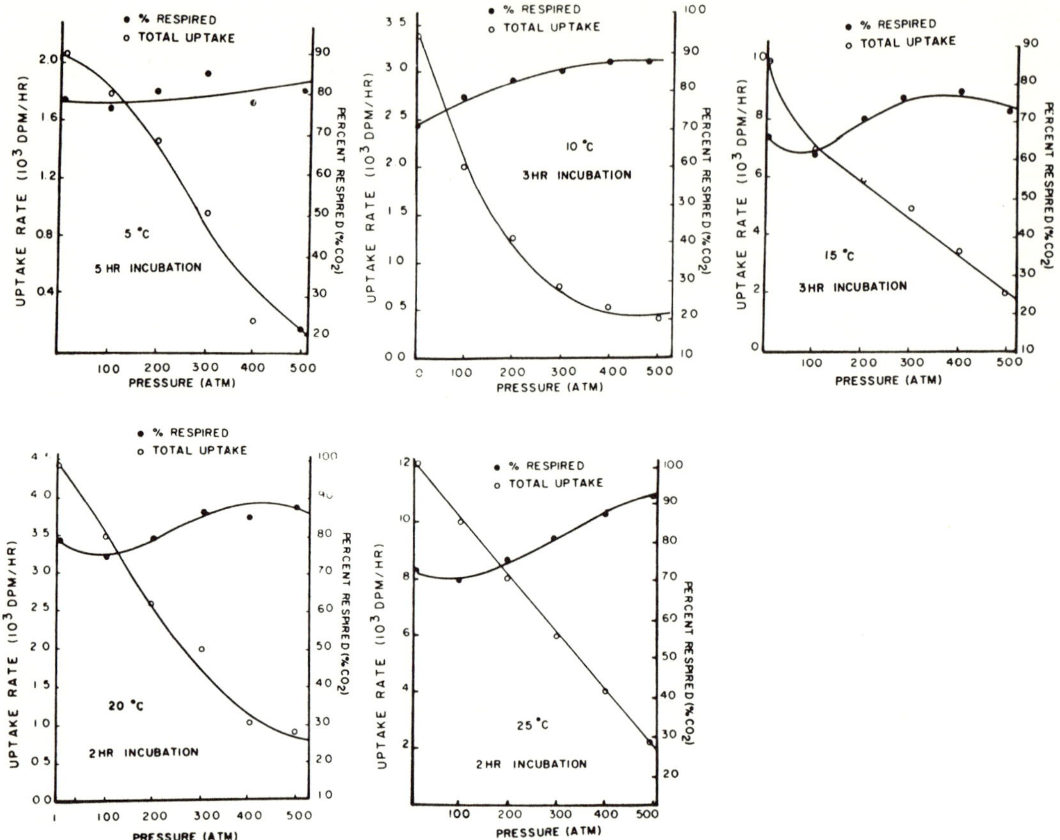

FIG. 5. *Uptake and respiration of* [14]*C-glutamic acid in artificial seawater at 5, 10, 15, 20, and 25 C. Cell concentration was approximately 6.0 × 10^5/ml; glutamic acid concentration was 200 μg/liter with an activity of 0.02 μCi/ml in 10 ml.*

affinity which would permit more effective competition for substrate on the part of the respiratory enzymes.

Carbon dioxide production may not be as affected by pressure as incorporation is, because of a possible intrinsic ability of respiratory enzymes to endure the molecular volume decrease effected by high pressure. If these enzymes, or a rate-limiting enzyme, undergo only a small molecular volume increase on forming the enzyme-substrate complex, the effects of pressure may be minimized. As the temperature is increased, the effect of pressure should be further minimized and finally offset. This may have been the case for glutamate uptake between 400 and 500 atm. At the lower temperatures, 500 atm may have been sufficient to inhibit respiration by preventing molecular volume increase. At 25 C, the effects of temperature may have compensated for the effects of pressure, permitting continued or increased activity of respiration enzymes.

Although the proportion of CO_2 increases, the actual CO_2 production decreases, as does the actual amount of glutamate incorporated. The actual amount of glutamate respired or incorporated decreases with pressure, because the total amount of glutamate taken up decreases with increasing pressure.

It is unlikely that glutamate uptake was influenced by amino acid pool levels. The high proportion of respired substrate indicates that the glutamate was rapidly utilized and spent little time in the pool. Using similar amino acid concentrations, Kay and Gronlund (7) found that pool levels did not influence amino acid uptake for *Pseudomonas aeruginosa*, also because of rapid utilization.

Figure 6 shows the uptake of glutamate as a function of temperature at 1 and 400 atm pressure. The optimal temperature of growth is lower than the optimum and maximum of the transport system, neither of which has been reached at 25 C. The temperature-pressure optimum effect discussed by Johnson (5) cannot be completely demonstrated with the data, since the optimal temperature at atmospheric pressure was not attained. However, it can be seen that, as the optimal temperature is approached by increasing the incubation temperature, the effect of pressure is decreased. The values for uptake at low temperatures are more depressed as compared with atmospheric values than are those at temperatures closer to the optimum for the glutamate transport mechanism. The inhibiting effect of pressure should be minimized at the optimal temperature; at higher temperatures, pressure should accelerate transport and protect transport proteins from heat damage (5, 15).

Studies on the transport of glycine, phenylalanine, and proline were also done at 5 C (Fig. 7). As with glutamate, increased pressure reduced the total amount of amino acid taken up by the organism. However, the respiration patterns at 5 C are different for all four amino acids. Where glutamate respiration increased up to 400 atm and then decreased slightly, glycine respiration was depressed strongly at higher pressure, proline respiration increased, and phenylalanine-CO_2 production was relatively unaffected. Since pressure affected amino acid incorporation the same amount for each amino acid, the differences in respiration patterns are probably due to intrinsic differences in the respiration enzymes for each amino acid.

The percentage of total glycine that was respired increased little with increasing pressure and was depressed strongly at higher pressures, pro-amounts for incorporation and respiration may not be altered, because both systems are equally pressure-sensitive up to 400 atm.

Phenylalanine respiration also occurred at a fairly constant proportion of the total take-up. This may indicate that the relative sensitivities of the metabolic pathways using phenylalanine were maintained under pressure. It is possible that the respiratory enzymes were functioning at their maximum capacity at each pressure level with respect to the total substrate concentration, a pseudosaturation condition.

Explanations for the increase in proline respiration are the same as those for glutamate, with some exceptions. If pressure inhibited the enzymes of proline respiration less than those of other pathways, available proline may have been metabolized at all pressures, indicating a higher degree of pressure resistance at this temperature than found for the glutamate enzymes. A similar curve would be expected if pressure stimulated proline respiration, either by increasing enzyme-substrate affinity or enzyme activity. Figure 8 indicates the differences in the response of MP-38 to these four substrates. It is possible that the four amino acid transport and metabolic systems have different temperature, pressure, and concentration optima, varying pool levels, and that any

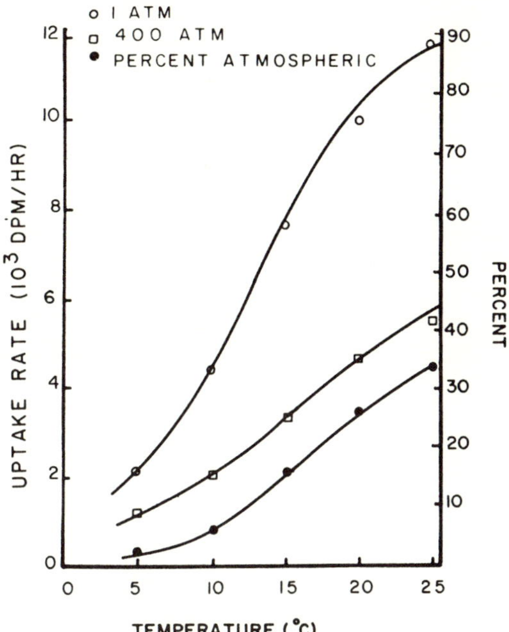

FIG. 6. *Total uptake of ¹⁴C-glutamic acid by marine bacterium MP-38 at 1 atm and 400 atm over the temperature range of 5 to 25 C. Uptake at 400 atm is plotted as a percentage of the uptake at 1 atm at the same temperature.*

given amino acid may have more than one transport mechanism.

These figures, taken together, indicate that the essential effect of high pressure is to reduce the organism's supply of nutrients at low temperature by inhibiting its transport system. Such an effect would have tremendous ecological ramifications for nutrient material in the deep sea, where the combined effects of cold and pressure could essentially prevent bacterial activity.

Figure 9 gives the reversibility patterns for each amino acid. All four transport systems exhibit reversible pressure inhibition with differing degrees of pressure damage. The proline transport system appear to be completely reversible in that the post-pressurization uptake rate is approximately the same as that at 1 atm without pressure.

However, there is a substantial decrease in the proportion of CO_2 respired compared to atmospheric conditions. The percentage respired is also lower at 1 atm than at 500, indicating the relative increase in CO_2 production with pressure for proline. The decrease of CO_2 production relative to a nonpressurized sample may indicate that some irreversible damage was done to the respiration pathway.

Glutamate, glycine, and phenylalanine transport systems all show impairment to the uptake

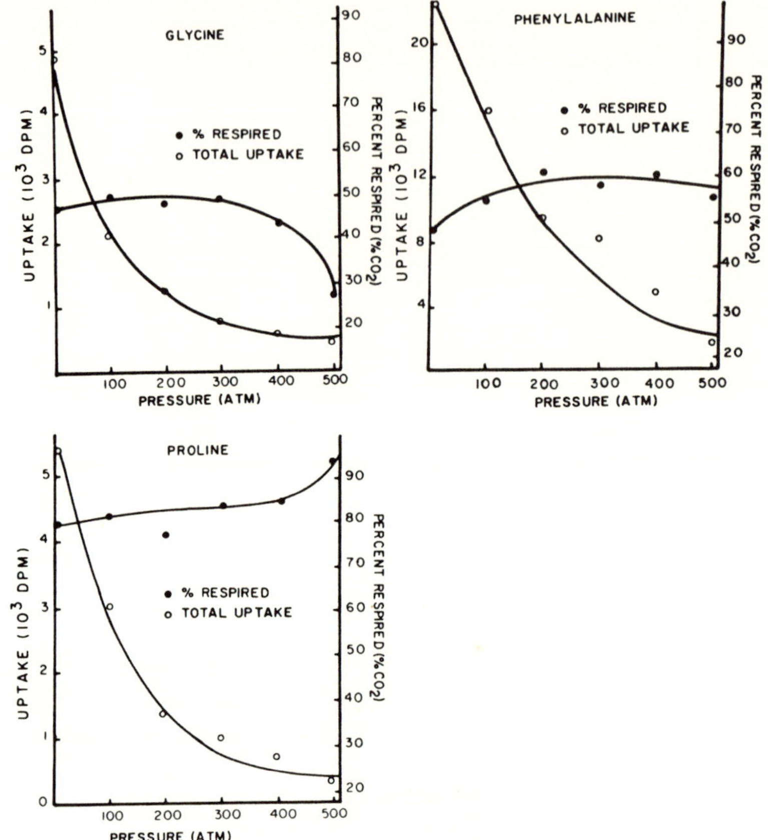

FIG. 7. *Uptake and respiration of ^{14}C-glycine, ^{14}C-phenylalanine, and ^{14}C-proline by marine bacterium MP-38 at 5 C, 5 hr of exposure. Cell concentration was approximately 6.0×10^5/ml in 10 ml of artificial seawater. Concentration of substrate was 200 μg/liter with an activity of 0.02 μCi/ml.*

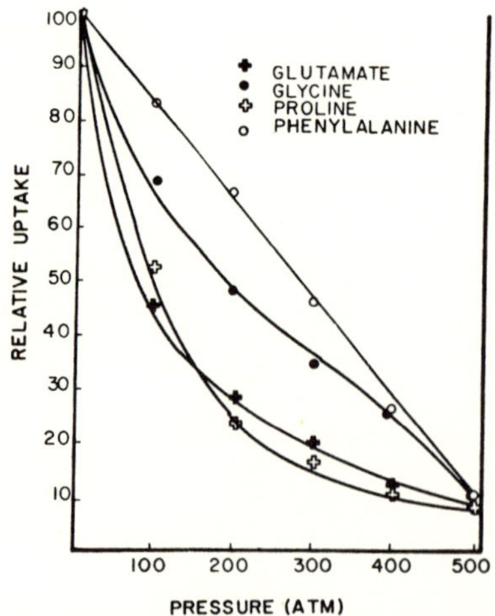

FIG. 8. *Relative uptake of four amino acids by marine bacterium MP-38, expressed as percentage of uptake at atmospheric pressure, at 5 C, 5 hr of exposure.*

mechanism, as evidenced by lowered uptake rates. The increased CO_2, relative to nonpressurized samples (percentage respired), indicates that the respiratory mechanisms are still functional and relatively undamaged after pressure release. Only glutamate transport exhibited a lag or recovery time after release of pressure.

Figures 10 and 11 show the result of exposing cells to substrate at 1 atm for 3 hr before pressurization. Counts from samples taken after exposure to 500 atm for given amounts of time indicate that both proline and glutamate were able to protect their transport proteins from pressure inhibition. The uptake rate at 500 atm without preincubation was 200 dpm/hr for glutamate and 85 dpm/hr for proline. With preincubation, the rates at 500 atm were 980 dpm/hr for glutamate

Cell concentration was approximately 6.0×10^5/ml. Substrate concentration was 200 μg/liter with an activity of 0.02 μCi/ml in 10 ml of artificial seawater.

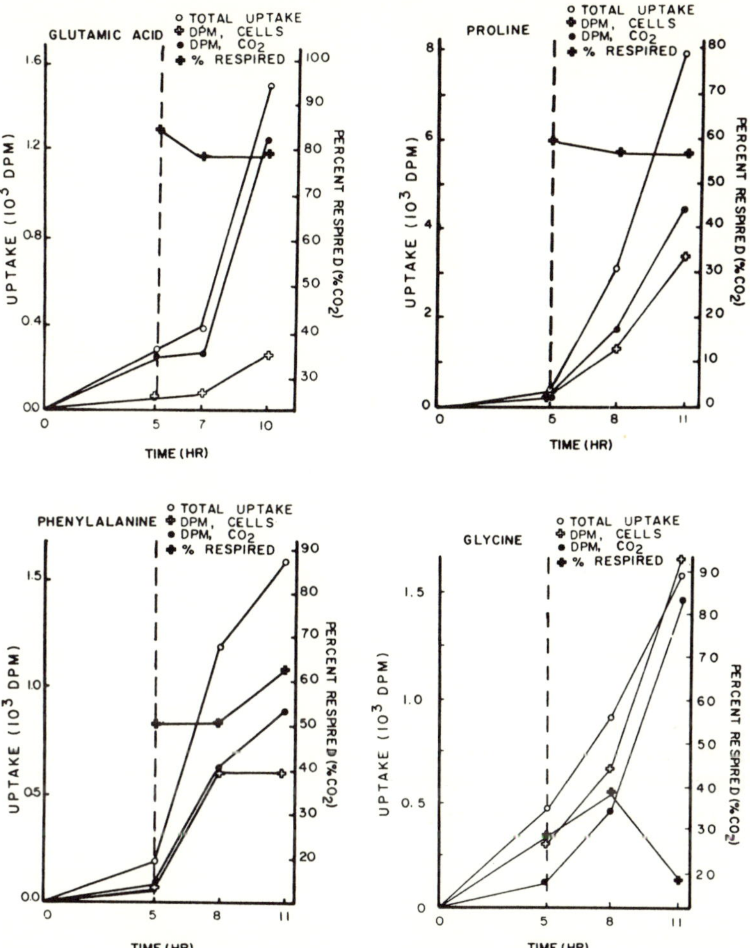

FIG. 9. *Reversibility of pressure inhibition of amino acid uptake after 5 hr of exposure to 500 atm pressure at 5 C. Cells were incubated for 6 hr after release of pressure. Cell concentration was approximately 6.0 × 10^5/ml; amino acid concentration was 200 μg/liter with an activity of 0.02 μCi/ml in 10 ml of artificial seawater.*

and 280 dpm/hr for proline. It appears that, with a large amount of available glutamate or proline, a site on the transport protein may remain charged with substrate and protected from pressure inhibition. The presence of cofactors combined with the protein in its active configuration may also help protect the transport mechanism (13).

Comparison of proline uptake and incorporation for MP-38 at 5 C, 500 atm, with and without preincubation, indicates the effect the transport system can have on data from incorporation studies. At atmospheric pressure at 5 C, MP-38 took up and incorporated into cellular material approximately 2 × 10^{-13} mg of proline per cell per hr. At 500 atm, this figure was reduced to 2 × 10^{-15} mg per cell per hr. If cells were preincubated with proline, the incorporation at 500 atm as a trichloroacetic acid precipitate was approximately 1.6 × 10^{-13} mg per cell per hr, equivalent to the amount found in whole cells (Fig. 4). The protection of the transport system by preincubation substantially changed the picture of incorporation under the same conditions. One would have to allow substrate to enter the cell before pressurization to see what effect pressure has on its use, not its uptake. Temperature variation, which also affected amino acid transport (Fig. 6) would be reflected in incorporation data. Cellular uptake of glutamate was approximately 2 × 10^{-12} mg per cell per hr at 20 C, 1 atm. At 5 C, the value was only 3 × 10^{-13} mg per cell per hr. Comparison of data for whole cells and that from trichloroacetic acid precipitates might help indicate whether transport or incorporation was the more sensitive process.

Studies of pressure effects on cellular systems using whole cells must take into consideration

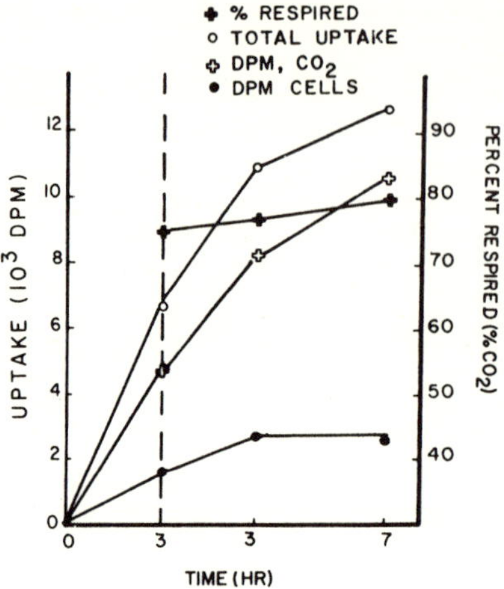

FIG. 10. *Uptake and utilization of* ^{14}*C-glutamic acid under pressure by marine bacterium MP-38 at 5 C. Cells were pressurized to 500 atm after being exposed to substrate for 3 hr at 1 atm. Cell concentration was approximately 6.0* × *10*^{5}*/ml. Glutamic acid concentration was 200 μg/liter with an activity of 0.02 μCi/ml in 10 ml of artificial seawater.*

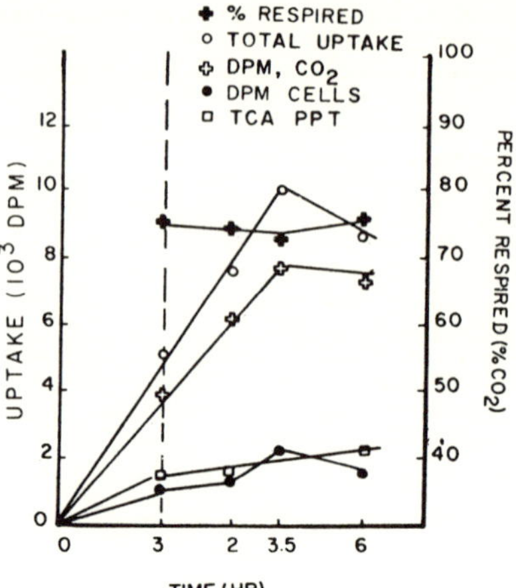

FIG. 11. *Uptake and utilization for* ^{14}*C-proline under pressure by marine bacterium MP-38 at 5 C. Cells were pressurized at 500 atm after being exposed to substrate for 3 hr at 1 atm. Cell concentration was approximately 6.0* × *10*^{5}*/ml; proline concentration was 200 μg/liter with an activity of 0.02 μCi/ml in 10 ml of artificial seawater.*

the effect of pressure on the transport system of compounds used for measuring incorporation and intracellular enzyme activity. Protective effects of substrates and cofactors should also be taken into account in intracellular and cell-free enzyme studies.

The data presented here suggest that the uptake of amino acids is inhibited by high pressure and low temperature; but the respiratory enzymes, necessary for subsequent amino acid metabolism, are not affected to any great degree. This naturally implies that we are dealing with a membrane phenomenon. Pressure and temperature have been shown to influence hydrophobic bonding (8). Hydrophobic groups of membrane proteins should also be influenced by both pressure and temperature changes. This should result in a conformational change of membrane proteins. This change is reflected in our studies by the lack of the ability of cells to take up amino acids.

The average pressure of the ocean has been calculated to be about 380 atm (18). This pressure has been shown to be strongly inhibitory to transport in many cases. It is interesting to speculate that, before the organism suffers a breakdown of its intracellular machinery due to pressure, it is already suffering from starvation because of an inability to obtain raw materials from outside the cell. If this is, in fact, the case, then a primary reason for the limitation of growth of organisms under high hydrostatic pressure in the ocean is the breakdown in the systems for transporting specific substrates to the metabolic machinery inside the cell.

ACKNOWLEDGMENTS

We are grateful to Joe Hanus for his valuable technical assistance.

This investigation was supported by the Oceanography Section, National Science Foundation, NSF grant GA-28521 and by grant NGR 38-112-017 from the National Aeronautics and Space Administration.

LITERATURE CITED

1. Albright, L. J., and R. Y. Morita. 1968. Effect of hydrostatic pressure on synthesis of protein, ribonucleic, and deoxyribonucleic acid by the psychrophilic marine bacterium, *Vibrio marinus*. Limnol. Oceanogr. **13**:637-643.
2. Haight, R. D., and R. Y. Morita. 1962. Interaction between the parameters of hydrostatic pressure and temperature on aspartase of *Escherichia coli*. J. Bacteriol. **83**:112-120.
3. Hill, E. P., and R. Y. Morita. 1964. Dehydrogenase activity under hydrostatic pressure by isolated mitochondria obtained from *Allomyces macrogynus*. Limnol. Oceanog. **9**:243-248.
4. Hobbie, J. E., and C. C. Crawford. 1969. Respiration correction for bacterial uptake of dissolved organic matter in natural waters. Limnol. Oceanog. **14**:528-532.
5. Johnson, F. H. 1970. The kinetic basis of pressure effects in biology and chemistry, p. 1-44. *In* A. Zimmerman (ed.), High pressure effects on cellular processes. Academic Press Inc., New York.
6. Johnson, F. H., H. Eyring, and M. J. Polissar. 1954. The kinetic basis of molecular biology. John Wiley & Sons, Inc. New York.
7. Kay, W. S., and A. F. Gronlund. 1969. Amino acid pool formation in *Pseudomonas aeruginosa*. J. Bacteriol. **97**: 282-291.
8. Kettman, M. S., A. H. Nishikawa, R. Y. Morita, and R. R. Becker. 1966. Effect of hydrostatic pressure on the aggregation of poly-L-valyl ribonuclease. Biochem. Biophys. Res. Commun. **22**:262-267.
9. Laidler, K. J. 1951. The influence of pressure on the rates of biological reactions. Arch. Biochem. **30**:226-236.
10. Landau, J. V. 1966. Protein and nucleic acid synthesis in

Escherichia coli: pressure and temperature effects. Science **153**:1273 1274.

11. Landau, J. V. 1970. Hydrostatic pressure on the biosynthesis of macromolecules, p. 45 70. *In* A. Zimmerman (ed.), High pressure effects on cellular processes. Academic Press Inc., New York.
12. Morita, R. Y. 1957. Effect of hydrostatic pressure on succinic, formic, and malic dehydrogenases in *Escherichia coli*. J. Bacteriol. **74**:251 255.
13. Morita, R. Y. 1967. Effects of hydrostatic pressure on marine microorganisms. Oceanogr. Mar. Biol. Annu. Rev. **5**:187 203.
14. Morita, R. Y., and R. R. Becker. 1970. Hydrostatic pressure effects on selected biological systems, p. 71 80. *In* A. Zimmerman (ed.), High pressure effects on cellular processes. Academic Press Inc., New York.
15. Morita, R. Y., and R. D. Haight. 1962. Malic dehydrogenase activity at 101 C under hydrostatic pressure. J. Bacteriol. **83**:1341 1346.
16. Oppenheimer, C. H., and C. E. ZoBell. 1952. The growth and viability of sixty-three species of marine bacteria as influenced by hydrostatic pressure. J. Mar. Res. **11**:10 18.
17. Pollard, E. C., and P. K. Weller. 1966. The effect of hydrostatic pressure on the synthetic processes in bacteria. Biochim. Biophys. Acta **112**:573 580.
18. ZoBell, C. E. 1961. Importance of microorganisms in the sea, p. 107 132. *In* Proceedings of a low temperature microbiology symposium, Camden, N.J., 1961. Campbell Soup Company, Camden, 1962.
19. ZoBell, C. E., and C. H. Oppenheimer. 1950. Some effects of hydrostatic pressure on the multiplication and morphology of marine bacteria. J. Bacteriol. **60**:771 781.

24

Reprinted from *Biochim. Biophys. Acta,* **238**(2), 347–354 (1971)

HYDROSTATIC PRESSURE EFFECTS ON THE TRANSLATION STAGES OF PROTEIN SYNTHESIS IN A CELL-FREE SYSTEM FROM *ESCHERICHIA COLI*

ROBERT M. ARNOLD AND LAWRENCE J. ALBRIGHT

Department of Biological Sciences, Simon Fraser University, Burnaby 2, British Columbia (Canada)

(Received November 30th, 1970)

SUMMARY

Hydrostatic pressure inhibits the binding of aminoacyl transfer RNA to ribosomes in a cell-free system from *Escherichia coli*, and has an even greater inhibitory effect on peptide bond formation. Evidence that pressure decreases the stability of the messenger RNA–ribosome complex is also presented.

INTRODUCTION

Studies of the effects of hydrostatic pressure on macromolecular synthesis in *Escherichia coli* and in marine bacteria[1–3] have indicated that protein synthesis in whole cells is inhibited at pressures which do not markedly affect nucleic acid synthesis. The mRNA translation system appears to be particularly sensitive to pressure, and LANDAU[2], using whole cell preparations of *E. coli*, has shown that translation is totally inhibited at 670 atm although it is unaffected at 265 atm.

Palmer and Albright (unpublished data) have studied the phenylalanyl-tRNA synthetase reaction in cell-free preparations of *E. coli* B/r and *Vibrio marinus* MP-1 and concluded that this reaction is not the limiting step in protein synthesis at pressures up to 600 atm.

This communication describes the use of a cell-free system from *E. coli* B/r in determining the effects of hydrostatic pressure on the stability of the mRNA–ribosome complex, the binding of aminoacyl-tRNA to the mRNA-ribosome complex, and the subsequent formation of peptide bonds between adjacent amino acid residues.

MATERIALS AND METHODS

Culture and growth medium

E. coli B/r was obtained by courtesy of Dr. D. J. Clark, Department of Microbiology, University of British Columbia and grown in medium described by LITTAUER AND EISENBERG[4].

11 l of this medium were inoculated with 1 l of an overnight culture of *E. coli* B/r. The cells were grown at 37° with vigorous aeration, harvested in early log phase

by centrifugation, and washed twice with 0.01 M Tris–HCl buffer, pH 7.5, containing 0.01 M magnesium acetate and 0.006 M 2-mercaptoethanol (buffer I). The washed cells were stored at −15° as a paste.

Preparation of a ribosome suspension

The frozen cells were ruptured by grinding with alumina as previously described by ALLENDE *et al.*[5] and an S-30 fraction, freed of endogenous mRNA by incubation was prepared by the method of NIRENBERG AND MATTHAEI[6]. Ribosomes were sedimented by centrifugation at 150 000 × *g* for 150 min, and the supernatant solution (S-150) was retained as a source of transfer factors. The ribosome pellet was suspended at 4° in 0.1 M Tris–HCl buffer, pH 7.8, containing 0.5 M NH_4Cl, 0.01 M $MgCl_2$, and 0.006 M 2-mercaptoethanol and collected by centrifugation. The suspension and sedimentation procedures were repeated three times, and the ribosome pellet was finally suspended at a concentration of approx. 20 mg/ml in Buffer I, and stored in liquid nitrogen.

Separation of transfer factors

Transfer factors were separated from the S-150 supernatant solution by ammonium sulphate precipitation, as described by RAVEL[7]. The resulting 40–65 % $(NH_4)_2SO_4$ fraction contained about 20 mg/ml of protein, determined by the method of LOWRY *et al.*[8], and was stored at −15°.

Assay for binding of phenylalanyl-tRNA

Bound [^{14}C]Phe-tRNA was measured by a modification of the method of NIRENBERG AND LEDER[9]. The reaction mixture contained, in a total volume of 0.9 ml: Tris–HCl buffer, 0.05 M, pH 7.7; 2-mercaptoethanol, 0.012 M; NH_4Cl, 0.08 M; KCl, 0.08 M; $MgCl_2$, 0.012 M; poly U (Sigma Co.), 3 μg; washed ribosomes, 0.1 mg; transfer factors, 50 μg protein from the 40–65 % $(NH_4)_2SO_4$ fraction; GTP, 0.1 mM; L-[U-^{14}C]phenylalanyl-tRNA (New England Nuclear Corp.), 0.1 mg, charged with 60.5 pmoles of L-[U-^{14}C]phenylalanine.

Reaction tubes were kept on ice and [^{14}C]Phe-tRNA was added last to initiate binding. After a 10 min incubation at 20° the assay tubes were placed on ice and the reaction mixtures were diluted tenfold with 0.05 M Tris–HCl buffer, pH 7.7, containing 0.16 M NH_4Cl and 0.012 M $MgCl_2$ (Buffer II). The diluted reaction mixtures were filtered immediately through Millipore filters (pore size 0.45 μm) which were then washed with three 3-ml portions of ice-cold Buffer II. The filters were dried at 60° for 15 min and dissolved in 15 ml of a toluene-based scintillation cocktail containing 2,5-diphenyloxazole (0.4 %), 1,4-bis-(5-phenyloxazolyl-2)-benzene (0.01 %), and 2-methoxyethanol (6 parts per 10 parts toluene–2,5-diphenyloxazole–1,4-bis-(5-phenyloxazolyl-2)-benzene). Counting was done with a Beckman LS-250 scintillation counter.

Assay for the binding of Phe-tRNA under pressure

Culture tubes (6 mm × 15 mm) containing the reaction mixtures described above were stoppered with neoprene stoppers (No. 124 S-1, West Co., Phoenixville, Pa.), and the contents mixed. Pressurization, using apparatus similar to that described by ZOBELL AND OPPENHEIMER[10] occurred 90 sec after addition of the [^{14}C]Phe-tRNA; depressurization after incubation for 8 min at 20° was followed, 35 sec later, by 10-fold dilution of the reaction mixtures using ice-cold Buffer II. The diluted reaction mixtures were then treated as described above.

Control experiments were conducted to assay the binding which occurred during the 90-sec period before pressurization and during the 35-sec period after depressurization, since the reactions could be neither initiated nor terminated while under pressure.

Assay for dipeptide formation under pressure

The reaction mixtures, as described above, were incubated under pressure, diluted with Buffer II, and ribosomes were collected on Millipore filters and washed as previously described. After drying, the filters were placed in 0.5 ml of 1 M NH_4OH containing 50 μg of unlabelled L-phenylalanine and 50 μg of unlabelled L-phenylalanyl-L-phenylalanine and agitated vigorously using a Vortex mixer. The filters were then removed, and the remaining solutions were evaporated to dryness over phosphorus pentoxide. The residues were then chromatographed as described by RAVEL[7]. Unlabelled-L-phenylalanine and unlabelled-L-phenylalanyl-L-phenylalanine controls gave R_F values of 0.52 and 0.83 respectively. The areas of the paper containing the radioactive samples were cut into 2 cm × 2 cm squares which were counted in a Beckman liquid scintillation counter.

Assay for the stability of the mRNA–ribosome complex under pressure

The reaction mixture contained, in a final volume of 0.9 ml: Tris–HCl buffer, 0.05 M, pH 7.7; 2-mercaptoethanol, 0.012 M; NH_4Cl, 0.08 M; KCl, 0.08 M; $MgCl_2$, 0.012 M; [^{14}C]poly U (Schwarz), 6 μg (equivalent to 136.4 nmoles of polynucleotide phosphorus); unlabelled poly U (Sigma Co.), as indicated.

The unlabelled poly U was added 2 min after binding was initiated with [^{14}C]-poly U, and the reaction mixtures were then incubated under pressure for 8 min as described above. Reaction mixtures were diluted with Buffer II and ribosomes were collected on Millipore filters and counted as previously described.

RESULTS

Binding of Phe-tRNA to ribosomes under pressure

A marked inhibition of binding of Phe-tRNA in response to increasing pressure is observed as shown in Fig. 1. The data are not corrected for binding which occurs during the 90-sec period before, and the 35-sec period after pressurization. Within these time intervals, 8.5 and 3.5 pmoles respectively of phenylalanine are bound to ribosomes. The total, 11.9 pmoles, is greater than the observed binding at pressures in excess of 600 atm, suggesting that these pressures cause dissociation of Phe-tRNA which bound to the ribosomes in the 90-sec period before pressurization took place.

A study of kinetics of binding of Phe-tRNA (Fig. 2) shows that the reaction proceeds at a rapid rate during the first 10 min of incubation at 1 atm. Thereafter it proceeds more slowly towards an equilibrium value which is reached after about 30 min (not shown on diagram). The values obtained for binding under pressure

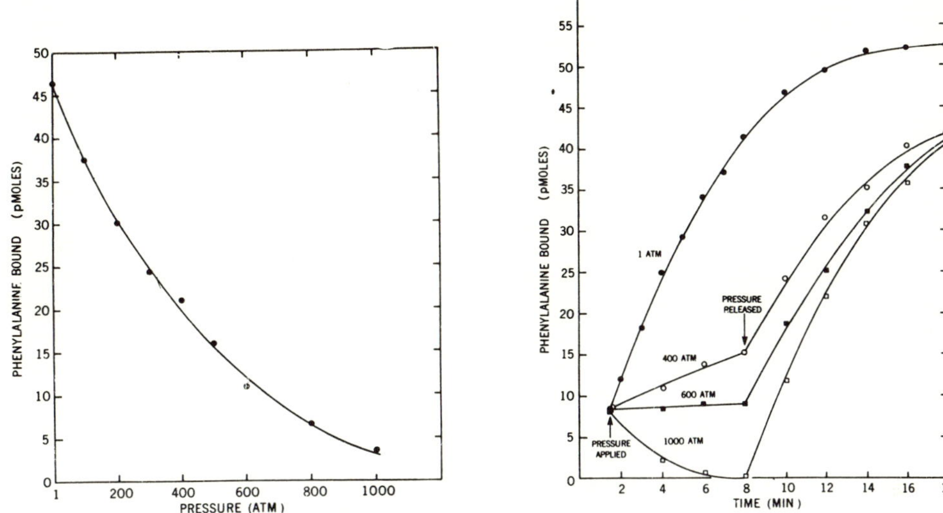

Fig. 1. The effect of hydrostatic pressure on the poly U-directed binding of [^{14}C]Phe-tRNA to *E. coli* ribosomes. Each point is the mean value obtained from 6 separate reaction mixtures.

Fig. 2. The effect of hydrostatic pressure on the rate of binding of [^{14}C]Phe-tRNA to *E. coli* ribosomes. Reactions were pressurized after 1.5 min at 1 atm and depressurized at intervals up to 6.5 min later. Values for binding under pressure have been corrected for binding which occurs in the 35-sec period between depressurization and reaction termination. Each point is the result of a duplicate assay.

have been corrected for binding which occurs in the 35 sec period between depressurization and termination of the reaction. Following depressurization, in the case of all pressures tested, binding appears to return immediately to a rate comparable to that at 1 atm. The rate of binding is markedly reduced at 400 atm, and approaches zero at 600 atm. A pressure of 1000 atm causes a rapid and complete dissociation of Phe-tRNA which binds in the 90-sec interval before pressure can be applied.

Dipeptide formation under pressure

In the experiments described above, bound phenylalanine was assayed as the sum of the contributions from single amino acid units, dipeptides, tripeptides, *etc.* In this section results are described which differentiate between the effect of pressure on the binding of Phe-tRNA to ribosomes and the subsequent effect on peptide bond formation between adjacent phenylalanine residues. It is assumed that peptides having a chain length in excess of two phenylalanine residues are present in such small quantities that their contributions may be neglected.

Both the binding of Phe-tRNA to ribosomes and the subsequent dipeptide formation are inhibited by pressure (Fig. 3). The ratio between bound phenylalanine and phenylalanylphenylalanine at pressures up to 600 atm was determined from Fig. 3, and the results are shown in Table I. Peptide bond formation is clearly more sensitive to pressure than are the aminoacyl-tRNA binding reactions. Essentially no dipeptide formation is observed at 800 and 1000 atm.

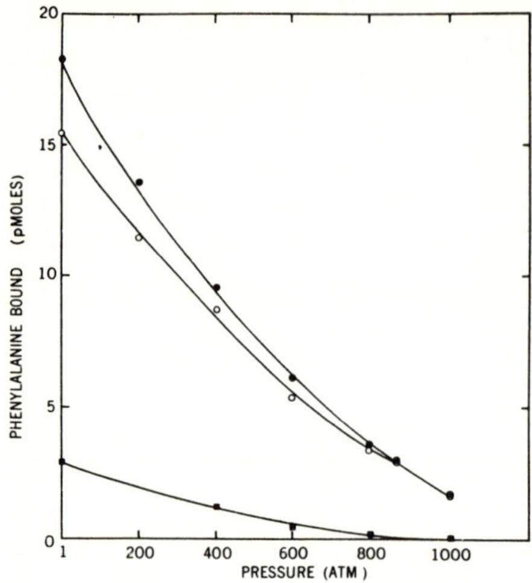

Fig. 3. The effect of hydrostatic pressure on the binding of tRNA carrying phenylalanine (○) and phenylalanylphenylalanine (■) to *E. coli* ribosomes. Total phenylalanine binding (●) is expressed as the sum of the other two curves.

TABLE I

THE EFFECT OF HYDROSTATIC PRESSURE ON THE RATIO OF PHENYLALANINE BOUND TO PHENYLALANINE POLYMERIZED IN AN *E. coli* RIBOSOMAL SYSTEM

Pressure (atm)	*Phenylalanine bound / Phenylalanine polymerized*
1	5.3
200	6.2
400	7.7
600	10.2

The effect of pressure on the total binding of phenylalanine (*i.e.* single amino acid units *plus* dipeptides) is also shown in Fig. 3. The inhibitory effect of increasing pressures closely parallels that described in Fig. 1, although the actual values in Fig. 3 are lower. This discrepancy is probably due to the inefficiency of the elution procedure used to extract the amino acid units and dipeptides from the tRNA–ribosome complex bound to the Millipore filters, and to losses of radioactive material during the chromatography and related procedures.

Stability of the messenger RNA–ribosome complex under pressure

When the technique of adding [^{14}C]poly U to a ribosomal suspension in reaction buffer, followed by pressurizing, depressurizing, and Millipore filtering was attempted, it was found that the quantity of poly U which bound to the ribosomes was constant at all pressures up to 1000 atm. It was also found that the time interval between depressurization and collection of ribosomes on Millipore filters was sufficient to allow almost complete binding of poly U to the ribosomes. These two observations suggest that either the poly U–ribosome complex is stable under pressure, or, if unstable, its dissociation is completely reversible following depressurization.

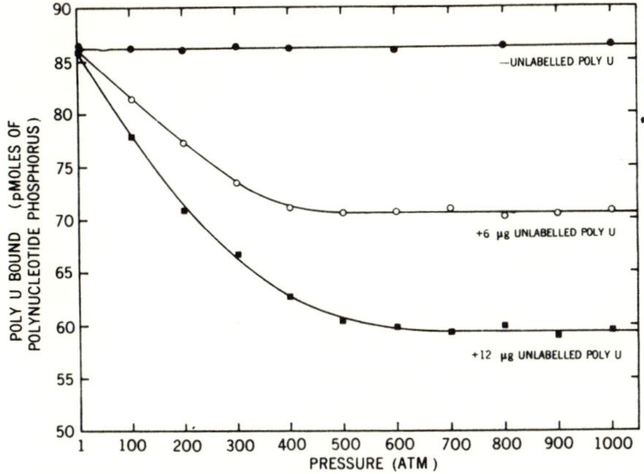

Fig. 4. The effect of hydrostatic pressure on the stability of the poly U–ribosome complex using a cell-free system from *E. coli*. The reaction mixture described in MATERIALS AND METHODS was supplemented with unlabelled poly U in the amounts indicated.

To distinguish between these alternatives, the modified procedure described in MATERIALS AND METHODS was used. The data in Fig. 4 indicate that increasing pressure causes a dissociation of poly U from the ribosomes, the maximum value of which depends on the concentration of unlabelled poly U. Control experiments in which the binding of [^{14}C]poly U at 1 atm was assayed at various time intervals following the addition of unlabelled poly U showed that there was no observable exchange between the two species of poly U during the 8 min incubation period. It is assumed that, at depressirization, there is competition between labelled and unlabelled poly U molecules for ribosomal binding sites resulting in a dilution of the label in proportion to the amount of disssociation which occurs under pressure. The observation that an increase in the concentration of unlabelled poly U results in a greater dilution of the label incorporated (Fig. 4) lends support to the assumption of dissociation under pressure, since the unlabelled species clearly acts as a better competitor for ribosomal binding sites at depressurization when present at a higher concentration.

A further set of experiments was carried out in which unlabelled poly U was added to a suspension of ribosomes in reaction buffer, followed by pressurization. After 8 min the pressure was released, and the ability of the reaction mixture to bind [^{14}C]Phe-tRNA was tested by assaying for binding at time intervals between 1 and 15 min after addition of the labelled tRNA. No difference was detected in the ability of the ribosome–poly U complex to bind Phe-tRNA at any pressure from 1 to 1000 atm.

Similar experiments showed that ribosomes subjected to various pressures from 1 to 1000 atm for an 8 min period subsequently bound [^{14}C]poly U with an equal efficiency in each case.

DISCUSSION

The results presented here indicate that pressure exerts its effect on protein synthesis by inhibiting at least three steps in the process.

Firstly, pressure decreases the stability of the mRNA–ribosome complex, causing at least partial dissociation of messenger molecules. The full extent of this dissociation is not known because, at depressurization, some of the reassociating messenger molecules will be labelled ones. This will reduce the value of the apparent dissociation as measured by the assay procedure.

The effect of pressure on the actual binding of mRNA to ribosomes has not been determined. The rate with which this reaction occurs[11] is such that it has virtually reached completion by the time pressure can be applied to the system. For this reason, emphasis has been placed on the stability of the complex under pressure, the results of which help to explain the inhibition of cell-free protein synthesis by pressure application after the start of the reaction at 1 atm.

The binding of aminoacyl-tRNA to ribosomes is more amenable to study using the techniques described here, since this reaction proceeds more slowly[11]. A similar consideration applies to peptide bond formation, for this occurs at a rate dependent on the rate of binding of the aminoacyl-tRNA molecules.

A more detailed study is needed to determine the precise nature of the effects described in this communication. It is possible that the observed inhibition of binding of Phe-tRNA under pressure (Fig. 1) may be mainly, if not entirely, due to the pressure-induced instability of the poly U–ribosome complex. However, the observation that dipeptide formation is strongly inhibited under conditions in which the availability of Phe-tRNA bound to the ribosomes is not limiting lends support to the hypothesis that pressure affects *in vitro* protein synthesis at more than one step in the synthetic sequence. A second possibility is that pressure may inhibit the activities of the transfer factors, F-I, which catalyses the binding of aminoacyl-tRNA to ribosomes[12], and F-II, which is required for the peptide bond forming reaction catalysed by peptidyl transferase[7,12].

Dissociation of the mRNA–ribosome complex under pressure may also explain the observation that pressures in excess of 600 atm cause dissociation of Phe-tRNA which binds to the ribosomes at 1 atm before pressure can applied. In the time interval between depressurization and reaction termination the reassociation of tRNA will lag behind that of messenger. Because about 13 % of the supplied tRNA binds in the 90-sec period before pressure can be applied, the effect described here can only be detected at pressures that appreciably shift the equilibrium in favour of dissociation of tRNA. For this reason it is not known whether increasing pressures between 1 and 600 atm cause less binding or whether they shift a binding–dissociation equilibrium increasingly in favour of dissociation.

There is no evidence from these results that pressures within the 1–1000 atm range applied for an 8 min period cause any permanent denaturation of ribosomes prepared from actively growing cells. It is of interest, however, to note that ribosomes isolated from cells grown to late log and stationary phases are more sensitive to pressure, as shown by the observation that a ribosome–poly U system using these ribosomes bound [^{14}C]Phe-tRNA at a reduced rate following pressurization above 500 atm, although no effect was detected below this pressure.

Comparison of the results described here with those of LANDAU[2] in which whole cell preparations of *E.coli* were used emphasizes the difficulties implicit in any attempt to explain the observed behaviour of intact cells in terms of results obtained with a cell-free system. Whereas protein synthesis in whole cells is reported to be unaffected by an applied pressure of 265 atm, this pressure is seen to cause an inhibition of the binding of Phe-tRNA to ribosomes and of dipeptide formation, and to decrease the stability of the mRNA–ribosome complex in the cell-free system described here. In both systems, however, a complete inhibition of protein synthesis is evident at pressures in excess of 600 atm.

ACKNOWLEDGEMENT

We gratefully acknowledge the financial support of the National Research Council of Canada for this project.

REFERENCES

1 L. J. ALBRIGHT, *Can. J. Microbiol.*, 15 (1969) 1237.
2 J. V. LANDAU, *Biochim. Biophys. Acta*, 149 (1967) 506.
3 E. C. POLLARD AND P. K. WELLER, *Biochim. Biophys. Acta*, 112 (1966) 573.
4 U. Z. LITTAUER AND H. EISENBERG, *Biochim. Biophys. Acta*, 32 (1959) 320.
5 J. E. ALLENDE, R. MONRO AND F. LIPMANN, *Proc. Natl. Acad. Sci. U.S.*, 51 (1964) 1211.
6 M. NIRENBERG AND H. MATHAEI, *Proc. Natl. Acad. Sci. U.S.*, 47 (1961) 1588.
7 J. M. RAVEL, *Proc. Natl. Acad. Sci. U.S.*, 57 (1967) 1811.
8 O. H. LOWRY, N. H. ROSEBROUGH, A. L. FARR AND R. J. RANDALL, *J. Biol. Chem.*, 193 (1951) 265.
9 M. NIRENBERG AND P. LEDER, *Science*, 145 (1964) 1399.
10 C. E. ZOBELL AND C. H. OPPENHEIMER, *J. Bacteriol.*, 60 (1950) 771.
11 C. SANDOR AND H. MATTHAEI, *FEBS letters*, 2 (1969) 293.
12 J. M. RAVEL, R. L. SHOREY, S. FROEHNER AND W. SHIVE, *Arch. Biochim. Biophys.*, 125 (1968) 514.

Part V

ELECTROKINETIC PROPERTIES

Editor's Comments on Papers 25 Through 28

The microscopic method (Uber, 1950) is particularly useful for gaining information about the electrokinetic properties of microorganisms and cell surfaces. Electrophoresis was discovered in 1808 by Reuss but only came into wide use approximately 40 years ago. Quantitative studies by the microscope method considerably preceded the moving boundary techniques described by Tiselius in 1937. The microscopic method remains the method of choice for the study of particles or surfaces of cells or membranes.

Several different microscopic electrophoresis instruments have been described. The first practical instrument used a flattened glass capillary for the electrophoresis cell (Northrop and Kunitz, 1925). This instrument was modified and improved greatly by fusing two glass microscope slides to form the electrophoresis cell and by using all-glass construction (Abramson, 1929a, b). The improved instrument had two great advantages in that it was easily cleaned, and there was no vibration due to rubber connections between parts. It used nonpolarizable copper, copper sulfate electrodes. A standard method for microscopic electrophoresis was suggested in 1936 by Moyer and has been used widely since that time. Abramson, Moyer, and Voet (1936) published a design for a vertical cell for microelectrophoresis which could be used for particles that did not remain in suspension. The Abramson microelectrophoresis apparatus has been the most frequently used, although other modifications have been made. The greatest improvement in design to date has been a microelectrophoresis cell that allows submersion in a water bath, thus providing accurate temperature control (Gittens and James, 1960). The modern cell adopted the silver, silver chloride electrode.

Measurement of the electrophoretic mobility of particles involves measurement of the time required to move a set distance as indicated by a calibrated ocular micrometer. The mobility of the particle must be determined at the 0.21 or 0.79 level of the cell in order that electroosmosis will not increase or decrease the velocity of the particle (Abramson, 1934a). The field strength must be determined carefully, and the mobility of the particles is recorded as either $cm^2/V \cdot s$ or as $\mu m/s/v/cm$. Although much emphasis has been placed upon calculation of the zeta potential from Henry's equation (Henry, 1931) and the net charge per unit area, mobility measurements can reveal a considerable amount of information about the particles. Particles migrating with equal mobilities are considered to possess equal zeta potentials and charge densities (Uber, 1950). It cannot be assumed that particles having the same mobility possess the same shape, size, and conductivity (Abramson and Michaelis, 1929).

Abramson (Paper 25) in 1931 reviewed the available experimental data on the influence of size, shape, and conductivity on electrophoretic mobility. The theory for the electrophoretic velocity of a particle in a given medium, initially developed by Helmholtz and later extended by Smolouchowski, was discussed in terms of the Debye–Hückel theory. Müller contributed a section to the paper to explain observed variations in electrophoretic mobility with the radius of the particle. The electrokinetic potential zeta (ζ) was demonstrated to depend on the charge density σ, the thickness of the Helmholtz double layer λ, and the dielectric constant of the medium D: $\zeta = 4 \pi \sigma D^{-1} \lambda$. Observed variations in cataphoretic velocity were due to a dependence of either σ or λ on the curvature of the surface. The relationship of the absolute electrophoretic mobility of particles to their position in the electrophoresis cell is discussed in terms of experimental results. Abramson concludes that the electrophoretic mobilities of erythrocytes are probably representative of the chemical makeup of the surface rather than of variations in the curvature of the surface.

Abramson (1934a) has discussed most of the literature. In his book he developed one of the most commonly used equations for the calculation of charge density. In a separate paper (1934b), the equation for the charge density was related to the zeta potential as follows:

$$\sigma = 2\sqrt{\frac{NDkT}{2000\pi}}\sqrt{c}\,\sinh Z\frac{e\zeta}{kT}$$

where N = Avogrado's number
k = Boltzmann's constant
T = absolute temperature
Z = valence
e = electronic charge
D = dielectric constant for the medium
c = salt concentration

Abramson, Gorin, and Ponder (Paper 26) in 1940 considered the chemistry and molecular geometry of the surface of particles and the associated effects on electrophoretic mobility. A net surface charge can occur in three ways: (1) Separation of charges may occur when ionizable and dissociable groups on the surface of the cell break down to form a residual charge if the dielectric constant of the medium, the pH, and other factors are favorable. (2) Ion-pair formation between various groups on the proteins and buffer molecules, such as phosphate, results in a shift of the net charge. (3) A process due simply to adsorption of ions to the surface.

The development of electrophoretic theory was limited to particles with perfectly smooth surfaces, yet on the surfaces of living cells there is always the possibility of "molecular" hills and valleys. However, surfaces of adsorbed protein appear to be electrokinetically identical with that of the unadsorbed, dissolved protein molecule (see Abramson, Gorin, and Moyer, 1939; Moyer and Abramson, 1938). This suggests that the bumps on the surface of the adsorbed film have the same physical significance as the curved surfaces of the free protein molecules. Abramson, Moyer, and Gorin (1942) should be consulted for a complete discussion of the electrophoresis of proteins and the chemistry of cell surfaces.

Staining techniques have been used extensively to predict the isoelectric points for bacterial cells. The apparent isoelectric point was interpreted as the pH at which the cells retained equal amounts of basic and acidic dyes. The results of such studies have been cited to demonstrate a basic difference in the isoelectric points of gram-positive and gram-negative bacteria. Harden and Harris (Paper 27) in 1953 determined the isoelectric points of 31 species of bacteria and compared their results with those in the literature. The results revealed no fundamental differences in the range of values for gram-positive and gram-negative bacteria. Apparently, the staining and electrophoretic methods do not measure the same properties. Harris (1953) found that treatment of both gram-positive and gram-negative bacteria with ribonuclease resulted in quantitative decreases in the adsorption of crystal violet by the cells and in the electric charge of the cells.

Carstensen et al. (1965) reported that the influence of environmental conductivity on the apparent conductivities of *Escherichia coli* and *Micrococcus lysodeikticus* can be explained in terms of cell wall properties. The evidence indicated that at higher medium conductivity ions from the medium invade the cell wall, thus increasing conductivity. The resultant model for this effect suggested the usefulness of dielectric techniques for the study of intact cell walls. The resistance of the bacterial cytoplasmic membrane was too great to account for the effective conductivity of *E. coli* or a *Micrococcus* species (Carstensen, 1967). Einolf and Carstensen (1967) found the surface charge of *Micrococcus* to be nearly independent of ionic strength when bacterial conductivity was considered. The conductivity of isolated cell walls of *M. lysodeikticus* was 0.40 S/m (Carstensen and Marquis, 1968). Observations of *M. lysodeikticus* protoplasts by Einolf and Carstensen (1969) showed that removal of the cell wall decreased the dielectric constant by two orders of magnitude. The conductivity of the protoplast was 0.001 S/m as compared with 0.045 S/m for the intact cell. Ou and Marquis (Paper 28) found that the cell walls of *M. lysodeikticus* expand and contract in response to changes in pH and ionic strength. The changes in volume were related to change in electrostatic interactions among the fixed ionized groups in the cell wall polymers.

REFERENCES

Abramson, H. A. 1929a. Modification of the Northrop-Kunitz microcataphoresis cell. *J. Gen. Physiol.* *12*(3): 469–472.

———. 1929b. Electrokinetic phenomena. I. The adsorption of serum proteins by quartz and paraffin oil. *J. Gen. Physiol.* *13*(2): 169–177.

———. 1934a. *Electrokinetic Phenomena and Their Application to Biology and Medicine*. Chemical Catalog Co., New York.

———. 1934b. The relation of the potential and charge of bacteria to their agglutination. *Trans. Electrochem. Soc.* *66*: 153–162.

———, and L. Michaelis. 1929. The influence of size, shape, and conductivity of microscopically visible particles on cataphoretic mobility. *J. Gen. Physiol.* *12*(4): 587–598.

———, M. H. Gorin, and L. S. Moyer. 1939. The polar groups of protein and amino acid surfaces in liquids. *Chem. Rev. 24*(2): 345–366.

———, L. S. Moyer, and M. H. Gorin. 1942. *Electrophoresis of Proteins and the Chemistry of Cell Surfaces.* Van Nostrand Reinhold, New York.

———, L. S. Moyer, and A. Voet. 1936. A vertical microelectrophoresis cell with non-polarizable electrodes. *J. Amer. Chem. Soc. 58*(12): 2362–2364.

Carstensen, E. L. 1967. Passive electrical properties of microorganisms. II. Resistance of the bacterial membrane. *Biophys. J. 7*(5): 493–503.

———, and R. E. Marquis. 1968. Passive electrical properties of microorganisms. III. Conductivity of isolated bacterial cell walls. *Biophys. J. 8*(5): 536–548.

———, H. A. Cox, Jr., W. B. Mercer, and L. A. Natale. 1965. Passive electrical properties of microorganisms. I. Conductivity of *Escherichia coli* and *Micrococcus lysodeikticus. Biophys. J. 5*(3): 289–300.

Einolf, C. W., Jr., and E. L. Carstensen. 1967. Bacterial conductivity in the determination of surface charge by microelectrophoresis. *Biochim. Biophys. Acta 148*(2): 506–516.

———, and L. Carstensen. 1969. Passive electrical properties of microorganisms. IV. Studies of protoplasts of *Micrococcus lysodeikticus. Biophys. J. 9*(4): 634–643.

Gittens, G. J., and A. M. James. 1960. An improved microelectrophoresis apparatus and technique for studying biological cell surfaces. *Anal. Biochem. 1*(1): 478–485.

Harris, J. O. 1953. Electrophoretic behavior and crystal violet adsorption capacity of ribonuclease treated bacterial cells. *J. Bacteriol. 65*(5): 518–521.

Henry, D. C. 1931. The cataphoresis of suspended particles. Part I. The equation of cataphoresis. *Proc. Roy. Soc. (London) A133*(821): 106–129.

Moyer, L. S. 1936. A suggested standard method for the investigation of electrophoresis. *J. Bacteriol. 31*(5): 531–546.

———, and H. A. Abramson. 1938. Electrokinetic aspects of surface chemistry. V. Electric mobility and titration curves of proteins and their relationship to the calculation of radius and molecular weight. *J. Biol. Chem. 123*(2): 391–403.

Northrop, J. H., and M. Kunitz. 1925. An improved type of microscopic electrophoresis cell. *J. Gen. Physiol. 7*: 729–730.

Tiselius, A. 1937. A new apparatus for electrophoretic analysis of colloidal mixtures. *Trans. Faraday Soc. 33*: 524–531.

Uber, F. M. 1950. *Biophysical Research Methods.* Wiley (Interscience Division), New York.

SELECTED READINGS

Abramson, H. A. 1929. The cataphoretic velocity of mammalian red blood cells. *J. Gen. Physiol. 12*(5): 711–725.

———. 1930. Electrokinetic phenomena. II. The factor of proportionality for cataphoretic and electroendosmotic mobilities. *J. Gen. Physiol.* *13*(6): 657–668.

———. 1932. Electrokinetic phenomena. V. A small but constant source of error in measurements of viscosity. *J. Gen. Physiol.* *15*(3): 279–281.

———. 1932. Electrokinetic phenomena. VIII. Surface conductance of cellulose and the theory of Smoluchowski. *J. Phys. Chem.* *36*: 2141–2144.

———. 1940. Microscopic method of electrophoresis and its application to the study of ionogenic and non-ionogenic surfaces. *Trans. Faraday Soc.* *36*(1): 5–15.

———, and E. B. Grossman. 1931. Electrokinetic phenomena. IV. A comparison of electrophoretic and streaming potentials. *J. Gen. Physiol.* *14*(5): 563–573.

———, and L. S. Moyer. 1937. Some recent developments in electrokinetic methods and their application to biology and medicine. *Trans. Electrochem. Soc.* *71*: 135–152.

———, F. Furchgott, and E. Ponder. 1939. The electrophoretic mobility of rabbit erythrocytes and ghosts. *J. Gen. Physiol.* *22*(4): 545–553.

———, D. H. Moore, and H. H. Gettner. 1972. A general electrophoretic pattern in extracts of pollens causing hay fever. *J. Asthma Res.* *9*(3): 153–163. [Editor's note: Several other articles by Abramson are included in the same issue of this journal: 9(3): 153–187.]

Booth, F. 1948. Surface conductance and cataphoresis. *Trans. Faraday Soc.* *44*: 955–959.

Carstensen, E. L., R. E. Marquis, and P. Gerhardt. 1971. Dielectric study of the physical state of electrolytes and water without *Bacillus cereus* spores. *J. Bacteriol.* *107*(1): 106–113.

Davies, J. T., D. A. Haydon, and E. Rideal. 1956. Surface behavior of *Bacterium coli*. I. The nature of the surface. *Proc. Roy. Soc. (London) B145*(920): 375–383.

Dozois, K. P. 1936. Variations in the electrophoretic mobilities of *Escherichia, Aerobacter* and "Intermediate" strains. *J. Bacteriol.* *31*(2): 211–215.

Dyar, M. T., and E. J. Ordal. 1946. Electrokinetic studies on bacterial surfaces. I. The effects of surface-active agents on the electrophoretic mobilities of bacteria. *J. Bacteriol.* *51*(2): 149–167.

Eckert, R. 1972. Bioelectric control of ciliary activity. *Science* *176*(4034): 473–481.

———, and Y. Naitoh. 1972. Bioelectric control of locomotion in the ciliates. *J. Protozool.* *19*(2): 237–243.

———, Y. Naitoh, and K. Friedman. 1972. Sensory mechanisms in paramecium. I. Two components of the electrical response to mechanical stimulation of the anterior surface. *J. Exptl. Biol.* *56*(3): 683–694.

Fricke, H. 1955. The complex conductivity of a suspension of stratified particles of spherical or cylindrical form. *J. Phys. Chem.* *59*(2): 168–170.

———, and H. J. Curtis. 1936. The determination of surface conductance from measurements on suspensions of spherical particles. *J. Phys. Chem.* *40*(6): 715–722.

Gortner, R. A. 1940. Electrokinetics. XIII. Electrokinetics as a tool for the study of the molecular structure of organic compounds. *Trans. Faraday Soc.* *36*: 63–68.

Hannan, P. J. 1961. Electrophoretic properties of spores of *Aspergillus niger*. *Appl. Microbiol.* *9*(2): 113–117.

Harris, J. O. 1949. The combination of dyes with growing, resting, and killed bacterial cells. *Trans. Kan. Acad. Sci.* *52*(3).

———. 1951. A study of the relationship between the surface charge and the adsortpion of acid dyes by bacterial cells. *J. Bacteriol.* *61*(6): 649–652.

———. 1956. Electrophoresis of bacterial variants. *Appl. Microbiol.* *4*(4): 161–163.

———, and R. M. Kline. 1956. Electrophoresis of motile bacteria. *J. Bacteriol.* *72*(4): 530–532.

Haydon, D. A. 1956. Surface behavior of *Bacterium coli*. II. The interaction with phenol. *Proc. Roy. Soc. (London)* *B145*(920): 383–391.

———. 1961. The surface charge of cells and some other small particles as indicated by electrophoresis. I. The zeta potential-surface charge relationships. *Biochim. Biophys. Acta* *50*(3): 450–457.

———. 1961. The surface charge of cells and some other small particles as indicated by electrophoresis. II. The interpretation of the electrophoretic charge. *Biochim. Biophys. Acta* *50*(3): 457–462.

Henry, D. C. 1948. The electrophoresis of suspended particles. IV. The surface conductivity effect. *Trans. Faraday Soc.* *44*(12): 1021–1026.

Marquis, R. E., and E. L. Carstensen. 1973. Electric conductivity and internal osmolality of intact bacterial cells. *J. Bacteriol.* *113*(3): 1198–1206.

Müller, H. 1933. The theory of ionic adsorption. *Cold Spring Harbor Symp. Quant. Biol.* *1*: 34–38.

Naitoh, Y., and R. Eckert. 1973. Sensory mechanisms in *Paramecium*. II. Ionic basis of the hyperpolarizing mechanoreceptor potential. *J. Exptl. Biol.* *59*(1): 53–65.

Offen, R. J., and A. M. Roberts. 1973. The relations between membrane potential and parameters of ciliary heat in free-swimming *Paramecium cadatum*. *J. Exptl. Biol.* *59*(3): 583–593.

Peniston, F. L., and J. O. Harris. 1954. The effect of lysozyme on the electrophoretic mobility of certain cocci. *Exptl. Cell. Res.* *6*(2): 307–310.

Roberts, A. M. 1970. Motion of paramecium in static electric and magnetic fields. *J. Theor. Biol.* *27*(1): 97–106.

Shibley, G. S. 1924. Studies in agglutination. II. The relationship of reduction of electric charge to specific bacterial agglutination. *J. Exptl. Med.* *40*(4): 453–466.

Sumner, C. G., and D. C. Henry. 1931. Cataphoresis. Part II. A new experimental method and a confirmation of Smoluchowski's equation. *Proc. Roy. Soc. (London)* *A133*(821): 130–140.

Thayer, D. W., and J. O. Harris. 1965. Electrokinetic properties of *Entamoeba invadens*. Rodhain 1934. *J. Protozool.* *12*(1): 144–145.

Wintersteiner, O., and H. A. Abramson. 1933. The isoelectric point of insulin. Electrical properties of adsorbed and crystalline insulin. *J. Biol. Chem.* 99(3): 741–753.

Wyatt, P. J. 1972. Dielectric structure of spores from differential light scattering. In *Spores V,* pp. 61–67. American Society for Microbiology.

25

Reprinted from *J. Phys. Chem.*, 35(1), 289–308 (1931)

THE INFLUENCE OF SIZE, SHAPE, AND CONDUCTIVITY ON CATAPHORETIC MOBILITY, AND ITS BIOLOGICAL SIGNIFICANCE. A REVIEW

BY HAROLD A. ABRAMSON

I. The Cataphoresis of Sub-microscopic and Microscopic Particles

Introduction. The characterization of the surfaces of microscopically visible particles by means of electrical mobility measurements is achieving more importance with the development of methods designed to measure accurately cataphoretic mobilities and streaming potentials. Thus the

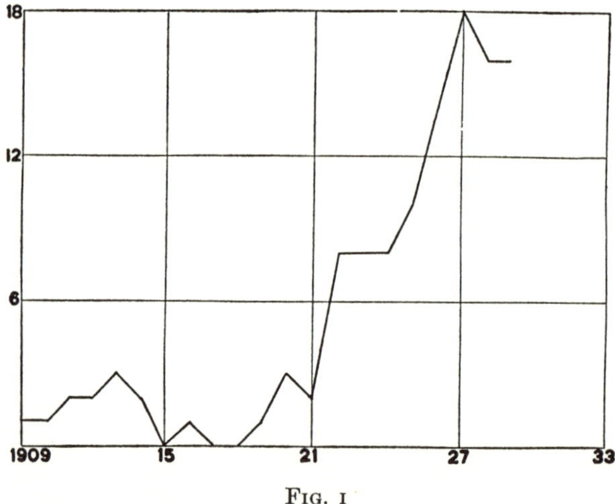

FIG. 1

The abscissa is marked off in years from 1909 to 1930. The ordinate values are the number of the references given under the heading *cataphoresis* in the yearly index of Chemical Abstracts during this period. This chart indicates the general increase in the number of investigations dealing with electrokinetic phenomena.

electrokinetic properties of cellulose particles, latex particles, oil droplets, blood cells, bacteria, inert surfaces covered more or less completely with proteins, protein particles, particles of clay and of fine metal wires have been of particular interest to the physicist, the chemist, and the biologist (Fig. 1). Indeed, the general chemical composition of the surfaces of particles suspended in liquids can be approached most satisfactorily by the study of electrokinetic phenomena.

Before satisfactory conclusions can be drawn, however, regarding the chemical composition of the surfaces of microscopic particles suspended in liquids, the influence of the size, shape, and conductivity of the particle on electrical mobility must be ascertained. It is the purpose of this communication to review data relative to these questions which the author has obtained or considered since the Symposium of 1928. Certain problems then presented are here brought to a more satisfactory solution.[1]

Theoretical. According to the theory developed first by von Helmholtz and subsequently extended by Lamb, Perrin, Pellat, and von Smoluchowski,

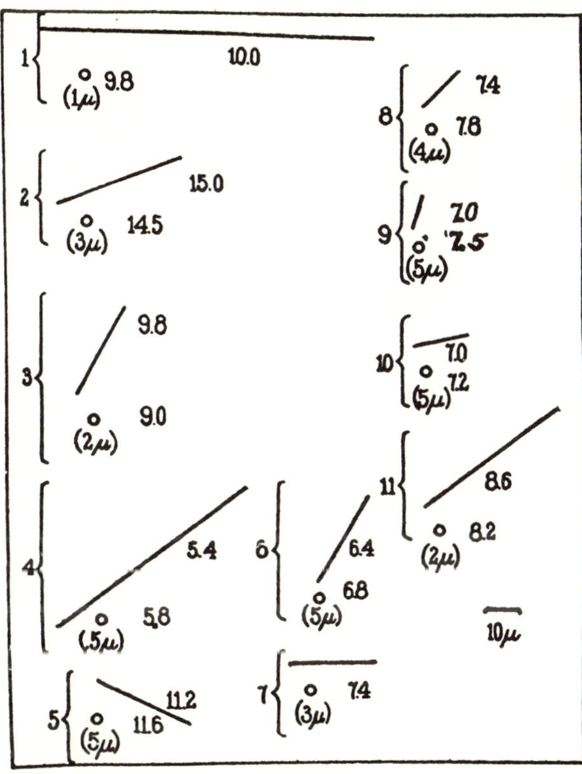

FIG. 2

The brackets indicate needles and globules studied in pairs. Nos. 1-3 are m-aminobenzoic acid crystals and mastic globules covered with gelatin. Nos. 4-10 are asbestos needles and paraffin oil globules covered with egg albumin. No. 11 is a m-aminobenzoic acid crystal and a paraffin oil globule covered with gelatin. The oil globule and mastic particles are all drawn the same size but their diameters are given in parentheses near the particle. The needles were drawn to the scale in the lower right hand corner. Each pair was studied at the same level in the cell. The numbers following each particle are the relative speeds. It is evident that when particles which vary as widely as these in size and shape have similar protein surfaces, there is no difference in cataphoretic velocity. (From the Journal of General Physiology).

the equation for V_p, the cataphoretic velocity of a particle relative to a given medium is

$$V_p = 1/4\pi \frac{XD\zeta}{\eta} = C\frac{XD\zeta}{\eta} \qquad (1)$$

(X = field strength; D = dielectric constant of the medium; ζ = electrokinetic potential; η = viscosity of the medium; all units c.g.s. electrostatic.)

[1] Abramson: Colloid Symposium Monograph, **6**, 115 (1928).

Equation (1) predicts that: (a) cataphoretic mobility should be independent of the size and shape of the particle, and (b) for similar surfaces (ion atmospheres), V_E, the electroendosmotic velocity of a liquid past the flat surface, should be equal to V_p, the velocity of the particle relative to the liquid.

Debye and Hückel,[2] on the other hand, have maintained on theoretical grounds that the constant, $1/4\pi$, in equation (1) was valid only for the cataphoresis of cylindrical particles. For spherical particles the factor $1/6\pi$ was substituted. In other words, according to this theory, the constant, C, in equation (1) is dependent upon the shape of the particle.

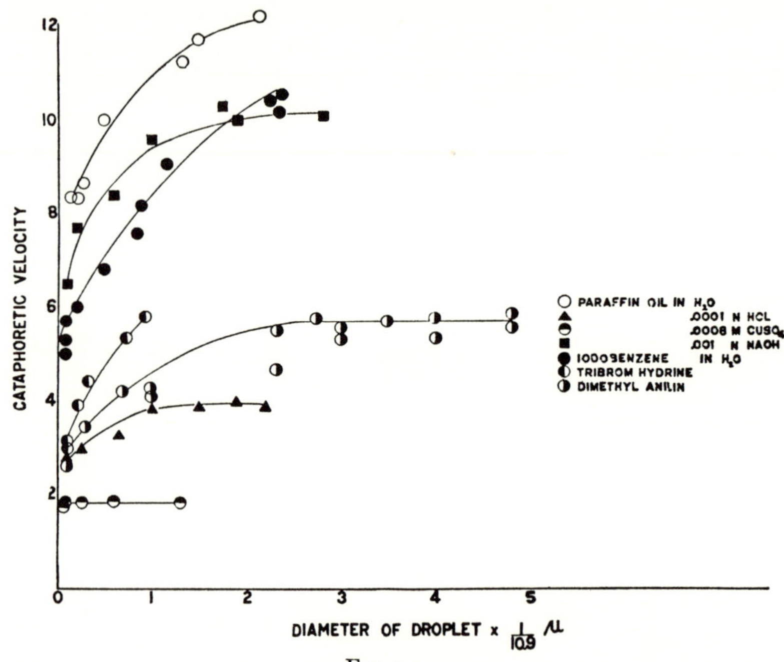

FIG. 3

Data of Mooney. Relative cataphoretic mobilities of the droplets of various emulsions in water and dilute electrolytes.

The Influence of Size and Shape of Particle on Cataphoretic Mobility. Table I summarizes the types of particles and the experiments in which the electrical mobility of the particle has been found to be independent of size and shape. The data there assembled are particularly striking if one recalls that the velocity of a spherical particle moving in a liquid in a gravitational field is proportional to the *square of the radius* and that the frictional resistance encountered by an uncharged spherical particle moving in a liquid is proportional *to the radius*. Fig. 2 illustrates a striking experiment comparing the mobilities of small spheres and very long cylinders.

Of particular importance in the conduct of experiments designed to study the influence of size and shape on electrical mobility is the technic that has

[2] Physik. Z., 25, 49 (1924).

TABLE I

A Summary of Particles that have Cataphoretic Mobilities independent of Size and Shape as determined by Experiment

Kind of Particle	Size Range (approximate)	Migration Velocity Independent of	Medium	Authority
Protein Sols	Differed sub-microscopically	Size	Acetic Acid	Hardy[3]
Silver Sols	Differed sub-microscopically	Size	Water	Burton[4]
Gas Bubbles	60μ to 160μ	Size	Water: Electrolytes	McTaggart[5]
Paraffin Oil Droplets	18μ to 28μ	Size	0.001 M NaOH	Mooney[6]
	33μ to 44μ	Size	0.0001 M NaOH	
	1μ to 16μ	Size	0.0008 M $CuSO_4$	
	16μ to 24μ		0.0001 M HCl	
Dimethyl Aniline Droplets	33μ to 50μ	Size	Water	Mooney[6]
Polymorphonuclear Leucocytes and their Aggregates	5μ to 10μ	Size and Shape	Serum; Plasma	Freundlich and Abramson[7]
Blood Platelets and their Aggregates	2μ to 15μ	Size and Shape	Plasma	Abramson[8]
Single Red Cells and their Aggregates	4μ to 30μ	Size and Shape	Serum	Freundlich and Abramson[7]
Air Bubbles in Gelatin Gels	2μ to 50μ	Size	Dilute Electrolytes	
Quartz, Glass, Clay	3μ to 15μ	Size and Shape	Dilute Electrolytes	Abramson[9]
			Distilled Water Sugar Solutions	Freundlich and Abramson[7]
Nujol	0.5μ to 30μ	Size	{ Water-Ethyl Alcohol and Phosphate Buffer	Abramson and Michaelis[10]
Castor Oil	1μ to 10μ	Size		
Benzyl Alcohol	1μ to 45μ	Size	HCl Solution and Gelatin }	
Cocoa Butter	1μ to 8μ	Size		
Nujol	1μ to 25μ	Size	HCl Solution and Gelatin	Abramson[10,11]

TABLE I (Continued)

A Summary of Particles that have Cataphoretic Mobilities independent of Size and Shape as determined by Experiment

Kind of Particle	Size Range (approximate)	Migration Velocity Independent of	Medium	Authority
Asbestos Needles	5μ to 50μ	Length	Distilled Water	Abramson and Michaelis[10]
Tyrosine Needles	2μ to 100μ	Length	Dilute HCl	Abramson[11]
Dissolved Crystallized Egg Albumin	5×10^{-7} cm (approx.)	This protein has been studied by these investigators in N/50 acetate buffer from pH 3.6 to 5.4 The mobilities of all three kinds of particles of egg albumin are very nearly alike.		Svedberg and Tiselius[12]
Gold Particles covered with Crystalline Egg Albumin	5×10^{-6} cm (approx.)			Prideaux and Howitt[13]
Quartz Particles covered with Crystalline Egg Albumin	0.5μ to 10μ			Abramson[11]
Paraffin Oil	1μ to 5μ (droplets)	When these particles which vary widely in size and shape are covered with gelatin, electrical mobility is independent of size and shape. See Fig. 2.		Abramson and Michaelis[10]
Asbestos Needles				
m-Amino-Benzoic Acid Needles	12μ to 150μ (needles)			

[3] J. Physiol., **29**, xxvi (1903).
[4] "Physical Properties of Colloidal Systems," 2nd ed., 144 (1921).
[5] Phil. Mag., **27**, 297 (1914).
[6] Phys. Rev., **23**, 396 (1924).
[7] Z. physik. Chem., **128**, 25 (1927); **133**, 51 (1928).
[8] J. Exptl. Med., **41**, 445 (1925); **48**, 677 (1928).
[9] J. Am. Chem. Soc., **50**, 390 (1928).
[10] J. Gen. Physiol., **12**, 587 (1929).
[11] Unpublished data.
[12] J. Am. Chem. Soc., **48**, 2272 (1926).
[13] Proc. Roy. Soc., **126**, A, 126 (1929).

been developed to insure chemical uniformity of surface. In the case of solid particles like quartz, glass particles, asbestos needles, and tyrosine needles electrical mobility is independent of size and shape in the size range indicated[14] regardless of the medium. Emulsions behave differently. Mooney[6] found that the migration of paraffin oil droplets increased with increasing radius. The addition of a trace of gelatin or other protein which is adsorbed by the droplets to these systems gives all the droplets, regardless of size, the same migration velocity. The gelatin, in all probability, gives the oil droplets similar surfaces. That in these instances the protein surfaces behave during cataphoresis very much like the dissolved protein has been pointed out previously. This point will be amplified further on.

Electrolytes have the same effect as the proteins, for in their presence the differences between large and small oil droplets become less or disappear.

The Cataphoretic Mobility of Oil Droplets. Figure 3 summarizes the data Mooney has obtained for the cataphoretic velocities of the droplets of various emulsions. The electrical mobility, as Fig. 3 demonstrates, increases with increase in radius. It seems likely that this variation in mobility with size must be related to changes in surface rather than to a purely frictional phenomenon. Even though it has been shown in the previous section that mobility is independent of size and shape for certain microscopically visible particles, and even though there does not seem to be an easily measureable difference between dissolved protein of radius about 10^{-6} cm and adsorbed protein of particle radius to about 10^{-2} cm this can by no means be extended to all particles below the size range investigated in particular.

The meaning of the differences in behavior between the particles listed in Table I and the oil droplets investigated by Mooney has been discussed with Professor H. Müller of the Massachusetts Institute of Technology. Before our discussion with Professor Müller we had believed that the differences were primarily due to a variation in charge density, σ, with curvature of the surface of the droplets. This viewpoint is discussed in section 2 below. Professor Müller has contributed the viewpoint presented in section A following. It is with much pleasure that we are able to incorporate his treatment here.

A.[15] According to Debye and Hückel[16] the cataphoretic velocity of a colloidal particle of spherical form is

$$V_p = \frac{1}{6\pi} \frac{\cdot XD\zeta}{\eta} \tag{2}$$

[14] Tyrosine particles less than 2μ in length may move more slowly than the larger needles, according to the results of a preliminary investigation. If one could obtain tyrosine needles smaller than 2μ, but still perfectly formed crystals, it seems likely that the mobility of these small needles would be like that of the larger ones. Pulverization of a crystal changes the surface energy and consequently, also, the ions adsorbed. It would be of interest to study the solubility of finely pulverized tyrosine. The amphoteric nature of this substance may lead to an unusual increase in solubility.

[15] Section A is a personal communication from Müller, Massachusetts Institute of Technology, May 28, 1930.

[16] Physik. Z., 25, 49 (1924); Hückel: 25, 205 (1924).

To explain the observed variation of the mobility with the radius we note that in this formula only the electrokinetic potential, ζ, may depend on the radius. This potential is due to the electric double-layer around the particle. It therefore depends on the charge density, σ, *i.e.*, the electric charge per cm,^{2}. and on λ, the thickness of the double layer. If we consider the double layer as a rigid system in the sense of a Helmholtz double layer, we would have

$$\zeta = \frac{4\,\pi\,\sigma}{D}\lambda. \tag{3}$$

There are consequently two possibilities to explain the observed variation in cataphoretic velocity according to this formulation. It may be due to a dependence of σ or of λ on the curvature of the surface.

In fact we can show that either quantity may be responsible for the observed variation.

1. Dependence of the thickness of the double layer on the radius of the particles.

According to Gouy,[17] Debye and Hückel, etc. we must assume that the electrical double layer is not rigid, but that the outer layer is "diffuse," *i.e.*, it is formed by those ions in the solvent which are electrostatically attracted to the surface. Debye shows that within this ionic atmosphere the potential, φ, decreases with the distance, r, from the center of the ion as

$$\varphi = K\,\frac{e^{-\kappa r}}{r}, \tag{4}$$

where

$$\kappa^2 = \frac{4\,\pi\,N\epsilon^2}{1000\,D\,k\,T}\,\Sigma\,\gamma_i\,z_i^2$$

N = Avogadro's number; ϵ = electronic charge; k = Boltzmann constant: γ = concentration of ions of the "ith" type in mol/liter having valence of the "zth" type.

For water at 0°C

$$\kappa = 0.229\sqrt{\Gamma}\cdot 10^8\ \text{cm}^{-1} \tag{5}$$

$\Gamma = \Sigma\,\gamma_i\,z_i^2$ = ionic strength of the electrolyte.

The constant K depends on the density of the adsorbed charges and is determined by the condition that on the surface r = R. (R = radius of particles)

$$\left(\frac{d\varphi}{dr}\right)_{r=R} = -\frac{4\pi\,\sigma}{D}. \tag{6}$$

The electrokinetic potential is the value of φ on the surface, hence

$$\zeta = K\frac{e^{-\kappa R}}{R}. \tag{7}$$

Differentiating φ and introducing in (6) gives

[17] J. Phys., (4) **9**, 457 (1910); Ann. Phys., **53**, 239 (1917).

$$\zeta = \frac{4\pi\sigma}{D}\frac{1}{\kappa}\cdot\frac{\kappa R}{1+\kappa R}. \qquad (8)$$

Comparison with (3) shows that this diffuse double layer is equivalent to a rigid double layer of the thickness

$$\lambda = \frac{1}{\kappa}\cdot\frac{\kappa R}{1+\kappa R}. \qquad (9)$$

This formula shows that, in general, the thickness of the double layer depends on the radius R of the particle. This dependence can, however, be neglected if $\kappa R >> 1$ for then $\lambda = 1/\kappa$, approximately. The dependence of λ on R will only manifest itself if κR is small. In order to find how small R has to be in order to produce any effect we plot in Fig. 3a the function $f(\kappa R) = \kappa R/1 + \kappa R$.

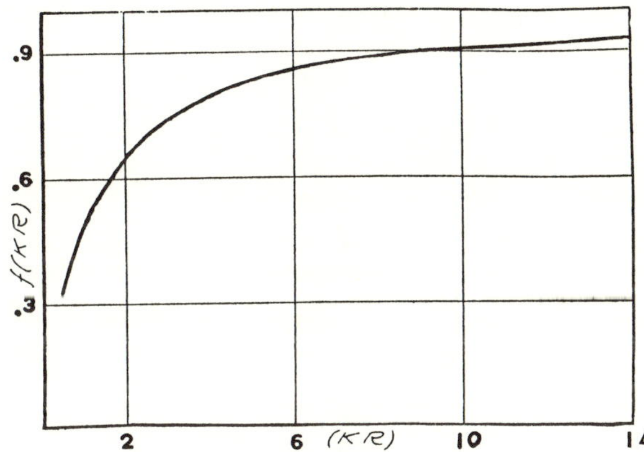

FIG. 3a
See text for explanation of this figure.

This diagram shows that the thickness of the double layer shows an observable variation with size of the particle only if

$$\kappa R \leqq 5$$

or if

$$R \leqq \frac{5}{\kappa} = \frac{2.2}{\sqrt{\Gamma}} 10^{-7}\ \text{cm} \qquad (10)$$

(for particle in water at 0°C).

Qualitatively, this theory gives, indeed a decrease of the thickness of the double layer, and hence also a decrease of the potential, ζ, and of the mobility, V_p, with decreasing radius of the particle. This is in good agreement with the observed facts. The theory also accounts for the fact that with an addition of any electrolyte the size-effect is shifted to smaller radii or disappears in the range investigated. An addition of electrolyte increases the ionic strength, Γ, hence according to equation (10) the radius, R, has to be correspondingly smaller to produce any variation.

Quantitatively however, the theory does not check very well. According to equation (10) the concentration of the electrolyte would have to be smaller than one micromol/liter in order to produce a measurable effect for particles of microscopic size ($1\mu - 10\mu$) while the effect is observed on even larger particles in solutions with probably stronger concentrations.

If the effect were due to the variation of the thickness of the double layer, it would only depend on the concentration of the electrolyte and not on the character of substance on the surface of the particle. Since the experiments indicate that the surface material is an important factor for the existence of the effect, it is more probable to find the reason for its existence in the dependence of the surface density, σ, on the curvature of the surface. (This is the end of Professor Müllers communication).

2. Dependence of the adsorbed charge density on the radius of the particle.

It is well known that the vapor pressure of very small liquid droplets is greater than the vapor pressure of large ones. The thermodynamic relationship between the vapor pressure of droplets and their radii has been given by Thomson[18]

$$\frac{RT}{M}\ln\frac{p_2}{p_1} = \frac{2S}{\rho}\left(\frac{1}{r_2} - \frac{1}{r_1}\right) \tag{11}$$

(R = gas constant; T = absolute temperature; M = molecular weight of the liquid; S = the surface energy (interfacial energy); ρ = density of the liquid; p = vapor pressure of the droplet; r = radius of the droplet.)

In equation (11) S is unknown and may be influenced by the differences in adsorption by droplets of different sizes and by the change in charge incidental to this adsorption. Similarly the vapor pressure of the liquid drops of different radii is also unknown. It is apparent, however, that the smaller droplets and solid particles can have higher vapor pressures (solubility) than the larger ones. If the adsorption of ions decreases with increasing vapor pressure, the decrease in velocity of the smaller droplets can be explained at least qualitatively as follows. If we assume S to remain constant for droplets of different radii in solutions of the same ionic strength, and the velocity of the droplet to vary inversely proportional to its vapor pressure, then, we may state,

$$\frac{RT}{M}\ln\frac{V_1}{V_2} = \frac{2S}{\rho}\left(\frac{1}{r_2} - \frac{1}{r_1}\right)$$

approximately.

It seems likely from Mooney's data that large droplets approach a limiting value, V. In that case $r_1 \rightarrow \infty$ and $V_1 \rightarrow V$. We would then have

$$\frac{RT}{M}\ln\frac{V}{V_2} = \frac{2S}{\rho r_2}. \tag{12}$$

Since Mooney's data are qualitative the validity of equations (12) can not as yet be exactly tested. From (12) it follows that

$$-\ln V_2 \; \alpha \; 1/r_2.$$

[18] Cited by H. Freundlich: Colloid Chemistry, p. 45 (1926).

TABLE II

The Bulk Conductivity of Microscopic Particles covered with Protein Films does not influence their Cataphoretic Mobility

Particle	Time to migrate given distance sec.	Medium
B*	9.4	N/100 HCl plus a trace of gelatin
Q	9.0	
Q	9.4	
B	9.0	
Q	8.8	
Q	9.6	
B	9.2	
Q	9.4	
B	9.1	
P	6.4	In equilibrium with N/50 Acetate Buffer and trace of gelatin
A	6.4	
A	6.4	
A	6.6	
P	6.7	
A	7.0	
A	6.8	
A	6.5	HCl plus trace of hemocyanine
A	6.4	
A	6.8	
P	7.2	
A	7.0	
A	7.0	
{P	7.8	N/100 HCl plus trace of gelatin
{A	7.0	
{A	7.0	
{P	6.0	
{C	6.0	
{P	5.8	
{C	5.8	
{P	4.5	
{C	4.8	

* B = benzyl alcohol, Q = quartz, A = agar particles in equilibrium with medium, P = paraffin oil, C = carbon.

That is, the negative logarithm of the droplet velocity should be inversely proportional to the radius of the droplet.

The proteins studied, as mentioned previously, hardly change their mobilities when the radii of the particle range[19] from approximately 10^{-6} cm to 10^{-3} cm. These substances acquire their charge in the systems studied not primarily by adsorption of ions but most probably by a stoichimetrical reaction. They possibly represent a special case. And it may be anticipated that when solid particles of pulverized quartz or tyrosine needles are investigated in the ultramicroscopic region, differences in mobility incidental to diminution in radii may be encountered. It is doubtful if the analogy with oil droplets is altogether strict. The pulverization of crystals produces changes in interfacial energy because of changes in the type of crystal surface made available to the solvent. What difference in mobility would exist between perfectly formed tyrosine needles 100μ and 0.01μ in length is unknown.

The influence of the bulk conductivity on cataphoretic mobility. Equation (1) is only strictly applicable to particles that are insulators. Particles homogeneous as a whole cannot be compared in order to study the effect of the conductivity of the particle on electrical mobility for the specific conductivity of the particle cannot be changed with varying the chemical composition of the phase boundary. Particles having identical surface films but varying in the specific conductivity of the enclosed bulk can be obtained by suspending various particles in dilute protein solutions. Table II summarizes experiments performed with Michaelis.[10] Particles of about the same size of quartz, paraffin oil, benzyl alcohol, carbon, and swollen agar covered with protein films, as indicated in the table, migrate independent of the bulk conductivity of the particle. In Table I the electrical mobilities of dissolved egg albumin, of submicroscopic gold particles, and of microscopic quartz particles covered with the same protein are similarly seen to be apparently unaffected by the bulk conductivity of the particle. Zakrzewski[20] has observed in a somewhat different type of experiment, however, that the streaming potentials set up in silvered glass capillaries are a function of increasing thickness of the silver film. The conditions of Zakrzewski's experiments are somewhat different from ours. But they emphasize that a good deal of similar experimentation is necessary before the problems dealing with bulk conductivity and electrokinetic phenomena are solved.

The Factor of Proportionality for Cataphoretic and Electroendosmotic Mobilities[21]

Theoretical. The data presented in the foregoing sections make it likely that microscopic particles having identical surfaces migrate independent of size and shape. This is, as previously stated, contradictory to the theory

[19] It will be shown in the next section that the range of radii of curvature over which a given protein does not change its mobility is from about 10^{-6} cm to 10^{a} cm where $a \rightarrow \infty$.

[20] Physik. Z., **2**, 146 (1900).

[21] This section has since appeared in J. Gen. Physiol., **13**, 657 (1930).

postulating that the factor C in equation (1) is a function of the shape of the particle. While these experiments deal with rather extreme variations of size and shape they are not definitely a test of that boundary condition of Debye and Hückel's theory which assumes that the radius of curvature of the cylinder is very large in comparison with the thickness of the ion atmosphere.

An experimental investigation including a test of the boundary condition involves the measurement of cataphoretic mobility of particles in a given medium simultaneously with the electroendosmotic mobility of the medium relative to a flat surface[22] having an ion atmosphere identical with that of the particle. Substituting the values of Debye and Hückel[2] for C in equation (1) and solving for R, the ratio of E_E and V_p, we obtain

$$R = V_E/V_p = 1.5$$

In other words, according to this theory electroendosmotic mobility must be 50 per cent greater than cataphoretic mobility.

Historical. Mooney[6] appears to be the first investigator to attempt to evaluate R. He found that the mobility of oil droplets was independent of size in $CuSO_4$ solutions (Fig. 3). Taking advantage of this fact, Mooney wet the inside of a round capillary tube with a paraffin oil and studied in this system the cataphoretic velocity of the oil droplets and the electroendosmotic velocity of the liquid against a surface presumably covered with oil. In one system V_p was very nearly equal to V_E.

The data of van der Grinten,[23] obtained in a flat cataphoresis cell, are in contrast to the finding of Mooney that R = 1.0 (approximately) for a round capillary. Van der Grinten studied the cataphoresis of small glass particles in distilled water, the particles composed of the same glass coverslips from which his flat cataphoresis cell had been assembled. He thus assumed that the surfaces of the particles of glass powder obtained by breaking up his coverslips were the same as that of the flat uninjured coverslips. Van der Grinten interpreted his data to give a mean value of R = 1.59, thus apparently confirming fairly well the theory of Debye and Hückel. Abramson[1] powdered pyrex glass and repeated the experiments of van der Grinten with a cell made of the same pyrex glass. This author found that for a given cataphoretic cell, R varied from 1.27 to 3.2 as a function of the nature of the medium. This cell of pyrex glass was not of uniform rectangular cross-section. The values obtained for R were consequently not considered absolute but rather pointing to the fact that a complete reinvestigation of the value of R was necessary under known hydrodynamic conditions and where the flat surface and surface of the particle were chemically identical.

The Movement of Liquids produced by Electroendosmosis in Flat Cataphoresis Cells. The movement of liquids in flat cataphoresis cells has been previously considered adequately for cells of various types by numerous investigators.

---

[22] An absolutely flat surface is, of course, not realizable experimentally.

[23] J. Chim. phys., **23**, 14 (1916).

Since the recalculations to be made here of certain data depend upon the movement of liquids in cataphoresis cells we shall briefly review the facts pertinent to our subsequent recalculations and investigations.

The theories of Ellis and von Smoluchowski (based upon an old observation of Quincke) have made possible the quantitative measurements of cataphoretic mobility. Ellis assumed that for a closed flat cataphoresis cell of depth x, the observed cataphoretic velocity of a particle was, at any level in the cell, (for a system with no turbulence)

$$V_{obs.} = V_p + V_w \qquad (13)$$

(V_p = absolute mobility relative to the liquid due to the charge, constant at all levels; V_w = the velocity of the liquid). The velocity of the liquid, as is well known, may vary from level to level so that if the electroendosmosis be in one direction, the return flow in the midregions of the closed cell must be in the opposite direction in closed systems like those considered. $V_{obs.}$ is, therefore, a function of the liquid streaming. The absolute velocity of a particle is, then, the mean velocity, M, of the particle within the cell,

$$M = 1/x_1 \int_0 V_{obs.}\, dx. \qquad (14)$$

Substituting (13) in (14)

$$M = 1/x_1 \int_0^{x_1} (V_p + V_w)\, dx = V_p + 1/x_1 \int^{x_1} V_w\, dx. \qquad (15)$$

For a closed cell $1/x_1 \int_0^{x_1} V_w dx = 0$ and since V_p is a constant for a given field strength,

$$M = 1/x_1 \int_0^{x_1} V_{obs.}\, dx = V_p \qquad (16)$$

TABLE III

Recalculation of van der Crinten's Data

Curve No.	$V_{1/5}$ μ/sec.	V_E* μ/sec.	$V_{(3/4\, V_{1/6} + 1/4\, V_{1/2})}$	V_E μ/sec.	$V_{Graphical\ Integration}$ μ/sec.	V_E μ/sec.	Mean. R
1 (Fig. III)	2.7	6.6	2.8	6.6	2.8	6.4	2.4
2 (Fig. III)	2.7	6.2	2.8	6.0	2.9	5.8	2.15
3 (Fig. III)	2.7	7.2	3.0	6.6	2.9	6.8	2.4
Fig. 4 Van der Grinten p. 228	2.5	7.2		2.5	7.4	6.8	2.8

* V_E is calculated by means of equation (4).

By measuring V at various levels, V_p may be calculated from the analytic expression relating $V_{obs.}$ to x or V_p is readily obtained by graphical integration.

Von Smoluchowski simplified the method adopted by Ellis by proposing that

$$V_p = \tfrac{3}{4} V_{1/6} + \tfrac{1}{4} V_{1/2} = V_{1/5} = V_{4/5}, \text{ very nearly,} \quad (17)$$

where the small sub-numerals represent level $\left(\text{level} = \frac{\text{depth}}{\text{total depth}}\right)$ in the cataphoresis cell. From the foregoing it can also be readily shown that

$$V_E = 2(V_{1/2} - V_p). \quad (18)$$

(V_E = mobility of the liquid relative to the wall of the cataphoresis cell.) The data of Ellis and of Svedberg and Anderssen have amply confirmed von

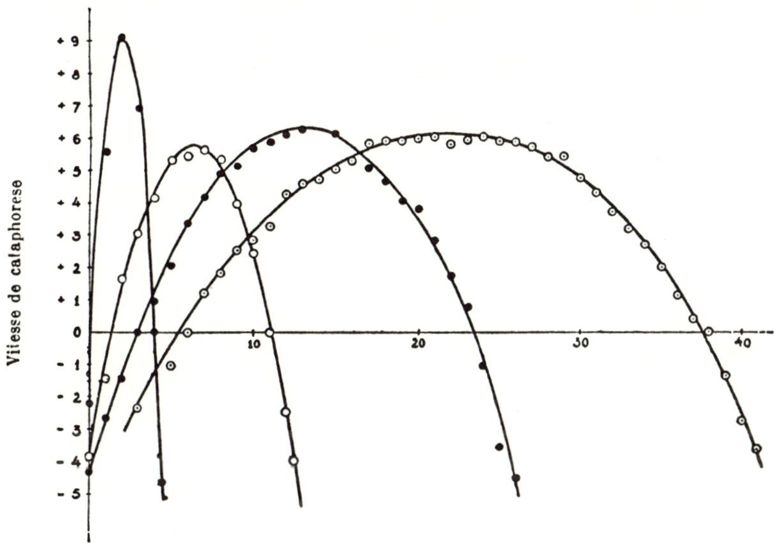

FIG. 4
Data of van der Grinten. (From J. Chim. phys.)

Smoluchowski's theory for the simple type of systems used by Ellis and by Svedberg and Anderssen. Their experiments are in accord with equation (17) in that

$$V_{1/5} = V_{4/5} = V_p = 1/x_1 \int_0^{x_1} V_{obs}\, dx \quad (19)$$

for flat cells from 50μ to about 1.0 mm thick. This means that *in cataphoresis cells of this type and depth the mobility is constant at a given level. And conversely if flat cells of different depths have mobilities in agreement with equations (18) and (19), then the absolute mobilities of the particles measured are governed by equations (18) and (19).* This is of the utmost importance in the recalculation of van der Grinten's data.

Recalculation of van der Grinten's data. It has been mentioned that van der Grinten found R = 1.5, approximately. The data submitted by van der Grinten are of the type given in Fig. 4 which is reproduced from the paper of this author. Curves 1, 2, 3, and 4 in Fig. 4 demonstrate that when van der Grinten's cataphoresis cells were more than 0.52 mm thick, the cataphoretic velocity of the particle (as well as the endosmotic velocity) remained practically constant for the same level in cells of different thicknesses. This does not mean, as van der Grinten interpreted it, that V_p, the speed of the particle relative to the liquid is that found in the midregions of the cells. It is rather, as Table III demonstrates, a further confirmation of the theory of von Smoluchowski. The table gives the values of V_p and V_E calculated by equations (17), (18), and (19) from the curves of Fig. 1 and another curve of van der Grinten's not reproduced here. It is evident that R is much greater than 1.5 in these systems and varies between 2.15 and 2.8. These data so calculated are in agreement with the author's previous experiments where similar high values of R for glass particle-glass surface systems similar to those of van der Grinten's were obtained. If one considers the data submitted by Lachs and Kronman[24] one could postulate *a prioi* that to determine R by means of a flat glass surface and glass particles would be impossible. Lachs and Kronman, after a series of careful streaming potential measurements on glass and quartz surfaces, concluded that no true electrokinetic equilibrium was reached. Furthermore, consideration of the well known sensitiveness to stresses of metal surfaces as determined by measurements of thermodynamic potentials makes it not unlikely that localized changes in surface energy due to pulverization of the glass leads to the high value of R. To determine R, a stable system was here sought where R would be independent of the electrolyte content of the medium, and where, with a reasonable degree of certainty, the particle surface and the flat surface were the same.

The Determination of V_E/V_p *for Flat Surfaces.* It has been shown by Davis,[25] Abramson,[26] and Freundlich and Abramson[27] that surfaces of quartz and glass are practically completely coated with certain proteins when in contact with dilute solutions of these proteins on either side of the isoelectric point,—the particles then acting very much like the native protein in cataphoresis experiments. The most varied substances in addition to quartz and glass coat themselves with gelatin and egg albumin. Thus cystine crystals,[11] menthol,[11] camphor,[11] oil droplets, agar, charcoal, zinc oxide powder and air bubbles behave in this fashion. Briggs[27] has also found that glass capillaries coat themslves with proteins. By suspending glass, quartz and other particles in a protein solution both particles and flat surface of the cataphoresis cell therefore can be coated with the same substance fulfilling the condition of chemical similarity of glass and particle surface.

In the experiments to be reported, R was determined in two different flat cataphoresis cells of uniform cross-section. One of the cells was a cemented cell, similar in arrangement to that described by Northrop[28] and constructed

[24] Ext. Bull. l'acad. Pol. Sci. Lettres (A), **1925**, 289.

[25] J. Physiol., **58**, xvi (1923).

[26] J. Am. Chem. Soc., **50**, 390 (1928); J. Gen. Physiol., **13**, 169 (1929).

[27] Briggs has obtained streaming potentials with protein-coated glass capillaries remarkably similar to those obtained by Abramson by the method of cataphoresis. The concentration and kind of electrolyte however differed in these experiments. Until the experiments are repeated under similar conditions of ionic strength and ionic species, *R* cannot be evaluated from these data; J. Am. Chem. Soc., **50**, 2358 (1928).

[28] J. Gen. Physiol., **4**, 629 (1923).

in the fashion previously described.[28] The approximate dimensions of this cell were length 7.0 cm.; thickness 0.1 cm; width 1.0 cm. The second cell, of fused glass, was the modification of the Northrup-Kunitz cell described by Abramson.[29] The approximate dimensions of this cell were: length 3.5 cm; thickness 0.08 cm; width 0.9 cm.[30]

It has been demonstrated for this type of flat cataphoresis cell that "the movements of the water and particle within the cell follow the theory of von Smoluchowski. . . . When the curve of particle velocity at different levels is parabolic, the curve of velocity as plotted against level is the same near the fused ends of the cell itself as in the middle. The stream lines of the liquid throughout the cell are therefore uniform."[29] The value of R may, therefore, be readily calculated by means of equations (17), (18), and (19).

Table IV gives the value of R for 16 experiments performed with various protein covered particles and the flat glass surfaces of the cataphoresis cells covered with the same proteins. These experiments were performed with two kinds of proteins on both sides of the isoelectric points of the proteins and in

TABLE IV

Experiments to determine R. G. = Fused Glass Cell; C = Cemented Cell

Exp. No.	Cell	Nature of System	V_P μ/sec.	V_E μ/sec.	$R=\frac{V_E}{V_P}$
1	G	Glass of cell, powdered. pH 3.6 N/50 Acetate Buffer + 0.1% Gelatin	11.1	11.8	1.08
2	C	0.004 N HCl Quartz Powder + 0.1% Gelatin	6.4	6.2	0.97
3	G	0.004 N HCl Quartz Powder + 0.1% Gelatin	8.7	8.3	0.95
4	G	Benzyl Alcohol + 0.2% Gelatin	6.0	5.4	0.90
5	G	As above	Data misplaced		0.97
6	G	Powdered Glass in Distilled Water	7.3	20.4	3.3
7	G	Quartz in M/150 pH 7.4 Phosphate Buffer + 1.3% Gelatin	1.23	1.28	1.12
8	C	As above	8.6	7.8	0.91
9	C	As above	9.6	9.1	1.06
10	C	As above but in Acetic Acid	10.0	7.8	0.78
11	C	As above	1.00	9.6	0.96
12	C	As above	10.5	11.4	1.08
13	G	0.1% Egg Albumin Quartz in M/50 Acetic Acid	3.2	3.3	1.06
14	G	As above	10.3	11.8	1.14
15	G	As above but in Phosphate Buffer	8.75	11.1	1.14
16	G	1/3% Gelatin + N/200 H_2SO_4	6.15	5.9	0.96
17	G	As above	12.0	13.9	1.16

For protein coated surfaces (except No. 10) Mn. R = 1.01 ± 0.088. Probable error ± 0.02.

[29] J. Gen. Physiol., 2, 469 (1928).

[30] The diameter of the side tubes connecting cataphoresis cells and stopcocks were large in comparison with the thickness of the cells themselves.

the presence of different cations and anions. The field strengths were also varied. The values of R for 15 of these experiments varied between 0.90 and 1.16. The sixteenth value was 0.78. The mean (excluding value 0.78) was equal to 1.01 ± 0.088 with the probable error of the mean equal to ± 0.02. These data point clearly to the conclusion that, under the given conditions, the ratio of cataphoretic to electroendosmotic velocity is very close to 1.00; and that the factor, C, in equation (1) is the same for V_p and V_E.[31]

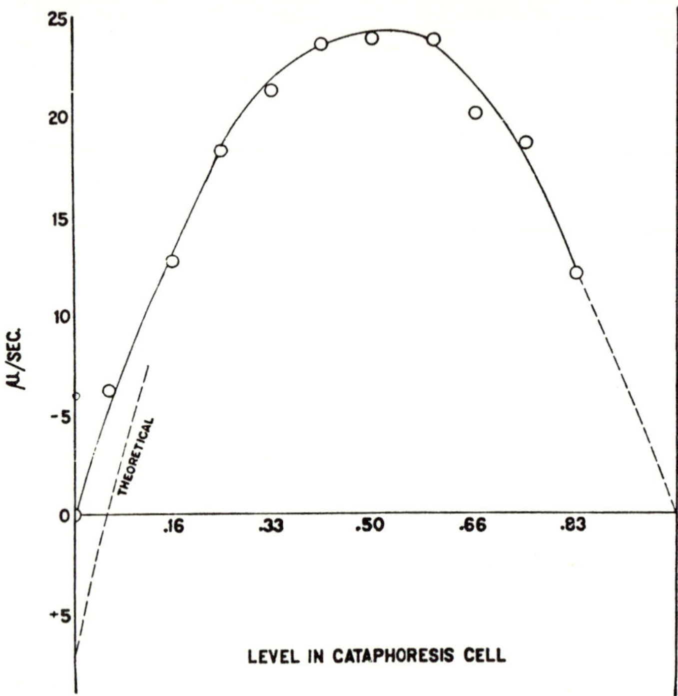

FIG. 5

The smooth curve extrapolated to $x = 1.0$ has been drawn through points of cataphoretic mobility of protein covered oil droplets. The curve passes through the origin. Since the origin ($x = 0$ or 1.0) is the wall of the cell it follows that $V_E = V_p$, or $\frac{V_E}{V_p} = 1.00$ approximately. The dotted line gives an idea of how the curve would look if $\frac{V_E}{V_p}$ were equal to 1.50. (From the Journal of General Physiology).

The Determination of V_E/V_p for a Round Surface. By proceeding as in the preceding section, V_E/V_p was determined in a round microcataphoresis cell by coating particle and glass surface with a protein, here gelatin. The curve drawn through the points in Fig. 5 is typical of the data of four similar experiments. The absolute mobility of the protein covered particle, V_p, is here at the level 0.15, approximately. The curve passes through the origin. This is only possible when

$$V_E = V_p \text{ or } V_E / V_p = 1.00$$

[31] This constant has not been determined experimentally.

The dotted line in Fig. 5 follows the approximate course of the curve calculated on the assumption that $V_E/V_P = 1.5$. That this ratio does not obtain under these conditions is obvious, confirming, therefore, the data for flat cells.

Significance of Data for Biological Objects

The Cataphoretic Mobility of Mammalian Red Blood Cells. Until more is known of the laws governing the cataphoresis of substances such as those ordinarily isolated from living cells, it is difficult to draw far-reaching conclusions concerning the data at present available dealing with the electrical mobilities of cells. There are, in fact, very few investigations which describe systematically, as *quantitatively* as the methods of cataphoresis permit, the comparative electrical mobilities of microscopic unicellular organisms. However, certain values given for the cataphoresis of mammalian red blood cells are of interest in view of the preceding discussion of the influence of size and shape on electric mobility. The data given in Table V show that very wide differences in the cataphoretic velocity of the red cells of different mammals exist.[32] Further, these values are independent of accidental impurities such as serum, etc. They are actually representative of the physicochemical make-up of the cell surfaces. But how much justification does there exist for the comparison of these values with one another?

If the surfaces of the cells were composed of protein exclusively, or of substances whose surfaces behaved like tyrosine needles, asbestos, etc., the problem would be simple; for it seems most probable that under such conditions curvature of the surface is not of much significance in determining cataphoretic mobility. Since it is most likely that fatty substances and other more complicated materials are chief components of certain red-cell surfaces, it is evident that curvature of the surface could affect the electric mobility of the red cells in exactly the same way that curvature affects the mobility of the oil droplets.

The following considerations are in favor of the tenability of the viewpoint that the electrical mobilities of the red cells studied are representative of chemical make-up of the surface rather than of variations in curvature of the surface:

a) There is too little known about the cataphoresis of organic substances in different states to accept curvature of the surface as a factor in determining red cell mobilities. The investigation of super-cooled emulsions and solid crystalline compounds of the type investigated by Mooney as well as the constituents of the red cell stroma must first be accomplished.

b) The "ghosts" (stromata) of red blood cells formed by hemolysis with dilute electrolytes can simultaneously be studied with single intact red cells. These hemoglobin-free lipoid portions of red cells undergo during hemolysis important changes in shape and stress incidental to swelling, besides losing

[32] J. Gen. Physiol., **12**, 711 (1929).

their hemoglobin. They have, in consequence, a new shape after hemolysis. The "ghosts" migrate with practically the same speed as normal intact red cells, in phosphate buffer.[11] The same has been found to be true for stroma debris.[33] Changes in curvature do not, therefore, seem to be of primary importance here.

c) Salts diminish the differences between droplets of the size of blood cells, particularly in the salt concentrations used.

d) There is no correlation between size and mobility. The diameter of rabbit cells are 7.16μ. Those of the dog are 7.20μ.[34] Yet these two types of cells have the lowest and highest values of mobility, respectively, among the mammals investigated. Similar examples can be cited for other cells of the group.

e) Red blood cells obtained from cases of primary and secondary anemias show very marked changes in size and curvature of the surface.[30] These differences in equivalent radius may be greater than 100 percent in suspensions from one patient. In suspensions from the same patient very little if any

TABLE V

The Cataphoretic Mobility of Mammalian Red Cells at pH 7.35 in M/15 Phosphate Buffer

Order	Animal	mn. μ/sec./volt/cm	Average Deviation
Primate	Man (White)	1.31	±0.02
	Man (Negro)	1.30	±0.05
	Monkey (*Macacus Rhesus*)	1.33	±0.02
Carnivor	Dog	1.65	±0.03
	Cat	1.39	±0.01
Ungulate	Pig	0.98	±0.03
Rodent	Rabbit	0.55	±0.05
	Guinea Pig	1.11	±0.02
	Mouse	1.40	±0.06
	Rat	1.45	±0.02
Edentate	Sloth (Two-toed)	0.97	One series of measurements
Marsupial	Opossum	1.07	±0.02

[33] One experiment.

[34] Tabulae Biological, Berlin, **2**, 459 (1925).

difference is found amongst the cells themselves. The values of mobility obtained for a series of anemia cases did not differ from the normal or differed very slightly, the difference being near the limits of experimental error. The cells, incidentally, also varied in hemoglobin content. This is excellent evidence that size and shape do not primarily invoke changes in red cell mobility.[35,36]

In conclusion we wish to re-emphasize that when data are available for other types of cells, a critique similar to that employed in the foregoing must be applied. The biologist in the meantime must wait upon the further acquisition of values of electrical mobilities of organic compounds in various states and in different suspending media before hoping to understand the chemical make-up of the surface and the many curious phenomena occurring at the phase boundaries of living and non-living microscopic particles and surfaces.

Department of Physical Chemistry in the Laboratories of Physiology,
Harvard Medical School,
Boston.

[35] It is not insinuated that no change in physico-chemical make-up occurs in the red cells in anemia. The changes that do occur cannot profoundly change the constitution of the surface.

[36] Since this paper was written Dr. G. Payling Wright and I have observed that foul red cells which vary in size, shape and cytoplasmic structure may have very much the same electrical mobilities.

26

Reprinted from *Cold Spring Harbor Symp. Quant. Biol.*, **8,** 72–79 (1940)

ELECTROPHORESIS AND THE CHEMISTRY OF CELL SURFACES*

HAROLD A. ABRAMSON, MANUEL H. GORIN, AND ERIC PONDER

For particles of the size of living cells suspended in salt solutions, the electrophoretic mobility depends only upon the nature of the cell surface and not upon size, shape, or orientation. Two factors determine the mobility of a particle of a surface at a given high ionic strength. The first is σ, the net charge per unit of surface area, and the second is the molecular geometry dealing with the average of the radii of curvature of all points on the surface. By microscopic curvature we mean the curvature of points on the surface, not the radius of the bulk of the particle. While the first concept, the charge density, has been used to a considerable extent, the second has not received as much attention, because of the experimental difficulties. Another factor preventing complete study is the unsatisfactory state of the theory bearing on the influence of the molecular geometry of the surface.

In practice, electrophoresis has been fruitful in biology in identifying the surface of the cell, particularly in the cases of immunological reactions at the surface (1). Even more important has been its application to the detection of changes in the cell surface during certain environmental variations. In some instances it is possible to compare the electrophoretic mobility-pH curve of a cell with that of known proteins believed to determine its surface composition. It is quite definite that, when the electric mobility-pH curve of a cell or particle is identical over a sufficient range with that of the protein, the outermost layer of the surface of the cell or particle is primarily made up of this protein. If the mobility cannot be identified with that of any known substance, we can only roughly classify the surface into main categories. Electrophoretic studies of the living cell have therefore been successful in scrutinizing changes or making comparisons; *e.g.*, the normal red cell and the ghost (2, 3), the red cell disc and sphere (3), red cells of different animals (4), surfaces of bacteria (5), the comparison of sensitized and unsensitized cells (2), interaction of cell surfaces with serum proteins and other substances (6, 7, 8), and variations in the surfaces of the blood cells of the horse (8).

It need hardly be pointed out that one great advantage of this method in studying the cell surface is that no essential injury of the surface occurs. Only small potential gradients are employed and these do not appreciably alter the surface of the cell. On the other hand, the method lacks sensitivity under certain conditions and it cannot be expected to detect small changes in the geometry or chemistry of the cell surface, *i.e.*, important physiological changes may occur without destroying a sufficient portion of the original surface to produce a detectable change in mobility.

The material to be presented will include (a) the meaning of surface charge and the boundary of shear, and general remarks on the origin of the surface charge and the classification of various surface types, (*b*) a short discussion of the difficult question of the microscopic surface of curvature, mainly in connection with the experimental evidence available with models rather than with living cells, and (*c*) the presentation of some of the experimental work on red cells, ghosts, leucocytes, etc. Finally an attempt will be made to sum up the results and to point out paths which future research could follow.

The Meaning of the Surface Charge and the Shear Boundary

Let us illustrate what is meant by the net charge of the surface of a cell by considering a single red cell, suspended in a very large volume of serum or isotonic salt solution. The distribution of electric charge in this system determines the mobility of the cell in an electric field. The total number of positive and negative charges of the whole system must be equal. The net charge of the system, therefore, is zero. It is only by having a charge distribution in which some of the charges will move with the red cell while other charges will move independently, just as if they were not associated with the cell, that a situation can arise in which the cell will move in an electric field. Thus, when n negative charges in excess of positive charges are associated with the red cell, there are p positive charges in excess of negative charges in the solution and n must equal p. These two sets of electric charges are independent of each other only in the same sense that the ions of a salt dissolved in polar solvents behave as if they were independent of one another. While sodium and chloride ions act as if sodium chloride were completely dissociated, their mutual effects can be predicted primarily by their electrostatic attraction. In the same way, the excess negative ions on the surface of the red cell behave as if they were independent of their partner ions in the solution. How may the surface of shear or the shear boundary be discussed in similar fashion?

The shear boundary is a hydrodynamic concept which may be essentially described in simple terms. A large particle like a red cell, moving

* This investigation was aided by a grant from the Josiah Macy, Jr. Foundation.

through water under the influence of a force of any kind, carries material which can be bounded by a surface that is called the surface of shear. Water molecules outside this surface of shear move as if they were essentially independent of the red cell and were part of the fluid in which the red cell is migrating. On the other hand, molecules inside the surface of shear act as if they were attached to the red cell and so travel with it. After having thus defined the surface of shear aside from electrical concepts, the concept of the surface charge may be properly introduced. If we consider a charge distribution in which there are n negative charges in excess inside the surface of shear, and p positive charges in excess outside the surface of shear, the cell migrates electrically as if it had a net charge of n units or a net surface density of charge given by the ratio of n to the surface area of the cell.

How does a distribution of charge occur which results in a net surface charge on the red cell? There are several ways which can be differentiated qualitatively (9). One of these involves the separation of charges. This is a process analogous to that which occurs when a crystal of sodium chloride is dissolved in a highly polar solvent like water (strong salts, bases and acids; weak salts, bases and acids). Ionizable and dissociable groups on the surface of the cell break down to form a residual surface charge when the dielectric constant of the medium, the pH, and other factors are favorable. More specifically, protein surfaces correspond essentially to the second type. At least two other processes in addition to ionogenic mechanisms are considered to be of importance: (a) adsorption and (b) ion pair formation (10). A rigid distinction between the two is difficult in an actual case, although in the ideal system it is very clear. Ion pair formation at a surface may be thought of as similar to ion pair formation in a solution. This is typified by the case of salts consisting of polyvalent ions (10). Ion pair formation requires points of charge on the surface to which ions opposite in sign may attach themselves. This mechanism is primarily independent of the net charge density but depends, rather, upon the total number of charged points capable of reacting specifically. Thus at the isoelectric point of an egg albumin molecule there are approximately 30 positive and 30 negative charges on its surface. With each charge, which may be considered as independent of the others, ion pair formation might take place to an extremely small extent. When, however, the effect on a single charge is multiplied by the total number of charges of opposite sign on the protein, an appreciable change in the net charge of the molecules may occur. In the case of a simple salt like NaCl there appears to be a greater tendency for chloride to bind to protein than for sodium. This is probably due to the smaller size of the chloride ion, but experimental information is still very meagre concerning the effects of salts of different types. With salts which consist of a polyvalent and a univalent ion, ion pairs are always formed preferentially by the polyvalent ion with charged points on the surface. Thus, relatively great changes in the isoelectric points of proteins occur with aluminum, thorium, ferricyanide, and even $HPO_4^{=}$ when these ions are combined with univalent ions of opposite sign. Polyvalent positive ions increase the pH of the isoelectric point whereas polyvalent negative ions shift it toward lower pH values. For simple uni-univalent electrolytes the effect is very small, but becomes relatively more important at higher concentrations. Moreover, it is nearly independent of pH in the region close to the isoelectric point.

The possible effect of ion pair formation on the mobility of hemoglobin as calculated (11) from the data of Davis and Cohn (12) may be seen in Fig. 1. When the mobility-pH data for four different ionic strengths are properly corrected for the effect of ionic strength, the curves become parallel to one another (dashed lines). There are important shifts in isoelectric point with ionic strength. These shifts in isoelectric point are probably caused by ion pair formation between the phosphate ion and positive groups on the protein,

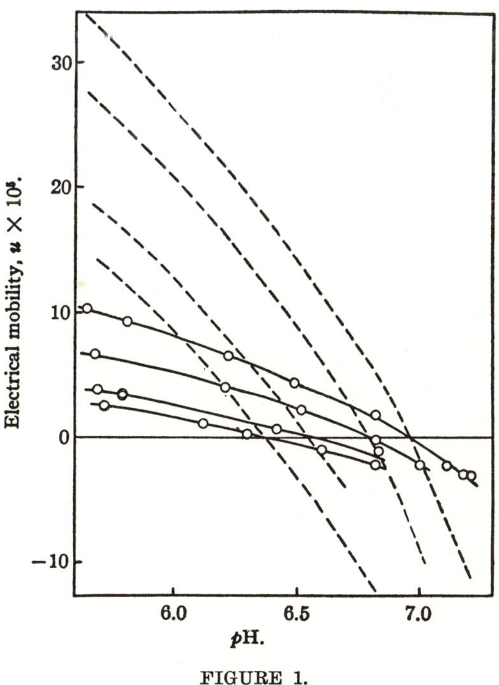

FIGURE 1.

the effect increasing with the concentration of buffer.

The second process usually comes under the general heading of adsorption. It is possible for an ion to be fixed at a surface even though there are no points of permanent charge to which it can be attached. An extreme example of this kind of charging process is the behavior of a droplet of paraffin oil or a particle of paraffin wax (also air bubbles). Chemically there are no points of residual charge to which attachment of ions in solution can be attributed. Still, particles of the types described acquire a surface charge when they are suspended in salt solutions. The forces responsible for adsorption of this kind are not completely understood theoretically. However, it is fairly definite that they arise chiefly from the polarization of the surface by the ion to be adsorbed. In this way, an induced charge is set up which corresponds somewhat to the permanent charge in the case of ion pair formation. It is interesting to point out that this second method of charging is probably more efficient and more important, especially at low concentrations of salt. The reason for this is not difficult to understand qualitatively. A residual charge on the surface is probably protected from charges in the solution by hydration or solvation in the same way that a sodium ion is protected from a chloride ion in water by a hydrated shell. Except for secondary effects which need not be considered here, a residual charge on the surface is no more apt to undergo ion pair formation than is an ordinary ion like a sodium ion. On the other hand, a nonpolar or an extremely weakly polar surface like paraffin is very poorly hydrated compared to either a polar surface or one that actually consists of points of residual charges such as amino acid crystals or protein. Here then, polarization forces can be most effective and, indeed, paraffin surfaces may become as highly charged, perhaps, as any other known surface. It is pertinent to our discussion to emphasize that when polarization forces alone are responsible for the charging of the surface, the charge is invariably negative in the case of uni-univalent electrolytes. There are, therefore, more negative ions attached to the surface than positive ions. This is probably caused by a smaller energy of hydration of the negative ion. For instance, in a crystal of KCl, the potassium ion is 0.6 A smaller than the chloride ion (13) but in aqueous solution they are very nearly the same size (14). This difference between the dissolved and crystal radii must be caused by hydration. Hydration is an important factor because adsorption would depend on the ease with which the ion can lose its shell of water. Another factor, the importance of which is difficult to estimate, is the polarizability of the ions themselves. This is a question which has not been investigated extensively, but qualitatively, the negative ions are much more polarizable than the positive ions. Whether this polarization factor is of direct influence or operates indirectly through its effect on the energy of hydration is not understood. A third possibility is that it has no effect at all. On the other hand, it would probably be of considerable importance in other methods of charging the surface, *i.e.,* in the formation of ion pairs.

The Molecular Geometry of the Surface

The development of electrophoretic theory has been limited to cases of particles with smooth surfaces. By a smooth surface we mean a surface which has no irregularities greater in magnitude than that corresponding to an unhydrated small ion (no more than 1 A). Even in these cases the theory is simplified because of the necessary assumption that the distribution of charge is uniform rather than discrete. The uncertainty due to this second assumption is difficult to estimate. However, the agreement between experiment and theory based on the uniform distribution of charge in the case of protein molecules is so good that this assumption seems to cause only a minor hiatus in the predicted electrokinetic behavior of the surface. Assuming a smooth surface, the theoretical investigations of Smoluchowski (15), Gouy (16), Henry (17), Debye and Hückel (18), and Gorin (14) have allowed the prediction with considerable accuracy of the behavior of particles ranging in size from that of a sodium or lithium ion to that of a protein molecule.

On the surfaces of living cells there is always the possibility that there are molecular hills and valleys which produce bumps much greater in magnitude than those which were set as a limit for smooth surfaces, *i.e.,* 1 A in radius. The bumps might correspond to the protein molecules or other structures which give essentially regions of much greater curvature than the bulk curvature of the surface. To approach this problem, we have at our disposal only the experimental evidence (19, 20) that particles coated with protein may have the same mobility as the given protein molecule in solution, within the limits of error. Proteins for which this is found to be true have radii of about 20-25 A and should behave quite differently than larger particles of the same charge density. To arrive at the source of this unexpected identity in mobility of dissolved and adsorbed protein we might ask what can happen when a protein molecule is adsorbed. Perhaps the most likely answer is that the comparatively large protein molecule is simply attached to the surface at a few points and essentially maintains its shape and chemical characteristics. If this is so, we are dealing with a type of surface for which the electrokinetic equations have not been solved, for there are no equations for the irregular surface of the kind (protein film) postulated above. The fact that this surface appears to be electrokinetically identical with that of the unadsorbed, dissolved protein molecule, suggests that the bumps on the surface of the adsorbed film have the same physical significance as the curved surfaces of the free protein molecules themselves. Let us accept the criterion of electrokinetic identity of molecules and bumps on the surface, for the time being, and look forward to further theoretical investigations of their behavior. Another possible explanation of the identity in behavior of the protein film and free molecules is that the charge density of the surface decreases by a factor of nearly three upon adsorption; *i.e.*, the surface area covered by a single protein molecule would have to be tripled. It is unlikely that the surface area can expand to so great an extent and still keep the net charge constant. Furthermore, if this should occur it would be difficult to imagine that the charge density could compensate so exactly for the change in surface area to give the same electric mobility.

Theoretical Considerations

For particles of smooth surface and given orientation the general equation is:

$$v = \frac{\sigma}{\eta}\left(\frac{1}{\kappa} + r_i\right)\chi \qquad (1)$$

where χ is a function of the size and shape of the particle and the ionic strength; v is the electric mobility; r_i is an average radius of the other ions in the solution; η is the viscosity of the medium; and κ is the well-known Debye-Hückel ionic strength function. The function χ depends upon the model chosen and the orientation of the particle. The function has been worked out for a sphere (14, 17, 18), for which, of course, the curvature parameter is the radius, r, and for an elongated cylinder (21, 22, 23) for which the parameter is the radius a. The function could, in addition, be derived with difficulty for ellipsoids of revolution, or even for the generalized ellipsoid.

For a sphere χ is equal to

$$\chi = \frac{\kappa r}{1 + \kappa r + \kappa r_i}$$

If r or a is sufficiently large, χ is independent of the orientation of the particle and of r or a; it then becomes equal to 1. This is the basis of the statement that the electrophoretic mobility of microscopically visible particles in salt solutions is independent of their size, shape, and orientation and is given by the equation

$$v = \frac{\sigma}{\eta}\left(\frac{1}{\kappa} + r_i\right) \qquad (2)$$

For platelets of flat surface, equation (2) should also apply.

In Table I, the calculated value of χ is given for a sphere of radius 23.6 A and for cylinders of the same volume but of various degrees of elongation.

It may be seen from Table I that χ is less than 1/3 at $\mu = 0.005$, and even at $\mu = 0.20$ it is about 1/2. If, then, a protein-coated surface has the same charge density as that of the unadsorbed protein molecule and the surface of the particle

TABLE I

Electrophoretic Shape Factors for Protein Molecules of Various Shapes (Molecular Wt. 45,000)

μ	χ	$1/\chi$
Sphere (radius 23.6 A)		
0.005	0.287	3.48
0.02	0.340	2.94
0.10	0.475	2.10
0.20	0.532	1.88
10.00	0.830	1.20
Cylinder ($l/a = 10$; $l = 121$ A; $a = 12.1$ A)		
0.005	0.295	3.39
0.02	0.320	3.12
0.10	0.427	2.34
0.20	0.463	2.16
Cylinder ($l/a = 45$; $l = 330$ A ; $a = 7.33$ A)		
0.005	0.241	4.15
0.02	0.273	3.66
0.10	0.381	2.62
0.20	0.426	2.35
Cylinder ($l/a = 90$; $l = 530$ A ; $a = 5.82$ A)		
0.005	0.211	4.74
0.02	0.242	4.13
0.10	0.336	2.97
0.20	0.378	2.65

behaves like a smooth plane surface, the mobility of protein-coated particles should be from 2 to 4.5 times as great as that of the unadsorbed molecule. Since no such differences are observed, it is likely that protein-coated surfaces are not smooth plane surfaces but rather are irregular. This forms the theoretical basis for the conclusions reached in the foregoing.

Electrophoresis of Red Cells

The most striking phenomenon encountered in the electrophoresis of red cells is the tenacity with which the cells maintain the general nature of their outer surfaces (2, 3). Important changes in shape (3), variation of hemoglobin content (24), the presence of lysins in the membrane (2, 3), the presence of the anti-sphering factor in the membrane (25), and the pathology of the cell (24) have not been found to necessarily cause a change in the electrophoretic mobility of the red cell if salts are present. Changes under other conditions have been observed (26, 27). These data will not be discussed here. Furthermore, it has not been possible, except in the cases of specific immunological reactions, to alter the surface of the red cell by putting it into contact with solutions of protein, like serum proteins, gelatin, and hemoglobin. When the cells have been injured by acids this rule does not obtain (28). This behavior in itself has been found only for protein surfaces studied recently by Moyer and Moyer (6, 29) and Moyer and Gorin (7). Thus surfaces made up of one of the serum proteins may or may not be changed when the particles on which they are coated are placed in a solution of another of the serum proteins. Collodion particles coated with globulin will not adsorb horse serum albumin A or B (30) while albumin A on quartz will not be replaced or coated over by globulins. When quartz particles are placed in horse serum, they become coated with a film of albumin A, while collodion particles adsorb globulin from serum. Casein particles in the neighborhood of the isoelectric point adsorb gelatin only over a limited pH range. Outside this range, casein surfaces are not altered appreciably in their electrophoretic mobility by gelatin (Fig. 2).

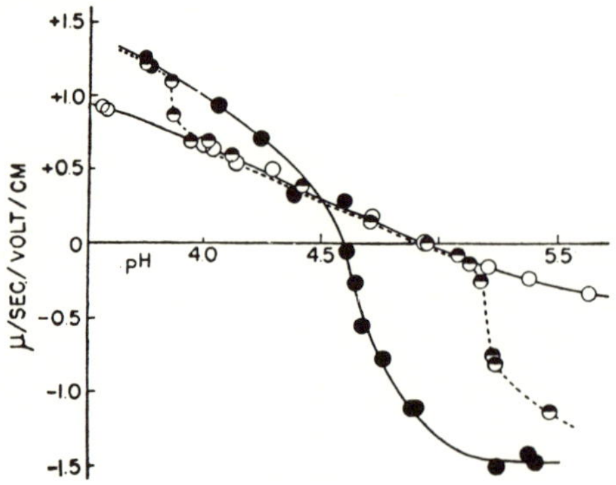

FIGURE 2.

From the limited data available for lipids it cannot be said that they show this exceptional behavior. As far as is known, cholesterol adsorbs proteins readily (31). All that can be said is that the behavior outlined above is not inconsistent with the idea that the red cell membrane is made up of a phospholipid-protein complex.

It was shown (28) that when sheep erythrocytes were sensitized with minute amounts of rabbit anti-sheep serum, the electric mobility of the sheep cells was not changed. If complement was added to these sensitized cells which had, at the most, less than 2 p.c. of their surface altered, hemolysis occurred. This hemolytic process in all probability was restricted to certain sensitized spots on the surface, which were too few in number to be detected by a change in electric mobility. The "key spot" hypothesis postulates that lysis takes place at isolated points on the surface rather than by a uniform change in membrane permeability. The hypothesis of "key spots" does not necessarily infer that these spots are structural units arranged in some pattern and picked out by the lysin because of their vulnerability. The complement fixation process is probably of a special type, in which the taking up of the complement is predetermined by the special positions of the antigen-antibody complexes on the surface of the cell.

Another point of interest is the function of the anti-sphering factor of Furchgott and Ponder in maintaining the normal shape of the red cell (25). The discovery that 0.8 gm. of crystalbumin (albumin B) per 100 cc. of cells can be washed out of the membrane without changing the electrophoretic mobility demands unequivocally that the material is either in the interior of the membrane or concentrated in isolated regions thereof and plays no significant role in the formation of that particular surface, the nature of which determines the mobility. This amount of crystalbumin is probably sufficient to form several monolayers on the cell surface if it were uniformly distributed. It is evident that a uniform film of this kind is not found; the mobility of the cell is quite different from that of a crystalbumin surface and does not change when the crystalbumin is washed out of it. Furthermore the normal erythrocyte is negatively charged down to pH 3.7 and shows no real isoelectric point even there (28).

Another conclusion can be drawn from the apparent identity in mobility of the normal red cell disc and the saponin sphere. Unless an improbable compensation occurs, it can be concluded from this evidence that the charge densities of the surface of the sphere and of the disc are identical. Considering it superificially, considerable change in surface area would be expected in going from a disc to a sphere of the same volume. Since the surface charge density does not change, it is probable that the disc-sphere transformation results in a folding of the surface which may be invisible microscopically. This is in agreement with the observations of Ponder from studies of osmosis (33).

All the results mentioned so far were obtained in salt solutions or salt solutions plus buffers. Freundlich and Abramson (34) have investigated mixtures of horse erythrocytes, polymorphonuclear leucocytes, small lymphocytes, and quartz particles in serum and in oxalated plasma. In this medium the red cell has a mobility quite different from that of the other particles, two of which, the quartz particles and the polymorphonuclear leucocytes, have identical mobilities. Similar results were also obtained in a gelatin-serum gel. The identity in behavior of different particles in the same electrophoresis cell is a sensitive criterion of their identity in surface composition, for two particles side by side at the same level in the cell can be observed to follow one another exactly. Such an observation is more accurate than the actual establishment of a numerical value for the electrophoretic mobility. We now know that quartz particles in horse serum are coated with serum albumin A (6). The conclusion can therefore be drawn that polymorphonuclear leucocytes are also coated with serum albumin A when in horse serum or plasma. Since the blood platelets of the horse have the same mobility as leucocytes they also must have a surface composed primarily of serum albumin A.

Ponder (35) recently found that ghosts produced by hemolysis in water retain hemoglobin in considerable amounts which can be washed out of the ghosts only with difficulty. Calculations demonstrate that from four to fourteen monolayers would be formed by the hemoglobin remaining. Since the mobility of the cell is not changed, it can be concluded that this hemoglobin is not adsorbed on the outer surface of the cell. This evidence supports the suggestion that there is some kind of structure inside the cell which is responsible for the retention of such large amounts of hemoglobin.

REFERENCES

1. Abramson, H. A. Electrokinetic Phenomena, Chemical Catalog Co., New York (1934). Chapt. XI.
2. Abramson, H. A. See Ref. 1, page 264.
3. Abramson, H. A., Furchgott, R. F., and Ponder, E. J. Gen. Physiol. *22,* 545 (1939).
4. Abramson, H. A. See Ref. 1, page 256.
5. Moyer, L. S. J. Bact. *32,* 433 (1936).
6. Moyer, L. S. and Moyer, E. Z. J. Biol. Chem. *132,* 373 (1940).
7. Moyer, L. S. and Gorin, M. H. J. Biol. Chem. *133,* 605 (1940).
8. Abramson, H. A. See Ref. 1, page 266.
9. Abramson, H. A. See Ref. 1, Chap. IV, V, VI.
10. Fuoss, R. M. and Kraus, C. A. J. Am. Chem. Soc. *55,* 476, 1019, 2387 (1933); *57,* 1 (1935). Kraus, C. A. Trans. Electrochem. Soc. *66,* 179 (1934).
11. Gorin, M. H., Abramson, H. A., and Moyer, L. S. J. Am. Chem. Soc. *62,* 643 (1940).
12. Davis, B. D. and Cohn, E. J. J. Am. Chem. Soc. *61,* 2092 (1939).
13. Pauling, L. The Nature of the Chemical Bond. Cornell University Press, Ithaca (1939). Chap. XI.
14. Gorin, M. H. J. Chem. Phys. *7,* 405 (1939).
15. Smoluchowski, M. In Graetz Handbuch der Elektrizität und des Magnetismus *2,* 366 (1921).
16. Gouy, G. -L. J. Phys. *9,* 457 (1910).
17. Henry, D. C. Proc. Roy. Soc. London, A, *133,* 106 (1931).
18. Debye, P. and Hückel, E. Phys. Z. *24,* 185 (1923); *24,* 49 (1924). Hückel, E. Phys. Z. *25,* 204 (1924).
19. Abramson, H. A. J. Gen. Physiol. *15,* 575 (1932).
20. Moyer, L. S. J. Biol. Chem. *122,* 641 (1938).
21. Abramson, H. A., Gorin, M. H., and Moyer, L. S. Chem. Rev. *24,* 345 (1939).
22. Gorin, M. H. Unpublished results.
23. Gorin, M. H. J. Phys. Chem. (In press, 1940).
24. Abramson, H. A. See Ref. 1, page 265.
25. Furchgott, R. F. and Ponder, E. J. Exp. Biol. *17,* 117 (1940).
26. Byler, W. H. and Rozendaal, H. M. J. Gen. Physiol. *22,* 1 (1938).
27. Stephens, J. G. J. Physiol. *95,* 92 (1939).
28. Abramson, H. A. J. Gen. Physiol. *14,* 163 (1930).
29. Moyer, L. S. and Moyer, E. Z. J. Biol. Chem. *132,* 3557 (1940).
30. Kekwick, R. A. Biochem. J. *32,* 552 (1938).
31. Abramson, H. A. J. Gen. Physiol. *12,* 711 (1929).
32. Ponder, E. The Mammalian Red Cell and the Properties of Haemolytic Systems, Gebrüden Borntraeger, Berlin (1934).
33. Ponder, E. J. Exp. Biol. *14,* 267 (1937).
34. Freundlich, H. M. J. Exp. Med. *47,* 677 (1928); Z. phys. Chem. *128,* 25 (1927).
35. Ponder, E. Unpublished results.

DISCUSSION

Dr. Jacobs: Have you any suggestion to make about the cause of the very striking differences in the mobilities of the red cells of different species?

Dr. Gorin: The differences in the mobilities of cells of different species can be related to differences in the numbers and kinds of polar groups at the shear boundary and their orientation. No correlation has been made so far between the electric mobility and chemical analysis of the membrane material for various species.

Dr. Schmitt: Have you made any studies of the electrophoretic mobilities of protein extracted from red cell stromata?

Dr. Furchgott: Haurowitz and Sladek reported that the extracted red cell stroma protein has no electrophoretic mobility, I would guess that their technique was at fault.

Dr. Abramson: There is an interesting paradox in that statement. Proteins must have an electrical mobility. It is only by the most delicate procedure that you can get protein to the isoelectric point.

Dr. Gorin: We do not know whether cephalin and lecithin adsorb protein.

Dr. Furchgott: On the basis of the amino acid analyses, the stroma protein would certainly not have a very low isoelectric point.

Dr. Davson: I think the main problem is really to find a surface which will not take on protein. It seems incredible that the erythrocyte should not take on some protein. It may be that it has a protein, and that the protein properties are obscured by the presence of acid and basic groups of the lipid itself.

Dr. Gorin: If the lipid surface actually adsorbs protein the mobility must change very markedly and approach, or become equal to that of a surface coated with the proteins used.

Dr. Abramson: In the presence of sufficient acid, red cells adsorb protein and act as though they had surface films of protein; apparently only injured cells behave in this way. I have observed that serum obtained from experimentally normal rabbits may lower the electric mobilities of sheep red cells if sufficient serum is added. The electric mobility, however, is not lowered very much. I believe this reaction is due to a specific reaction rather than an adsorption phenomena.

In this connection Porter's work on red cell ghosts is of some interest. She has found that the ghosts of human red cells hemolysed by distilled water may readily be agglutinated by their iso-agglutinating serum. In other words, not only is the electric mobility of red cells unchanged by this type of hemolysis but also certain groups responsible for specific immunological reactions are preserved.

Dr. Bull: There is evidence that in the case of egg albumin there are two layers of protein molecules at the air-solution surface.

Dr. Gorin: It may require more than one layer of adsorbed protein at the solid liquid interface to give a mobility identical with that of the dissolved molecules.

Dr. Schmitt: Red cells are readily adsorbed on a barium stearate surface made hydrophilic with thorium. However, they apparently also adhere to a ferric stearate surface and even to a clean glass surface. Neither these facts nor data concerning the relative hydrophilic or oleophilic properties of the surface of the dried envelopes necessarily tell whether the outermost layer of envelope is protein or lipide. That the protein and lipide constituents are not uniformly distributed in the envelope is shown by the discontinuities of protein thickness revealed by the analytical leptoscope.

Dr. Gorin: All this evidence is in agreement with the idea that red cell membrane must be inhomogeneous in nature. The evidence from electrokinetics can be interpreted in no other way.

Dr. Schmitt: In your opinion, do the electrophoretic data require that the outer layer of the red cell be composed of lipide?

Dr. Gorin: I think that all we can say is that we have no objection on the basis of the electrokinetic evidence to an outermost surface containing lipid. The orientation of the lipid molecules, however, must be of a very specific kind in order to result in a surface which does not adsorb gelatin.

Dr. Abramson: Very few experiments have been done on the reaction of red cells with protein. Although the experiments referred to have demonstrated that the types of red cells which have been investigated do not adsorb protein non-specifically, the subject is by no means closed. Red cells not yet investigated may have surfaces which adsorb protein and other proteins.

Dr. Moyer: In regard to the possibility of cephalin being on the surface, I think we must keep in mind the probability of the orientation of the molecules. Cephalin oriented with non-polar groups sticking out toward the aqueous phase would present a surface which should be similar in adsorption properties to mineral oil, as far as proteins are concerned. It is known that mineral oil droplets tend to adsorb egg albumin and gelatin, which are not adsorbed by the red cell. Furthermore, mineral oil droplets in serum become coated with a film of globulin which then does not appear to interact with the dissolved albumin present. The high isoelectric point of globulin would speak against its being the outer surface of the red cell. These facts speak against the first type of orientation.

If the cephalin were oriented the other way it would probably have a very high maximal density of net charge and a low isoelectric point. Although Abramson and I find that the red cell charge density is not unusually high, Abramson finds that red cells are still negative down to pH 3.6. It would be an interesting experiment to see if cephalin surfaces with polar groups toward the aqueous phase can adsorb proteins. So far, the only surfaces we have produced which cannot adsorb a protein have been particles coated with another protein. Horse serum contains two albumins called by Kekwick A and B. It has been found that quartz particles put in serum become coated with a film of albumin A, as shown by the quantitative agreement of their mobility-pH curve with the curve for particles coated with pure albumin A. Mineral oil and collodion become covered with globulin when immersed in serum. Although an inert particle has not been found which selectively adsorbs albumin B from serum, the presence of B is inferred by the results of Furchgott with the antisphering factor. If these quartz and collodion particles are each coated with albumin A and then treated with globulin, Gorin and I find that the collodion particles release their albumin and take up globulin. The albumin coated quartz, however, retains its surface and is not affected to a measurable extent by the presence of globulin. Likewise, gelatin and casein do not interact at surfaces to any marked extent except in the range of pH close to the isoelectric point of casein. It appears, therefore, that one protein may be unable to adsorb another. Whether cephalin will behave in this way with proteins remains to be seen.

27

Reprinted from *J. Bacteriol.*, **65**(2), 198–202 (1953)

THE ISOELECTRIC POINT OF BACTERIAL CELLS[1]

VIRGINIA P. HARDEN AND JOHN O. HARRIS[2]

Department of Bacteriology, Kansas State College, Manhattan, Kansas

Received for publication July 11, 1952

This study was undertaken to compare the isoelectric points of gram positive and gram negative bacterial cells by the method of microelectrophoresis. Within recent years a number of workers have referred to differences in the isoelectric point of the gram positive and the gram negative groups of bacteria. This concept arose from the studies of Stearn and Stearn (1924, 1925, 1928), Tolstoouhov (1929), and others who have used differential staining procedures at different pH levels. The apparent isoelectric point was taken to be the pH at which the cells retained equal amounts of basic and acidic dyes. Recent studies of the staining mechanism by Knaysi (1951) indicate that the basic dyes do not combine with the cell wall but stain the cytoplasmic membrane. The dye method would indicate then the properties of the cytoplasmic surface rather than the actual cell surface. It could be expected that the protoplasm of the gram positive cells with the high ribonucleic acid content would exhibit greater acidic properties than that of gram negative cells. However, the isoelectric point of the cytoplasmic membrane would correspond to the point at which equal amounts of basic and acidic dyes combined only if the dyes used possessed opposite charges of the same magnitude and if no other factors other than electrical charge influenced the combination of dyes with the cells. It has been demonstrated that in a series of structurally related acidic dyes, adsorption depends to a great extent in the configuration of the dye molecule (Harris, 1949). Furthermore the dye method as applied in the past depends upon the ability of the investigator to judge by microscopic examination when cells have been stained equally by the two dyes. It is apparent that considerable variation would result from such a procedure.

The electrophoretic behavior of colloidal particles is largely a function of the surface in contact with the surrounding environment. It is not known how great an influence the cytoplasmic membrane would have on the electrical properties of the cell surface, but the effect would be expected to be small in comparison with that of the cell wall or slime layer directly in contact with the suspending liquid. Methods of microelectrophoresis offer a precise means for the determination of the isoelectric point of colloidal particles. Although many factors, such as type and concentration of the ions of the suspending fluid and age of the bacterial cells, influence the electrophoretic mobility of bacteria, most of these factors can be so controlled as to result in the measurements being made under a standard set of conditions.

It was the purpose of this investigation to determine whether actual differences exist in the pH values of the isoelectric point of gram positive and gram negative bacteria. A survey of the literature revealed that a number of isoelectric values for both groups of bacteria have been indicated by other workers. These will be tabulated later and compared with the results of this investigation.

EXPERIMENTAL METHODS

Young actively growing cells of various species of bacteria were obtained by growing cultures in Roux flasks containing 100 ml of the following medium: 0.2 per cent K_2HPO_4, 0.1 per cent KH_2PO_4, 0.5 per cent Difco peptone, 0.1 per cent commercial sucrose, and 1.7 per cent agar in distilled water. Exceptions to the use of this medium were *Mycobacterium tuberculosis* grown on Corper's egg medium, *Azotobacter* species grown on Ashby's mannitol agar, and the anaerobic species grown in deep bottles of the liquid medium without added agar. All cultures were transfers from the departmental stock collection.

[1] Contribution no. 273, Department of Bacteriology, Kansas Agricultural Experiment Station, Manhattan, Kansas.

[2] Portion of a thesis presented by the senior author as partial fulfillment of the requirements of the degree Doctor of Philosophy in Bacteriology at Kansas State College.

Those obtained originally from the American Type Culture Collection are labeled with the American Type Culture Collection number.

Cultures were incubated at 30 C except for the few requiring 37 C. Incubation was usually for 24 hours, though in a few exceptions longer periods were necessary for the slow growing types, such as *M. tuberculosis*. Cells were removed from the growth medium and washed four times by centrifuging and resuspending in a balanced salts solution having the following composition: K_2HPO_4, 14.0 g; KH_2PO_4, 5.6 g; NaCl, 1.6 g; $MgSO_4$, 1.6 g; $CaCl_2$, 0.16 g; glass-distilled water, 8 liters. This solution was sterilized and filtered just before use. Although a major effort was made

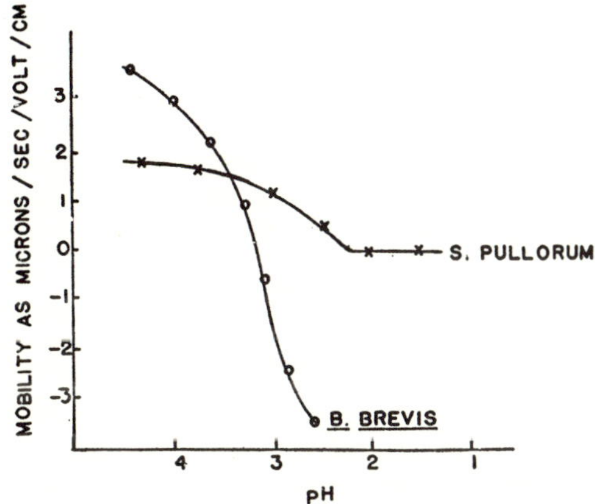

Figure 1. The electrophoretic mobility of *Salmonella pullorum* and *Bacillus brevis* at different pH values.

to maintain uniform procedures for comparative purposes, the highly pathogenic types were heat-killed by steaming for 10 minutes; all others were examined in a living condition. Concentrated cellular suspensions of the washed bacteria were stored at 4 C for not over 8 hours prior to the electrophoretic measurements.

Since the isoelectric point of colloidal particles is influenced by the kind and concentration of electrolytes present as well as the pH of the suspending menstruum, a dilute buffer system of monovalent salts was chosen to avoid any variable effects of multivalent ions upon the different bacteria being studied.

The buffer system was composed of various combinations of sodium acetate, acetic acid, and hydrochloric acid to give a series of buffer solutions of different hydrogen ion concentrations but a constant ionic strength (0.01 μ). One-half milliliter of the concentrated cellular suspension

TABLE 1

Isoelectric points for 31 bacterial species determined by using acetate-HCl buffer

GRAM POSITIVE SPECIES	ISO-ELECTRIC POINT (pH)	REVERSIBILITY OF CHARGE
Mycobacterium tuberculosis (human strain)	4.15	+
Bacillus sphaericus	3.99	+
Bacillus terminalis	3.57	+
Bacillus cereus	3.55	+
Rhodospirillum rubrum	3.46	+
Bacillus brevis	3.20	+
Bacillus polymyxa	3.12	+
Erysipelothrix rhusiopathiae	2.91	+
Clostridium sporogenes	2.75	+
Lactobacillus casei, strain 9595	2.45	+
Leuconostoc mesenteroides, strain 8042	2.25	+
Sarcina lutea	2.20	+
Bacillus subtilis	2.19	−
Micrococcus citreus	1.90*	−
Streptococcus faecalis, strain 9790	1.90*	−
Bacillus alvei	1.85*	−
Bacillus megaterium, strain 8245	1.80*	−
Bacillus pumilis	1.75*	
GRAM NEGATIVE SPECIES		
Mycoplana bullata, strain 4278	3.65	+
Moraxella bovis	3.47	+
Alkaligenes faecalis, strain 8749	3.28	+
Pseudomonas cyanogenes, strain 795	3.25	+
Erwinia carotovora, strain 495	2.99	+
Pseudomonas convexa	2.95	+
Proteus vulgaris	2.67	−
Klebsiella pneumoniae	2.48	−
Aerobacter aerogenes	2.42	−
Serratia marcescens	2.17	−
Pseudomonas aeruginosa	2.17	−
Salmonella pullorum	2.12	−
Azotobacter chroococcum	2.07	−

* It was necessary to use one-tenth normal acids to obtain pH values below 2.0; hence, these solutions have an ionic strength of 0.1 μ.

was placed in 100 ml buffer and allowed to equilibrate for 20 minutes. Electrophoretic measurements were made following the standard procedure suggested by Moyer (1936) using the

Abramson modification of the Northrup-Kunitz flat type microelectrophoresis cell. Comparable results were obtained throughout the study with the two cells used. Experiments were carried out at 25 to 27 C. Measurements of the mobility of the cells were made at the upper stationary level of the cell.

TABLE 2

Isoelectric points of various bacteria obtained from data published by other workers who have used electrophoretic measurements

NAME OF THE ORGANISM*	ISOELECTRIC POINT (pH)	REFERENCE
Gram positive		
Diplococcus pneumoniae	3.5 to 4.7	Thompson (1932)
Bacillus subtilis	3.6	Yamaha and Abe (1934)
Staphylococcus aureus	3.4	Yamaha and Abe (1934)
Bacillus anthracis	3.1	Yamaha and Abe (1934)
Bacillus cereus	3.0	Winslow, Falk, and Caulfield (1923)
Diplococcus pneumoniae	2.7 to 3.3	Falk and Jacobson (1926)
Sarcina lutea	2.6	Yamaha and Abe (1934)
Clostridium flabelliferum	2.5	Winslow and Upton (1926)
Lactobacillus acidophilus	2.5	Winslow and Upton (1926)
Mycobacterium smegmatis	2.5	Winslow and Upton (1926)
Mycobacterium leprae (R)	2.2	Reed and Gardiner (1932)
Diplococcus pneumoniae	2.0 to 3.0	Mudd (1933)
Bacillus pseudodiphtheriae	1.8	Yamaha and Abe (1934)
Mycobacterium leprae (S)	1.2	Reed and Gardiner (1932)
Staphylococcus sp.	0.7	Verwey and Frobisher (1940)
Gram Negative		
Hemophilus influenzae	3.0 to 4.0	Mudd (1933)
Brucella abortus	3.0 to 4.0	Mudd (1934)
Bacillus caryocyaneus	3.5	Lasseur, Dupaix-Lasseur, and Fribourg (1933)
Phytomonas stewartii	3.5	Frampton and Hildebrand (1944)
Erwinia amylovora	3.3	Frampton and Hildebrand (1944)
Bacillus pyocyaneus	3.1	Yamaha and Abe (1934)
Pseudomonas pyocyaneus	2.5	Winslow and Upton (1926)
Bacterium coli	2.5	Winslow and Upton (1926)
Diplococcus gonorrhoeae	2.5	Yamaha and Abe (1934)
Bacillus pertussis	2.5	Yamaha and Abe (1934)
Brucella abortus	2.3 to 3.3	Stearns and Roepke (1941)
Bacillus proteus	2.1	Yamaha and Abe (1934)

* The name of the organism is that as used in the original publication.

RESULTS

Immediately following the electrophoretic measurements, the specific conductivities and the pH values of each suspension were determined. The mobilities (calculated as microns/second/volt/centimeter) were plotted against the pH values, and the isoelectric value was found by interpolation from the graph. Typical mobility-pH curves are shown in figure 1. Measurements were made over a range of hydrogen ion concentrations sufficiently acidic to result either in no movement by the bacteria as shown by *Salmonella pullorum* or in reversal of charge and movement in the opposite direction as exhibited by *Bacillus brevis* in figure 1.

Thirty-one different bacterial species were studied by the procedure described. The isoelectric points for each species are shown in table 1 together with the gram staining reaction of the organism. The isoelectric point in the gram positive group varied from pH 1.75 to 4.15 and in the gram negative group from pH 2.07 to 3.65.

These results indicate that the isoelectric point value for gram negative cells does not vary over as wide a range as for gram positive cells. It is apparent, however, that the isoelectric point of the gram negative group does not occur at more alkaline reactions than that of the gram positive bacteria.

Also recorded in table 1 are data indicating any alteration in charge at pH values below the isoelectric point. These data show a reversal of charge only in those organisms possessing the more alkaline isoelectric points. This reversibility of charge has been interpreted by some workers as indicating a protein-like surface. It should be pointed out that the data in table 1 represent values for a given set of experimental conditions and should not be taken as absolute values for the organism.

While most previous electrophoretic studies have been directed toward detecting fundamental differences in cell surfaces which might be related to pathogenicity, variability, or some other cell characteristic, a number of references contain data revealing the isoelectric point of the particular species under investigation. Such values for a number of different bacteria are shown in table 2.

DISCUSSION

Although many factors, such as age of cells, methods of growing, and lack of uniform procedure by early workers, would tend to cause variation in these results, the values recorded in the literature agree well with those of the present study. There was no relationship between the gram staining reaction and the isoelectric point of the bacterial cells.

Since the data obtained by the electrophoresis studies show no fundamental differences in the isoelectric behavior of gram positive and gram negative cells, it is apparent that misleading conclusions can result from attempts to interpret information regarding cellular isoelectric points from data obtained by the dye method. It is likely that the two methods measure the properties of two entirely different cellular surfaces. The results of the electrophoretic studies give indirect evidence to substantiate the theory of Knaysi (1951) that the basic dyes combine with the cytoplasmic surface rather than the cell wall. In the interest of accuracy, discussions of the isoelectric properties of bacteria should distinguish clearly whether the surface is that of the cytoplasmic membrane or that of the outermost layer of the cell (cell wall or slime layer) and whether the properties are those that effect the behavior of the cells as colloidal particles or the behavior of certain cellular components.

SUMMARY

Electrophoretic determination of the isoelectric points of 31 bacterial species in a dilute monovalent buffer system showed no fundamental differences in the range of values for gram positive and gram negative bacteria. These results are compared with data obtained by other workers using differential staining procedures. It is concluded that the two methods do not measure the same properties of the cell and that the staining method has led to inaccurate interpretations regarding the colloidal behavior of bacteria.

REFERENCES

Falk, I. S., and Jacobson, M. A. 1926 Studies on respiratory diseases. XXVIII. Electrophoretic potential, acid and serum agglutination of pneumococci. J. Infectious Diseases, **38**, 182–187.

Frampton, V. L., and Hildebrand, E. M. 1944 Electrokinetic studies on *Erwinia amylovora* and *Phytomonas stewartii*. J. Bact., **48**, 537–545.

Harris, J. O. 1949 Combination of certain fluorane derivative dyes with bacterial cells at different hydrogen ion concentrations. Stain Technol., **24**, 217–221.

Knaysi, G. 1951 *Elements of bacterial cytology*. Second edition. Publishing Company, Inc., Ithaca, New York.

Lasseur, P. H., Dupaix-Lasseur, A., and Fribourg, R. 1933 Fixation des colorants par les elements microbiens. IV. Cataphorese des plastides bacteriennes. Trav. lab. microbiol. fac. pharm. Nancy, **6**, 48–49. (Original not seen—from Biol. Abstracts, **8**, 18347, 1934.)

Moyer, L. S. 1936 A suggested standard method for the investigation of electrophoresis. J. Bact., **31**, 531–546.

Mudd, S. 1933 Agglutination. Cold Spring Harbor Symposia Quant. Biol., **1**, 65–76.

Mudd, S. 1934 Sensitization of bacteria with normal and immune human serum. J. Immunol., **26**, 447–454.

Reed, G. B., and Gardiner, B. G. 1932 Studies

in variability of T. B. bacilli. V. Acid agglutination and electrophoretic potential in *M. leprae*. Can. J. Research, **6**, 622–631.

STEARN, E. W., AND STEARN, A. E. 1924 Chemical mechanism of bacterial behavior. J. Bact., **9**, 463–477.

STEARN, E. W., AND STEARN, A. E. 1925 A study of the chemical differences of bacteria. J. Bact., **10**, 13–23.

STEARN, A. E., AND STEARN, E. W. 1928 Studies in physico-chemical behavior of bacteria. Univ. of Missouri Studies, **3**, 3–84.

STEARNS, T. W., AND ROEPKE, M. H. 1941 Electrophoresis studies on *Brucella*. J. Bact., **42**, 411–430.

THOMPSON, R. L. 1932 Electrophoresis of normal and chemically treated strains of pneumococcus. Am. J. Hyg., **15**, 712–725

TOLSTOOUHOV, A. V. 1929 Detailed differentiation of bacteria by means of a mixture of acid and basic dyes at different pH values. Stain Technol., **4**, 81–89.

VERWEY, W. F., AND FROBISHER, M. 1940 Electrophoresis of *Staphylococcus*. I. Factors affecting electrophoretic velocity. Am. J. Hyg., **32**, 55–62.

WINSLOW, C. E. A., AND UPTON, M. F. 1926 Electrophoretic migration of various types of vegetative cells. J. Bact., **11**, 367–392.

WINSLOW, C. E. A., FALK, I. S., AND CAULFIELD, M. F. 1923 Electrophoresis of bacteria as influenced by hydrogen ion concentration and the presence of sodium and calcium salts. J. Gen. Physiol., **6**, 177–200.

YAMAHA, G., AND ABE, S. 1934 Weiteres über den isoelektrischen Punkt der Bakterien. Sci. Repts. Tokyo Bunrika Daigaku, Sect. B, **1**(21), 221–229.

28

Reprinted from *J. Bacteriol.*, **101**(1), 92–101 (1970)

Electromechanical Interactions in Cell Walls of Gram-Positive Cocci

LI-TSE OU AND ROBERT E. MARQUIS

Department of Microbiology, University of Rochester School of Medicine and Dentistry, Rochester, New York 14620

Isolated cell walls of *Staphylococcus aureus* and *Micrococcus lysodeikticus* were found to expand and contract in response to changes in environmental *p*H and ionic strength. These volume changes, which could amount to as much as a doubling of wall dextran-impermeable volume, were related to changes in electrostatic interactions among fixed, ionized groups in wall polymers, including peptidoglycans. *S. aureus* walls were structurally more compact in the hydrated state and had a higher maximum charge density than *M. lysodeikticus* walls. However, they were less responsive to changes in electrostatic interactions, apparently because of less mechanical compliance. In media of nearly neutral *p*H, *S. aureus* walls had a net positive charge whereas *M. lysodeikticus* walls had a net negative charge. These charge differences were reflected in Donnan distributions of mobile ions between wall phases and bulk medium phases. Cell walls of unfractionated cocci also could be made to swell and contract, and wall tonus in intact cells appeared to be set partly by electrostatic interactions and partly by mechanical tension in the elastic structures due to cell turgor pressure. The experimental results led to the conclusions that bacterial cell walls have many of the properties of polyelectrolyte gels and that peptidoglycans are flexible polymers. A reasonable mechanical model for peptidoglycan structure might be a sort of three-dimensional rope ladder with relatively rigid, polysaccharide rungs and relatively flexible polypeptide ropes. Thus, the peptidoglycan network surrounding cocci appeared to be predominantly an elastic restraining structure.rather than a rigid shell.

Most bacterial and plant cells are surrounded by a thick, porous, polymer meshwork called a cell wall. This structure, which may occupy as much as 50% of total cell volume, has a major influence on the nutrition and general physiology of these cells. In recent years, knowledge of the chemical structure of cell walls has increased tremendously, so much so that it is now reasonable to attempt to interpret many of the physical properties of walls in terms of their chemical structures.

It is well known that bacterial cell walls have shape-retaining properties and that isolated walls, especially those of gram-positive bacteria, tend to retain the general shapes of cells from which they were derived. Because of this shape-retaining tendency, bacterial walls have been described as rigid structures. However, Knaysi (13) pointed out that the type of rigidity exhibited by bacterial walls is of the same type as that exhibited by an inflated rubber inner tube or balloon, both of which retain their general shapes when deflated. Wámoscher (26) demonstrated clearly that surface structures of a variety of bacterial cells are extensible, flexible, and elastic, in that they can be deformed easily by micromanipulation and then quickly regain their original shapes when deforming forces are removed. Further, a number of investigators (13, 16, 19) have concluded that cell turgor pressure produces elastic stretching of bacterial walls, and that relief of this pressure during plasmolysis results in contraction of the walls, even though plasmolysis vacuoles may form between walls and protoplast membranes. Knaysi (13) also observed that bacterial walls collapse and flatten when intact cells are dried on collodion films.

The polymers most often implicated as being responsible for the shape-retaining properties of bacterial walls are peptidoglycans. If peptidoglycans are damaged enzymatically, they become water-soluble and lose their capacity to serve as mechanical supporting structures. This damage may involve cleavage of either glycan or peptide portions of the molecules or of links between muramic acid and peptides. In fact, it appears that the insolubility of peptidoglycans is similar to that of certain other biological gels in being related to degrees of polymerization and to the numbers of cross-links among polymer stands. In unfractionated cell walls, peptidoglycans may have other polymers, such as teichoic acids, covalently bonded to them, but usually removal of these attached polymers does not sufficiently reduce the degree of polymerization of the complex to render it water-soluble.

In an earlier study (16), it was found that *Bacillus megaterium* walls undergo major changes in volume as a result of changes in electrostatic interactions among fixed, charged groups in wall polymers, especially amphoteric peptidoglycans. Enhanced electrostatic attraction resulted in wall contraction whereas enhanced electrostatic repulsion resulted in wall swelling. In other words, the walls had many of the properties of polyelectrolyte gels. In this paper, we describe an extension of this previous work to a consideration of electrostatic volume changes in walls isolated from *Staphylococcus aureus* strain Duncan and *Micrococcus lysodeikticus* strain ATCC 4698. These organisms were chosen because the former has compact, highly cross-linked peptidoglycans and a net positive wall charge, whereas the latter has less compact, less extensively cross-linked peptidoglycans and a net negative wall charge.

MATERIALS AND METHODS

Bacteria. *S. aureus* Duncan and *M. lysodeikticus* ATCC 4698 were grown at 30 C on a rotary shaker in a medium prepared by dissolving 30 g of Oxoid tryptone (Colab Laboratories, Inc., Chicago Heights, Ill.), 10 g of glucose, 1 g of Marmite (a commercial yeast extract from Marmite Ltd., London, England), and 0.5 mmole of phosphoric acid in 1 liter of distilled water. The *p*H of the medium was adjusted to 7.0 with NaOH before autoclaving. *S. aureus* cultures were harvested by centrifuging them in the cold after they had been cultivated for 44 to 48 hr. These cultures were in the early stationary phase; they contained about 10^{10} cells per ml and about 2.2 mg of cells (dry weight) per ml. *M. lysodeikticus* cultures were harvested similarly after 68 to 72 hr of growth when they were in the phase of decreasing growth rate; they contained about 1.2×10^{10} cells per ml and about 2.9 mg of cells per ml. Harvested cells were used immediately or frozen until needed.

Wall isolation. Cells were disrupted in small batches by sonic disintegration, as described previously (5), or in large batches by high-speed stirring with a Sorvall Omnimixer (Ivan Sorvall Inc., Norwalk, Conn.), as described by Sharon and Jeanloz (24). For both procedures, glass beads were added to suspensions to aid in cell disruption, and the suspensions were chilled with ice or circulating iced water during treatment. Wall fragments were separated from other cell debris by repeated differential centrifugation; they were heated at 90 C for 10 min and were then treated with trypsin to remove proteins (23).

A number of criteria were used to evaluate the extent of contamination of wall preparations by other cell components. A reduction of over 95% in the optical density (700-nm light) of *M. lysodeikticus* wall suspensions was recorded after addition of lysozyme. A reduction of over 80% in the optical density of *S. aureus* wall suspensions was recorded after addition of lysostaphin (kindly supplied by Walter Zygmunt of the Mead Johnson Research Center). Absorbance-scattering spectra (220 to 700 nm) of suspensions containing walls showed no absorption peaks. Acid-hydrolyzed (4 M HCl at 105 C for 24 hr) wall preparations showed little reactivity with the Folin-Ciocalteau reagent; 1 mg (dry weight) of *S. aureus* walls had the same reactivity as 0.04 mg of bovine serum albumin, and 1 mg of *M. lysodeikticus* walls had the same reactivity as 0.05 mg of serum albumin. In addition, wall preparations were routinely monitored during purification and use by phase-contrast microscopy.

Isolated walls were used shortly after preparation or were frozen until needed.

Table 1 presents a description of the main polymers that have been isolated and characterized from *M. lysodeikticus* and *S. aureus* walls. Walls isolated from *M. lysodeikticus* cells usually are composed mainly of loosely cross-linked peptidoglycans to which small amounts (less than 10% of total wall dry weight) of teichuronic acids (polymers of glucose and 2-acetamido-2-deoxy-mannuronic acid) are linked via phosphodiester bonds. However, there is a recent description (12) of peptidoglycan-poor *M. lysodeikticus* walls which contain as much as 260 μmoles of glucose per g. Analyses of the walls we used (Table 2) indicated very low glucose or phosphate content and, therefore, minimal teichuronic acid deposition in the structures. *S. aureus* walls appear to be composed mainly of highly cross-linked peptidoglycans to which polyribitol-phosphate teichoic acids are linked via phosphodiester bonds. It should be appreciated that the views of wall structure presented in Table 1 are rather rough ones strictly applicable to only certain strains of the species. They are at variance with our findings (presented in Table 2) and those of Salton (22) that *S. aureus* Duncan walls contain an excess of cationic groups relative to anionic groups. Preliminary experiments to be described elsewhere indicate that this excess is due to unacetylated muramyl residues in walls prepared from stationary-phase cells. The content of free carboxyl groups in our wall preparations (Table 2) also was rather high, and it suggested that as many as 50% of the carboxyl terminal alanyl residues might not have been involved in cross-linking.

Determination of wall volume. The volumes occupied by cell walls in suspensions were assessed by measuring dextran-impermeable volumes of the suspensions by means of the commonly used space or thick-suspension technique. For these measurements, high molecular weight dextrans (Dextran 2000 with average molecular weight of ca. 2 million obtained from Pharmacia Inc., New Market, N.J.) were used. These dextran molecules are too large to enter wall

TABLE 1. *Previously described polymers of coccal cell walls*

Type of polymer and main characteristics	Organism[a]	
	S. aureus	*M. lysodeikticus*
Peptidoglycan		
Predominant muramyl-substituted peptide	N^{α}-(L-alanyl-D-isoglutamyl)-L-lysyl-D-alanine	N^{α}-(L-alanyl-γ-(α-D-glutamyl-glycine)-L-lysyl-D-alanine)
Predominant cross-linking peptides	(Glycine)$_5$ (Glycine + serine)$_5$	Direct links between lysyl and alanyl residues or bridging polymers of above pentapeptide
Percentage of muramic acid residues, with attached peptide	ca. 100	ca. 40
Percentage of lysyl or N-terminal glycyl amino groups free to accept protons	7–25	80
Extent of acetylation of muramyl amino groups	Complete	Incomplete
Polymers covalently linked to peptidoglycan	Ribitol teichoic acid with *N*-acetylglucosamine and ester-linked alanyl substituents	Copolymer of equimolar amounts of glucose and 2-acetamido-2-deoxymannuronic acid (plus some phosphate groups)

[a] Polymer structures given for *S. aureus* have been determined mainly with walls of the Copenhagen strain; pertinent references are 1, 9, 10, 11, and 25. Pertinent references to work on *M. lysodeikticus* wall polymers are 12, 14, 18, and 21.

pores (8), and so the dextran-impermeable volume of a wall suspension includes the volume of wall polymers plus the volume of fluid within wall pores. When walls contract, fluid is squeezed out of the pores, and it can then act as a solvent for dextran.

Our exact experimental technique and procedure for calculating dextran-impermeable volumes have been described in detail elsewhere (16). In essence, the technique used was as follows. Water-washed wall suspensions were centrifuged at 30,000 × *g* for 40 min to obtain well-packed wall pellets weighing from 3 to 5 g (wet weight). Then measured volumes (approximately equal to the wall pellet volume) of 1% dextran solutions containing desired concentrations of acids, bases, or salts were added to the pellets, the walls were resuspended, and the suspensions were incubated for 40 min at ice temperature before being centrifuged. Supernatant fluids were carefully decanted, and samples were pipetted into cellophane dialysis bags. The bags were sealed and placed in large beakers of cold distilled water, which was changed frequently until all dialyzable material had been removed and the electrical conductivity of the dialysate was the same as that of distilled water. The dialyzed dextran solutions were transferred, with thorough washing of the dialysis bags, to tared weighing pans. The pans were placed in an oven at 100 C for 24 hr and then were weighed to determine the amount of dextran in the sample.

Acid-base titrations. Titrations of ionizable cell wall groups were carried out in two ways. For some titrations, no attempt was made to control ionic strength, which changed from ca. 0 to ca. 0.15 during the course of titrations. For other titrations, ionic strength was held nearly constant at 0.2 by adding KCl to suspensions. The exact titration procedure has been described previously (16).

Other analytical procedures. Total numbers of anionic groups were estimated from the capacities of cell walls to bind cationic safranine dye. The procedure described by Fraenkel-Conrat and Cooper (7) was followed. Total free amino groups were estimated from capacities of walls to react with fluorodinitrobenzene. The procedure applied by Ghuysen and Strominger (10) to *S. aureus* walls was used. The phosphate contents of walls were estimated by first hydrolyzing them (4 N HCl at 105 C for 20 hr) and then assaying the hydrolysates for inorganic phosphate by means of the Fiske-SubbaRow (6) method. Glucose was assayed with glucose oxidase (Worthington Biochemical Corp., Freehold, N.J.). When necessary, components of hydrolyzed cell walls were separated by paper chromatography or by thin-layer chromatography on silica gel.

Removal of teichoic acid from S. aureus walls. Ester-linked D-alanyl residues were removed from teichoic acids of isolated *S. aureus* walls by treating them with 0.1 N NH_4OH [1 g (wet weight) of walls plus 20 ml of 0.1 N NH_4OH] at room temperature for 5 to 6 hr. This procedure removed 0.53 mmoles of alanine per g (dry weight) of walls. No other ninhydrin-positive compounds that could be detected on paper chromatograms were removed by NH_4OH treatment.

Ammonia-treated walls were washed and resuspended in 0.1 M sodium acetate buffer, *p*H 5.6, containing sufficient sodium metaperiodate for maximal oxidation of teichoic acid diols. The reaction mixtures were incubated at ice temperature in the dark for 20 hr. This treatment was found to remove approximately 85 to 90% of the wall phosphate.

RESULTS

Table 2 presents pertinent chemical and physical characteristics of the cell walls we used to study electrostatic volume changes. Measurements of wall dextran-impermeable volumes indicated that *S. aureus* walls were structurally more compact than *M. lysodeikticus* walls, and this compactness appeared to be related to the extensive peptide cross-linking that is characteristic of *S. aureus* peptidoglycans (9). A 1-g quantity (dry weight) of extensively washed *S. aureus* walls suspended in distilled water occupied a volume of 5.1 ml, compared with 10.9 ml for the same weight of *M. lysodeikticus* walls. *S. aureus* walls had fewer dissociable groups per gram (dry weight) than did *M. lysodeikticus* walls; again, this finding could be related to the high degree of peptide cross-linking in *S. aureus* peptidoglycans. Of the anionic groups in *S. aureus* walls, more than 50% were teichoic acid phosphates, and nearly 50% of the cationic groups in the walls were teichoic acid D-alanyl amino groups. Since the number of phosphates in the walls was essentially equal to the number of D-alanyl residues that could be removed by mild alkaline hydrolysis, it appeared that, on the average, each ribitol residue in teichoic acids had one esterified D-alanyl substituent. If one assumes that each ribitol also has one *N*-acetyl-glucosamine substituent (1), then the phosphate analyses indicate that approximately 25% of *S. aureus* wall dry weight was accounted for by teichoic acids.

M. lysodeikticus walls had much lower phosphate contents (0.9 μmoles per g, dry weight), and nearly all the anionic groups in them appeared to be peptidoglycan carboxyl groups, including muramic acid carboxyl groups lacking peptide constituents. The *M. lysodeikticus* walls we used were relatively poor in teichuronic acids, as indicated by their low phosphate content and low glucose content (36 μmoles per g, dry weight).

The results presented in Table 2 indicate that cell walls of *S. aureus* Duncan suspended in media of nearly neutral *p*H should have a net positive charge, whereas *M. lysodeikticus* ATCC 4698 walls should have a net negative charge. This difference in net charge was reflected in the *p*H values of water suspensions of the walls. Since the structures are not soluble in water, suspensions have two phases—a wall phase with solvation shell and a bulk medium phase. An asymmetric Donnan distribution of mobile ions develops between these two phases. Hence, the bulk-phase *p*H (measured with a glass electrode) of water suspensions of well-washed cationic *S. aureus* walls was below 7, owing to wall-phase retention of hydroxyl ions as counterions for fixed cations. The *p*H of water suspensions of *M. lysodeikticus* walls was greater than 7, owing to retention of hydronium ions as counterions for fixed anions.

TABLE 2. *Chemical and physical characteristics of isolated coccal cell walls*

Characteristic	Organism: *S. aureus*	Organism: *M. lysodeikticus*
Amino groups free to react with fluorodinitrobenzene (meq/g, dry wt)	1.17	1.45
Amino groups of esterified D-alanyl residues (meq/g, dry wt)	0.53	0
Anionic safranine-binding groups (meq/g, dry wt)	0.90	2.09
Phosphate groups (meq/g, dry wt)	0.52	<0.01
Carboxyl groups[a] (meq/g, dry wt)	0.38	2.09
Dextran-impermeable volume (ml/g, dry wt)	5.1	10.9
Total maximum charge density: meq/g, dry wt	2.07	3.54
Total maximum charge density: meq/ml	0.406	0.325
Minimum net charge density: meq/g, dry wt	+0.27	−0.64
Minimum net charge density: meq/ml	+0.053	−0.059
*p*H of water suspension of well-washed walls	6.6	7.6

[a] Carboxyl groups were assessed from the difference between total anionic groups and phosphate groups; the estimate is probably somewhat too high because some phosphates may have been doubly anionic.

Although *S. aureus* walls had appreciably fewer dissociable groups per gram (dry weight) than did *M. lysodeikticus* walls, they had a higher maximum charge density in the hydrated state because of being structurally more compact. However, there is a problem in assigning values for charge density when considering structures that contain flexible amphoteric polyelectrolytes such as peptidoglycans, because there may be extensive internal neutralization of charge. In a prior study, Carstensen and Marquis (5) found that the charge densities of *M. lysodeikticus* walls, assessed by measuring capacities of the structures to conduct electrical current, were close to, but significantly more than, net charge densities calculated from acid-base titration results. Electrical current is carried through cell walls essentially only by mobile counterions, such as Na^+, and not by the relatively immobile charged groups of wall polymers. In other words, *M. lysodeikticus* wall polymers are sufficiently flexible to allow most, but not all, wall cations to serve as counterions for wall anions.

Changes in wall volume associated with acid-base titration. Electrostatic interactions appeared to play a significant role in determining the compactness of hydrated bacterial cell walls and, as shown in Fig. 1 and 2, isolated walls expanded and contracted in response to changes in environmental *p*H and ionic strength. Figures 1A and 2A show volume changes for *M. lysodeikticus* and *S. aureus* walls, respectively, and Fig. 1B and 2B show proton-binding curves for the walls. The *p*H values indicated in the abscissae of the figures are bulk-phase *p*H values measured with a glass electrode. Because of Donnan effects, wall-phase *p*H may be higher or lower than bulk-phase *p*H, depending on the net charge of the walls, and this disparity is reduced in media of higher ionic strength. These complicating factors made it difficult to interpret proton-binding curves in a strictly quantitative manner, at least with current electrolyte theory.

There were also difficulties in estimating total numbers of ionizable groups from titration curves.

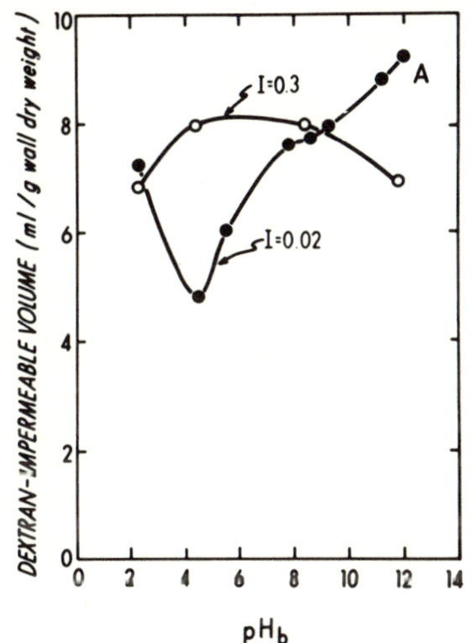

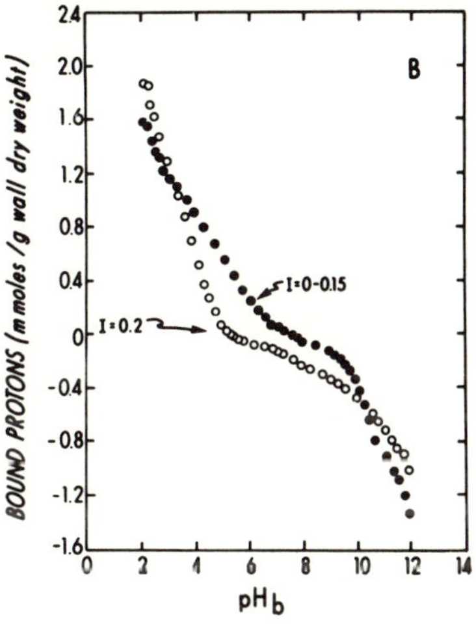

FIG. 1. *Acid-base induced volume changes of M. lysodeikticus cell walls. (A) Dextran-impermeable volumes of walls suspended in media of constant ionic strength, 0.02 or 0.3, plotted in relation to bulk-phase pH (pH_b) of the suspensions. (B) Titration curves for walls in media of constant or variable ionic strength.*

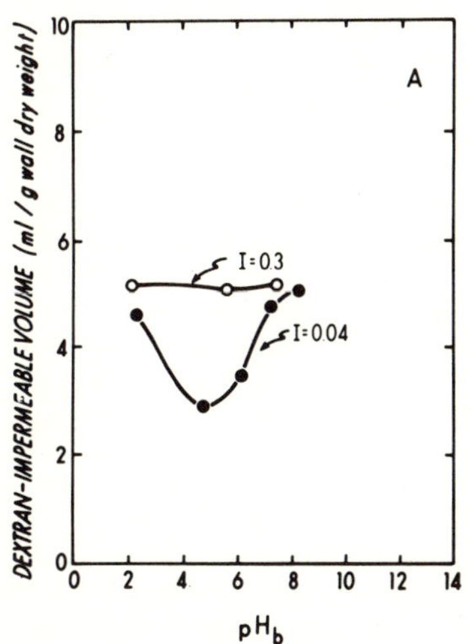

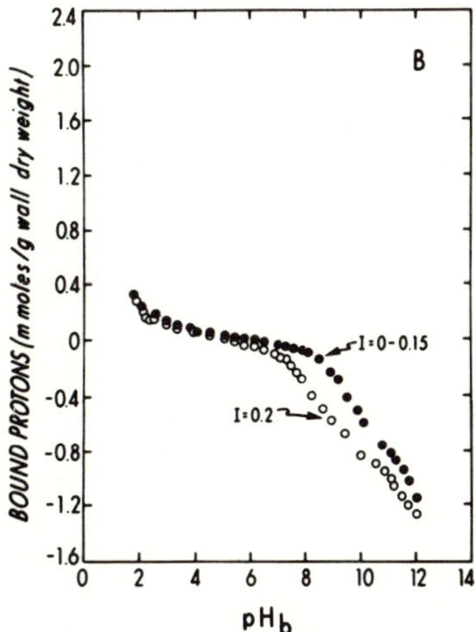

FIG. 2. *Acid-base induced volume changes of S. aureus cell walls. (A) Dextran-impermeable volumes of walls suspended in media of constant ionic strength, 0.04 or 0.3, plotted in relation to bulk-phase pH. (B) Titration curves for walls in media of constant or variable ionic strength.*

One of the main problems was the difficulty in accurately assessing isoelectric and isoionic *p*H values for insoluble polyelectrolytes, and we arbitrarily chose initial bulk-phase *p*H values of wall suspensions as zero points for considering addition or loss of protons. Also, it was not always possible to obtain titration end points, especially for phosphate groups, because of unavoidable electrode errors at extremes of *p*H. In all, we considered that the chemical characteristics (safranine-binding capacity, phosphate content, and fluorodinitrobenzene reactivity) gave more reliable estimates of total numbers of ionizable groups than did acid-base titration.

Despite these problems, it is possible qualitatively to relate titration behavior to volume changes for isolated cell walls. For example, the results presented in Fig. 1 indicate that *M. lysodeikticus* walls suspended in media of relatively low ionic strength (0.02) and high *p*H (11 to 12) were expanded polyanions. Most amino groups were uncharged, and all carboxyl (and phosphate) groups were negatively charged. When environmental *p*H was lower, ϵ-amino groups of peptidoglycan lysyl residues (average intrinsic *p*K of ca. 10.4) were protonated, and they electrostatically attracted anionic carboxyl groups so that the walls contracted. In media of still lower *p*H, more amino groups (including any α-amino groups with average *p*K of ca. 7.8) became protonated, and the walls contracted further. *M. lysodeikticus* walls were most compact in low ionic strength media of *p*H ca. 4.5, a value close to the predicted isoelectric *p*H for the walls. In more acid media, protonation of carboxyl groups (average *p*K of ca. 3.8) was sufficient to convert the walls to polycations so that transfer to still more acid media resulted in expansion of the structures.

Electrostatic repulsions seemed to predominate in *M. lysodeikticus* walls suspended in media with *p*H above ca. 9. When environmental ionic strength was raised from 0.02 to 0.30, the walls shrank due to the shielding effects of mobile counterions. The opposite effect occurred for walls in media with *p*H between 2.5 and 9; increased environmental ionic strength induced swelling because of a predominance of electrostatic attractions in the walls.

The curves for *S. aureus* walls can be interpreted qualitatively in much the same manner. However, we found that *S. aureus* walls exposed to media of *p*H 9.5 or higher lost appreciable amounts of esterified alanyl groups and could not be made to regain their original compactness. Nevertheless, *S. aureus* walls in low ionic strength media of mildly alkaline *p*H appeared to be expanded, and protonation of amino groups led to contraction (Fig. 2A). The walls were most compact in media of *p*H ca. 4.8, and further reduction in environmental *p*H converted them to polycations so that protonation of carboxyl or phosphate groups led to swelling. Electrostatic attractions seemed to predominate in *S. aureus* walls in media of *p*H 2 to 8 since increased environmental ionic strength induced swelling.

Salt-induced volume changes. Isolated cell walls suspended in distilled water could be made to contract by adding NaCl to the suspensions. Examples of NaCl-induced contraction of *M. lysodeikticus* and *S. aureus* walls are presented in Fig. 3A and 4A, respectively. The maximal contraction (per cent decrease in dextran-impermeable volume) that could be induced in this manner was ca. 30% for *M. lysodeikticus* walls, compared with only about 12% for compact *S. aureus* walls. Maximally effective NaCl concentrations were approximately 1.0 and 0.5 M, respectively. Transfer of walls to more concentrated solutions resulted in less contraction rather than more. It should be noted that the concentrations (ionic strengths and osmolalities) indicated in Fig. 3 and 4 are initial ones; final concentrations were about one-half the initial ones because of dilution by water in wall pellets. NaCl-induced wall contraction was reversible, and appeared to be electrostatic rather than osmotic in nature because it could be induced with other salts but not with sucrose.

Figures 3B and 4B show that whole cells of *M. lysodeikticus* and *S. aureus* also contracted when transferred from distilled water to NaCl solutions. Whole-cell contraction in nonplasmolyzing solutions more dilute than ca. 0.2 to 0.4 M appeared to be due mainly to electrostatic wall contraction. Thus, for example, *M. lysodeikticus* cells contracted by some 9% when transferred from distilled water to 0.3 M NaCl solution, whereas isolated walls contracted by some 25% following similar transfer. Previously reported results (5) obtained with *M. lysodeikticus* cells grown in this laboratory indicated that the wall occupies about 50% of the cell volume. Therefore, electrostatic wall contraction should have caused the cells to contract by about 12.5%. The finding that the measured volume change was somewhat less than expected suggests that walls of intact cells were more resistant to contraction than isolated walls. This conclusion does not seem unreasonable in view of the impossibility of isolating cell walls without mechanically damaging them. Also, it is known that bacterial cells have relatively high turgor pressures, so that there is a tendency for the cell protoplast to press against the wall and stretch it with development of mechanical tension.

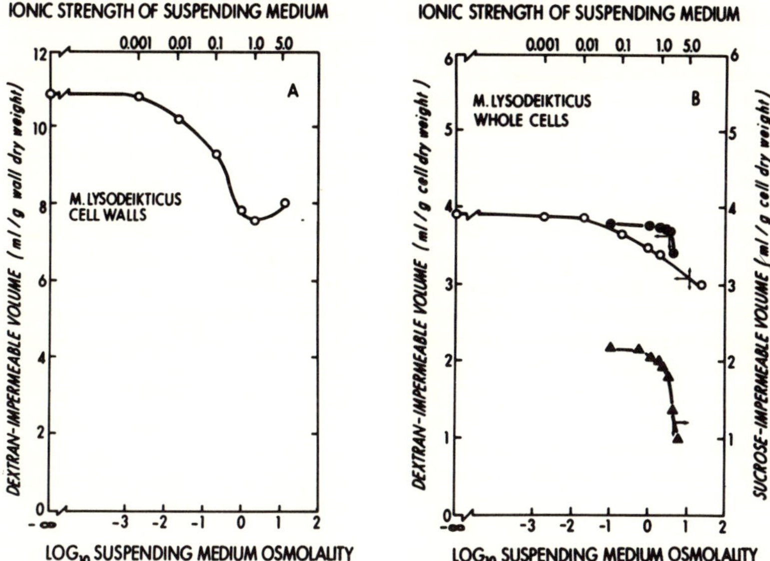

FIG. 3. *NaCl-induced contraction of M. lysodeikticus cell walls or whole cells and sucrose-induced contraction of whole cells. (A) Dextran-impermeable volumes of isolated walls in NaCl solutions (○). (B) Dextran-impermeable volumes of cells in NaCl (○) or sucrose (●) solutions and sucrose-impermeable volumes of cells in sucrose solutions (▲). Arrows on curves indicate the pertinent axes.*

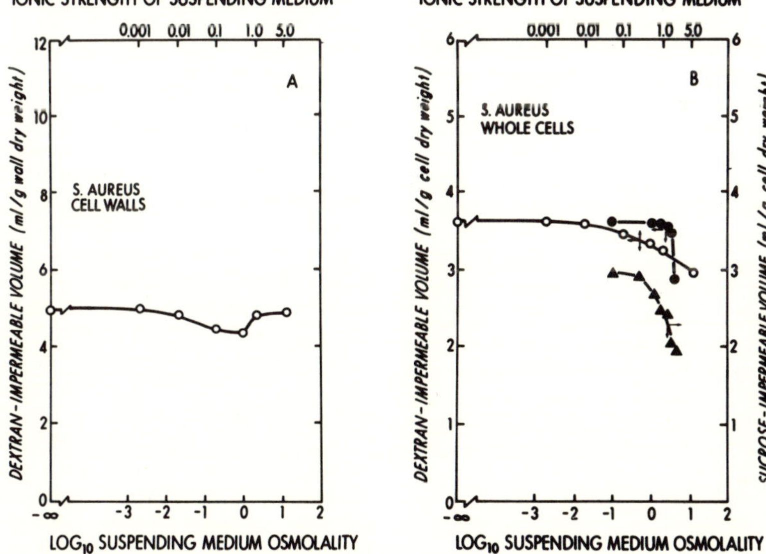

FIG. 4. *NaCl-induced contraction of S. aureus cell walls or whole cells and sucrose-induced contraction of whole cells. (A) Dextran-impermeable volumes of isolated walls in NaCl solutions (○). (B) Dextran-impermeable volumes of cells in NaCl (○) or sucrose (●) solutions and sucrose-impermeable volumes of cells in sucrose solutions (▲). Arrows on curves indicate the pertinent axes.*

In very concentrated salt solutions, osmotic effects were compounded with electrostatic effects, and, as shown in Fig. 3B and 4B, cells shrank even when sucrose rather than NaCl was used as solute. Cells in concentrated sucrose solutions did not release appreciable quantities of electrolytes, and so purely osmotic effects could be measured when sucrose solutions were used. The marked decreases in sucrose-impermeable volumes, relative to dextran-impermeable volumes, of cocci transferred to sucrose solutions more concentrated than ca. 0.4 osmolal indicated that the cells had become plasmolyzed, even though plasmolysis vacuoles could not be seen with a phase microscope. (Here, the sucrose-impermeable volumes of cells were considered to be essentially equal to protoplast volumes, whereas dextran-impermeable volumes were considered to be equal to whole cell volumes.) Presumably, many small plasmolysis vacuoles, rather than one or two large ones, formed in each cell.

The contraction of whole coccal cells in hypertonic sucrose solutions appeared to be similar to the type of contraction measured previously (16) for plasmolyzed *B. megaterium* cells. Retraction of the turgid protoplast from the cell wall of an intact cell reduces mechanical tension in the wall and leads to elastic contraction. This type of contraction seemed to overshadow electrostatic volume changes for cells suspended in plasmolyzing salt solutions. For example, intact *S. aureus* cells shrank by about 11% when transferred from 0.5 to 5.0 M NaCl solution, even though isolated walls swelled about 9% as a result of similar transfer.

Salt-induced displacement of protons from isolated cell walls. Salt-induced volume changes of coccal cell walls appeared to be mechanistically more complex than volume changes accompanying *p*H shifts in constant ionic strength media. Many salts, including NaCl, KCl, $MgCl_2$, and $CaCl_2$, displaced protons from isolated walls, and, as shown in Fig. 5 and 6, addition of these salts to water suspensions containing *M. lysodeikticus* or *S. aureus* walls resulted in decreased suspension *p*H. Part of this displacement for *M. lysodeikticus* walls can be attributed to shifts in the Donnan equilibrium with substitution of salt cations for hydronium ions as counterions for fixed wall anions. But disturbance of the Donnan equilibrium for cationic *S. aureus* walls should have just the opposite effect—suspension *p*H should increase when salts are added.

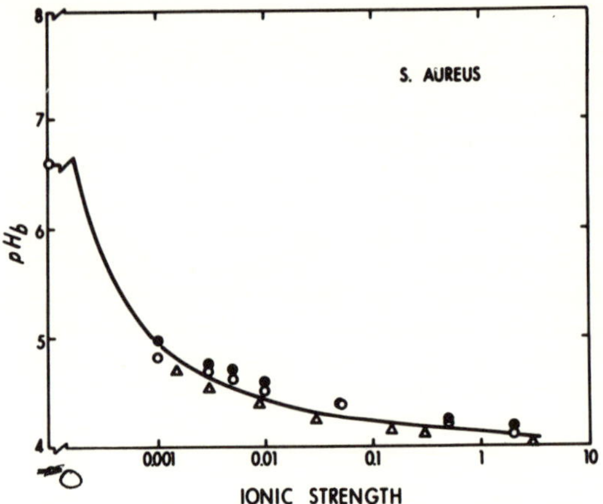

FIG. 6. *Proton displacement from isolated S. aureus cell walls (10.6 mg of walls, dry weight, per ml) by NaCl (○), KCl (●), or $CaCl_2$ (△).*

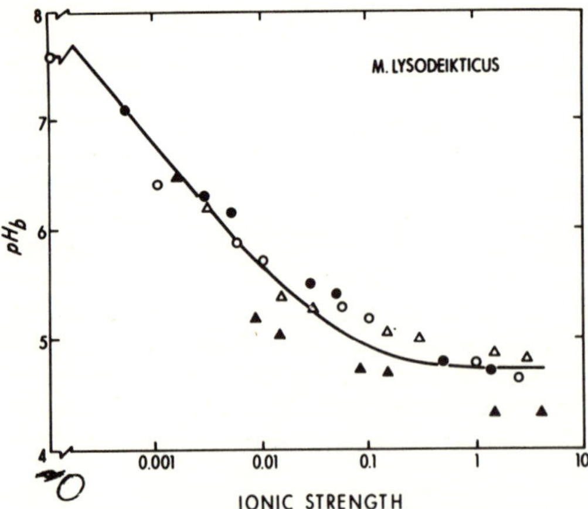

FIG. 5. *Proton displacement from isolated M. lyodeikticus cell walls (10.3 mg of walls, dry weight, per ml) by NaCl (○), KCl (●), $CaCl_2$ (△), or $MgCl_2$ (▲).*

It is well known that charged forms of dissociable compounds are favored in media of increased ionic strength; for example, the apparent *p*K values of protein carboxyl groups are lower in media of higher ionic strength. Thus, if NaCl were added to a suspension of *S. aureus* walls in water with an initial *p*H of 6.6, any carboxyl groups that were protonated should have an increased tendency to give up their protons to produce ion pairs. The curves of Fig. 1B and 2B show that walls of both types of cocci were more highly protonated in media of low ionic strength than in media of high ionic strength in the pertinent *p*H range from 4 to 8.

Even though NaCl-induced *p*H shifts were marked—amounting to 1 *p*H unit or more of change—only a relatively small percentage of the wall carboxyl groups, less than 0.4% for *M. lysodeikticus* walls and less than 0.7% for *S. aureus* walls, would have to dissociate to produce the measured acidification of the suspensions. Apparently the *p*K-lowering effect of increased environmental ionic strength for wall carboxyl groups overshadowed shifts in Donnan equilibria, because addition of even small amounts of NaCl to *S. aureus* wall suspensions led to decreased suspension *p*H.

Role of teichoic acids in electrostatic volume changes of S. aureus walls. Removal of teichoic acid alanyl groups from *S. aureus* walls caused the walls to swell, as indicated by a rise in dextran-impermeable volume from 5.1 to 10.1 ml per g. Subsequent removal of 85 to 90% of teichoic acid phosphates from the walls by means of periodate treatment resulted in a reduction of dextran-impermeable volume to a value of 6.2 ml per g. (The weights used here are dry weights of the degraded walls rather than that of the original walls.) These volume changes are probably due mainly to changes in electrostatic interactions, but may also be related to reduction in degree of polymerization of the wall complex.

DISCUSSION

The experimental results presented in this paper and in a previous one (16) led us to conclude that bacterial peptidoglycans are flexible polyelectrolytes. Thus, the major structural polymer of bacterial walls appears to be much more flexible than cellulose or chitin of plant cell walls. Cellulose and chitin are both mechanically rigid polymers, owing, at least in part, to hindered rotation around β-glucosyl bonds (20). The glycan portions of peptidoglycans should be relatively rigid also, but they are short chains, 12 to 65 disaccharide units in length, compared with cellulose or chitin chains which have polymerization degrees in the range of 1,400 to 8,000. A reasonable mechanical model for peptidoglycans might be a sort of three-dimensional rope ladder with relatively rigid polysaccharide rungs and relatively flexible polypeptide ropes. In other words, the flexibility of peptidoglycans is related mainly to the known flexibility of polypeptide chains. Compounds such as teichoic acids may be covalently linked to peptidoglycans, and, as our work with *S. aureus* walls indicated, these substituent polymers may also contribute to wall mechanical properties.

Bacterial peptidoglycans appear to be less densely packed in the hydrated state than are celluloses and chitins. Measured values for the density of cellulose in water range from about 1.5 to 1.6 g per ml (15). Bacterial walls have densities well below these values. The *M. lysodeikticus* walls used in this study had a density of about 1.06, as measured by a method similar to the one used by Black and Gerhardt (2), and *S. aureus* walls had a density of about 1.07. *M. lysodeikticus* walls were composed almost entirely of peptidoglycan, and it seems reasonable to conclude that bacterial peptidoglycans have a much more open, loosely knit structure than do celluloses and chitins. In fact, both cellulose and chitin form relatively large, condensed crystallites into which water molecules cannot readily diffuse. Because of the looseness of the peptidoglycan matrix, one would expect little weak interchain bonding to occur. Also, the peptide chains are too short for extensive helices to form.

Our major aim in studying electrostatic interactions of bacterial cell walls is to be able to interpret wall functioning in terms of polymer chemistry. Bacterial walls are considered to be supporting structures for bacterial cells, but, as indicated previously, it might be more appropriate to think of them as ensheathing elastic, restraining structures with sufficient mechanical compliance to allow for moderate cell swelling and bending. Compliance of elastic net structures is reduced by mechanical tension, so bacterial walls should be less compliant when stretched by cell turgor pressure and more compliant when turgor pressure is reduced.

Bacterial cell walls act also as heteroporous molecular sieves (8) to exclude macromolecules from cells. Presumably, electrostatic shrinkage of walls would result in contraction of wall pores and lowering of the molecular exclusion threshold. However, experimental tests are needed to substantiate this view, since pore shape may, in some circumstances, be as important as pore size in the molecular sieving effect.

Bacterial cell walls behave also as ion-exchangers in that they retain exchangeable counterions in association with fixed, charged groups of wall polymers. Relatively little detailed information is available regarding the specificity and other characteristics of ion exchange in cell walls. Carstensen (4) found that the concentration of mobile counterions in walls of intact *M. lysodeikticus* cells may exceed that in the cytoplasm. Previously, Britt and Gerhardt (3) found that cationic lysine molecules were bound to *M. lysodeikticus* walls, and Marquis and Gerhardt (17) showed that passive uptake of nonmetabolized α-aminoisobutyrate by *B. megaterium* cells can be almost entirely accounted for by wall uptake of the amino acid. The role of this ion retention by walls in cell economy is still largely unknown, although it is easy to see how binding of nutrients such as amino acids and organic acids could be important to a cell with intermittent nutrient supply. Certainly, it is reasonable to assume that the general physical structure of bacterial peptidoglycans and the variants in structure found among different types of bacteria reflect adaptations of the organisms to their natural habitats.

ACKNOWLEDGMENTS

We thank Aaron Morse for skillful and diligent technical help. This investigation was supported by Public Health Service research grant RO1 AM 08990 from the National Institute of Arthritis and Metabolic Diseases and by National Science Foundation grant GB-6573.

LITERATURE CITED

1. Baddiley, J. 1962. Teichoic acids in walls and cells of gram-positive bacteria. Fed. Proc. **21**:1084–1088.
2. Black, S. H., and P. Gerhardt. 1962. Permeability of bacterial spores. IV. Water content, uptake, and distribution. J. Bacteriol. **83**:960–967.
3. Britt, E. M., and P. Gerhardt. 1958. Bacterial permeability. Total uptake of lysine by intact cells, protoplasts, and cell walls of *Micrococcus lysodeikticus*. J. Bacteriol. **76**:288–293.
4. Carstensen, E. L 1967. Passive electrical properties of micro-organisms. II. Resistance of the bacterial membrane. Biophys. J. **7**:493–503.
5. Carstensen, E. L., and R. E. Marquis. 1968. Passive electrical properties of microorganisms. III. Conductivity of isolated bacterial cell walls. Biophys. J. **8**:536–548.
6. Fiske, C. H., and Y. SubbaRow. 1925. The colorimetric determination of phosphorus. J. Biol. Chem. **66**:375–400.
7. Fraenkel-Conrat, H., and M. Cooper. 1944. The use of dyes for the determination of acid and basic groups in proteins. J. Biol. Chem. **154**:239–246.
8. Gerhardt, P., and J. A. Judge. 1964. Porosity of isolated cell walls of *Saccharomyces cerevisiae* and *Bacillus megaterium*. J. Bacteriol. **87**:945–951.
9. Ghuysen, J.-M. 1968. Use of bacteriolytic enzymes in determination of wall structure and their role in cell metabolism. Bacteriol. Rev. **32**:425–464.
10. Ghuysen, J.-M. and J. L. Strominger. 1963. Structure of the cell wall of *Staphylococcus aureus*, strain Copenhagen. I. Preparation of fragments by enzymatic hydrolysis. Biochemistry **2**:1110–1119.
11. Ghuysen, J.-M. and J. L. Strominger. 1963. Structure of the cell wall of *Staphylococcus aureus*, strain Copenhagen. II. Separation and structure of disaccharides. Biochemistry **2**:1119–1125.
12. Ghuysen, J.-M., E. Bricas, M. Lache, and M. Leyh-Bouille. 1968. Structure of the cell walls of *Micrococcus lysodeikticus*. III. Isolation of a new peptide dimer, N^{α}-[L-alanyl-γ-(α-D-glutamylglycine)]-L-lysyl-D-alanyl-N^{α}-[L-alanyl-γ-(α-D-glutamylglycine)]-L-lysyl-D-alanine. Biochemistry **7**:1450–1460.
13. Knaysi, G. 1951. Elements of bacterial cytology. Comstock Publishing Co., Ithaca, N.Y.
14. Leyh-Bouille, M., J.-M. Ghuysen, D. J. Tipper, and J. L. Strominger. 1966. Structure of the cell wall of *Micrococcus lysodeikticus*. I. Study of the structure of the glycan. Biochemistry **5**:3079–3090.
15. Mark, R. E. 1967. Cell wall mechanics of tracheids. Yale University Press, New Haven, Conn.
16. Marquis, R. E. 1968. Salt-induced contraction of bacterial cell walls. J. Bacteriol. **95**:775–781.
17. Marquis, R. E., and P. Gerhardt. 1964. Respiration-coupled and passive uptake of α-aminoisobutyric acid, a metabolically inert transport analogue, by *Bacillus megaterium*. J. Biol. Chem. **239**:3361–3371.
18. Mirelman, D., and N. Sharon. 1967. Isolation and study of the chemical structure of low molecular weight glycopeptides from *Micrococcus lysodeikticus* cell walls. J. Biol. Chem. **242**:3414–3427.
19. Mitchell, P., and J. Moyle. 1956. Osmotic function and structure in bacteria, p. 150–180. *In* E. T. C. Spooner and B. A. D. Stocker (ed.), Bacterial anatomy. Cambridge Univ. Press, Cambridge, England.
20. Morawetz, H. 1965. Macromolecules in solution. Interscience Publishers, New York.
21. Perkins, H. R. 1963. A polymer containing glucose and aminohexuronic acid isolated from cell walls of *Micrococcus lysodeikticus*. Biochem. J. **86**:475–483.
22. Salton, M. R. J. 1961. Studies of the bacterial cell wall. VIII. Reaction of walls with hydrazine and with fluorodinitrobenzene. Biochim. Biophys. Acta **52**:329–342.
23. Salton, M. R. J. 1964. The bacterial cell wall. Elsevier Publishing Co., New York.
24. Sharon, N., and R. W. Jeanloz. 1964. A procedure for the preparation of gram-quantities of bacterial cell walls. Experientia **20**:253–254.
25. Tipper, D. J., and M. F. Berman. 1969. Structures of the cell wall peptidoglycans of *Staphylococcus epidermidis* Texas 26 and *Staphylococcus aureus* Copenhagen. I. Chain length and average sequence of crossbridge peptides. Biochemistry **8**:2183–2192.
26. Wámoscher, L. 1930. Versuche über die Struktur der Bacterienzelle. Z. Hyg. Infektionskr. **111**:422–460.

Part VI
SOUND

Editor's Comments on Papers 29, 30, and 31

In their natural environments microorganisms are not subjected to sound waves of sufficient intensity or frequency for sound to be of any real importance. However, high-frequency sound has proved of value in sterilization and disintegration of cells for extraction of enzymes and antigens from microorganisms without denaturization. Strangely, the entire field of study developed from 1917 efforts to detect submarines by high-frequency sound waves. Langevin had observed that small fish were killed by high-frequency sound waves during the initial military studies. The description of both physical and biological effects of high-frequency sound waves (Paper 29) in 1927 by Wood and Loomis led to an incredible number of applications of the process, so that by 1951, in a time span of only 24 years, over 580 papers had been published on the effects of sonic and ultrasonic vibration. The early bibliography from 1900 to 1950 has been published by Naimark, Klair, and Mosher (1951).

El'piner (1964) describes two stages in ultrasonic chemical reactions both of which are induced by cavitation. The term "cavitation" refers to the formation of cavities in the liquid and their subsequent collapse.

> State I. According to Frenkel, photo and electrochemical effects occur in the cavitation bubble when the pressure in the bubble is very low. The conditions created at this stage of development of the cavitation bubble are characteristic of a low-pressure electric discharge in the presence of gases and substances in the vapor state. In these conditions the gases present in the cavitation bubble undergo ionization or activa-

> tion. The formation of activated or ionized particles obviously terminates in electronic breakdown of the bubble.
>
> State II. The further development of the cavitation bubble is accompanied by an increase in pressure in it and the loss of its lenticular shape. In the compression phase the bubble collapses. The relatively long lived active radicals and atoms formed pass into the surrounding medium. In addition, the collapse of the cavitation bubble gives rise to a shock wave.

Other effects occur because of the sharp temperature increases due to vibration.

In Paper 29 the physical and biological effects of high-frequency sound waves are described. The ultrasonic sound was generated with a crystal quartz piezoelectric oscillator driven by a 50,000-V current at 300,000 Hz. The authors observed intense heating effects at the end of glass rods used to transmit the sound. Sufficient heat was generated to burn holes in wood or glass. Sufficient pressure was developed by the radiation in oil above the quartz vibrator to support 150 g on an 8-cm^2 glass plate. Maximum pressure existed when the distance between the surface of the plate and the quartz oscillator was an integral number of half-wavelengths. The radiation supported a 10-cm mound of oil over the oscillator! The authors observed the emulsification of immiscible liquids and the breakup of erythrocytes, spirogira, and paramecia. Bacteria survived the effects of the radiation. A mouse was killed by ultrasonic radiation, its body cavity reaching a temperature of over 40°C. The authors suggested that internal heating may be one of the most important biological effects of ultrasonic irradiation.

Schmitt, Olson, and Johnson (1928) used a 250-W oscillator at 750,000 Hz to study the effect of ultrasonic sound on protozoa. They found that paralysis accompanied by coagulation of protoplasm preceded death. Harvey, Harvey, and Loomis (1928) built a high-frequency sound generator directly on the stage of a microscope. This allowed ultrasonic sound effects on erythrocytes, *Elodea* leaf cells, and bacteria to be observed directly. The bacteria were violently agitated but were not destroyed. Irradiation of the two-cell-layer *Elodea* leaves caused the protoplasm to rotate, and with increased intensity of radiation fragments were torn loose. The most striking effect in living organisms, besides heating, was the intracellular stirring which Harvey and Loomis (1928) found was affected by various factors, such as air bubbles, amount of water, interference patterns, and wave frequency. Harvey and Loomis (1929) observed the effects of sound (400,000

Hz) on the luminous bacteria *Bacillus fischeri (Vibrio fischeri)*. The suspensions were cooled first to eliminate or reduce the effects of heating. The bacterial luminescence quickly diminished upon irradiation and the cells were broken up. Johnson (1929) proposed that the expulsion of dissolved gas from an aqueous medium is responsible primarily for the destruction of protozoa or erythrocytes. Destruction was prevented by application of sufficient external pressure or by evacuation. Williams and Gaines (1930) found a frequency of 8800 Hz to be bactericidal for *Escherichia coli* and to lyse erythrocytes. The lethal effect was attributed to violent agitation set up within the cells by the sound waves.

Schmitt and Uhlemeyer (Paper 30) in 1930 reported the results of experiments in which they extended the observations of Johnson. Many microorganisms were tested. A pressure of 60 to 80 lb was sufficient to prevent damage to cells during irradiation. The concept that cavitation bubbles reflect the sound from their surface and thus concentrate the radiation was discounted by two experiments. Suspensions of cells were irradiated in the presence of infusorial earth, and in a second experiment paramecia that had ingested Chinese black were irradiated. The increase of surface area either in or outside of the cells did not aid in producing the lethal effects. Evidence was obtained indicating that the plasma membrane was more permeable after sonic irradiation. The author's experiments with artificial plasma membranes led them to propose that the lethal effect is due to a rupture of the plasma membrane by a chemical or a physiochemical effect produced by cavitation in the water immediately surrounding the cell.

Harvey (1930) reviewed the results of 23 papers and listed seven major effects of sonic irradiation: (1) Heating of media that absorb waves. (2) Movement of the particle into the mode of the standing wave pattern. (3) Floculation of particles above a critical size. (4) Dispersion at liquid–gas, liquid–liquid, and liquid–solid interfaces. (5) Cavitation resulting from expulsion of dissolved gas from solutions. (6) Compression and expansion of media through which the sound waves pass. (7) Acceleration of chemical reactions. Harvey and Loomis (1931) were able to photograph the destruction of *Arbacia* eggs by sonic waves with a camera system capable of 1200 pictures/s. The disintegration occurred in even less time.

Applications of high-frequency sound waves for sterilization of liquids such as milk followed. Chambers and Gains (1932) developed a device for sterilization of a continuous flow of milk.

The breakup of cells by sonic vibration suggested its possible use for preparation of cellular extracts. Chambers and Flosdorf (1936) utilized sonic vibration at 8900 Hz to extract antigens without loss of activity from *Salmonella typhosa* and *Streptococcus hemolyticus.*

Stumpf, Green, and Smith in 1946 (Paper 31) made the first systematic study of ultrasonic disintegration for the extraction of bacterial enzymes. A 700-W generator with a frequency range of 200 to 1000 Hz was utilized for the studies. This device allowed both the frequency and the intensity to be varied. Prolonging the irradiation beyond certain limits did not achieve further cellular disintegration. The thickness or viscosity of the suspensions affected the results. A very thick paste of packed cells was not affected by the radiation, but diluted suspensions disintegrated. When the base of the container was placed at an odd multiple of one-half the wavelength, reinforcement of the wave occurred and maximum disruption was obtained. The degree of disintegration was independent of frequency and dependent only on intensity and amplitude. If the power output was raised beyond a certain range, denaturation of labile proteins occurred.

REFERENCES

El'piner, J. E. 1964. *Ultrasound—Physical, Chemical, and Biological Effects.* Consultants Bureau, New York. (Translated from Russian by F. L. Sinclair.)

Chambers, L. A., and E. W. Flosdorf. 1936. Sonic extraction of labile bacterial constituents. *Proc. Soc. Exptl. Biol. Med. 34*(5): 631–636.

———, and N. Gaines. 1932. Some effects of intense audible sound on living organisms and cells. *J. Cell. Comp. Physiol. 1*(3): 451–473.

Harvey, E. N. 1930. Biological aspects of ultrasonic waves, a general survey. *Biol. Bull. 59*(3): 306–325.

———, and A. L. Loomis. 1928. High frequency sound waves of small intensity and their biological effects. *Nature 121*(3051): 622–624.

———, and A. L. Loomis. 1929. The destruction of luminous bacteria by high frequency sound waves. *J. Bacteriol. 17*(5): 373–376.

———, and A. L. Loomis. 1931. High speed photomicroscopy of living cells subjected to supersonic vibrations. *J. Gen. Physiol. 15*(2): 147–153.

———, E. B. Harvey, and A. L. Loomis. 1928. Further observations on the effect of high frequency sound waves on living matter. *Biol. Bull. Marine Biol. Lab. 55*(6): 459–469.

Johnson, C. H. 1929. The lethal effects of ultrasonic radiation. *J. Physiol. 67*(4): 356–359.

Naimark, G. M., J. Klair, and W. A. Mosher. 1951. A bibliography on

sonic and ultrasonic vibration: biological, biochemical and biophysical applications. *J. Franklin Inst. 251*: 279–299.
Schmitt, F. O., A. R. Olson, and C. H. Johnson. 1928. Effects of high frequency sound waves on protoplasm. *Proc. Soc. Exptl. Biol. Med. 25*(8): 718–720.
Williams, O. B., and N. Gaines. 1930. The bactericidal effects of high frequency sound waves. *J. Infect. Dis. 47*(6): 485–489.

29

Reprinted from *Phil. Mag.*, Ser. 7, **4**, 417–436 (1927)

THE PHYSICAL AND BIOLOGICAL EFFECTS OF HIGH-FREQUENCY SOUND-WAVES OF GREAT INTENSITY

R. W. Wood and Alfred Lee Loomis

Introduction.

IN the present paper we shall give an account of a preliminary survey of what appears to be a wide field for investigation, opened up by the study of the very surprising and remarkable effects obtained with sound-waves of high frequency and great intensity generated in an oil-bath by a piezo-electric oscillator of quartz operated at 50,000 volts and vibrating 300,000 times per second.

The radiation pressure exerted against a glass disk 8 cm. in diameter amounts, under certain circumstances, to 150 grams, and when operating against the free surface of the oil (from which the radiation is totally reflected) raises it in a mound 7 cm. in height, surmounted by a fountain of oil drops, some of which are projected to an elevation of 30 or 40 cm.

The waves can be transmitted along a glass thread 0·2 mm. in diameter and a metre or more in length, and the end of the thread, if squeezed between the thumb and finger, burns a groove in the skin. A tapering glass rod, 0·5 mm. in diameter at the tip, can be thrown into vibration of such ntensity that a pine chip smokes and emits sparks when pressed against the tip, the rod burning its way rapidly through the wood, leaving a hole with blackened edges. If a glass plate is substituted for the chip, the rod drills its way through the plate, throwing out the displaced material in the form of a fine powder or minute fused globules of glass.

If the waves are passed across the boundary separating two liquids such as oil and water or mercury and water, more or less stable emulsions are formed. Chemical reactions are accelerated, crystallizations started, and other remarkable effects produced by these very intense super-sonic vibrations.

Preliminary experiments with interference fringes formed between a vibrating plate and one at rest indicate that the amplitude of the vibration is of the order of magnitude of a wave-length of light, yet an enormous amount of energy is delivered. The mean energy and acceleration are both proportional to the square of the frequency, and we are here dealing with what Prof. C. V. Boys very tersely describes as "*All* acceleration and *no* motion."

The method employed in the generation of the waves is essentially the one developed by Professor Langevin in 1917 for the purpose of locating submarines by the echo of a narrow beam of high-frequency sound-waves. Shortly thereafter experiments along similar lines were inaugurated by the British and American navies.

In Langevin's original apparatus the vibrations of the piezo-electric quartz plate were excited by a Poulsen arc in connexion with suitable condensers and coils. Voltages as high as thirty or forty thousand were applied to the plates, and the amplitude of the waves raised to such a degree that small fish were killed by the radiation and pain of considerable severity was experienced when the hand was thrust into the water in the tank. The Poulsen arc proved troublesome, however, owing to its instability, which made it impossible to keep the electrical vibration in tune with the natural frequency of the quartz plate, and it was speedily supplanted by the vacuum tube, which is used exclusively at the present time.

Since the coefficient of viscosity increases with the square of the frequency, comparatively low frequencies only can be employed when beams of sound are to be projected under water over long distances for signalling or other purposes. Voltages of one or two thousand at frequencies of from 30 to 40 thousand are used for the most part in work of this nature.

In our experiments this limitation is not imposed, since

absorption of the radiation by the medium does not interfere to any great degree with the study of effects close to the source, and we operate usually with voltages in the vicinity of 50,000 at frequencies ranging from 200,000 to 500,000.

Description of the Apparatus.

(See Plate VII. fig. 1.)

The apparatus employed in the present work was built in the Research Laboratory of the General Electric Co. at

Fig. 1.

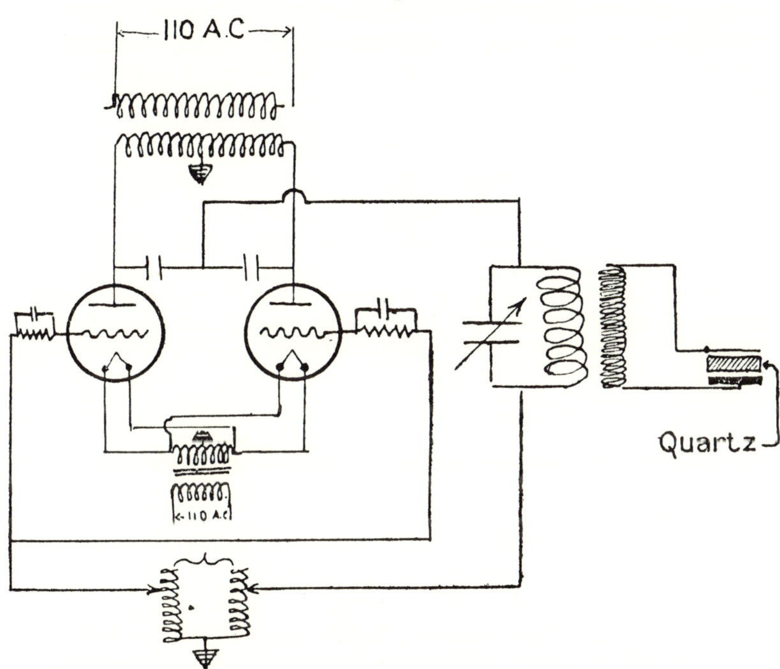

Schenectady. It consists of a two kilowatt oscillator, designed originally for an induction furnace, a bank of oil condensers, giving capacities up to 0·1 microfarad, a large variable air condenser, and several pairs of coaxial coils for raising the voltage. The primary or outer coil consisted of from 7 to 20 turns of Litzendraht cable, the coils varying from 16 to 24 cm. in diameter. The secondary coils were wound on glass-cylinders (100 to 250 turns) and mounted within the primaries. Fig. 1 shows in conventional manner the wiring of the various parts. The use of several coils was found to be necessary as we employed quartz plates varying in thickness from 7 to 14 mm., with which we obtained waves with frequencies ranging from 100,000 to 700,000 cycles per second. The quartz plates were circular disks, and when in operation, one of them rested on a disk of sheet lead at the bottom of a dish of transformer oil. The other electrode consisted of a disk of very thin sheet brass resting on the upper surface of the quartz. The coils for raising the potential and the glass dish with oil, in which the quartz oscillator is immersed, are shown on Plate VIII.

Pressure due to the Radiation.

We have already mentioned the pressure developed against the free surface of the oil above the quartz vibrator as a result

Fig. 2.

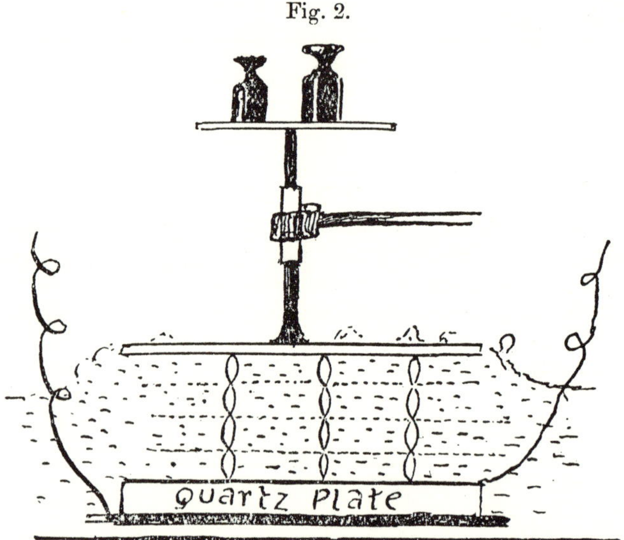

of the reflexion of the radiation. In the case of reflexion from plates of glass or metal the magnitude of the effect can be measured. We found that a glass disk 8 cm. in diameter attached to a glass rod and supported as shown in fig. 2 would support a weight of 150 grams. The pressure is a maximum when the distance between the under surface of the plate and the upper surface of the quartz oscillator is a whole number of half wave-lengths. Under this condition the reflected wave strikes the oscillator when its phase is such as to reflect the wave back to the plate. The energy is thus imprisoned by multiple reflexions between the vibrator and the plate, and the amplitude rises to a very high value, for the same reason that the amplitude of vibration of the stationary waves on a thread attached to a vibrating tuning-fork may be twenty or thirty times the amplitude of the fork when the length of the thread is properly adjusted.

If the rod which supports the glass plate is held in the fingers and the plate pushed down gradually into the oil a strong resistance is encountered periodically, as if the plate were breaking its way through a series of resisting films. In the positions of maximum pressure the energy is reflected back and forth between the oscillator and plate. In the positions of minimum pressure the wave reflected down from the plate meets the oscillator when it is in such a phase as to transmit the reflected wave. That such is the case was shown by the following experiment.

The quartz oscillator was mounted vertically in a large oil bath and a metal vane hung by a bifilar suspension at a short distance to one side of it. With the oscillator functioning, this vane was deflected to one side by the pressure of the radiation. If now a glass plate was immersed in the oil on the opposite side of the oscillator, the deflexion of the vane increased periodically as the glass plate was moved towards the oscillator, *i. e.* it swung back and forth.

This shows us that, to get the maximum amount of energy from the oil into a bath of some other liquid (or into a solid) immersed in the oil, the distance between the bottom of the bath and the upper surface of the oscillator must be so adjusted that the energy builds up between the two by multiple reflexions. A beaker of water, for example, is heated much more rapidly when the above condition is fulfilled. This operation will be referred to in future as adjusting for energy density.

The height of the oil mound raised above the surface over the plate depends in the same way upon the depth of the oil. We have never obtained a smooth uniform mound, as might be expected with low amplitude if the plate were simply expanding and contracting as a whole. With the oscillator operated at low voltage, a number of humps appear on the surface which shift their position with every alteration in the capacity of the condenser. When the frequency of the electrical oscillation is tuned exactly to the natural frequency of the crystal plate the oil mound rises to a height of 7 cm., its summit erupting oil drops like a miniature volcano. Further increase of the power gave a mound 10 cm. in height, but the quartz plate broke into fragments. A photograph of the oil fountain taken against a very bright background with an exposure of 1/500 second is reproduced on Plate VII. fig. 2.

Stationary waves on tubes

If a glass tube a metre long and 2 or 3 cm. in diameter, closed at the bottom like a test-tube, is coated on the inside with a layer of a heavy oil, the oil gathers itself together in rings about 3 mm. apart, which line the tube from top to bottom, as soon as the lower end is dipped into the vibrating oil over the quartz plate. The rings appeared to be interrupted in places by another system of waves, and a permanent record of the pattern was secured by substituting paraffin, coloured with aniline red, for the oil and using it in a warm tube. A photograph of a portion of the tube with the rings cut across into a pattern of regularly spaced dots is reproduced on Plate IX. fig. 3.

The wave-length of the oblique system is about 1·5 times that of the horizontal system.

We at first attributed the rings to a stationary system of compressional waves formed by interference between disturbances reflected down from the top of the tube with those coming up from below, but the velocity deduced from the wave-length and frequency was much less than the velocity of sound in glass. The waves turned out to be transverse vibrations, the wave-length being but a small fraction of the diameter of the tube. If a similar glass tube was used without the oil, and the outer surface heated to the softening point in the flame of a blast-lamp while the vibrations were running up and down the tube, the stationary system was permanently recorded in the glass, and could be made visible by casting a shadow of the tube with sunlight (or light from any concentrated source) on a sheet of paper held at a distance of 10 or 15 cm. from the tube. A shadow photograph, made in this way, is represented on Plate IX. fig. 4.

We have made no careful study of the modes of vibration of a tube for high frequencies, and have no explanation for the oblique system of greater apparent wave-length. It seems evident, however, that the velocity of propagation is greater for the waves forming this system than for the others. It was observed also, as the tube cooled down, that the paraffin remained fluid longer in those portions of the tube where the double system registered than at other places, indicating that the internal heating of glass was greater here than elsewhere.

Transverse waves on glass plates, rods and threads.

If a glass rod is cemented with sealing-wax to the centre of a circular glass disk, dusted with lycopodium, a beautiful system of concentric circular rings forms on the plate as soon as the lower end of the rod is brought into contact with the vibrating oil-bath.

These rings are formed at the nodal lines of a system of stationary waves in the plate, formed by the interference of the waves reflected from the rim with those radiating from the centre. If the disk is thicker at the centre than at the rim (we used the base and stem of a broken wine-glass) the distance between the rings is less at the rim than at the centre, from which the inference can be drawn that the waves are transverse vibrations, which is to be expected considering the arrangement of the rod and disk. The velocity is higher at the thick than at the thin portions, consequently the rings are further apart.

If the rod is cemented to the disk at a point situated at a small distance from the centre, we obtain the complicated pattern reproduced as a negative (*i. e.* the lycopodium lines black) on Plate X. fig. 5. Here we evidently have the waves reflected from the rim coming to a focus, which becomes a second source of radiation on the side of the centre opposite to the rod, and a system of radiating interference fringes is formed, as with two similar sources of light. The pattern in the immediate vicinity of the sources is of especial interest.

A beautiful system of circular rings of variable spacing is produced by dusting the inner surface of a champagne glass with lycopodium, and touching the base to the surface of the vibrating oil. At the rim the rings are closer together than near the centre, where the glass is thicker.

Applying the lycopodium method to a glass rod which has been drawn down in a flame to a long tapering point, the diameter varying from 7 to 0·5 mm., gives us a system of rings, the separation of which decreases rapidly as we pass from the thick to the thin portion. This shows that the velocity of propagation is a function of the diameter of the rod, which will of course be true for transverse, but not for longitudinal disturbances. If the rod terminates in a fine point no rings are formed, since in this case the reflexion from the end is negligible and the stationary wave system is not formed.

More permanent rings better suited for wave-length measurements were made by the following method. A small ball of soft red wax was stuck on the point of the rod held vertically over the oil. This melted and slid down the rod as soon as the lower end was dipped in the oil, owing to heat developed by friction between the vibrating glass and the wax. The wax solidifies in rings above the ball as it descends leaving a permanent record of the wave-length, as shown on Plate X. fig. 6.

In this particular case the rod is drawn down from a glass tube closed at the bottom. The energy abstracted from the oil and thrown into the rod is greater than when a large solid rod forms the collector. Remarkable calorific effects obtained with this type of collector will be described presently.

Heat developed at Contact-point between vibrating rods and matter.

This type of heating was first accidentally observed when taking the temperature of the oil in the erupting mound over the vibrator. Though the mercury registered only 25°, the thermometer tube became so hot at the point where it was held between the thumb and finger that it had to be released. The heat of course is developed by friction between the vibrating glass stem and the skin of the fingers, or rather by the rapid *pounding* of the transverse vibrations, and becomes unbearable only when the glass is squeezed tightly between the thumb and finger. This same heating is observed when any object such as a rod, tube, beaker or flask is held by the fingers and dipped into the vibrating oil-bath. If a glass rod is drawn out into a long thread of the diameter of a horsehair, terminating in a pear-shaped bead, the heat developed, when the end of the thread is squeezed between the fingers and the bead dipped into the oil fountain, is so great that a groove with seared edges is left in the skin. A week later bright red spots similar in appearance to blood-blisters developed, which did not disappear for several weeks. These were perhaps due to an effusion of blood from capillaries deep down in the skin, which were ruptured by the vibration. Still more powerful effects were obtained with rods 0·5 mm. in diameter drawn out from the top of an Erlenmeyer flask the neck of which had been closed by fusion in a blast-lamp. Shown at extreme right of Plate VIII., which also shows rack-and-pinion stand for adjusting a flask of water containing a frog for maximum energy density. A side tube was fused to the flask, by which it was supported in a clamp-stand, furnished with a rack-and-pinion movement, by which the distance between the flat bottom of the flask and the vibrator could be accurately adjusted for the position necessary for securing the maximum density of the radiation imprisoned between the surfaces. With this condition fulfilled a dry pine-chip, pressed against the top of the glass rod, smoked and emitted an occasional spark, while the rod rapidly burned its way through the wood, leaving a hole with charred edges. The heating, of course occurs only at the point of contact, the remainder of the rod being quite cold.

If a plato of glass is pressed lightly against the top of the rod, the surface of the plate is etched at the point of contact, the microscope showing a curious scalloped pattern. If the pressure is increased the rod drills its way rapidly through the plate, and the microscope shows small globules of molten glass, and finely powdered material.

This method of conducting the vibration along threads of glass yields a valuable technique for the investigation of the biological effects of the high-frequency vibration, which can thus be applied at a small point on a living organism, egg or embryo under the microscope. We have felt the heat at the end of a thread a metre long and 0·2 mm. in diameter. With the flask form of collector, with proper adjustment for the system of stationary waves between the top of the quartz vibrator and bottom of the flask, the energy thrown into the glass thread is often so great that the thread breaks into pieces.

Another form of collector by which energy can be abstracted from the oil and conducted into a rod or thread is shown in fig. 6, Plate X.—a tube of glass closed with a round bottom and drawn down to rod or thread at the opposite end. The sloping wall of the bottom appears to facilitate the production of transverse waves and it was found also, in experimenting with a flat collector made of a thin plate of glass, drawn off into a rod at one corner, that the most vigorous vibrations occurred in the rod when the plate was immersed in the oil in a slightly oblique position and not when it was parallel to the surface of the quartz plate.

Velocity and Dispersion of Transverse Waves in Solids.

By the methods just indicated it is possible to measure the velocity of propagation of the transverse waves if the frequency is known. As has been said, the velocity is a function of the diameter of the rod or the thickness of the plate. Micro-photographs of the rings on rods of 0·15, 0·5 and 1 mm. in diameter are reproduced on Plate XI. fig. 7. Observations made with waves generated by quartz vibrators of different thicknesses showed that the velocity was also a function of the frequency, as is the case with light traversing a dispersing medium. We investigated the phenomenon of dispersion employing disturbances of four different frequencies—441, 405, 350, and 285 thousand vibrations per second—forming the rings of red wax or lycopodium on rods and threads of glass varying in diameter from 6 mm. to 0·1 mm. The results are shown graphically in the curves reproduced in fig. 3. The velocity, which at

Fig. 3

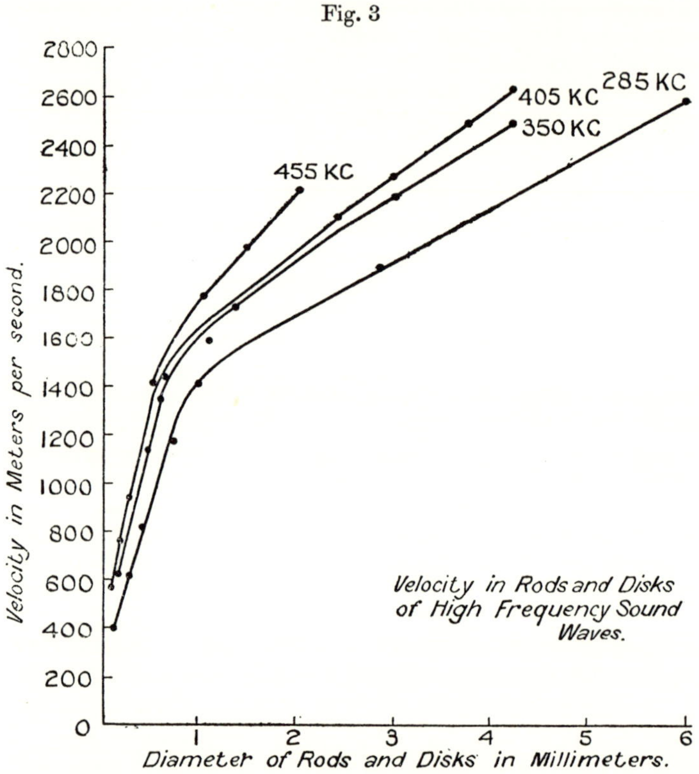

285 kilocycles is 2600 metres per second in a glass rod 6 mm. in diameter, falls off to 400 metres in the case of a glass thread 0·1 mm. in diameter.

As the diameter of the rod is increased it becomes increasingly difficult to form the rings, and for values of the order of magnitude of the half wave-length it becomes impossible to obtain any record of the transverse waves, even by the lycopodium method, which is the more sensitive of the two. With very large energy input we several times obtained under these conditions indications of waves of considerably greater wave-length than the transverse ones which we at first believed represented the longitudinal disturbance. Measurements, however, were not in agreement with the known velocity of sound in glass, which is in the neighbourhood of 5000 metres per second.

The dispersion, for a rod of any given diameter, is given by taking the ordinates from the four curves corresponding to the same abscissa, the velocity of propagation increasing with the frequency at first rather rapidly, then more slowly, and finally rapidly again. These values are preliminary only and do not represent a very high degree of accuracy.

Sonic Interferometer and the Velocity of the Waves in Liquids.

Another method was developed for the determination of the velocity in liquids, depending upon the formation of a system of stationary waves between the vibrator and a reflector. We have alluded to the variable periodic pressure exerted upon a plate pushed down through the oil over the vibrator. By counting the number of resisting planes as the plate is lowered through a measured distance, the wave-length can be determined. It is obvious that the accuracy of this method depends upon the precision with which the points of maximum or minimum pressure can be determined. It was found that the best results were secured by employing electrical means for registering the reaction of the reflected waves upon the piezo-electric vibrator, as employed by Professor Pierce of Harvard, and a very compact instrument of low power has been developed in collaboration with Professor J. C. Hubbard, now of Johns Hopkins University, who has already made a series of very satisfactory observations of the velocity of sound in various liquids, solutions, and liquid mixtures. It is possible to obtain results of great accuracy with only a few cubic centimetres of liquid. The results of this work will be reported in a subsequent paper.

As is apparent this method is analogous to the employment of the Fabry and Perot interferometer for determining the wave-length of light, and the apparatus may be termed a Sonic Interferometer. With the plate in the position for maximum pressure, and multiple to-and-fro reflexion of the waves, the condition is similar to that obtaining with a Fabry and Perot instrument illuminated by *parallel rays* at normal incidence, with its plates at such a distance as to secure the maximum transmission of light. If the wave-length of light or the distance between the plates is slightly altered, the transmission falls to a very low value. Transmission in this case corresponds to the entrance of the energy into a beaker of water tuned for maximum energy density, as previously described. In the usual treatment of the Fabry and Perot interferometer one is apt to overlook the circumstance that under certain conditions transmission of the light is refused, as it is customary to illuminate the instrument by an extended luminous source such as a flame—in which case the amount of light transmitted is independent of wave-length, but transmission is permitted in specified directions only, this limitation giving rise to the rings.

Heating of Liquids and Solids.

The kinematic coefficient of viscosity increases as the square of the frequency. At frequencies from 300 to 400 K.C. the heating of liquids is very pronounced. Thus a test-tube filled with water and immersed in a beaker containing water and cracked ice heats rapidly when the beaker is lowered over the vibrating quartz disk, showing that the energy of the sound-waves (which pass into the water in the test-tube after traversing the intervening glass walls and the ice-water) is converted into heat by absorption, the water in the test-tube rising rapidly in temperature notwithstanding the circumstance that it is surrounded by a layer of water at 0°. The rise of temperature may be as great as one degree every three seconds, the rate depending upon whether the depth of the water and the distance between the vibrator and the bottom of the beaker are adjusted for multiple reflexions of the radiation, as previously described. With 250 c.c. of water the heat developed amounted to about 900 calories per minute—with 150 c.c., 750 cals.; with 100 c.c. ; 700 cals.; and with 50 c.c. (in a test-tube) 430 cals.

These results show that though the temperature rise is higher in the case of small volumes of liquid, the total energy abstracted from the radiation increases with the volume of liquid employed.

We have not as yet made any precise determinations of the heating of various liquids by the radiations. Three determinations with 45 c.c. of ethyl alcohol gave 4°·1, 3°·6, and 4°·5 as the temperature elevation for an exposure of 20 seconds. These observations were, however, made before the necessity of accurate adjustment of the containing vessel had been realized. In work of this nature it will be necessary to devise some method by which it will be possible to throw the same amount (or a measured amount) of energy into the fluids under investigation.

A few observations have also been made of the internal heating of solids. A block of newly formed ice (distilled water frozen in a beaker by ice and salt) was subjected to the action of the sound-waves for two minutes in a beaker of ice water containing numerous small fragments of ice which kept the temperature of the water at 0°. Adjustment for maximum energy density was secured by watching the block of ice which was elevated above its normal position in the water by the radiation pressure. At the end of the exposure the block of ice, on being squeezed between the thumb and fingers, broke up into small fragments showing that liquefaction had taken place throughout the mass, as in the case of so-called "rotten ice" after exposure to the sun's rays.

We found that this experiment could not be duplicated with natural ice (*i. e.* pond ice), and believe that this may be due to the circumstance that in this case we are dealing with a single crystal, whereas in the case of the artificial ice we have a mass of interlocking crystals, the heating taking place at the crystal interfaces.

To eliminate effects due to air bubbles in the ice, distilled water, thoroughly boiled to remove air, and coloured with fluorescein, was covered with a thin layer of paraffin and frozen. The outer portions of the ice block were perfectly transparent, while the central portion was yellowish in colour, somewhat cloudy and devoid of fluorescence, showing that the fluorescein was not in solution. On exposing this block to the radiation for thirty seconds and examining it in sunlight against a black background, it was found to be traversed by innumerable interlacing planes of green fluorescence, caused by the internal melting and consequent solution of the dye. A somewhat analogous effect was noted in one of our earlier experiments, in which a bit of candle (probably stearic acid) was melted on the surface of water in a test-tube and allowed to solidify. Water was then introduced above the solid plug, and the lower end of tube subjected to the vibration. No trace of the radiation appeared in the water above the plug the *under surface* of which melted rapidly and was thrown down into the water as a white emulsion, showing the powerful absorption of the radiation by the solid stearic acid.

It was during an attempt later on to duplicate this experiment with paraffin that we obtained the apparent crystallization of the substance referred to later on in this paper.

Formation of Emulsions and Fogs.

If two non-miscible liquids such as oil and water are simultaneously subjected to the radiation in the same beaker, an emulsion or colloidal solution is formed as a result of the forces acting at the interface between the liquids. This phenomenon was first observed in an earlier arrangement of our apparatus, in which the quartz oscillator operated in a thick layer of oil floating on water. White clouds of finely divided oil were thrown down into the water and occasionally, probably when exact tuning happened to be secured, large masses of oil were projected, with almost explosive violence, down into the water, as a shower of large drops. Stable emulsions resembling milk can be made of stearic acid or paraffin and water. A beaker of water with a layer of mercury at the bottom, under the influence of the vibrations becomes first milky, then brown, and finally black. At the end of twenty-four hours most of the mercury has settled, but a sufficient amount remains in suspension to make the water slightly turbid.

Colloidal solutions of the low melting-point alloys have also been made.

At a liquid-air interface, in the case of less viscous liquids the forces brought into play drive the liquid into the air in the form of a spray of minute droplets, forming a fog. This atomization (to use the popular term) of a liquid by the sound-waves is best shown by pouring a little benzol into a beaker and lowering the beaker into the oil, tuning it for energy density. The beaker fills rapidly with a cloud of white smoke, a benzol fog, the surface of which is in tumultuous motion. A photograph of this phenomenon is reproduced on Plate XII. fig. 10. The beaker was illuminated by sunlight (reflected from a mirror) against a black background, the camera pointing as nearly as possible towards the direction from which the light was coming, to secure the maximum illumination. The filaments rising from the surface are larger droplets moving too rapidly for the camera shutter. Fogs can also be formed over water, but in this case the droplets are larger and settle rapidly.

Prof. C. V. Boys has drawn my attention to the analogy between this experiment and a phenomenon produced by the explosion of a "depth-charge." The first indication of the explosion seen at the surface is the sudden development of a great cloud of fine spray which is projected to a height of ten or fifteen feet. This is followed immediately by the rising mound of water and the great fountain, lifted by the expanding gases. The spray is due to the shock of a "pulse"-wave of almost instantaneous pressure. The spray is never seen following the explosion of a mine, which always contains a large volume of air. This acts as a cushion, and prevents the development of the instantaneous pressure.

A fog of extremely small droplets of heavy transformer oil can be formed by a collector of special construction. This is perhaps the most spectacular experiment of all. A glass tube of about 2·5 cm. diameter is closed at one end, and drawn down to a diameter of about 7 mm. at the other end. The tube is clamped to the rack and pinion stand, and the rounded bottom lowered into the oil, adjustment being made for energy density. This form of collector for compressing the radiation into small volume (if we may use this expression) is the most efficient thus far found. The constricted portion of the tube heats rapidly by internal friction, and if touched with the finger becomes unbearably hot. To distinguish between heating by internal friction and the heat developed when another body is pressed against the vibrating glass it is only necessary to operate the tube for a few seconds, shut off the power and touch the thin constricted portion with the finger. If now, with the tube adjusted for maximum heating and the oscillator working at full power, we apply a little oil with a medicine dropper to the outside of the tube above the narrowed portion, a very surprising thing happens. The oil spreads over the surface and is thrown out in jets of spray resembling smoke and a dense cloud gathers about the tube. If a match flame is brought gradually up to this cloud brilliant flashes of countless small scintillations occur resembling the sparks of the Japanese fireworks, and if the flame is brought closer, the whole cloud goes up in a grand burst of flame and the top of the tube continues to burn fiercely like a torch until the oil supply is exhausted. For some reason not quite clear the oil is unable to run down the thin portion of the tube, gathering in a ring of greater or less width at the top. If the power on the oscillator is reduced or the tube thrown out of adjustment by an up or down movement of a fraction of a millimetre the ring crawls down the tube, rising again as soon as the intensity of the vibration is increased. This driving of the oil layer up the tube is doubtless due to the fact that the energy of the radiation travelling up the tube is greater than the energy reflected down. It is obvious that the walls of the tube are vibrating as stationary waves, since close inspection shows that the oil in the ring has gathered in more or less regularly arranged dots, and that the jets of spray shoot out from these dots (fig. 4).

Fig. 4.

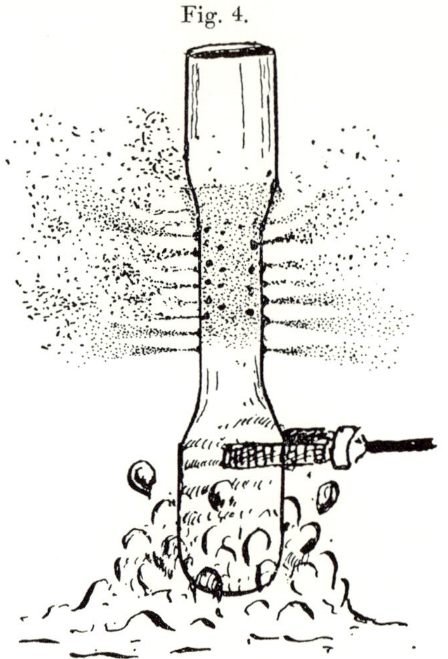

A photograph of the tube in action is reproduced on Plate XIII., the clouds of spray and the jets being illuminated by sunlight and photographed under the same conditions as with the benzol fog in the beaker. Twenty or more of the small jets of spray are visible in the original photograph shooting out from the tube just below the cloud, and three vertical rows of the oil dots referred to above can be seen without difficulty between the jets.

A fog of metallic mercury was also formed with this type of collector.

With a collector of this type the amplitude of the vibration at the constricted portion frequently becomes so great that the tube is fractured in a curious manner, small irregular pieces of glass breaking away from the tube. A photograph of a tube fractured in this way is reproduced on Plate XI. fig. 8.

Flocculation of Suspended Particles in a Liquid.

In the case of particles exceeding a certain size and of a specific gravity not much greater than that of the liquid in which they are suspended, flocculation occurs the moment the liquid is traversed by the waves, the particles rushing together to form clusters which presently gather into a single dense mass just under the surface. We first noticed this effect when studying the action of the radiation on the unicellular organism paramecium, and at the moment interpreted it as a biological effect, but we presently duplicuted it with fine sawdust which had soaked until water-logged.

The phenomenon is probably the result of radiation pressure combined with shielding perhaps, or analogous to the attractions observed and studied by Bjerknes in pulsating liquids.

Effects of the waves on Chemical Reactions and Crystallization.

The effects of these high frequency radiations on Chemical reactions is under investigation by Dr. W. T. Richards of Princeton University. In work of this sort it is very necessary to distinguish between effects due to the heating of the liquid as a whole and effects due to the vibration. The liberation of dissolved gases from water has been a matter of common observation by all who have worked with super-sonics. The bubbles appear the moment the vibration is started, long before any sensible rise of temperature is observed, and instead of rising with a uniform velocity they remain suspended in the nodal planes moving up in an irregular manner, with frequent pauses. Distinct evidence has been found that certain chemical reactions are accelerated by the vibrations, the most striking case being the so-called "clock-reaction" in which the termination of the reaction is marked by the sudden change from a clear transparent solution to a deep blue one. In attempting to repeat the experiment, previously made with stearic acid, with paraffin wax, we obtained what appeared to be a crystallization of the paraffin induced by the vibration. The melted paraffin was allowed to solidify on the surface of hot water in a beaker. When the whole was quite cold, the bottom of the beaker was dipped into the vibrating oil. Small opaque white spots immediately appeared in the layer of translucent paraffin, which increased in size, forming irregular clusters. A photograph of the sheet of paraffin (natural size) by transmitted light is reproduced on Plate XI. fig. 9. Under a Zeiss binocular stereoscopic microscope the white spots appeared to be nodules covered with protruding points which suggested a crystalline structure. They have not yet been carefully examined however. Sir W. Bragg, on seeing the photograph, remarked that paraffin crystallized in two modifications, one at the solidifying point, and another at a temperature a few degrees lower.

Dr. Richards failed to get conclusive results on the crystallization of super-saturated solutions of sodium hyposulphate by subjecting the entire amount to the vibrations, but we made one interesting experiment in which the super-sonic waves were carried to the surface of the solution by a bent glass thread. Crystallization immediately started around the tip of the thread, and also around a minute crystal which we dropped on the surface at a distance from the thread, the *type of crystallization* and *rate of growth* being different in the two cases. At the request of the Bureau of Soils, Dept. of Agriculture, we made some experiments on the dispersion of colloids from soils. In soil analysis it is often a long and tedious process to disperse the colloids adsorbed on the soil grains. It was found that the colloids of a "very difficult" soil sample sent for examination were completely dispersed into the water by a few minutes treatment to the super-sonic vibration, while by the usual methods the process of shaking violently and centrifuging has to be repeated twenty or thirty times before the colloid is completely dispersed.

Biological Effects.

Though the effects of these waves upon living matter might more properly be discussed elsewhere, it may not be out of place to mention briefly a few of the observations which we have made as they have some bearing on the physical processes involved.

In marked contrast to the flocculation, or driving together of small particles of suspended matter, which has been mentioned, we have fragmentation, or the tearing to pieces of small and fragile bodies. Filaments of living spirogyra were torn to pieces and the cells ruptured. Small unicellular organisms such as paramecium were rendered immobile by a short treatment to vibration of moderate intensity, subsequently recovering, but were killed by a longer exposure, many of them being torn open. The circumstance that all are not treated alike is doubtless due to the fact that those which manage to keep out of the nodes of the stationary wave system are less roughly handled by the vibrations. Bacteria apparently are able to survive owing to their small size, for the fragmentation of larger bodies is due to the fact that the forces applied to their surfaces vary in magnitude and direction at different points of the body, while in the case of a bacterium the whole body is subjected to the same treatment.

Red blood corpuscles in physiological salt solution are rapidly destroyed, the turbid liquid becoming as clear as a solution of a red aniline dye.

With vibrations of less intensity the destruction is less complete, a blood count made at the end of each 15 seconds of exposure showing that the percentage destroyed decreases, a point being reached at which no further destruction occurs unless the intensity of the radiation is augmented. This means of course that some of the corpuscles, the recently formed ones perhaps, are more hardy than those of greater age. Small fish and frogs are killed by an exposure of one or two minutes, an observation also made by Langevin at Toulon with his Poulsen arc oscillator (see Plate VIII.) Mice are less sensitive, a twenty-minute exposure not resulting in death, and though at the end of the treatment the animal was barely able to move, the recovery was fairly rapid. Blood counts made with a mouse during exposure showed a diminishing number of corpuscles, until a stationary state (about 60 per cent. normal) was reached. The biologists inform us, however, that the blood count of a mouse is affected by fear, the corpuscles hiding in the liver until tho danger is over! We made the count with drops taken from the tip of the tail.

We have not yet determined the cause of death in the case of the fishes and frogs. They were protected against rise of temperature as much as possible by ice fragments, dropped into the water from time to time, but this does not shield them from internal heating, which may be the cause of death, as in the case of small animals introduced into a high-frequency electric field. In the case of a mouse killed by an exposure of two minutes between the plates of an air-condenser operated at about 1000 volts with a frequency of 100 million, we found that the temperature of the body-cavity was over 113° F.

With distilled water or a fairly strong solution of salt in a test-tube between the plates of the condenser, little or no heating occurred; but for small concentrations the heating was very marked, the maximum being for ·8 per cent., which is very nearly the amount found in mammalian blood. At lower frequencies the heating appears to be greater for distilled water, at least with high voltages. We found that one terminal of our 60,000 volt coil could be held in the fingers without the production of any sensation, but if dipped into the open end of a glass tube a metre long and filled with distilled water, caused the water to boil in less than 10 seconds. The introduction of a small amount of salt into the water, prevented the heat entirely in this case, which explains why no thermal discomfort was felt when the wire was held in the hand. The wire must be seized, however, before the current is turned on, otherwise a very vicious arc jumps to the finger producing a burn which is very slow in healing.

Fig. 1.

Two 1000 Watt tubes. Oil condensers. Variable air condenser and coils.

Fig. 2.

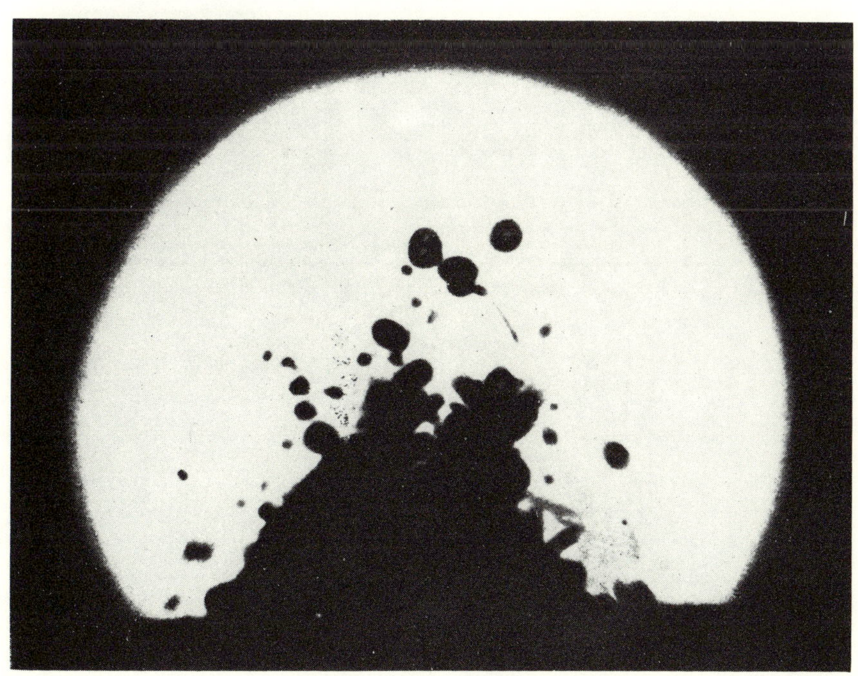

FIG. 3.

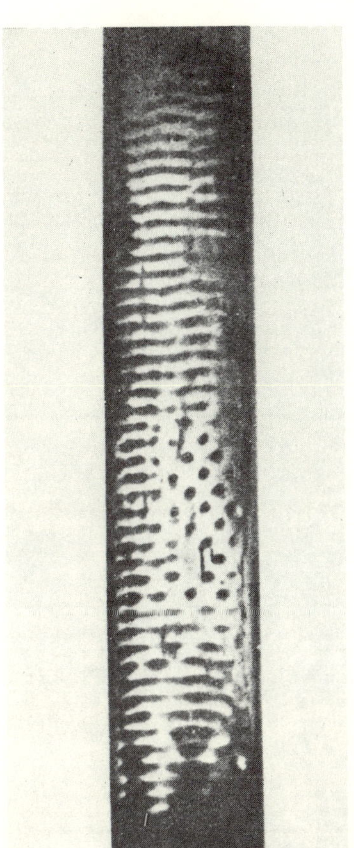

FIG. 4.

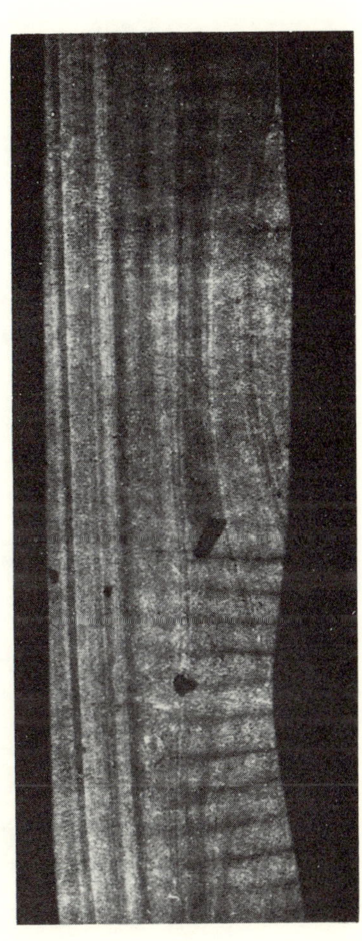

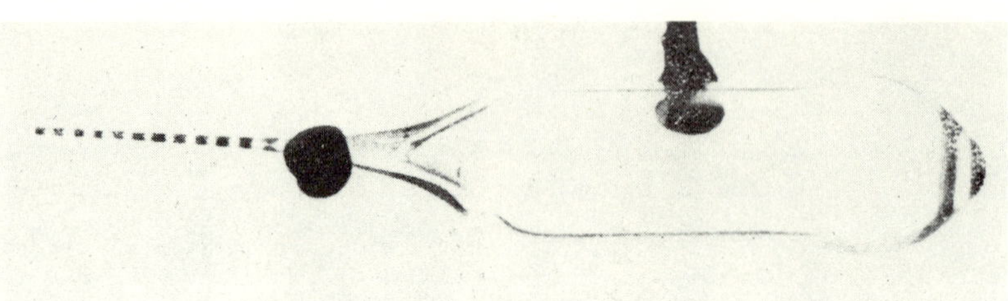

Fig. 6.

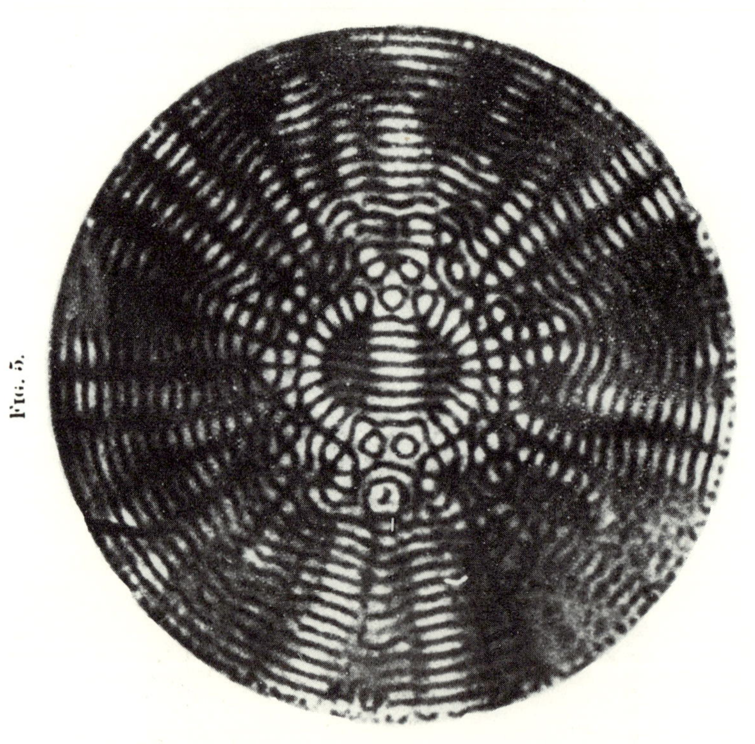

Fig. 5.

Fig. 7.

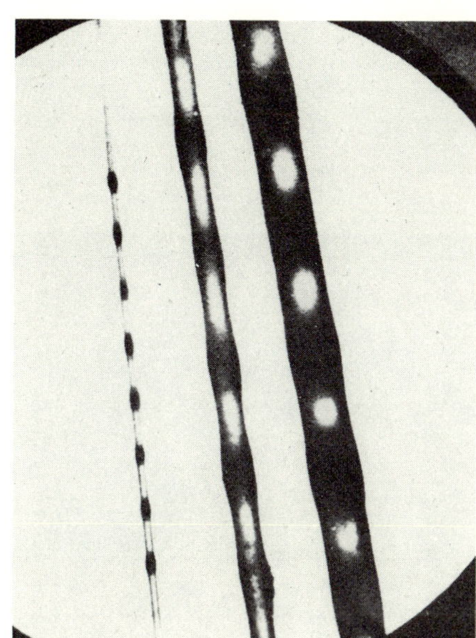

Fig. 8.

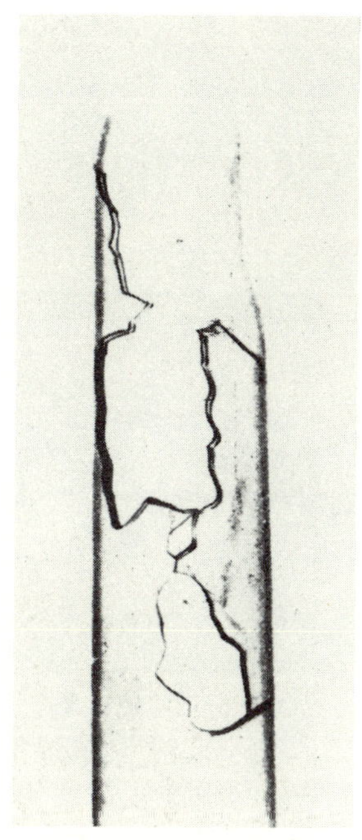

Fig. 9.

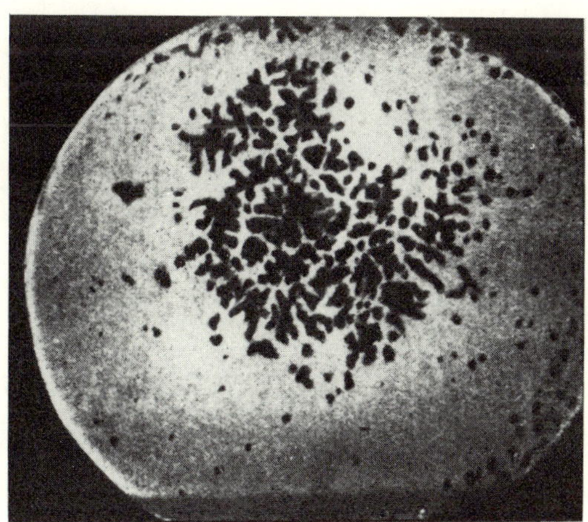

Fig. 10.

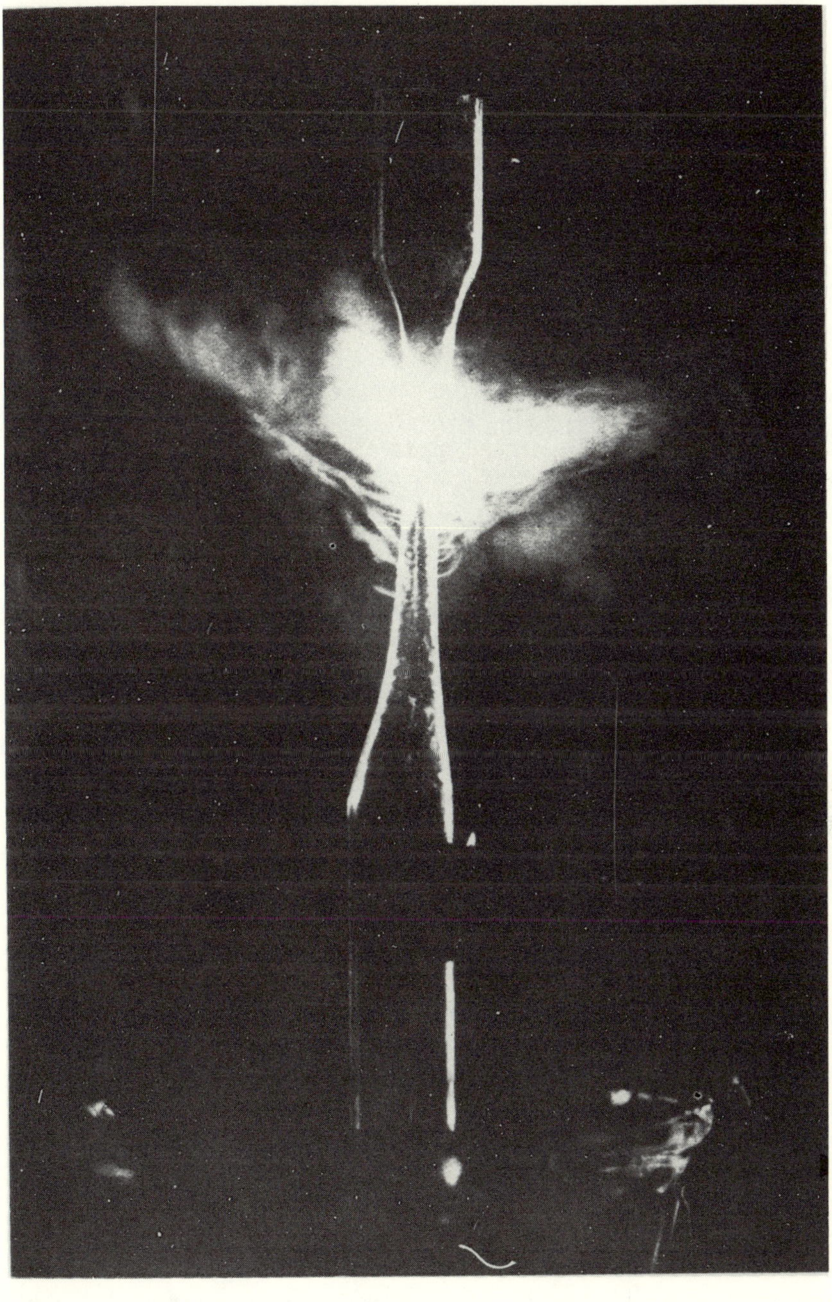

30

Reprinted from *Proc. Soc. Exptl. Biol. Med.*, **27**(7), 626–628 (1930)

The Mechanism of the Lethal Effects of Ultrasonic Radiation.

FRANCIS O. SCHMITT AND BERTHA UHLEMEYER.

From the Department of Zoology, Washington University, St. Louis, Mo.

That the lethal effects of ultrasonic waves on protozoa and other single cells could be traced to the cavitation of dissolved gas, has recently been discovered by Johnson.[1] Any gas suffices, apparently; there is no specificity of cavitated oxygen as was found for acceleration of chemical reactions by Schmitt, Johnson and Olson.[2] We have repeated the experiments of Johnson and are able to confirm his results fully. The present communication extends these observations and examines the mechanism of the lethal effect. A large number of microorganisms were tested, including various types of protozoa, rotifers, copepods, Daphnia, etc. In addition to these a number of eggs and embryos of frogs and snails were treated. About

[1] Johnson, C. H., *J. Physiol.*, 1929, lxvii, 356.

[2] Schmitt, F. O., Johnson, C. H., and Olson, A. R., *J. Am. Chem. Soc.*, 1929, li, 370.

10 cc. of the fluid containing the material to be tested was placed in a glass tube and radiated at atmospheric pressure, and at pressures of 60-80 lb. per square inch. After the minimum lethal dose for those radiated at atmospheric pressure was determined, it was found that an exposure many times this dosage had no effect upon cells radiated under 60-80 lb. pressure. Radiation of frog embryos yielded results which seem to indicate that radiation affects the head region rather preferentially, and that there is a definite selective interference in organogenesis. A study of these effects will be reported later.

To test the possibility that the function of the cavitated bubbles is to reflect the waves from their surfaces and thus tend to concentrate the radiation locally, two types of experiments were performed. In the first, cells were radiated in a suspension of infusorial earth, the suspended particles ranging from very small to relatively coarse. In the second, paramecia were allowed to ingest Chinese black which packed the food vacuoles with solid granules. Radiation of the cells in either case, with pressure and without, gave no results which would lend support to the theory that such an artificial increase of surface, inside or outside the cell, aided in producing the lethal effects concerned. It seems, therefore, that the lethal effect is produced at the surface of the cavitated gas bubbles. Johnson is of the opinion that the effect on cells is external rather than internal, and to his reasons for so believing, we may add the fact that, owing to the greater viscosity of the protoplasm, there will be considerably less tendency for cavitation inside the cell than outside. Thus, for example, when cells are radiated in a viscous medium, the minimal lethal dose is greatly increased. This retarding effect of viscosity is observable also in the promotion of chemical reactions by ultrasonic radiation.

To test the view that the lethal effect is due to a chemical or physical change at the plasma membrane, Spirogyra filaments were treated with a radiation strength insufficient to rupture the cellulose wall, or to cause visible disorganization. The strands were stained with neutral red, and then placed in a mixture of NaOH, NaCl, and $CaCl_2$. Radiated material gave evidence of a much more rapid penetration of the alkali than unradiated controls. Although adequate temperature control is difficult there seems little doubt but that the radiation alters the membrane so as to make it more permeable.

In further testing this view, another interesting and suggestive experiment was performed. Artificial cells were made by mixing a chloroform solution of lecithin with a protein solution according to the directions of Harvey.[3] After the chloroform has diffused out of the "cells", an aqueous mixture of lecithin surrounded by a denaturized protein membrane presumably remains. Radiation of these "cells" gave very curious results. Sufficiently long exposures, of course, destroy the "cells". A dose of 10 seconds, however, sufficing to kill all paramecia present, results only in the accumulation of a bubble due to the cavitation of air inside the "cell"; successively longer radiations cause the appearance of protrusions, which in reality, are myelin forms resulting from the escape of lecithin into the water through ruptures in the cell wall. Microscopic examination occasionally reveals minute breaks in the membrane. That radiation may almost instantly break certain unstable emulsions, such as oils in water, and cause a separation of phases, has been demonstrated by Schmitt.[4] This seems highly significant in view of the Clowes theory of the emulsion structure of the membrane. It has also been discovered[4] that types of chemical reactions other than oxidations may be promoted by radiation in the presence of water and cavitated gas. Thus the hydrolysis of carbon tetrachloride and other halogens requires only cavitation; for this, pure nitrogen suffices.

The membranes of artificial cells may not be ruptured even though cavitation has taken place on the inside and the outside by radiation strong enough to kill all protozoa present; one seldom, if ever observes a cavitated bubble inside of radiated protozoa; very energetic chemical reactions may be promoted by radiation, when cavitation is permitted, even in the absence of oxygen; unstable emulsions may be instantly cracked by the radiation. These facts render plausible our tentative position that the lethal effect is due to a rupture of the plasma membrane by a chemical or a physical chemical effect produced by cavitation in the water immediately surrounding the cell.

---

[3] Harvey, E. N., "Laboratory Directions in General Physiology," 1913, 31.

[4] Schmitt, F. O., in press.

31

Reprinted from *J. Bacteriol.*, **51**(4), 487–493 (1946)

ULTRASONIC DISINTEGRATION AS A METHOD OF EXTRACTING BACTERIAL ENZYMES[1]

P. K. STUMPF, D. E. GREEN, AND F. W. SMITH, JR.

Departments of Medicine and Biochemistry, College of Physicians and Surgeons, Columbia University, New York

Received for publication December 15, 1945

In recent years several techniques have been developed for the purpose of disintegrating bacterial cells. These have depended upon different principles ranging from autolytic rupture of the cell to grinding with powdered glass (Lipmann, 1938; Koepsell and Johnson, 1942; Wiggert *et al.*, 1940; Kalnitsky *et al.*, 1945; Booth and Green, 1938). Although each method has merit for some specific problem, it frequently proves inadequate for other applications.

Of the several methods employed in this laboratory, that of disintegrating bacterial cells by exposure to an ultrasonic field[2] has proved of most value when limited quantities of bacteria were available and when the possibility of preparing cell-free enzyme extracts had to be explored. Further, since irradiation of bacteria by ultrasound can be carried out under completely aseptic conditions, it becomes possible to analyze pathogenic bacteria with respect to their endotoxins, enzymes, polysaccharides, and other substances which can be extracted from cells only after disintegration. The method, however, has two drawbacks: (1) the construction of the apparatus is expensive, and (2) not all bacteria can be disintegrated by ultrasound.

APPARATUS

As a detailed account of the ultrasonic equipment employed in this laboratory has been published elsewhere (Smith and Stumpf, 1946), only a brief description is given here. An X-cut quartz crystal[3] ($1\frac{1}{2}$ inches in diameter and ground to a frequency of 600 kilocycles) is driven by a radio-frequency generator, operating at 2,000 volts with a rated output of 700 watts. The generator has a frequency range from 200 to 1,000 kilocycles and employs beam power tetrodes to eliminate neutralization. It has the usual crystal controlled exciter as well as a variable frequency exciter, which is stabilized and calibrated, and which facilitates exploring a wide range of radio frequencies. Figure 1 is a diagrammatic representation of the complete unit and the disintegrating chamber, respectively.

An efficient crystal mounting must fulfill several requirements, among which are low mechanical damping of the crystal by the holder, high insulation against

[1] This investigation was made possible by a grant from the John and Mary Markle Foundation.

[2] The application of sonic energy to biological problems is not new (cf. Chambers and Flosdorf, 1936), but previously no systematic effort has been made to explore the technique from the standpoint of extracting bacterial enzymes.

[3] We are indebted to the August E. Miller Laboratories and the Federal Telephone and Radio Corporation for quartz crystals.

voltage breakdown, and ready accessibility to permit rapid change of operating frequency. With these requirements in mind, we developed a crystal holder such

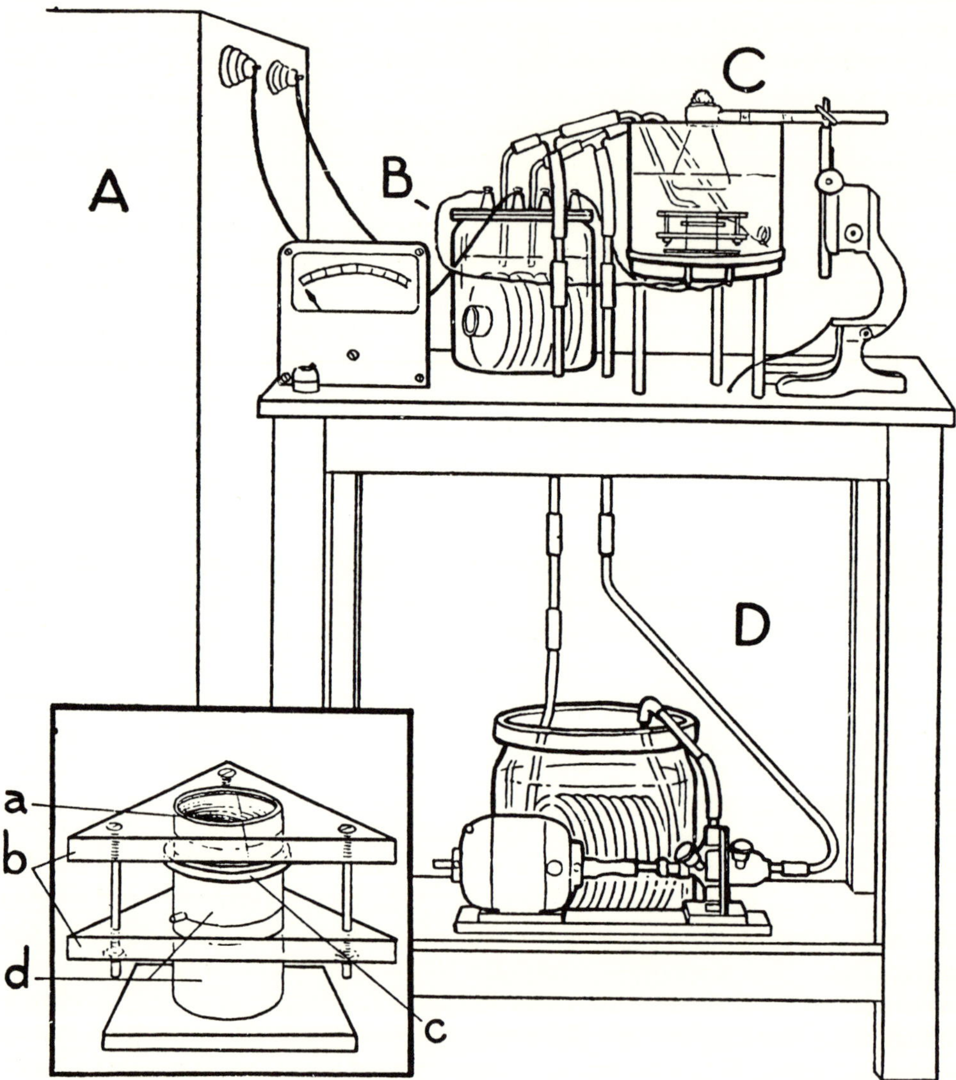

FIG. 1

A. Radio-frequency transmitter.
B. Coupling coil.
C. Disintegration chamber and rack and pinion arrangement.
D. Cooling arrangement for oil in circulating system.
a. Upper brass electrode.
b. Adjustable lucite plates for holding electrodes in place.
c. Quartz crystal.
d. Lower brass electrode.

as is shown in figure 1. It consists of a quartz crystal sandwiched between two brass rings held together by a lucite press device which permits changes in contact pressure of the brass electrodes on the quartz surfaces. Although it leads to some damping, this arrangement has the advantage that breakdown of the insulating medium is eliminated completely. With other types of holders which were constructed and tested, voltage breakdown of the transformer oil resulted in immediate and irreparable damage to the surface of the quartz crystal. This surface is silver-plated routinely every 6 months by Brashear's method (Hodgman, 1937). It was found that a silver-plated surface is superior to one of aluminum or silver foil cemented or spattered onto the quartz. To increase the voltage breakdown path, the upper and lower plating is not carried out to the extreme edge of the crystal. Lucite is used exclusively in the mounting and in the oil chamber since, as insulation against high voltage, it is superior to either neoprene or bakelite.

The crystal holder is completely immersed in transformer oil,[4] which serves at once as insulation medium, conducting medium for ultrasound, and circulating fluid for the cooling system. The oil is circulated continuously by a centrifugal pump from the ultrasound apparatus through a copper coil surrounded by an ice-water mixture. Much heat is formed during operation, and efficient cooling is necessary to prevent the temperature from rising above 30 C in the bacterial suspension.

When the crystal holder is immersed in the oil bath, the lower brass ring electrode automatically forms an air pocket, which almost completely reflects the sound waves at the lower side of the quartz plate. The reflected waves come automatically into the right phase with those radiated upwards and consequently increase their energy (Bergmann and Hatfield, 1938). Hence the total damping of the crystal becomes less, owing to the decrease in the radiation decrement.

The bacteria to be disintegrated were made up in a thin suspension containing about 20 mg dry weight per ml and were placed in a 50-ml Erlenmeyer flask, which was then lowered by a rack and pinion device into the oil chamber to a critical distance from the upper face of the quartz crystal. The optimum distance is reached when the cone of fluid in the flask is at maximum height during ultrasonic irradiation. A 10-minute exposure usually ensures satisfactory disintegration. The suspension is spun for 10 minutes at 20,000 rpm in the high-speed conical head of the refrigerated International centrifuge no. 1. After centrifugation, three layers are observed: a supernatant fluid which is faintly opalescent and pale yellow, and a sediment consisting of two layers, (a) a lower, hard-packed layer of intact cells and (b) an upper, gelatinous layer of ghost cells and debris. The supernatant fluid is removed with a pipette, care being taken not to disturb the sediment. The same technique is employed when pathogens are disintegrated. It is of course necessary to follow aseptic procedures in handling pathogenic material.

The supernatant fluid still contains a residual number of intact, viable cells, but the number is so small that these do not contribute significant enzymic activity to the juice. This conclusion is most easily demonstrated by considering the enzymic activity of the supernatant fluid of a suspension of disintegrated cells

[4] Wemco "C" transformer oil obtained from Westinghouse Electric Supply Co., New York.

of *Proteus vulgaris* with respect to pyruvic oxidase (Stumpf, 1945). Whereas the whole cell rapidly oxidizes pyruvic acid without addition of any factors, the supernatant juice from disintegrated cells is completely inert unless cocarboxylase is added (as indicated in figure 2). From this fact it follows that the number of intact cells is too low to account for any appreciable fraction of the observed enzyme activity. As a precaution against gross bacterial contamination, however, the extracted juice is centrifuged daily in the high-speed attachment before its use and is always kept chilled at about 4 C. In some cases it may be lyophilized and stored at 0 C.

THE FACTORS WHICH AFFECT THE EFFICIENCY OF ULTRASONIC DISINTEGRATION

A number of factors have been considered which determine the maximum utilization of ultrasonic energy for the disintegration of bacteria. Two factors

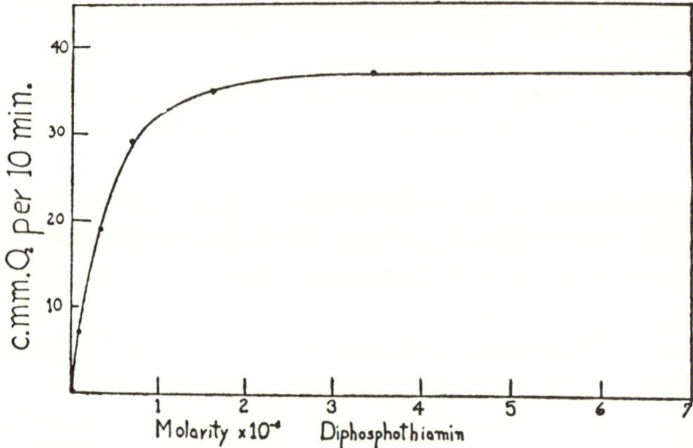

FIG. 2. REACTION VELOCITY OF CELL-FREE ENZYME PREPARATION OF PYRUVIC OXIDASE OF PROTEUS VULGARIS AS A FUNCTION OF DISPHOSPHOTHIAMINE CONCENTRATION

Each manometric cup contained 0.5 ml of enzyme, 0.5 ml of M/2 acetate buffer at pH 6.0, 0.1 ml of 0.2 per cent manganese sulfate, 0.5 ml of M/5 lithium pyruvate. Final volume 3 ml; NaOH in center pot, temp 38 C.

which come immediately to mind—namely, time of exposure and shape of container—can be treated briefly. As indicated in table 1, after 10 minutes' ultrasonic irradiation, the maximum degree of disintegration is reached. The suspension probably contains forms of the bacterial cell which are resistant to the disruptive forces, and hence complete disintegration is not possible. As for the second factor, it has been our experience that vessels such as test tubes or round-bottom flasks, because of their convex bases, reflect and thereby dissipate a large percentage of the ultrasonic radiation. An ideal container should therefore have a perfectly flat, thin glass base which would permit a maximum passage of disruptive energy. Routinely, we employ small Erlenmeyer flasks or volumetric flasks which have relatively thin bases.

As shown in table 2, the thickness of the suspension, or viscosity, is a factor of considerable importance. A thick paste of bacteria suffers little or no disintegration of cells. When the suspension is diluted, however, disintegration by ultrasound takes place. Evidently, when a suspension exceeds a certain viscosity, ultrasonic radiation which strikes such a suspension is largely converted to heat, which has no disruptive effect.

TABLE 1

Relation between time of exposure and degree of disintegration of E. coli

EXPOSURE TIME	MG OF NITROGEN PER ML OF ORIGINAL SUSPENSION	MG OF NITROGEN PER ML OF CELL-FREE "JUICE"	DEGREE OF DISINTEGRATION*
minutes			*per cent*
2.5	0.94	0.19	25
5	0.94	0.30	40
10	0.94	0.43	56
15	0.94	0.47	63

* $\frac{\text{mg of nitrogen per ml cell-free juice}}{\text{mg of nitrogen per ml original suspension}} \times \frac{100}{0.80}$. The theoretical limit for this ratio is approximately 80 per cent, since ghost cells and cellular debris which contribute the remaining 20 per cent of nitrogen are removed by centrifugation.

TABLE 2

Relation between concentration of bacteria and degree of disintegration of E. coli

Bacteria

A. MG BACTERIAL N PER ML OF SUSPENSION	B. MG BACTERIAL N PER ML OF CELL-FREE "JUICE"	DEGREE OF DISINTEGRATION 100B/0.80A
		per cent
2.18	0.44	25
1.37	0.37	34
0.63	0.20	39
0.32	0.13	50

Erythrocytes

DILUTION OF SUSPENSION	TIME FOR COMPLETE HEMOLYSIS
	seconds
Thick paste of undiluted centrifuged cells	No hemolysis
1:5 (dilution with isotonic salt solution)	75
1:10 (dilution with isotonic salt solution)	50
1:15 (dilution with isotonic salt solution)	25
1:20 (dilution with isotonic salt solution)	20
1:25 (dilution with isotonic salt solution)	7

When the distance between the upper face of the crystal and the base of the container is such that the suspension is mildly agitated by ultrasonic energy, the degree of disintegration is about half that which is obtained when the distance is critical and the suspension is violently agitated (table 3). Obviously, when the distance between the upper face of the quartz plate and the base of the container is a multiple of the whole wave length of the ultrasonic vibration, any reflection from the base of the container will cause partial interference. When, however, the distance traveled by the waves is an odd multiple of one-half the wave length, reinforcement of the reflected wave occurs and considerably more ultrasonic energy is carried into the suspension. As shown in table 4, a considerable number of cells readily yield to treatment, but others are completely refractory to ultrasound.

Finally, the relative degree of disintegration of bacteria seems to be independent of the frequency of the ultrasonic wave and dependent only on the intensity or amplitude of the wave. It is unfortunate, however, that, if the power output

TABLE 3

The position of the irradiating flask with reference to the quartz crystal as a factor in determining the degree of disintegration

DISTANCE OF FLASK FROM QUARTZ CRYSTAL	STATE OF BACTERIAL SUSPENSION DURING IRRADIATION	MG OF BACTERIAL NITROGEN PER ML OF SUSPENSION	MG OF BACTERIAL NITROGEN PER ML OF CELL-FREE "JUICE"	DEGREE OF DISINTEGRATION
				per cent
1 mm shift from the critical position	Gentle agitation	0.92	0.14	19
Critical position	Vigorous agitation	0.91	0.29	40

TABLE 4

Organisms exposed to ultrasonic irradiation

EASILY DISINTEGRATED		REFRACTORY
Hemophilus influenzae	*Proteus vulgaris*	*Sarcina lutea*
Salmonella typhi-murium	*Clostridium welchii*	*Micrococcus lysodeikticus*
Brucella abortus	*Escherichia coli*	*Acetobacter suboxydans*
Lactobacillus casei	*Pseudomonas pyocyanea*	*Saccharomyces cervesiae*
Lactobacillus delbrueckii	*Pseudomonas fluorescens*	(Chloroplasts)
Proteus morganii	*Staphylococcus aureus*	
	(Erythrocytes)	

is raised beyond a certain range, denaturation and inactivation of labile proteins rapidly takes place. Thus there is actually only a relatively narrow region of power intensity in which ultrasonic energy can be effectively used as a disintegrating agent.

SUMMARY

A method of disintegrating bacterial cells by ultrasonic radiation is described. The method has been employed to prepare cell-free enzyme extracts of bacteria. A number of factors which influence the degree of efficiency of ultrasonic disintegration are discussed.

REFERENCES

BERGMANN, L., AND HATFIELD, H. S. 1938 Ultrasonics, p. 37. John Wiley and Sons, New York.

BOOTH, V. H., AND GREEN, D. E. 1938 A wet-crushing mill for microorganisms. Biochem. J., **32**, 855–861.

CHAMBERS, L. A., AND FLOSDORF, E. A. 1936 Sonic extraction of labile bacterial constituents. Proc. Soc. Exptl. Biol. Med., **34**, 631–636.

HODGMAN, C. D. (Editor.) 1937 Handbook of chemistry and physics, **23**, 1953–1954.

KALNITSKY, G., UTTER, M. F., AND WERKMAN, C. H. 1945 Active enzyme preparations from bacteria. J. Bact., **49**, 595–602.

KOEPSELL, H. J., AND JOHNSON, M. J. 1942 Dissimilation of pyruvic acid by cell-free preparations of *Clostridium butylicum*. J. Biol. Chem., **145**, 379–386.

LIPMANN, F. 1938 An analysis of the pyruvic acid oxidation system. Cold Spring Harbor Symposia Quant. Biol., **7**, 248–259.

SMITH, F. W., JR., AND STUMPF, P. K. 1946 High ultrasonic generator and crystal holder design. Electronics. *In press*.

STUMPF, P. K. 1945 Pyruvic oxidase of *Proteus vulgaris*. J. Biol. Chem., 529–544.

WIGGERT, W. P., SILVERMAN, M., UTTER, M. F., AND WERKMAN, C. H. 1940 Preparation of an active juice from bacteria. Iowa State College J. Sci., **14**, 179–186.

AUTHOR CITATION INDEX

SUBJECT INDEX

About the Editor

DONALD W. THAYER is associate professor of microbiology at Texas Tech University, where he teaches courses in bacterial physiology and biochemistry, and general and industrial microbiology. Dr. Thayer received his B.S. and M.S. in bacteriology from Kansas State University. He completed the Ph.D. in microbiology at Colorado State University. During the 1966–1967 academic year, Dr. Thayer held a postdoctoral, resident research associateship in biochemistry at the Naval Medical Research Institute in Bethesda, Maryland. He was a research chemist at the Institute until coming to Texas Tech University in the fall of 1969. Dr. Thayer is conducting research on bacterial cellulases and the production of single-cell protein from cellulose.